Public Transportation Systems

Principles of System Design, Operations Planning and Real-Time Control

Public Transportation Systems

Principles of System Design, Operations Planning and Real-Time Control

Carlos F. Daganzo
University of California, Berkeley, USA

Yanfeng Ouyang
University of Illinois at Urbana-Champaign, USA

World Scientific

NEW JERSEY • LONDON • SINGAPORE • BEIJING • SHANGHAI • HONG KONG • TAIPEI • CHENNAI • TOKYO

Published by

World Scientific Publishing Co. Pte. Ltd.

5 Toh Tuck Link, Singapore 596224

USA office: 27 Warren Street, Suite 401-402, Hackensack, NJ 07601

UK office: 57 Shelton Street, Covent Garden, London WC2H 9HE

Library of Congress Cataloging-in-Publication Data
Names: Daganzo, Carlos, author. | Ouyang, Yanfeng, author.
Title: Public transportation systems : principles of system design, operations planning and real-time control / by Carlos Daganzo, University of California, Berkeley, USA and Yanfeng Ouyang, University of Illinois at Urbana-Champaign, USA.
Description: New Jersey : World Scientific, [2019] | Includes bibliographical references.
Identifiers: LCCN 2018048228| ISBN 9789813224087 (hc : alk. paper) |
ISBN 9789813224094 (pbk : alk. paper)
Subjects: LCSH: Urban transportation. | Local transit. | Transportation engineering.
Classification: LCC TA1205 .D34 2019 | DDC 388.068--dc23
LC record available at https://lccn.loc.gov/2018048228

British Library Cataloguing-in-Publication Data
A catalogue record for this book is available from the British Library.

For any available supplementary material, please visit
https://www.worldscientific.com/worldscibooks/10.1142/10553#t=suppl

Desk Editor: Herbert Moses

Typeset by Stallion Press
Email: enquiries@stallionpress.com

Preface

This book is an extended and updated version of a set of lecture notes for a graduate course on public transportation at U.C. Berkeley. The 2010 version of the notes, entitled "Public Transportation Systems: Basic Principles of System Design, Operations Planning and Real-Time Control" (C. F. Daganzo, Institute of Transportation Studies, UCB-ITS-CN-2010-1, 2010) has been published online and made freely available. Since 2013, the notes have also been the basis for a joint undergraduate–graduate course on public transit taught annually by one of the authors at the University of Illinois, Urbana-Champaign. This effort has led to additional improvements.

The notes were translated to Chinese in 2016 and used at some major Chinese universities such as Tsinghua University (Beijing). This interest was perhaps sparked because the notes espouse an integrated approach to design, operations, and management that is based on physics principles and yields holistic insights on transit performance vis-à-vis other transportation means. Rather than describing what exists and basing designs on what has been successful in similar situations, the notes show how to optimally design complete systems for arbitrary regions from scratch using as inputs only the basic economic and operational features of the technology to be used, e.g., the unit costs of the guideway, the vehicles and the stations; and the vehicles' speed and capacity, among other things.

This approach to design is useful because it allows designers to come up with concepts that have never been tried before. Notably, it was used by one of the authors to conceive and then help design a high-performance bus system for Barcelona, Spain. Called the Nova Xarxa (the "new network" in Catalan) it is by now almost completely deployed. The Nova Xarxa is different because unlike other bus systems, it stresses high frequency, few stops and complete coverage of the urban area with a single and convenient transfer. This design provides users an experience comparable to the subway at the cost of bus. Thanks to its unusual attributes, the Nova Xarxa is already attracting considerably more demand than expected. And about 30% of the trips on it involve one transfer. This is something unheard of for bus networks.

The success in Barcelona, as well as feedback from our classrooms, convinced the authors that there would be merit in polishing the notes and sharing them in a more visible way. We naively thought that this would be easy, but of course, we were wrong. We discovered many gaps, errors and numerous topics that the earlier notes did not address. Since a minor fix was not going to be satisfactory, we decided to produce a more complete product that could hopefully be used, not just for teaching, but also as a source of ideas for professionals that would complement existing books. This, at least, has been our goal.

Besides considerably updated and expanded planning chapters, the book now discusses the physics of emergent transit services such as crowd-sourced taxis, ride-sharing on the fly, and intelligent vehicle-sharing; the effect of driverless vehicles on both, transit and the future of cities; human and physical plant management; robust methods of real-time control; basic analysis methods including a section on dimensional analysis — a useful topic that is rarely covered in transportation books — and an epilog discussing demand elasticity, pricing, subsidies and economics. Given the supply-side emphasis of the book, this epilog is intended to highlight the role of supply side ideas in economic policy analyses.

The book is not a manual but it still attempts to be comprehensive and versatile. Rather than covering the state of the practice with overviews of current software packages and conventional wisdom tenets, it presents a general analysis approach that can be used to deepen the reader's understanding and help solve arbitrary real-world problems. The approach is illustrated with solutions and detailed formulas for simple idealized problems of various types. These formulas give a qualitative idea of what transit can accomplish in real applications. The formulas also reveal insights that help illuminate the big picture — for example, they are used in the book, both, to identify the societal and technological factors that most affect an application's success, and to unveil the proper niche of each transit technology.

Despite its many new additions, the book is organized much like the original, so it can still be used to teach advanced undergraduate and entry-level graduate courses. It has three main parts. First is an introduction (Chapters 1–3) that covers the basics of transit and the most important analysis tools — the latter should already be familiar to graduate students in engineering but are presented in an effort to reach a broader audience. This part is followed by a module on planning and design (Chapters 4–7).

This module is mode-abstract and self-contained; it explains how a transit systems' spatiotemporal coverage in terms of routes and schedules (what the public sees) should be organized. For pedagogical reasons the ideas are presented in order of increased complexity, using the results for simple systems as building blocks for the more complex. Chapter 4 considers shuttles; Chapter 5 considers corridors; and Chapters 6 and 7 explain complete systems covering two-dimensional regions. The third part of the book explains how to make the plans happen. It describes the behind-the-scenes actions that a transit agency must take to deliver the advertised services. Chapters 8 and 9 explain how to manage both the vehicle fleet and its human drivers to cover the plan while maximizing their productivity. These are medium-term decisions, usually revisited every few months. Chapter 10 explains the much shorter term decisions that an agency must make in real time from the control room to keep its operations on schedule despite random delays and other external disruptions. Although all three chapters describe the status quo, their main objective, as in the rest of the book, is explaining how things *can* be done rather than how they are done. The book ends with the aforementioned epilog (Chapter 11) that discusses economic ideas from a supply-side perspective.

To be helpful to students, each chapter ends with a set of references and homework exercises. The references are related readings that either support or expand the chapter's ideas and are not meant to be an exhaustive bibliography. The homework exercises are meant to solidify the concepts in the chapter. Some are direct applications of the chapter's ideas and others (marked with an asterisk) expand the ideas into new contexts. Four of the chapters (5, 6, 9, and 10) include a mini-project that students can undertake either alone or in groups to practice the creative activities discussed in the respective chapters. In our experience, students benefit from the homework exercises quite a bit, but only the mini-projects allow them to experience something resembling a professional activity. Our students' reception to this endeavor has been quite positive.

The book can be used to teach two types of courses. An undergraduate course on transit planning can be built around Chapters 1 through 7. We recommend spending about two weeks per chapter on average, and saving the last two weeks of the semester to raise the student's awareness about the "plumbing" aspects of transit. This could be done in a seminar format while students conclude term-papers or mini-projects. A graduate course on transit planning and operations covering Chapters 4–11 could also be offered, again using about two weeks per chapter. Chapters 1–3 could be assigned as independent readings, assumed as prerequisites, or covered very quickly. Alternatively, the most arcane methodological ideas could be briefly reviewed when the need arises. At the graduate level too, we recommend complementing the lectures with homework exercises and mini-projects.

Although we have tried to eliminate as many errors as possible, experience tells us that errors will likely remain. Therefore corrections will be posted online and periodically updated. The link to the website is: `https://www.worldscientific.com/worldsscibooks/10.1142/10553#t=suppl`. We recommend visiting this website before using the book as a teaching tool. This website also includes datasets for the mini-projects and an animation that pertains to Chapter 10. Solutions to the homework exercises and the mini-projects will not be posted online but will be made available to professors of accredited schools. To obtain a copy, please write to either author. Please include your university and affiliation, as well as a signed non-disclosure agreement affirming that the solution sets are for your instructional purposes only and that they will not be copied and/or shared with anyone.

This book has benefited from the input of friends, colleagues, and students. Above all, we wish to thank Michael J. Cassidy (the Horonjeff Professor of Transportation Engineering at U.C. Berkeley) and Josh Pilachowsky (DKS Associates) for their creative contributions to most of the mini-projects and for the many comments arising from their separate experiences teaching from the notes. Former U.C. Berkeley doctoral students Vikash Gayah (Penn State University), Eric Gonzales (U. Massachusetts, Amherst) and Josh Pilachowsky helped compile the 2010 version of the notes in their roles as Graduate Student Instructors. Subsequently, Drs. Yiguang Xuan, Juan Argote, and Paul Anderson suggested modifications and provided valuable feedback, also in their roles as U.C. Berkeley graduate student instructors. More recently, U.C. Berkeley lecturer Lewis Lehe (now at the University of Illinois) taught the course and provided valuable comments on the latest additions to the book. Graduate students and teaching assistants at the University of Illinois, Antoine Petit, Ruifeng She, Zhoutong Jiang, Laura Ghosh, and Liqun Lu have also provided many valuable comments and feedback on the text, mini-projects and homework exercises. We extend our warm thanks to all the named individuals.

Finally, and most importantly, both authors are immensely grateful to our respective spouses, Valery and Wei, for their patience and support. As non-engineers, they have been gracious listeners and helped provide real-world perspective — and not just on transportation matters. This book is dedicated to them with much love and appreciation.

About the Authors

Carlos F. Daganzo is a Chancellor's Professor of the Graduate School in Civil and Environmental Engineering at the University of California, Berkeley. He is also a member of the National Academy of Engineering and recipient of numerous honors. He co-founded *Via Analytics*, a California benefit corporation devoted to improving mobility through technology. He has served as a Convenor of the oldest international symposium on transportation and traffic theory (the ISTTT) and as an Associate Editor of *Transportation Research* (Part B, methodological) and *Transportation Science*. Daganzo has supervised 38 PhD students who hold faculty positions at top-ranked schools of engineering, business and management all over the world. He has authored *Fundamentals of Transportation and Traffic Operations* (Pergamon-Elsevier, 1997) — an internationally used and translated graduate-level textbook. Noted for his contributions to econometrics, logistics, freight operations, network theory, traffic flow, and transit operations, Daganzo has also written more than 180 papers in scientific journals and four sole-authored research books: *Multinomial Probit: The Theory and its Application to Demand Forecasting* (Academic Press, 1979), *Logistics Systems Analysis* (1st–4th editions, Springer, 1991, 1996, 1999, 2005), *A Theory of Supply Chains* (Springer, 2003) and *Public Transportation Systems: Basic Principles of System Design, Operations Planning and Traffic Control* (published online, 2010; translated to Chinese in 2016). The latter contains the blueprint for Barcelona's new high-performance bus system, which Daganzo co-designed. The Universitat Politecnica de Catalunya, Barcelona, Spain, bestowed on him an honorary doctorate, partly in recognition for this work. The ISTTT Symposium of 2011 was dedicated to him in recognition of his cumulative contributions to transportation science. In 2018, Daganzo was appointed as the Overseas High Talent Strategic Scientist to the mayor of Beijing with the charge of conceiving and overseeing a transit master plan for Beijing, and addressing other current transportation issues. Daganzo obtained his degree in Civil Engineering from the University of Madrid in 1972 and his Master of Science and PhD in Civil Engineering (Transportation) from the University of Michigan in 1973 and 1975, respectively.

Yanfeng Ouyang is George Krambles Endowed Professor in Rail and Public Transit, Paul F. Kent Endowed Faculty Scholar of Civil and Environmental Engineering, and Donald B. Willett Faculty Scholar in Engineering at the University of Illinois at Urbana–Champaign (UIUC). He also holds the Changjiang Chair Professorship from the Ministry of Education of P.R. China through affiliation with the Harbin Institute of Technology. His research mainly focuses on modeling transportation, logistics, and infrastructure systems, and various applications to energy, military, and agricultural industries. To date, Ouyang has published over 100 peer-reviewed articles in leading journals. He has also co-authored six book chapters and two translated books. He currently serves as a Department Editor of *IIE Transactions*, an Area Editor of *Networks and Spatial Economics*, an Associate Editor of *Transportation Science*, an Associate Editor of *Transportation Research Part C*, an Associate Editor of *Transportmetrica B*. His work has been recognized with numerous awards, including a Walter L. Huber Research Prize from the American Society of Civil Engineers, a High Impact Project Award from the Illinois Department of Transportation, and a Faculty Early Career Development (CAREER) Award from the U.S. National Science Foundation, among others. Ouyang obtained his undergraduate degree in Civil Engineering (*summa cum laude*) from Tsinghua University, Beijing in 2000. He received his postgraduate degrees in Civil and Environmental Engineering from the University of Washington in 2001 and in Industrial Engineering and Operations Research from the University of California, Berkely in 2005, respectively. He received his PhD in Civil and Environmental Engineering (Transportation) from the University of California, Berkeley in 2005.

Contents

Chapter 1

Transit Basics

The American Public Transit Association in its *Glossary of Transit Terminology* (1994) defines public transit, or public transportation, to be

> *"Transportation by bus, rail, or other conveyance, either publicly or privately owned, which provides to the public general or special service on a regular and continuing basis."*

And the Merriam-Webster Dictionary defines public transit as

> *"the system that is used for moving large numbers of people on buses, trains, etc."*

No matter how the definition is phrased, the basic goal of public transit is to enable the mass movement of passengers from their origins to their destinations in an efficient and economical manner. Like all transportation systems, public transit involves both a temporal dimension and a spatial dimension. It achieves its economy by collecting and consolidating passengers over time and space into shared vehicles. It achieves its efficiency by smart spatiotemporal operating plans that endow these passengers with high mobility, giving them access to the opportunities available in a city.

These spatiotemporal plans are a central topic of this book and will be examined in analytical detail. This chapter's goal is to set the context. As such, the chapter provides basic ideas and facts about transit. It discusses the role transit can play in a city, its stakeholders, the challenges it faces and the available technologies. The chapter is organized as follows: Section 1.1 explores transit's role with a simple example that explains how public transportation meets a fundamental need of dense cities; Section 1.2 discusses both the expectations all transit systems aim to fulfill for its various stakeholders, and the challenges that stand in the way; Section 1.3

reviews current technologies focusing on their key attributes and introduces terminology; and finally, Section 1.4 summarizes the pros and cons of these technologies.

1.1. Need for Public Transportation

The U.S. Census Bureau's 2010–2013 American Community Survey estimated that automobile ownership in the United States would reach 1.8 vehicles per household in 2013. In other words, that by 2013 there would be almost one car per adult. Still, despite the large number of cars today, not every traveler has access to one, or the ability to drive. This segment of the population, which includes the poor, disabled, some elderly, and the young is captive to public transit and needs it. Although fulfilling this need is an important reason for the provision of transit, it is not the only one, however.

Evidence is that the same survey also revealed that about 30% of the commuters in the central parts of the 10 largest U.S. metropolitan areas use public transit as the means of transportation. Since it is hard to imagine transit captives making up the bulk of this 30% in any U.S. city, something else must be going on. Is there another need transit is fulfilling? Why do these extra people use transit?

The extra use arises at least in part because transit is often subsidized by governments so it can function well and compete on cost with the automobile. For example, in November 2016, the voters of Los Angeles County approved a 0.5% of sales tax hike projected to raise $120 billion over the next 40 years, with much of the money going to expand the city's light rail and subways (Nelson and Poston, 2016). But why would governments do this in the first place?

One reason is that well-used transit systems are friendlier to the environment than cars — since transit vehicles can serve many passengers at the same time and therefore can do so with fewer vehicle-kilometers traveled — and governments may be interested in the environment. This is not all, however. There is an even more compelling reason. Governments know from experience that transit is *essential* for the functioning of large and dense cities because, without transit, the demand for travel would flood the streets with cars to unthinkable levels of congestion and create a parking nightmare.

Dwellers of these cities also know in their guts that cars alone can only serve a modicum of their transportation needs and therefore, they will often support the transit expenditures, as in the case of Los Angeles that we

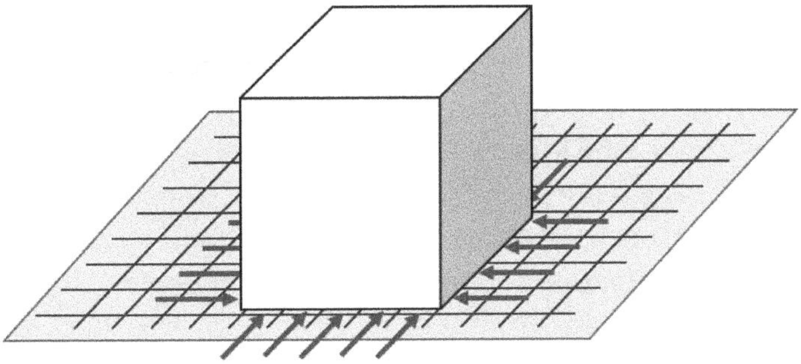

Figure 1.1. An idealized CBD.

have cited. The consequences of eliminating transit can also be understood systematically. The following example (Newell, 1994) shows what would happen during the morning rush in a monocentric city where all travel is done by car.

The city consists of a central business district (CBD) with no residences — think of the southern tip of Manhattan — and a peripheral zone where all the CBD workers reside. The workday starts approximately at the same time for everyone. Figure 1.1 depicts the CBD by means of a box, which is assumed to have a square footprint. During the rush, all employees access the CBD by car using the grid of city streets shown in the figure. We optimistically assume that once these commuters enter the CBD they can move freely and park. However, the CBD's perimeter could be a bottleneck.

To pursue this idea, let us first calculate the demand — the total number of commuters wishing to cross the perimeter. Let A [km^2] be the area of the CBD, p the fraction of this area occupied by buildings, n the average number of floors, and a [km^2] $\ll A$ the size of an average one-person office including its prorated share of common space such as stairwells, corridors, and lavatories. The total number of commuters can then be expressed as pnA/a.

Next, evaluate the perimeter's capacity — the maximum number of people who can enter the CBD in 1 h by car. To this end, let s be the spacing between adjacent city streets, q the maximum number of cars that a street can carry in one direction in 1 h, and k the average car occupancy. Now, since the CBD is a square, its perimeter length is $4\sqrt{A}$ [km] and the total number of streets crossing it is $4\sqrt{A}/s$. Therefore, the maximum number of people able to enter the CBD in 1 h is $qk(4\sqrt{A}/s)$.

We can now explore the balance between demand and capacity by assigning reasonable numbers to the variables in these formulas. Data from the U.S. Bureau of Labor Statistics (2016) reveal that the average U.S. office space is about 200 ft^2/person, or approximately $a = 2 \times 10^{-5}$ [km^2]. In most U.S. cities, the length of a city block is somewhere between 300 and 500 ft, or approximately $s = 0.16$ km. Reasonably, we assume that the streets are bidirectional with two lanes in each direction and that each lane can carry about 500 vehicles/h on account of the red phases of the traffic signals and the disruptions of left turns and pedestrians. Then, $q = 1000$ [veh/h]. In 2010, the average passenger occupancy in the U.S. was approximately 1.7 across all private vehicle types (www.BTS.gov), so we shall use $k = 2$ to be conservative. In the same spirit, we choose small values for the remaining parameters: $A = 2$[km^2], $p = 0.5$, and $n = 10$. Combining these numbers, we see that the demand would be $pnA/a = 5 \times 10^5$ (commuters) and the perimeter capacity $qk(4\sqrt{A}/s) = 7.1 \times 10^4$ [commuters/h]. Since these two estimates differ by a factor of 7, it follows that the rush hour would have to last about 7 h — an unrealistically large number. This demonstrates the impossibility of car-only commuting for cities with concentrated CBDs.

Given the trend toward urbanization, especially in the developing countries, the values of A and n representative of CBDs in some of the world's megacities should be considerably larger than those assumed in the above calculations. In such cases, the problem becomes even more severe, since as the formulas show the demand/capacity ratio is proportional to $n\sqrt{A}$.

What can we do? We cannot easily change a city's structure, captured by the parameters p, n, A, and a; nor can we keep building new roads (reducing s) or expanding existing lanes (increasing q). The easiest parameter to change in the example is k; the vehicle occupancy. This is fortunate because the demand/capacity ratio varies inversely with it. It is also fortunate because it can be changed by nearly two orders of magnitude with buses and by three with long trains. This can have a huge effect. For example, if 25% of the vehicles entering the city were buses filled with $k = 50$ passengers instead of two-occupant cars, then the average vehicle occupancy would be $k = 14$ instead of $k = 2$. It would increase by a factor of 7. On the other hand, buses may have a deleterious effect on the perimeter's vehicular capacity since they have a larger footprint. To account for this effect, we treat each bus as the equivalent of two cars when calculating capacity, i.e., we assume that the time needed by 100 vehicles (including 25 buses) to cross the perimeter is the same as the time needed by $25 \times 2 + 75 = 125$ cars. Thus, the time needed to process the same number of vehicles when 25% of them are buses increases to $125/100 = 5/4$ of the original. Conversely, the perimeter's capacity is reduced to 4/5 of the original. Now, combining the demand and capacity factors, we see that the demand/capacity ratio would decrease by a factor of $7 \times 4/5$. Thus, the introduction of buses would reduce the rush from 7 h to $5/4 = 1^1/_4$ h. Buses would make our hypothetical city viable, and benefit car commuters to boot.

Although the example is based on a very extreme scenario, it clearly illustrates the essential need for public transportation in densely populated metropolitan areas. The example also shows that things get worse for larger and denser cities. In fact, when A and n are larger, even extensive use of buses can result in unacceptably long rush periods. In these cases, a transit system with larger vehicles and operated on a separated right-of-way (ROW), e.g., an underground subway, is needed.

The reader may have noticed that in our example most of the commuters travel by bus (50 for every 6 that travel by car) and this is a much larger transit share than the 30% reported by the American Community Survey. The reason for the discrepancy is that the Survey result includes *all* commuting trips in cities and not just those destined for dense CBD's. But the transit share of the example is realistic. Take for example, the commute from San Francisco's East Bay to San Francisco itself. These two communities are joined by the San Francisco Bay Bridge, which conveys all the car traffic from the East Bay and distributes it to San Francisco through numerous off ramps; and by BART — a suburban rail line. During the rush, both systems work at capacity. With five lanes, the bridge can theoretically carry about 10,000 vehicles/h, but the entrances into San Francisco can only accommodate about 5,000, and the bridge gets congested every morning as a result. Therefore, 5,000 is the perimeter capacity of San Francisco. By comparison, BART runs long trains, which can easily carry well over 1,000 passengers each, every 3 min during the rush hour. This conservatively works to about 20,000 passengers per hour, i.e., *four* times more than the perimeter of San Francisco can absorb by car. (BART reports that its ridership during the whole rush is 60,000.) Thus, the real-world ratio of transit to car commuters into San Francisco from the East Bay must be about 4:1, which is much larger than that in the Survey. Imagine what would happen to communities like the San Francisco Bay Area, Manhattan, Paris, or Shanghai if transit was taken away.

As the above discussion illustrates transit is of benefit because it makes travel over long distances possible in dense cities, with fewer vehicles in operation and less congestion. It should be apparent and *stressed* that it benefits non-transit travelers immensely. This positive "externality" is fairly well recognized by the public, even in the car-centric U.S. According to the

Future of Transportation National Survey (2010), nearly 60% of Americans believed that improved public transportation systems coupled with better walking and biking environments are the most promising means toward a less congested and more sustainable urban future.

1.2. Expectations and Challenges

This section will explore what transit can do for the different population segments affected by it and the challenges that stand in the way.

1.2.1. *Stakeholders and Expectations*

A public transit system, just like any other transportation system, involves multiple stakeholders. Here we recognize four classes: passengers, operators, the community at large, and the government. Each of these groups has its own perspective and expectations.

Passengers normally care about the individual experience associated with their trips. Although they may have different tastes, most users prefer quick access, short waits, few transfers, fast movement, and safe and reliable travel. Surveys show that reliability (the ability to predict arrival times) is of paramount importance; even more than short travel times.

By operators, we mean the agencies that run the systems. They can be non-profit public agencies or for-profit franchisees. Both should be interested in running a tight ship; the former because it shows that management is effective and the latter because low costs maximize profit. And both should be interested in providing good service because this is the only way in which the operators can attract demand. To achieve both goals operators should deploy all the resources they control (thanks to subsidies and farebox revenue) in the best way possible to serve the public. And they must make value judgments in the process, with due consideration of public pressure. For example, an agency may want to weigh how much emphasis to place on the segments of the public receiving the worst service, as compared to the majority of users.

The community at large includes all people exposed to the social, economic, and environmental externalities of a transit system. These externalities include congestion reduction, land value changes, energy consumption, gas emissions, noise, and safety. Automobile drivers (and bus riders) desire less congestion; property owners the higher rents near popular stations; people who live along the ROW less noise, pollution and pedestrian accidents; and many in society a greener footprint.

Last but not least, the government is an important stakeholder and driver of policy. At the most basic level, government serves as a steward of the community by enacting and enforcing safety and other type of regulations for the industry. Government's involvement does not end there, however, because with rare exceptions transit is not financially self-sufficient. For this reason, the government nearly always subsidizes transit's capital costs and often also subsidizes its operating costs. This can turn out to be highly desirable for society as we shall see in Chapter 11. The example in Section 1.1 also supports this statement — it strongly suggests that car drivers in crowded cities would benefit by cross-subsidies that encourage other drivers to switch to transit, as then travel times for both groups would dramatically decline. This would be a win–win proposition for both groups and therefore highly desirable. Something similar happens with "value capture" taxes. Subsidies for station construction can be funded by slightly taxing the profits landowners make when stations are built near their properties, and again everyone wins. Thus, in most cases, subsidies make sense, so the government is both a regulator and funder of transit. In some countries, however, the government's role is even more extensive. Their governments, or proxies such as state-owned companies, may completely own and operate the system.

The above discussion shows that the desires of the different stakeholder classes are not perfectly aligned. Since conflicts may arise as a result, it is also interesting to review these objectives and see how they relate to the stakeholders. This review will also be an opportunity to introduce terminology. The following seven objectives are considered:

- *Accessibility*: Accessibility (to activities) is the ability to engage in a wide choice of spatially scattered activities. For a single person, accessibility can be measured by counting the number of potential activities (opportunities) that the person can reach from his/her home or workplace locations. For the system as a whole, it can be measured by averaging the accessibility of different locations across the city. To simplify matters, one sometimes assumes that opportunities are distributed in proportion to the population density (or uniformly) and then measures it by the percentage of the population (or area) that can be reached within a stipulated time budget from different points in the city (see e.g., Vuchic, 2005). Accessibility can be improved by increasing the spatial density of activities, and also by improving the performance of the transportation system in general and transit in particular. Accessibility is very important to transit users. It is, in fact, one of the most important factors influencing the public's perception of public transit. The Future of Transportation National Survey (2010), for example, revealed that the top reasons for transit non-users to avoid public transportation were "not available in my community" (47% of responding non-users) and "not convenient to my home or where I need to go" (35% of responding non-users). Both responses are markers for low accessibility. Governments and operators are also interested in accessibility as an objective and a highly relevant measure of performance.

- *Mobility*: We define mobility (of a transit service) as the average distance that passengers can travel in a unit time. Mobility is obviously related to accessibility since more mobility implies more accessibility. For example, in the special case where opportunities are uniformly distributed in space, accessibility is proportional to the square of mobility. Transit mobility is therefore important to users, operators, and the government. Transit's mobility is often compared to that of the automobile. Many factors influence the former. They include the service frequency which influences waiting and transferring times, the separation between stops which affects walking and riding times, and the circuity of the transit routes and the vehicles' cruising speed, which affect riding times. Like accessibility, higher transit mobility implies better service in a way that strongly affects passengers' willingness to ride transit. In a 2014 Mobility Attitudes Survey with 11,842 respondents (Transit Center, 2014), all age groups chose low travel times (i.e., high mobility) as the most important driver toward riding transit. Government and operators are also interested in high mobility for the reasons already stated. But in the case of operators, there is also a self-interest angle. When higher mobility is achieved by more direct routes with faster vehicle travel and fewer stops, vehicles can be turned around in less time. Thus, vehicle productivity improves (see below) and the cost of providing service declines.
- *Vehicle productivity*: We define vehicle productivity as the ratio of passenger-distance served (the service output) to the vehicle-time in operation (the resource input). This indicator has the physical units of speed multiplied by vehicle occupancy. It is of interest to operators and the government, as a measure of efficiency in the use of streets. Transit systems typically have a much larger productivity than personal cars because they can carry more passengers. High productivity is desirable because it ensures that vehicles use space efficiently. The ability to convey high numbers of passengers has important consequences. Recall from the example of Figure 1.1 how the duration of the rush hour, and consequently traffic congestion and all its environmental impacts, are

reduced significantly when high-productivity modes (such as buses or subways) are used.

- *Reliability*: Reliability, i.e., the predictability of travel time, is very important for transit users. The Mobility Attitudes Survey (Transit Center, 2014) showed that travel time reliability, together with total travel time, was ranked by all age groups as one of the top two most important mode-choice factors. Since traffic congestion can significantly affect the reliability of surface transit modes, passengers often resort to segregated modes such as metro or light rail. Traffic congestion is not the only factor affecting reliability, however: poor operational practices, e.g., a disregard for the schedule, slow passenger boardings, maintenance practices that lead to frequent equipment failures, and no planning for unexpected disruptions can severely impact reliability, as we will see later in the book. Disruptions due to equipment failures turn out to be a prime concern for transit agencies. A survey conducted by Smart *et al.* (2009) revealed that transit managers regarded these types of failures as the 5th most important metric of transit system performance. Of course, to achieve good overall reliability, a transit agency must not only maintain all its equipment and infrastructure in a state of good serviceability, but also train their operations personnel properly, have extra drivers and vehicles for times of need, and use resilient control strategies. The effort must go on continuously.
- *Safety, security, and comfort*: Transit service should be safe (accident-free), secure (crime-free), and comfortable during every part of a passenger's trip. Important to passengers, this applies not just to the vehicle ride but also to the parts of the trip that occur outside the vehicle: accessing the stations, waiting, boarding, alighting, and transferring. Our own informal surveys show that some people (especially women) prefer to transfer aboveground rather than in long underground tunnels at night for safety and security reasons, and this influences their mode choice. Good lighting, cameras, and security systems can help improve safety and security. Safety and security is also important for the transit agency's field employees. Comfort also affects mode choice. Comfort is closely associated with crowdedness and is often measured by the availability of space and seats. Crowdedness triggers not just physical but also psychological discomfort in the form of anxiety, stress, and feelings of low privacy. Surveys have shown that passengers start to feel uncomfortable when the probability of getting a seat drops below 40% (Vovsha *et al.*, 2014) and that a perception of discomfort affects passengers' route and mode choices (Li and Hensher, 2011; Tirachini *et al.*, 2013). Other factors affecting comfort are the smoothness

of the ride (Castellanos and Fruett, 2014), the amount of vibration (Sekulic *et al.*, 2013), and the ambient environment created by temperature, ventilation, and noise (Zhang *et al.*, 2014). Transit operating agencies can influence many of the factors affecting passenger safety, security and comfort, and governments can institute regulations to encourage follow through.

- *Affordability and equity*: Low transit fares are quite important to most transit passengers. According to Mattson *et al.* (2012), about 60% of the respondents to the 2009 National Household Travel Survey, including both auto drivers and transit riders, stated that the price of travel was a "big issue" and another 30% said it was a "moderate issue." This affordability perception obviously depends on a passenger's socioeconomic status, and sometimes on the cost of competitive modes. A survey conducted by Taylor *et al.* (2010) found that "inadequate transportation continues to be a greater problem for people with incomes of $15,000 or less, among those both with and without disabilities." In that survey, 26% of non-disabled respondents with annual household incomes at or below $15,000 identified their transportation as inadequate, compared to just 13% of those who earned $35,000 or more. Fares are important to operators in a different way because they are a source of income. This income is also important for the government because high farebox revenue reduces the need for subsidies. While both operators and the government have an interest in maximizing farebox revenue, transit users do not, and neither do the recipients of transit's congestion–reduction externality. This issue will be revisited in Chapter 11. Be that as it may, it should be clear that the decision of what fares to charge is a sensitive political matter. It is particularly delicate because the fare-setting body has to decide how equitably to treat disadvantaged social classes.
- *Sustainability*: As the global trend toward urbanization continues, public transit has become the tool to make urban transportation sustainable. This is of interest to users and non-users alike. For the same trip production, well organized public transportation systems can (i) reduce congestion; (ii) densify land use, especially if encouraged by transit-oriented development (TOD) policies; and (iii) reduce noise, greenhouse gas emissions, air pollution, and energy consumption (Federal Transit Administration, 2010). For example, in the United States, metro and light rail transit on average consume 1.7–1.8 MJ of energy for each passenger per kilometer of service; this is about 30% lower than the 2.3–2.6 MJ/passenger-km consumption rate of a passenger car or personal truck. In terms of air pollutants, transit buses emit 0.95 g of

nitrogen oxides, 1.89 g of carbon monoxide, and 0.12 g of hydrocarbon per passenger-km, in contrast to 1.28 g/passenger-km, 9.34 g/passenger-km, and 1.3 g/passenger-km, respectively, by single-occupant automobiles (Davis *et al.*, 2009). In addition, TOD policies have succeeded in densifying land use, not only conserving land but also decreasing the distances people need to travel to reach their destinations. These reductions are important because they improve accessibility without increasing the demand for travel. Transit-oriented development is frequently seen in European and Asian cities (e.g., Hong Kong) and also in some major U.S. cities (e.g., Boston, Chicago, New York, Philadelphia, Denver, and the San Francisco Bay Area). For a detailed discussion of TOD, see Cervero and Landis (1997), Casello (2007) and Ratner *et al.* (2013).

1.2.2. *Challenges*

The previous subsection argued that good public transportation is a necessity for well-functioning dense cities, and described what it can deliver for its various stakeholders. Here we explore some of the challenges that need to be overcome to deliver on this promise. Methods for achieving this objective are the focus of this book.

The first challenge is economic. Most transit systems in the world require some form of subsidy to be financially viable. Given the positive externalities of transit, subsidies make sense, at least in the abstract. However, determining and securing the desired amounts is always a practical challenge. This book will not explore funding mechanisms in detail as this is more appropriate for a book on finance and public policy, but it will examine how to estimate the costs needed to provide any desired level of service. This type of information is useful for budgeting, e.g., when considering funding options, because there is always a tradeoff between the quality of service and the operator expenditures. The information is important because transit is costly — after all, providing mobility for spatially and temporally distributed travel requests on a mass scale is an enormous and complex task.

The second challenge is low demand. Low demand is problematic because the cost of providing good transit service is high even if the vehicles are nearly empty. Understandably, societies may be reluctant to fund in a big way little-used transit systems. Unfortunately, when this happens reductions in funding beget cuts in service which further depress demand, making the problem worse. The end result is a poor transit system that is lightly used; or no transit system at all. Called the "Mohring effect" (1972), this vicious cycle is particularly strong in low-density urban regions of car-centric societies, which is logical because in this type of operating environment transit cannot come close to matching the automobile's performance. The U.S. is particularly vulnerable to the Mohring effect. Kenworthy *et al.* (1999) reported that in 1990 an average American drove about 16,045 km and rode transit for only 474 km, i.e., less than 3% of the total, whereas an average European traveled 6,602 km by car and 1,895 km by transit, i.e., about 23% of the total. In Japan and Hong Kong, with even denser urban areas, the transit percentages exceed 60% and 80%, respectively (Kenworthy and Laube, 1999). This pattern of lower transit ridership in the U.S. than in the rest of the world has persisted for decades.

Low ridership also has environmental implications. While well-used transit systems consume less energy and generate fewer emissions than cars per passenger-kilometer, low ridership can negate this benefit. This occurs in the U.S. With an average occupancy of about 10 passengers per vehicle, U.S. buses consume more energy per passenger kilometer (2.78 MJ/passenger-km) than an average passenger car or a personal truck (2.30 or 2.60 MJ/passenger-km) (see Davis *et al.*, 2009). A silver lining is that a slight increase in ridership could reverse this statistic, and that ridership is already larger in the densest urban areas where the environmental impacts are felt the most.

The next challenge is posed by the nature of the task (grouping people into shared vehicles) which requires both the passengers and the vehicles to adjust their travel paths. For this reason, an operating agency cannot provide unlimited mobility, even with unlimited resources. Indeed, the temporal and spatial dispersion of demand forces transit vehicles to make detours to visit the locations where passengers can access the system and to spend valuable time allowing these passengers to board and alight. These two necessities increase the time passengers must spend onboard to travel a given distance. As we will see in later chapters, there is a floor to this travel time that is fairly high, but with good planning, one can get close to it. Just to give you an idea, in the U.S. transit buses travel at about 17–40 km/h (including stops) and rail trains at about 30–44 km/h (American Public Transportation Association (APTA), 2016); and these figures do not include the delay due to detours. The onboard delays transit riders experience due to stops and detours explain why transit, no matter how well planned, can only be broadly attractive in situations when door-to-door automobile travel is itself retarded due to congestion and lack of parking.

Although it is in these situations where transit is most needed, it should always be planned and managed so it performs at its potential. Typical planning and management questions, which will be discussed extensively throughout the book, are the following: For a given budget, what mixture of resources should an agency employ? How should they be deployed in space and time? Should there be stations? Where should the stations be? Should there be routes? How should vehicles be routed? Should there be schedules? What should the schedules be? And given these choices, what will the passenger's experience be? How should specific buses be deployed to cover the service program that is advertised to the public? How should drivers be rotated to populate these buses? The list goes on, but the reader probably gets the idea. Planning and managing the system is complex. Therefore, a systematic approach is desirable. This is a main focus of this book.

The last challenge is operational and, like the previous one, also needs to be met by the transit agencies. Left alone, transit vehicles cannot provide the regular reliable service included in the plans because the vagaries of station dwell times and traffic prevent it. For this reason, most transit systems are operated under some form of real-time control. This can work well for rail systems operated in isolation from other traffic, but otherwise, service often deteriorates severely throughout the day despite an agency's best efforts. A reason is that if the control allows for significant unevenness in the separations between the route's vehicles then the transit system develops a resistance to the control and the unevenness worsens. (The resistance arises because vehicles at the tail ends of large gaps are slowed by the extra passengers they need to pick up, so the gaps grow larger.) Unfortunately, the forms of strict control required to overcome this problem have not worked their way into practice yet, so the problem persists. This hinders reliability and dampens demand. Given the importance of the issue, Chapter 10 will describe transit control in considerable detail, including some promising solutions that are in the offing.

1.3. Basic Typology of Transit Systems

Transit systems come in many different shapes and forms. This is fortunate since the physical, geographical, socio-economic, and environmental conditions under which service must be provided vary drastically around the world. A broad menu of transit types allows cities to better match systems to their needs. In view of its variety, it is important to know what the menu has to offer. For the purposes of this book, it suffices to categorize transit systems according to four attributes: the ROW, the infrastructure, the vehicle, and the type of service.

1.3.1. *ROW*

This attribute refers to the path (or way) in which transit vehicles (have the right to) travel. Of particular interest is the extent to which this right is shared with the automobile and other transportation modes such as pedestrians and bicycles because sharing critically affects speed and performance. In general, the more extensive the separation, the better the performance, but the higher the capital costs of infrastructure. Thus, determining the appropriate level involves a delicate balance. The stakes are so high that transit routes will sometimes use different separation levels in different parts of a city. Three levels are commonly recognized: (I) no separation, (II) partial separation, and (III) full separation. Level I refers to systems where vehicles operate on surface city streets, mixed with other traffic. This is the most common configuration for bus operations. Understandably, this type of ROW produces the lowest service quality, as vehicles are subject to interference from ambient traffic conditions.

Level II includes systems that operate on surface streets but enjoy some separation from the other modes. This separation can be provided both, by physical barriers and by the right to use exclusive lanes that are not physically separated from the rest. At intersections, where separation is not possible, transit vehicles may receive some form of priority. Physical separation allows for larger, less maneuverable vehicles, and the whole level II setup for faster travel. Therefore, vehicles can be considerably more productive. Two prime examples of level II systems are "light-rail transit" (LRT) and "bus rapid transit" (BRT). The former, also known as "streetcar," "tramway," "tram," or "trolley car," uses relatively lightweight vehicles on rails embedded in the pavement. The latter is the bus version of LRT; it uses large capacity buses with widely spaced stations. To further enhance productivity, LRT and BRT stations are typically outfitted with elevated platforms and ticket dispensing machines that allow users to embark and disembark quickly — without stepping up or down, or delaying everyone else while purchasing tickets. Of course, all the infrastructure and equipment improvements of level II systems require a considerable investment.

Finally, level III refers to systems with a totally separated ROW, either underground or above ground, and whose vehicles do not interact at all with the surface modes. This level includes monorails, subways (a.k.a. "metros"), automated guided transit (a.k.a. "personal rapid transit"), and commuter rail (a.k.a. "regional" or "suburban" rail). Obviously, systems at this level are very expensive but full separation allows designers to configure stations and guideways that can accommodate very long vehicles — trains that can carry many hundreds of passengers — and high speeds. If demand is high, this can increase the system's productivity enough to justify the expense. Figure 1.2 shows real-world examples of the three ROW types.

1.3.2. *Infrastructure*

This is the fixed part of a transit system. It includes the guideway, the stations, and other fixed operational facilities. The guideway is the structure that supports vehicles as they travel, and is the part of the infrastructure that differs most across transit systems. Although its primary function is to support vehicle motion in the vertical direction, some guideways also help control vehicle motion on the horizontal plane — only laterally across the direction of travel for the most part, but sometimes also longitudinally.

The most basic type of infrastructure only provides support — so vehicles must guide themselves laterally and longitudinally. This type includes street pavements and bodies of open water such as bays and estuaries. These are the cheapest guideways because they already exist and may only need to be upgraded for transit service. They are normally used with level I ROWs.

Also common are guideways that provide both, support and lateral guidance. Typically associated with levels II and III of ROWs, these include monorails, dual rail tracks, and even cables for aerial trams. Often, these guideways are outfitted with additional equipment to help propel the vehicles in the longitudinal direction. For level III ROWs, the propulsion aids can either be embedded in the guideway itself (e.g., as a high-voltage third rail for electric systems, a second cable for aerial trams, or propulsion coils for magnetic levitation applications) or be overhead in the form of electric cables. Embedding is typical for underground subway systems and overhead wires for above ground rail. Overhead wires are also used for ROWs of levels I and II involving trolleybuses and trams.

Stations are a very important part of the infrastructure. These are the locations where passengers enter and exit the vehicles to begin and end their trips, and also to transfer in the middle. Stations should provide access for transit vehicles and passengers, comfort and shelter for passengers, good information, and a safe and efficient interface for boarding to occur. In some cases, ticket dispensing systems are also provided.

(a) ROW I: Buses on Beichen West Road in Beijing, China.

(b) ROW II: Tram approaching the Blauwbrug Bridge in Amsterdam, Netherlands.

(c) ROW III: Metro train on West Lake St. in Chicago, USA.

Figure 1.2. Illustration of ROW levels.

Information is critical to passengers, especially those who are not familiar with the system. Essential information includes maps with routes and schedules and the location of boarding bays for specific routes. Dynamic vehicle information, e.g., with estimated arrival times, is also highly desirable as it reduces passenger anxiety and allows them to determine whether an alternate route is advisable. Rarely provided, but very inexpensive and clever for its simplicity is a Japanese invention for subways that helps reduce walking. Every boarding platform displays a table indicating which car a passenger should board for easier access to important landmarks at different destination stations — because on arrival at the destination station the chosen car will be closest to the best exit tunnel for the corresponding landmark. The table helps reduce unnecessary walking, saves time by allowing passengers to do some of the necessary walking while waiting at the origin, and reduces crowding by encouraging passengers to board different cars. In all applications, good information is relatively cheap, so it should *always* be provided.

Stations can range hugely in cost and complexity, and this is what differentiates them most. For level I ROWs they can be as simple as a post with a schedule and route number. For level III ROWs, they need to be whole buildings with escalators, gates, walkways, platforms, and all the components necessary for proper functioning. These buildings must be particularly large if they are to accommodate longer trains, multiple routes, and transfers. And burying these buildings underground increases the capital cost significantly. Large intermodal transfer stations can be larger than a city block and enormously expensive. In fact, for metro systems, the cost of building the stations can exceed the cost of the guideway. Obviously, these costs must be seriously and separately considered when deciding the number and size of the stations a system should have.

Other infrastructure facilities include depots where vehicles are stored and repaired, a control center and the agency's headquarters. These facilities do not differ substantially across agencies to deserve much discussion for the purposes of this book.

1.3.3. *Vehicles*

Vehicles are the moving parts of the transit system. They must offer comfort to passengers and have an ability to move in a controlled way. Three attributes affect productivity the most: their passenger holding capacity, their cruising speed, and the rapidity with which passengers can embark and disembark. All three will play an important role in this book, and all three can be achieved with different technologies. These are described below.

Most transit systems with level III ROWs, subways in particular but also LRTs, use steel wheels on steel tracks because of their low rolling resistance. Vehicles are typically propelled by electric motors, which enjoy better energy efficiency due to regenerative braking and the lack of gears, plus lower emissions where they operate even if the electricity is generated from dirty fossil fuels. Electric vehicles also produce less noise. Some level III systems are controlled by either automatic or semi-automatic guidance systems. Others, especially in hilly areas, may use rubber tires on a hard surface to benefit from the higher friction this combination produces. To facilitate embarkation, vehicles of level III ROWs are usually endowed with wide doors, level with station platforms.

Level I systems require vehicles to have more independence and for this reason, the internal combustion engine, sometimes in hybrid form, is a common option. Two exceptions are trolleybuses, which derive their power from overhead wires, and battery-powered electric buses. The latter present operational challenges because today's batteries do not last for a full day — as a result, transit agencies have to include recharging intervals into their plans, which generally reduce vehicle productivity. (This problem may be eliminated if in a not-too-distant future streets are wired-up to charge a moving vehicle's electric batteries over the air, or if vehicle batteries could be quickly changed.) Vehicles of level I systems, at least today, require human drivers for control. For the most part, they are slower and less capacious than level III vehicles.

1.3.4. *Type of Service*

The transit attributes described so far are physical characteristics of the system's hardware. The final attribute describes its software, i.e., how the system is organized to provide service. Although there are many possible organizations, one discriminating factor — whether or not fixed stations and routes are used — divides the universe into two very different classes. The two classes are so different that they require different analysis methods, and are covered separately in this book.

Most transit systems, especially in dense urban areas where individual transit vehicles carry many passengers (as they should) serve their customers using fixed stations, routes, and timetables, which advertise either a schedule or service intervals. We call them "fixed-route" systems. The reason for fixing the configuration is efficiency. To get many passengers into each vehicle, efficiency dictates that they should be picked up and dropped off in groups at preset locations — since doing it individually would delay too many people already onboard and reduce vehicle productivity. Fixed stations serve this consolidation function. They are where people congregate for pick-up, and where they can exit in groups. The fixed routes and timetables allow people to plan their trips.

By contrast, when the transit service area is not dense, so demand is light and transit vehicles lightly loaded, it can make sense to pick up and deliver passengers individually as the need arises, going where they are, and doing away with stations and timetables altogether. This often occurs in rural and suburban areas, and at night when the demand is low. It also makes sense when customers are willing to pay for automobile-like level of service. This type of operation, without stations, preplanned routes, or timetables, will be called "flexible transit." Examples of flexible transit include taxi service, ride-sharing services, and some forms of paratransit. Systems such as *Uber*, *Lyft*, and *Scoop* fall into this category.

Of course, the world is not black and white, and there are hybrids which incorporate features from both fixed-route and flexible transit. For example, there are systems that combine fixed stops with flexible routes. Examples are airport shuttles, which have one or more fixed stops at the airports they serve but undefined stops in their service regions, and paratransit services for commuter rail stations. When the latter are synchronized with the trains they also have a timetable, so this kind of hybrid exhibits not just a partly fixed route but also a partly fixed schedule.

There is even more variety. Some communities, especially in the developing world, are served by "hub-and-spoke" systems of "jitneys" or "shared taxis." These systems dispatch vehicles from a central terminus over a published route but according to a flexible schedule, and usually with flexible stops. A common rule is that vehicles are dispatched once they are full. Routes can span a whole city and passengers can use the terminus as a transfer point to reach points other than the terminus. More than one terminus can be used if the city is large and multi-centric.

1.4. Cost and Performance

It is useful to think for planning purposes that transit systems are made up of three basic building blocks (routes, stations, and vehicles) and that the purpose of planning is choosing and arranging these ingredients in the best way possible. All three have unit costs and other physical characteristics that affect performance. Obviously, these features need to be quantified, at least approximately, before one can make sound choices. This section gives a rough idea of the magnitudes involved.

As we will see in this book, three performance characteristics turn out to be most important for planning and design. The first is vehicular capacity. This feature is important because it directly influences vehicle productivity which, as explained in Section 1.2.1, helps match the required resource input (number of vehicles) to the desired output (served demand). Obviously, for a given number of vehicles, the larger the capacity is, the larger the demand that can be served.

The second important characteristic is the cruising speed achieved by the ROW–guideway–vehicle combination. As we explained in Section 1.2.1, cruising speed strongly influences passenger travel time, and therefore mobility. It also affects vehicle productivity.

The third and final characteristic is the dwell time of vehicles at stations. This characteristic depends on a number of factors such as a vehicle's acceleration and deceleration capabilities, but it is crucially influenced by the rate at which passengers can embark and disembark when a vehicle's doors are open at a station. Large values of such rates reduce station dwell times and therefore reduce overall travel times; also, improving the system's productivity and the passengers' mobility.

Before we review the specifics, one should keep in mind that there is always a tradeoff in selecting our ingredients: the better their performance the higher their cost. For example, level III ROWs exhibit better performance in terms of speed, capacity, and boarding rapidity than level I or II systems, but they are also costlier. On the other hand, a level I bus route will be slower, less reliable, and less productive than level II or III systems but will require little infrastructure investment and be less expensive. The only way in which this tradeoff can be resolved is with a planning process that properly accounts for the key ingredient features. Because these features depend very heavily on the particular city, the two subsections below can only give a rough idea of what is involved. Cost is discussed first, and then performance.

1.4.1. *Costs*

The costs for transit systems include not only the capital costs associated with the acquisition of the ROW, the fixed infrastructure of guideways and stations, and the vehicles, but also the recurring costs associated with day-to-day operations. The latter include employee salaries and benefits, as well as the cost of supplies and consumables, such as electricity and petroleum. According to the APTA almost 100% of transit's capital costs in the U.S., and about 64% of its operating costs are covered by subsidies from the federal, state, and local governments. The remaining 36% of the operating costs are covered by the agencies' revenue from passenger fares. These data are from the *2015 Public Transportation Fact Book* of the American Public Transportation Association (Neff and Williams, 2015).

Capital cost. Table 1.1 below lists the U.S. transit capital expenditure by major modes and by types of assets in the year 2013. All data are from the above-mentioned APTA's Fact Book, which defines capital costs as "expenses related to the purchase of equipment" that is "non-expendable," "tangible," and "having a useful life of more than one year."[1] First, note the distribution of the costs across the three subcategories, and how the costs for "Infrastructure and Facilities" dominate the total for all rail-based modes. This should not be surprising since rail-based modes are nearly always operated on expensive levels II and III ROWs with large specialized stations. In contrast, the vehicle costs for the "all bus" and "demand responsive" modes, which for the most part are operated on level I ROWs, with either very simple stations or no stations at all, constitute more than half of the total.

The table also breaks down the infrastructure cost by subcategory (e.g., guideways, stations, and supporting facilities). The row on "unit guideway cost" is of particular interest to us. It is the ratio of the total capital guideway cost in that particular year to the total number of kilometers of directional service line. Note how this unit cost is one to two orders of magnitude higher for rail-based modes than for bus-based modes. For example, we see that the annual capital cost of a unit of guideway is practically negligible (i.e., about $500 per km per year) for level I buses, but is on the order of $1 million per kilometer per year for light and heavy rail. The latter can further increase by several orders of magnitude when expropriations and/or tunnels are necessary, especially in expensive urban areas. For example, the upfront capital costs of typical LRT systems in Europe are about 20–30 million euros per km of guideway constructed, including stations, and considerably more for underground subways. The construction cost per kilometer of the latter can range from $100 million to more than a billion in London and New York (Davies, 2012) due to pricey real-estate and the high expenses associated with tunneling.

---

[1]These costs are reported in one fiscal year, so they should not be interpreted as the "upfront" total expenses for constructing the entire infrastructure in operation that year. Instead, they roughly represent the "prorated" capital expenses on a per-year basis.

Table 1.1. U.S. transit capital costs reported in 2013.

Item	All bus		Heavy rail		Regional rail		LRT & streetcar		Demand responsive	
Infrastructure & facilities										
guideway (million$)	215.6	5%	2,344.4	38%	1,276.9	42%	2,569.4	73%	0.0	0%
Stations (million$)	443.8	10%	1,718.5	28%	339.2	11%	307.7	9%	22.4	4%
Admin & maintenance facilities (million$)	756.6	17%	224.6	4%	190.5	6%	130.4	4%	83.9	14%
Subtotal (million$)	1,416.1	31%	4,287.4	70%	1,806.6	60%	3,007.6	86%	106.3	18%
Directional route length (km)	428,024		2,610		14,389		2,553		—	
Station count	0.5		898.1		88.7		1,006.3		—	
Unit guideway cost (1,000$/km)	—*		1,044		1,296		887		—	
Unit station cost (1,000$/station)	—		1,646.1		261.7		346.9		—	
Vehicle										
Subtotal (million$)	2,361.9	52%	441.1	7%	780.2	26%	312.1	9%	412.3	69%
Available vehicle count	71,139		10,380		7,369		2,387		68,559	
Unit vehicle cost (1,000$/vehicle)	33.2		42.5		105.9		130.7		6.0	
Average purchase price (1,000$/vehicle)**	486.8		2,254.4		2,444.6		3,300.0		77.6	
Other										
Subtotal (million$)	746.4	16%	1,428.4	23%	437.8	14%	195.1	6%	81.4	14%
Total	4,524.4	100%	6,156.9	100%	3,024.6	100%	3,514.8	100%	600.0	100%

Source: Neff and Dickens (2015).

* The APTA Fact Book reports a total of 1,627 bus stations in the U.S. in 2013, but this number does not include street stops which constitute the majority of bus stations. Hence, we omit the count of bus stations here to avoid confusion.

** These data are based on samples from 2012 to 2014 for, respectively, standard two-door bus with length $\geq 27'6''$, heavy rail car with one level and one cab, locomotive-hauled commuter rail car with two levels and 0 cab, single articulated light rail car with one level and two cabs, and demand response vehicle (minibus, van, car, SUV) with length $< 27'6''$.

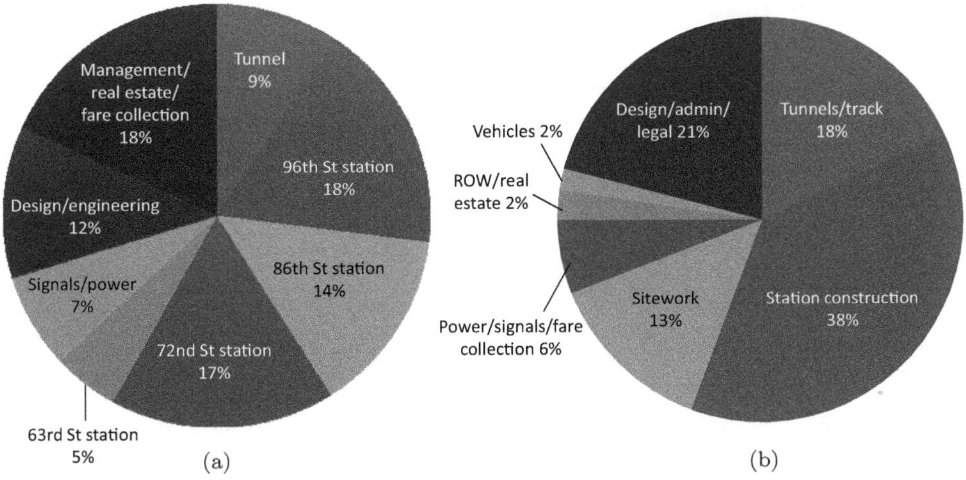

Figure 1.3. Cost breakdown for (a) 2nd Avenue Subway in Manhattan, phase I ($4.5 billion, 1.6 miles) and (b) San Francisco Central Subway ($1.6 billion, 4 stations, 1.7 miles).

What about stations? Their costs also vary significantly across modes. A web search reveals that a simple bus stop with a shelter costs about $10,000 in the U.S., although the actual figure can range widely from $5,000 to 30,000. The yearly maintenance cost is also on the order of $10,000, but this can be partly offset with advertising. For LRT and metro, station costs are much higher and more variable. Table 1.1 lists the capital costs for rail passenger stations in the U.S. as of 2013 on an amortized basis. They range from $0.26 to $1.65 million per station per year. These numbers, however, are very dependent on the specific country, the value of real estate and the size of the station. For example, the amortized cost of the San Francisco Transbay Station, which spans well over a city block, is an architectural landmark and reportedly cost about $4.5 billion, will easily exceed $100 million per year. The cost of stations for subways in large and dense urban areas often dwarfs other construction costs, see for example, the cost breakdowns in Figure 1.3.

Vehicles are easier to price because transit companies request bids from suppliers, which in the U.S. are public records. Table 1.1 reports average prices for representative transit vehicle types in the U.S. The purchase price of a city bus is on the order of $500,000 and that of a rail car ranges from $2.2 million to $3.3 million.

Operating cost. Table 1.2 lists APTA's 2013 U.S. transit operating costs by mode and asset type. These costs include "expenses associated with the

Table 1.2. U.S. transit operating costs reported in 2013 (million \$, unless specified otherwise).

Type	All bus		Heavy rail		Regional rail		LRT & streetcar		Demand responsive	
Vehicle operations	10,706.7	52%	3,220.4	39%	1,905.5	35%	654.6	38%	1,560.8	30%
Vehicle maintenance	3,514.4	17%	1,279.7	16%	1,109.5	20%	352.7	21%	297.5	6%
Other maintenance	883.5	4%	2,225.6	27%	849.3	16%	303.7	18%	54.7	1%
Administration	3,176.3	16%	1,390.8	17%	886.7	16%	313.9	18%	657.1	13%
Outsourcing	2,166.5	11%	56.6	1%	700	13%	93.9	5%	2,587	50%
Total	20,447.4		8,173.1		5,451		1,718.8		5,157.1	
\$/passenger-km	0.57		0.28		0.29		0.43		1.48	

Source: Neff and Dickens (2015).

operation of the transit agency and goods and services purchased for system operation." This includes driver salaries and benefits.

The table displays unit costs on the basis of passenger-kilometers carried. Curiously, lower costs are associated with the modes using larger vehicles. Although this makes sense, the statistic is not a good indicator of cost efficiency because it also reflects demand conditions — after all, very efficient but empty vehicles will have a high unit cost. And this seems to be the case in the table because all the statistics are comparable or higher than the \$0.26 figure corresponding to the use of private vehicles.[2] The statistics would be much lower in cities like New York, Paris or Tokyo where transit vehicles, especially subways, are full most of the time. As we will see later in the book, a more universal and useful statistic for design is the operating cost per *vehicle*-kilometer and/or *vehicle*-hour.

To get an idea of these statistics for buses, consider that their operation, including maintenance, depreciation, and fuel, costs about \$2 per km in the U.S.[3] And considering typical wages and bus speeds, a driver costs about \$3 per km; so the combined operating cost per vehicle is about \$5 per km. Alternatively, we can price drivers separately on an hourly basis

[2]According to the U.S. Department of Transportation (U.S. Department of Transportation, 2016) the average cost of owning and operating an automobile in the U.S. in 2013 was about \$0.368 per km; and with an average occupancy of 1.4 passengers per private vehicle, the average cost per passenger-km turns out to be about 0.26.

[3]A bus life cycle model published by the Volpe National Transportation Systems Center (Kay *et al.*, 2011) reveals that the costs of a bus per kilometer are on the order of \$0.4 for gas, \$0.5 for depreciation and \$1.0 for maintenance.

since they have to be paid whether they are moving or not. A reasonable range for their wages is \$11 to \$30 per hour in the U.S., plus fringe benefits that approximately equal the wages (see Bureau of Labor Statistics, 2016). Ballpark figures for all modes can also be obtained by multiplying the values of the last row of Table 1.2 by the average vehicle load in passengers, which is given on the seventh row of Table 1.4. The resulting operating cost per vehicle-kilometer is about: \$6 for buses; \$7.5 for heavy rail cars; \$10 for regional rail, LRT, and streetcar; and \$2.3 for demand responsive service. It should be remembered that these are averages, provided only for guidance, and that individual cases will vary. For example, the cost of the train conductors (if used) should be prorated to the number of cars in the train, so longer trains will generally exhibit lower costs per train.

We close this subsection with a final word about cars. Does the low operating cost per kilometer (\$0.26) mean that driving a car is cheaper than riding a bus or even a subway in the U.S.? The answer is no. First of all, the \$0.26 figure is an average that includes rural areas and off-peak travel on highways. Obviously, the cost must be higher for urban travel, which requires low speeds and stop–go maneuvers, especially during peak hours. Furthermore, the \$0.26 figure does not include the cost of parking during the day or the cost of garaging the car at night. And these costs can be quite high: in 2016, a covered parking spot in San Francisco cost about \$17,000 per year (estimated for a minimal 12×22 sq. ft. space, using a going rate of \$66 per square foot per year). So garaging alone can nearly triple the cost of operating a car in an urban area — to nearly \$0.76 per kilometer in San Francisco — and this easily exceeds the cost of transit. Furthermore, we have seen that transit users only pay as fares about 36% of an agency's operating costs, i.e., about \$0.19 per passenger-km for buses, or about \$0.09 per passenger-km for heavy rail. By considering these factors, we see that a transit user's out-of-pocket expense is in many cases a small fraction of the cost of owning and using an automobile. However, for transit to be really appealing its level of service must be competitive with the automobile. The following subsection reviews those features of our transit ingredients that most affect the level of service.

1.4.2. *Performance*

The three most important performance-defining features of a transit system are cruising speed, vehicular capacity, and boarding rapidity.

We define cruising speed as the speed that vehicles can sustain in their guideway and ROW environments without stopping at stations. For levels I and II ROWs, cruising speeds are determined by the ROW environment, including speed limits and traffic. As such, these speeds are usually on the order of 20–30 km/h. For level III systems, speeds are determined by the vehicle and guideway technology. Urban rail systems will typically achieve somewhere between 50 and 100 km/h; and level III LRT systems between 30 and 100 km/h.[4]

Vehicular capacity is on the order of 10–20 passengers for small vans used for demand responsive service, about 30–40 seats for buses and about twice that for articulated buses. Rail cars and streetcars have more efficient floor spaces and can accommodate from 50 up to well over 150 passengers, depending on vehicle length and width as well as the density of standing passengers. For example, BART cars can accommodate up to 200 passengers under "crush" loads. Table 1.3 lists typical capacities for various vehicle types in terms of seated and total passengers.

Boarding rapidity very much depends on the door configuration and its relation to the station. A passenger embarking and paying a fare will take about 5 s if climbing steps, and about half that time otherwise (Kittelson and Associates *et al.*, 2013). A disembarking passenger will take about 2 s. The process can be facilitated by using multiple doors and streamlining the passenger flow, e.g., by using different doors for boarding and alighting and/or using wide doors that enable multiple passengers to move in and out simultaneously. The process can be further accelerated if steps are eliminated and fares are paid outside the bus, as with BRT and LRT systems. Under these conditions, passengers can board at a rate easily exceeding 1 passenger per second per door if the doors are wide. This is common for subway systems.

To give an idea of current transit conditions in the U.S., we close this chapter with some overall transit statistics for 2013 by mode (see Table 1.4). The first three rows encapsulate the aggregate supply. A vehicle's service distance and time are also called "revenue" distance and time. For rail modes, a "vehicle" indicates a railcar. The third row shows typical numbers of vehicles in a train for rail modes. The next three rows characterize the passenger demand. Note from the fourth and fifth rows how the majority of this demand is served by the bus and heavy rail modes, especially the former. Perhaps not surprisingly, regional rail serves the longest trips and therefore also accounts for a non-trivial fraction of the total vehicle distance. The table's last three rows show performance measures that arise from the balance between supply and demand. The seventh row shows

[4]See http://www.reconnectingamerica.org/assets/Uploads/bestpractice175.pdf for more details.

Table 1.3. U.S. transit vehicle characteristics.

Performance metric	Standard transit bus	Articulated transit bus	Heavy rail car	Commuter rail car	Light rail car	Demand response
Vehicle size (seated/ total pax)	30–47/< 93	65/120	31–80/ 70–227	148–190/ 148–190	64/ 125–153	<30/<40
Typical boarding time (sec per pax)	1.6–8.4	0.9–1.2	1.1–2.6 (without steps) 2.9–4.0 (with steps)			—
Typical alighting time (sec per pax)	1.2–2.0	0.6–0.8	1.4–2.0 (without steps) 3.3–4.0 (with steps)			—

Source: Neff and Dickens (2015); Kittelson and Associates *et al.* (2003, 2013).

Table 1.4. 2013 U.S. transit service statistics.

	All bus	Heavy rail	Regional rail	LRT & streetcar	Demand responsive
Vehicle service distance (million km)	3,343.9	1,084.2	537.4	167.4	2,197.4
Vehicle service time (million hrs)	161.1	32.6	10.3	7.1	92.2
Average train length (vehicles)	—	7.28	6.05	1.92	—
Passenger trips (millions)	5,330	3,817	487	510	223
Passenger distance (million km)	35,646.9	28,976.2	19,225.2	3,994.4	3,493.9
Average trip length (km)	6.7	7.6	39.4	7.9	15.6
Average vehicle load (pax/veh)	10.7	26.7	35.8	23.9	1.6
Average commercial speed (km/h)	20.8	33.3	52.2	23.6	23.8
Vehicle productivity (pax-km/veh-hr)	221.3	888.8	1,866.5	562.6	37.9

Source: Neff and Dickens (2015).

that the number of passengers per vehicle is consistently low across all transit modes, especially for the two bus-based columns. This, however, is a function of the low-density environment in which a majority of the U.S. transit services operate, and cannot be generalized to dense cities and other countries. The eighth row lists the average commercial speed, i.e., the average speed passengers experience, including stops at the stations. This important variable depends on three system features: the vehicle cruising speed; the spacing between the stations, which is a design variable; and the rapidity of boarding. Thus, the actual value in an application is not intrinsic to its components but depends on the system design. As one might expect, it is higher for level III ROW modes than for the rest. The final row gives the actual vehicle productivity observed in the U.S. Not surprisingly, we see that the modes with the largest vehicle capacities exhibit the largest productivity. Productivity values are affected by both, the system design and the demand. As such, these values are merely snapshots of conditions in the U.S. as of 2013, and should not be viewed as targets for our designs.

References

Bureau of Labor Statistics (2016). "Occupational Employment and Wages: 53-3021 (Bus Drivers, Transit and Intercity), May 2016." https://www.bls.gov/oes/current/oes533021.htm, Accessed on September 30, 2017.

Casello, J. M. (2007). "Transit competitiveness in polycentric metropolitan regions." *Transportation Research Part A: Policy and Practice*, 41(1), 19–40.

Castellanos, J. C. and Fruett, F. (2014). "Embedded system to evaluate the passenger comfort in public transportation based on dynamical vehicle behavior with user's feedback." *Measurement*, 47, 442–451.

Cervero, R. and Landis, J. (1997). "Twenty years of the bay area rapid transit system: Land use and development impacts." *Transportation Research Part A: Policy and Practice*, 31(4), 309–333.

Davies, A. (2012). "Why do subways cost so much more here than elsewhere?" *The Urbanist*, Feb. 15, 2012. https://blogs.crikey.com.au/theurbanist/2012/02/15/why-do-subways-cost-so-much-to-build-here-than-elsewhere/, Accessed on October 20, 2017.

Davis, S. C., Diegel, S. W., and Boundy, R. G. (2009). *Transportation Energy Data Book*, Edition 28. U.S. Department of Energy.

Federal Transit Administration (2010). "Public Transportation's Role in Responding to Climate Change." https://www.transit.dot.gov/, Accessed on April 29, 2016.

Future of Transportation National Survey (2010). http://t4america.org/maps-tools/polling/2010survey/, Accessed on May 19, 2016.

Kay, M., Clark, M., Duffy, C., Laube, M., and Lian, F. S. (2011). *Bus Lifecycle Cost Model for Federal Land Management Agencies: User's Guide*. John A. Volpe National Transportation Systems Center (U.S.), United States Department of the Interior.

Kenworthy, J. R. and Laube, F. B. (1999). "Patterns of automobile dependence in cities: an international overview of key physical and economic dimensions with some implications for urban policy." *Transportation Research Part A: Policy and Practice*, 33(7), 691–723.

Kenworthy, J. R., Laube, F. B., Newman, P., Barter, P., Raad, T., Poboon, C., and Guia, Jr. B., (1999). *An International Sourcebook of Automobile Dependence in Cities 1960–1990*. University Press of Colorado, Boulder, CO.

Kittelson and Associates, Inc., KFH Group, Inc., Parsons Brinckerhoff Quade and Douglass, Inc., and Hunter-Zaworski, K. (2003) *TCRP Report 100: Transit Capacity and Quality of Service Manual, 2nd Edition.* Transportation Research Board, Washington DC.

Kittelson and Associates, Inc., Parsons Brinckerhoff, Inc., KFH Group, Inc., Texas A&M Transportation Institute, ARUP (2013) *TCRP Report 165: Transit Capacity and Quality of Service Manual, 3rd Edition.* Transportation Research Board, Washington DC.

Li, Z. and Hensher, D. A. (2011). "Crowding and public transport: A review of willingness to pay evidence and its relevance in project appraisal." *Transport Policy*, 18(6), 880–887.

Mattson, J. W. (2012). "Travel behavior and mobility of transportation-disadvantaged populations: Evidence from the National Household Travel Survey (No. DP-258)." Upper Great Plains Transportation Institute. North Dakota State University. Fargo, ND. http://www.ugpti.org/pubs/pdf/DP258.pdf, Accessed on January 20, 2016.

Mohring, H. (1972). "Optimization and scale economies in urban bus transportation." *American Economic Review*, 62(4), 591–604.

Neff, J. and Dickens, M. (2015). *APTA 2015 Public Transportation Fact Book.* American Public Transportation Association, Washington, D.C.

Nelson, L. J. and Poston, B. (2016). "L.A. voters just showed how frustrated they really are over traffic." *Los Angeles Times*, November 21, 2016. http://www.latimes.com/local/lanow/la-me-ln-metro-tax-vote-20161117-story.html, Accessed on October 22, 2017.

Newell, G. F. (1994). *Lecture Notes on Public Transportation.* University of California at Berkeley.

Ratner, K. A. and Goetz, A. R. (2013). "The reshaping of land use and urban form in Denver through transit-oriented development." *Cities*, 30, 31–46.

Sekulić, D., Dedović, V., Rusov, S., Šalinić, S., and Obradović, A. (2013). "Analysis of vibration effects on the comfort of intercity bus users by oscillatory model with ten degrees of freedom." *Applied Mathematical Modelling*, 37(18), 8629–8644.

Smart, M., Miller, M. A., and Taylor, B. D. (2009). "Transit stops and stations: transit managers' perspectives on evaluating performance." *Journal of Public Transportation*, 12(1), 59–77.

Taylor, H., Krane, D., and Orkis, K. (2010). "The ADA, 20 years later." *New York: Harris Interactive* (conducted for the Kessler Foundation and National Organization on Disability).

Tirachini, A., Hensher, D. A., and Rose, J. M. (2013). "Crowding in public transport systems: effects on users, operation and implications for the estimation of demand." *Transportation Research Part A: Policy and Practice*, 53, 36–52.

Transit Center, (2014). *2014 Mobility Attitudes Survey.* http://transitcenter.org/our-work/, Accessed on May 1, 2016.

U.S. Bureau of Labor Statistics, (2016). *Occupational Employment Statistics.* http://www.bls.gov/oes/, Accessed on January 20, 2016.

U.S. Census Bureau's 2010–2013 American Community Survey. http://www.census.gov/data/developers/data-sets/acs-survey-3-year-data.html, Accessed on January 28, 2016.

U.S. Department of Commerce, (2014). *Census Bureau, Population Division, Annual Population Estimates.* http://www.census.gov/popest. Accessed on March 15, 2015.

U.S. Department of Labor, (2014). *Bureau of Labor Statistics, American Time Use Survey 2013.* http://www.bls.gov/tus, Accessed on April 11, 2015.

U.S. Department of Transportation, (2016). *Bureau of Transportation Statistics, National Transportation Statistics*. http://www.bts.gov/publications/national_transportation_statistics/, Accessed on June 11, 2016.

U.S Department of Transportation, (2011). *National Household Travel Survey: Summary of Travel Trends*. http://nhts.ornl.gov/2009/pub/stt.pdf, Accessed on June 11, 2016.

Vovsha, P., Marcelo, G. S. O., and William, D. (2014). "Statistical analysis of transit user preferences including in-vehicle crowding and service reliability." *Proceedings of 93rd Transportation Research Board Annual Meeting*, Washington DC, USA.

Vuchic, V. R. (2005). *Urban Transit Operations, Planning and Economics*. John Wiley and Sons. ISBN 978-0471632658.

Zhang, K., Zhou, K., and Zhang, F. (2014). "Evaluating bus transit performance of Chinese cities: Developing an overall bus comfort model." *Transportation Research Part A: Policy and Practice*, 69, 105–112.

Homework

1.1 Pick one public transit system of your choice (e.g., a system in your hometown), and do the following:

(a) Collect basic information about: (i) the geography, population distribution and demographics of the system's service region, (ii) the type of the transit system (e.g., the ROW and vehicle type, the power source and the control method); (iii) the network geometry and other operational service characteristics (e.g., the total route-kilometers, annual cost, vehicle and passenger kilometers, revenue, and ridership).

(b) Has there been news media coverage about the transit system in this city/town? What are the major issues or concerns?

(c) Qualitatively discuss how you would improve the transit system in this town if it was up to you. What possible options you can think of? Be brief.

1.2 Many cities in the world have CBDs that are surrounded by ring-radial street networks. Assume one such city has a circular CBD with radius, $r > 0$, and that a fraction p of the CBDs area is occupied by skyscrapers with n floors. The average space per employee a, the street spacing s, the street capacity q, and the vehicle occupancy k are assumed to take the values of the example of Section 1.1. Given this information, estimate the commuter demand [commuters] and the perimeter roadway capacity [commuters/hr]. Then, show how the duration of the rush depends on r, n, and p.

1.3 The idea of autonomous (or automated) buses is becoming less and less utopian every day. How do you think such emerging technologies will change the current transit systems in terms of operation, cost and performance? Please discuss briefly. You may consider different scenarios, e.g., (i) a futuristic world where all vehicles are autonomous and connected, and (ii) an intermediate phase when autonomous vehicles and regular vehicles coexist.

Chapter 2

Analysis Tools

This chapter reviews four analysis tools that will be used throughout this book. The tools are fairly basic and are also used to study other transportation systems. Readers already familiar with these concepts may want to use this chapter as a reference, skipping familiar sections or even the whole chapter. Continuity will not be lost. To facilitate this kind of selective reading, the four techniques are reviewed in separate sections: Section 2.1 covers the time–space diagram, Section 2.2 the queuing diagram, Section 2.3 some optimization concepts, and Section 2.4 the basics of dimensional analysis.

2.1. The Time–Space Diagram

A graphical tool, the time–space diagram is used in the transportation field to study the motion of relevant objects such as vehicles and passengers along one-dimensional channels. In the following, the ideas are expressed in terms of vehicles but the procedures are general. The following subsections briefly discuss: (i) how the movement of a single vehicle (object) can be tracked and represented mathematically; (ii) how the collective motion and interactions of many vehicles can be described; and (iii) how the time–space diagram can be used to facilitate modeling and analysis.

2.1.1. *A Single Vehicle*

While studying transit systems, we often encounter vehicles that are constrained to a one-dimensional channel. This can occur either physically, like a train along its track, or logically, as when a bus is required to serve a particular route. In both cases, however, we can follow the location of

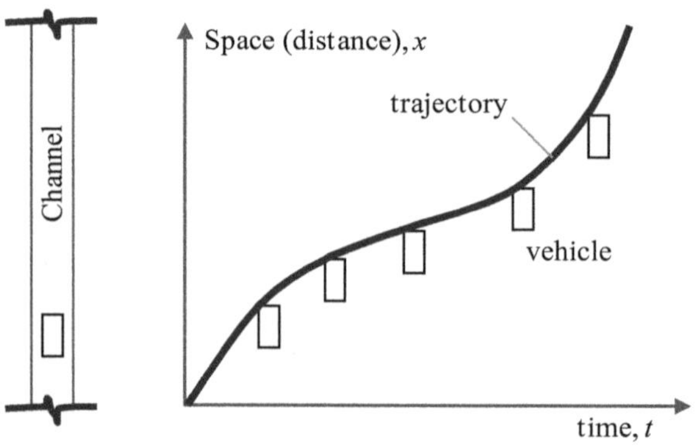

Figure 2.1. A vehicle's trajectory in a time–space diagram.

the vehicle over time by keeping track of the distance it has traveled along the channel. And since vehicles are solid bodies it suffices to record this information for a single specific point, such as the vehicle's front bumper. The result can be graphed as a curve in the time–space plane (see Figure 2.1). The curve is called a trajectory and the chart is a time–space diagram. The curve can be expressed as a mathematical function, $x(t)$, which returns the vehicle's (front bumper) position, x, for every moment of time t.

Vehicle trajectories condense all relevant information about a vehicle's movement. For example, the slope of the curve at any time t, given by the first derivative $dx(t)/dt$, is the instantaneous speed $v(t)$ along the channel; and the second derivative $dx^2(t)/dt^2$ is the acceleration, $a(t)$.[1] Figure 2.2 illustrates these ideas. Trajectory A depicts a vehicle traveling with a constant speed in the x^+ direction; trajectory B illustrates a vehicle accelerating in the x^- direction; and trajectory C illustrates a vehicle decelerating, first in the x^- direction to a full stop, and then accelerating in the x^+ direction. Trajectories are useful because they can be represented as functions and treated mathematically. It should be remembered, however, that not every curve is a trajectory; realistic trajectories must be continuous *functions* of time. For example, curves D and E are not valid trajectories. Curve D is discontinuous, unrealistically implying that the vehicle it describes would change position instantaneously. And curve E is not a

[1]Combined with the channel's geometry, i.e., with its three geolocation coordinates for each x, one can convert these scalar measures of speed and acceleration into velocity and acceleration vectors to get a full sense of the vehicle's three-dimensional motion.

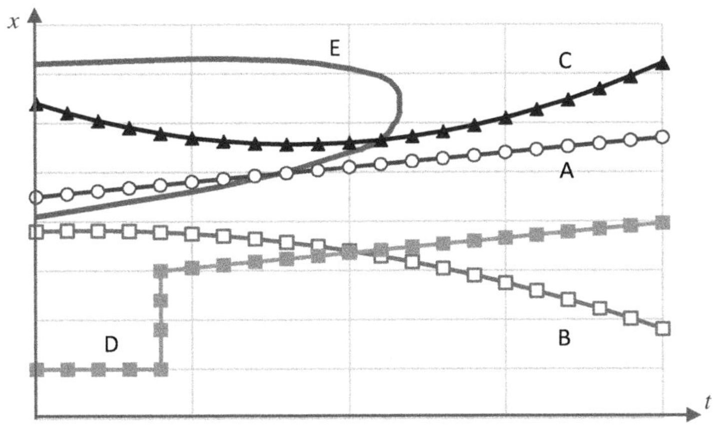

Figure 2.2. Examples of valid and invalid trajectories.

function because it does not return a single position for each time t, which is a physical impossibility.

In practical applications, vehicle trajectories sometimes have to be developed from partial information. For example, from an initial speed and a deceleration rate, we may look for the distance required to bring a train to a stop. These types of analyses are done using the laws of physics. Newton's second law, in particular, is used to describe vehicle acceleration as a function of vehicle mass, the engine's propulsion, or alternatively the resistance of braking, and other forces such as aerodynamic drag, rolling resistance, and the gravity pull due to grade. Because the characteristics of the vehicle, the driver, the roadway, as well as the ambient environment, may vary over time and space, and because some of the forces, e.g., the aerodynamic drag, depend on the vehicle's speed, the instantaneous acceleration can often be expressed in applications as a mathematical function of space, speed, and time: $a(x, v, t)$.

Once this function has been identified, kinematic differential equations can be used to find the vehicle's trajectory. For example, if in addition to the acceleration function, we are given the vehicle's initial speed and position at $t = 0$, $\{v(0), x(0)\}$, then the trajectory $x(t)$ can be found by solving the following pair of ordinary differential equations:

$$a(x, v, t) = d^2x(t)/dt^2, \; v(t) = dx/dt.$$

In all cases the solution can be found numerically by stepping through time and, when the acceleration function is simple, also analytically. Interested readers are referred to the relevant chapters in Mannering and Kilareski (1990) or Daganzo (1997) for more detail.

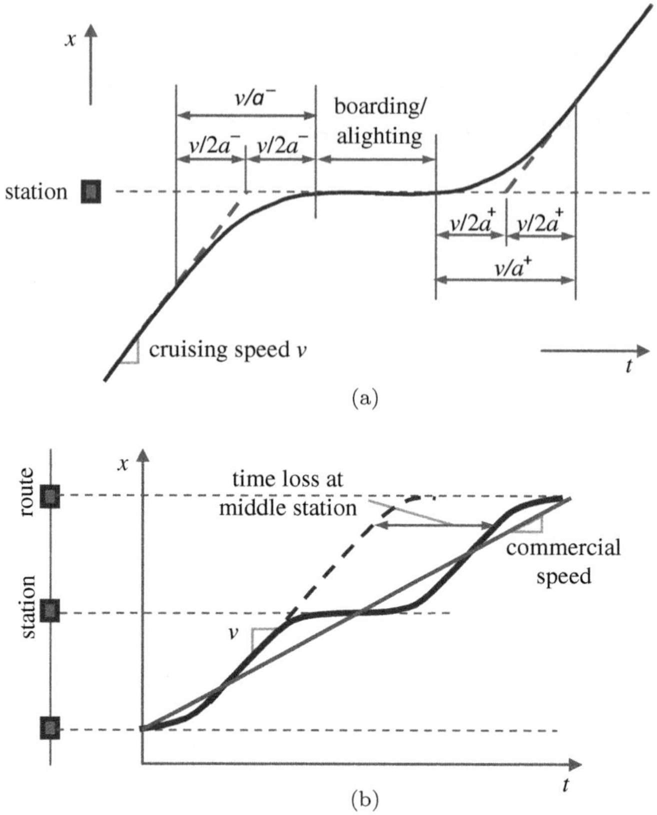

Figure 2.3. Transit vehicle trajectories: (a) near a station and (b) at multiple stations.

This kinematic equation approach can be used to derive a formula for the delay transit vehicles endure when stopping at a station. (By delay we mean the extra travel time the vehicle has to endure because it has to stop.) To see how this is done consider Figure 2.3(a), which shows a typical trajectory. The delay is the horizontal separation between the two dashed lines, and it includes three parts: time lost during deceleration, stopped time (boarding/alighting), and time lost during acceleration.

Analysis of the kinematic equations for an invariant acceleration reveals a well-known result: that if the cruising speed is v and the vehicle accelerates at rate a^+ then the acceleration maneuver lasts v/a^+ time units and spans $\frac{1}{2}v^2/a^+$ distance units. Now note that if the vehicle had been traveling at the cruising speed that distance would have been covered in $\frac{1}{2}v/a^+$ time units instead of v/a^+. Thus, the delay due to non-instantaneous acceleration is $v/(2a^+)$, as the figure shows. Similar considerations show that if the train decelerates at rate $a^- > 0$ then the delay due to deceleration is $v/(2a^-)$.

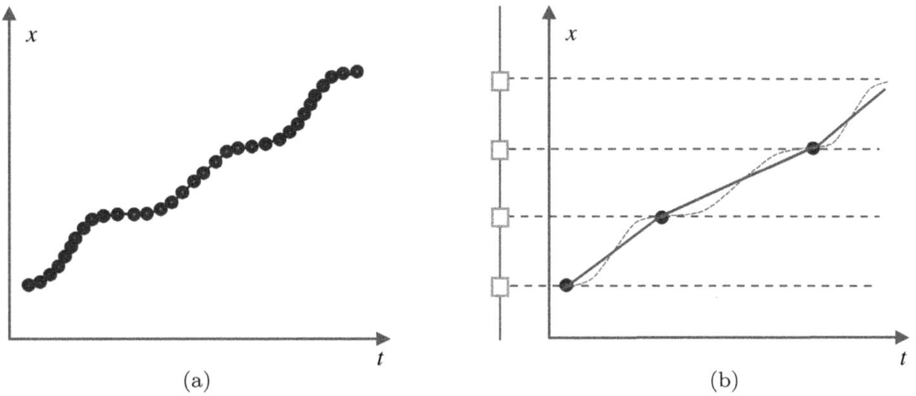

Figure 2.4. Constructing the trajectory of a transit vehicle from field observations: (a) with an AVL system and (b) with stationary observers.

Thus, the total station delay is $v/(2a^-) + v/(2a^+)$ + (stopped time). For rail systems, which exhibit high-cruising speeds, low accelerations, and quick boarding, the first two terms are usually comparable with the third. For bus systems, they are considerably smaller.

Along the entire route, the vehicle will experience an acceleration–cruising–deceleration–stopping cycle after each station departure, as shown in Figure 2.3(b). Each of these cycles introduces the aforementioned delay and adds to the vehicle travel time. Therefore, our formula will be used in later chapters to build models of bus travel times and predict their average or "commercial" speed.

In some cases, e.g. when diagnosing a system, we may just want to record what vehicles do without any models, i.e., to construct vehicle trajectories directly from field observations. This is easy when transit vehicles are equipped with automatic vehicle location (AVL) systems because these systems periodically report to a control center the position of each vehicle along with a time stamp, i.e., a point on the time–space diagram. Modern AVL systems can refresh every few seconds. Since the points obtained for a single bus are on its trajectory, and since the points are fairly closely spaced, we can approximate the trajectory by connecting the points with a continuous curve, e.g., with linear interpolation (see Figure 2.4(a)).

If onboard sensors are not available, one can approximate trajectories using stationary observers (or detectors) along the vehicle's path, e.g., at the stations along the route. These "observers" record the time at which each vehicle is observed and its ID. Time–space points for a particular ID are obtained using the location of the observer and the time stamp. By considering the data from all observers and then connecting all the points with the same ID in the direction of increasing time, e.g., with straight lines, one approximates the vehicle's trajectory. Figure 2.4(b) shows how this is done for a bus that the observers identified on the time–space diagram by a small dark circle. Although this low-tech method is less accurate, the trajectories it yields can effectively serve the needs for transit design and analysis.

2.1.2. *Multiple Vehicles*

Repeated across vehicles, the methods of the previous subsection can be used to obtain multiple trajectories which can be plotted on the same diagram to describe the vehicles' collective motion (e.g., as in Figure 2.5). This is very useful because the resulting picture encapsulates all relevant aspects of the collective motion in a simple chart. This all-encompassing simplicity can help us unveil and then quantify the complex vehicle interactions that drive the collective dynamics. As we shall see in later chapters, such interactions normally involve the spatiotemporal separations between consecutive vehicles, which are transparently displayed by the time–space diagram. This subsection shows how these separations and other key measures of performance can be identified and related to one another using the time–space diagram.

First, consider the point of view of a stationary observer, e.g., a passenger at a station. An important measure of performance from this point of view, i.e., a horizontal line on the time–space diagram, is the time-interval length between consecutive bus arrivals.[2] Refer to Figure 2.5 and note that except for the first and last, all buses have two intervals: one in the front and one in the back. To distinguish these two, we call the interval in front "headway" and the interval in back "backward headway." The headways labeled in the figure correspond to the bus with the thick trajectory. As we showed in later chapters, headways are important to transit systems as they directly impact the passenger's experience and the agency's operating cost.

Also from the standpoint of a stationary observer, we are often interested in the service frequency over a time interval. This is the ratio, n/T, of the observed number of buses n and the duration of the interval, T (see Figure 2.5). This ratio is also known as the "bus flow." If the observation

[2]In this chapter the word "bus" will be often used as shorthand for the longer phrase: "transit vehicle."

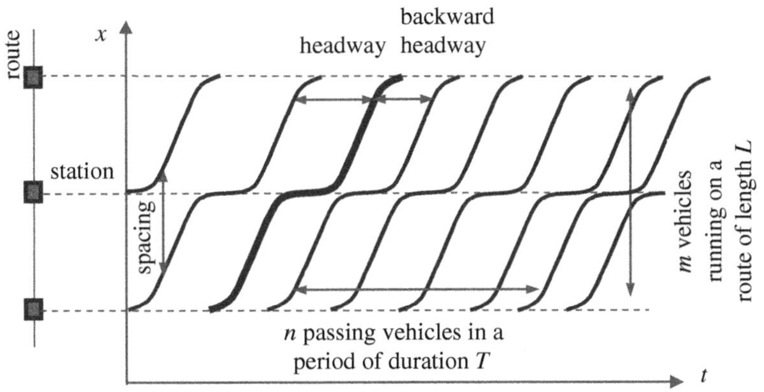

Figure 2.5. Headway, spacing, flow, and density.

interval begins and ends immediately before a bus's arrival, as in Figure 2.5, then T equals the sum of the observed vehicle headways. This means that the inverse of the service frequency, T/n, is the average headway. Consideration shows that this relationship is also true, albeit only approximately, if the observation period is not synchronized with the bus arrivals but contains many headways. The precision of this approximation is explored in homework exercise 2.1.

Also important for transit systems is the passenger flow, i.e., the number of passengers N observed in T. As should be intuitive, passenger flow is the product of the bus flow, n/T, and the average number of passengers in a bus, N/n. This should be clear as $(n/T)(N/n) = N/T$.

The time–space diagram also clearly displays measures of performance from other points of view. We can, for example, freeze time and look at the system as in a photograph. This corresponds to a vertical line on the diagram. In this case, the (vertical) separation between trajectories is a distance, which we call the spacing. The ratio, m/L, of the number of buses m in a (vertical) section of length L is called the "density" (see Figure 2.5). Density has units of buses/distance. Note that if the section is an entire route served by a single bus line, then m is the number of buses in service. And finally, as occurred with headways, the average spacing in a section is approximately equal to the inverse of the bus density.

The time–space diagram also reveals how the measures of performance arising from these two views are related. To see this consider, a transit route consisting of a closed loop, e.g., going from terminus A to terminus B at the other end of the route, and then back to A. One such route is conceptually illustrated in the time–space diagram of Figure 2.6. The route length is

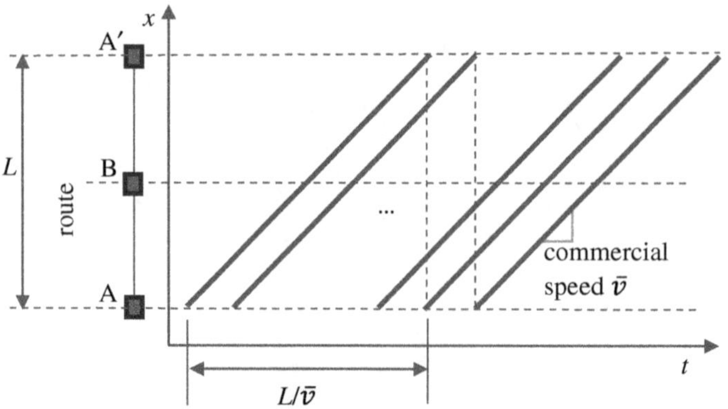

Figure 2.6. Vehicles running on a closed loop.

denoted L. Point A′ is a duplication of terminus A since the route returns to this terminus after L distance units. The figure also displays terminus B and omits intermediate stations.

The slanted lines in the diagram are bus trajectories with stops at intermediate stations smoothed out. The slope of these lines is the commercial speed, $\bar{v}$. The lines are parallel and equidistant on the assumption that all vehicles have the same commercial speed and the headways are uniform. Note every trajectory ends when its vehicle reaches A′ and then simultaneously re-emerges and continues from A. This is logical since reaching A′ means that the vehicle has returned to A.

The figure can be used to demonstrate how the average headway and frequency are related to average spacings, density, fleet size, route length, and commercial speed. For example, we see from the right triangle formed by the upper part of the second trajectory and the two dotted lines emanating from the upper end point of the first trajectory that (spacing) = (headway) $\times\ \bar{v}$. This can be restated as $(\text{headway})^{-1} = \bar{v} \times (\text{spacing})^{-1}$, or equivalently $(\text{flow}) = \bar{v} \times (\text{density})$, which some readers will immediately recognize as one of the fundamental relationships of traffic flow theory.

Now consider the right triangle formed by the entire first trajectory and the two dotted lines emanating from the toe of the fourth trajectory shown in the figure. The length of the base of this triangle is the duration of a vehicle's round trip, $L/\bar{v}$. Also, note from the figure that the first dispatch after the interval can be fulfilled by the first vehicle because it has just returned. Therefore, service can be provided with a fleet size equal to the number of dispatches included in the base of the triangle, i.e., the ratio of $L/\bar{v}$ to the headway. This result, which is of fundamental importance for transit

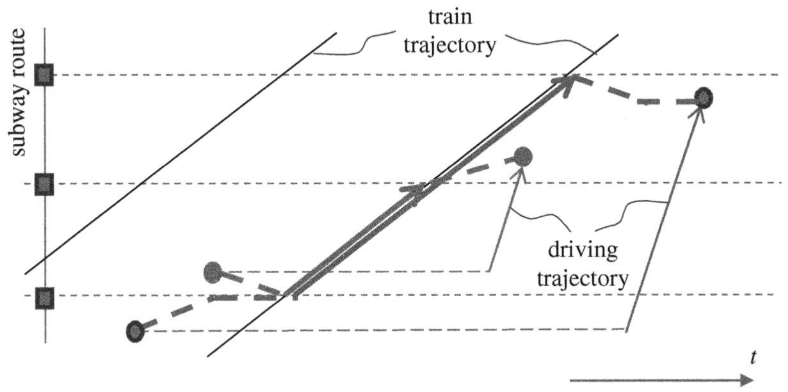

Figure 2.7. Auto travel vs. transit travel in a time–space diagram.

system design and operations, can be expressed in four different ways using the relations we have already unveiled:

$$\begin{aligned}\text{fleet size} &= L/(\bar{v} \times \text{headway}) = L/\text{spacing} \\ &= (\text{density}) \times L = (\text{frequency}) \times L/\bar{v}. \end{aligned} \tag{2.1}$$

A time–space diagram can also qualitatively illustrate ideas about accessibility. The Swedish geographer Torsten Hägerstrand, who is famous for introducing "time geography," used time–space diagrams to that end. In the same spirit, we now use the diagram to visualize some of the pros and cons of car driving vs. transit riding, enriching in this way the discussion of Section 1.1. The vertical axis of Figure 2.7 shows a portion of an imaginary linear city served by a subway and the stations in the section. The two parallel slanted lines are train trajectories. Passengers can access the subway trains only where these lines intersect the dotted horizontal lines corresponding to the stations, and this hinders their mobility.

To illustrate how, consider the two door-to-door passenger trajectories described by the thick piecewise linear curves of Figure 2.7. The passengers have appointments marked by the thick dots where the curves end, on the upper right. The thick dots on the bottom left show when and where they start their trips. The passenger trajectories include an in-vehicle portion denoted by a solid arrow spanning the gap between the stations they use and out-of-vehicle portions denoted by dashed lines. The horizontal portions of these dashed lines, if any, correspond to waiting and the slanting portions to walking — note the relatively flat slope of the walking parts.

Now compare this with the trajectories corresponding to driving. These are the thin bilinear curves connecting the same origin and destination dots. The solid part of each trajectory corresponds to in-vehicle travel, at a considerably higher speed, and the horizontal dashed part to out-of-vehicle waiting. (We assume for this illustration that parking is available at the origin and the destination so walking distance is zero.) The driving option is more convenient because, unlike in the case of transit, the out-of-vehicle waiting is very long and done at the origin location before accessing the vehicle. Thus, travelers can use this time for productive or enjoyable endeavors. Moreover, the automobile allows them to choose an earlier departure time along the dashed line, make an intermediate stop anywhere along the route and still meet the deadline. This flexibility makes automobile travel appealing. On the other hand, we should acknowledge that the above scenario is favorable for driving. If parking is limited and congestion rears its ugly head, as occurs in many cities, the advantages of driving greatly diminish — access to the car then requires considerable waiting (e.g., for parking) and walking, and travel by car can be slower than that by subway. Furthermore, with today's car technology auto drivers have to focus on driving whereas transit riders can use their in-vehicle time to do other things. Clearly, transit is better in some environments and the car option in others. The methodological point, however, is that the time–space diagram can help us visualize the role of the environment, parking regulations, station spacings, transit headways, and commercial speeds when comparing these two transportation modes.

2.1.3. *Application to Modeling and Analysis*

The time–space diagram is useful not only as a visualization tool but also to solve problems. It is particularly effective for exploring ways in which multiple vehicles should be scheduled over a common right-of-way. As an illustration of the process, this subsection considers an example in which passenger and freight trains share a common section of single-track without sidings (see Figure 2.8).

It is assumed that both types of trains travel in the same direction but passenger trains are faster and have high priority. Clearly, to avoid conflicts, trains of both types should be dispatched at the upstream end of the single-track section in a coordinated fashion. We now show how the time–space diagram can be used to devise a good schedule and to evaluate its effectiveness vis-a-vis that of segregated service with double tracks.

The section of shared track depicted in the figure has a length of $L = 60\,\text{km}$. Passenger trains travel at $v_1 = 120\,\text{km/h}$, and freight trains at $v_2 = 60\,\text{km/h}$. For safety, passenger trains must be at all times

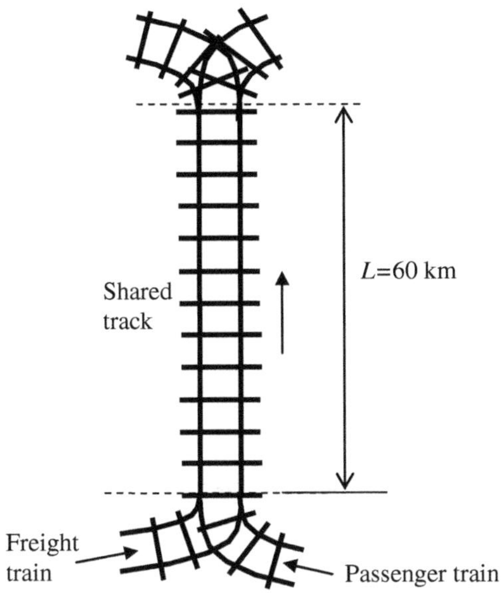

Figure 2.8. Shared railroad track for passenger and freight trains.

at least $s_1 = 22.5\,\text{km}$ behind another moving train; and freight trains at least $s_2 = 15\,\text{km}$ behind. (These spacings are hypothetical; in actual applications, they are based on train lengths, speed variations, reaction times, and deceleration distances.) We assume that passenger trains must be dispatched regularly every $H = 2\,\text{h}$ and freight trains can be dispatched flexibly. Below, we look for answers to the following questions:

(1) What is the maximum freight train flow if passenger trains operate on separate tracks?
(2) What is the maximum freight train flow without segregation?
(3) To increase the maximum freight train flow, is it beneficial to slow down or speed up the passenger trains? What is the freight train flow when v_1 is reduced to 90 km/h?

Let us start with question 1. If passenger trains operate on separate tracks the entire traffic in our section consists of freight trains. And, since they travel at the same speed, they can be spaced $s_2 = 15\,\text{km}$ apart. This translates to a headway of $s_2/v_2 = 15/60 = 0.25\,\text{h}$, which corresponds to a (maximum) freight train flow of 4 trains/h.

Now consider question 2. Since two train types are involved, we draw a time–space diagram (see Figure 2.9). The trajectories of the high-priority passenger trains (leaving at 0:00:00 and 2:00:00) are pre-determined, so

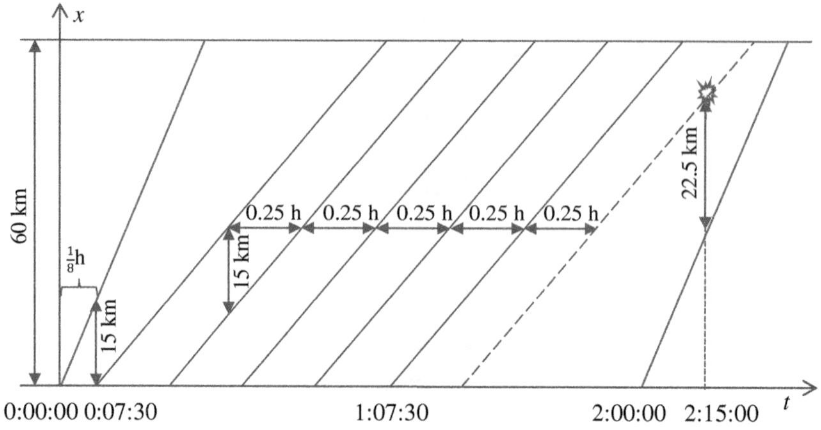

Figure 2.9. Time–space diagram for passenger and freight trains on a shared railroad track.

they are drawn ahead of time. Both have a slope equal to $v_1 = 120\,\text{km/h}$. Their horizontal separation is $H = 2\,\text{h}$. Next, we draw the trajectories of the freight trains. Since the first freight train travels more slowly than the preceding passenger train, the two are closest when the freight train first enters the shared track. Therefore, it can be dispatched as soon as the passenger train has traveled s_2 km, and the headway at that point is $s_2/v_1 = 0.125\,\text{h} = 7.5\,\text{min}$. Accordingly, the figure shows this first dispatch occurring at 0:07:30 h. The figure also shows that the following freight trains can be dispatched in intervals of 0.25 h, as occurred with question 1; and that the sixth freight train would violate the clearance requirement for the 2:00:00 passenger train before reaching the end of its trip. This means that only five freight trains can be dispatched. Since this limit reoccurs every 2 h, we see that the maximum daily freight train flow is now $5/2 = 2.5$ trains/h. This is 37.5% fewer trains per day than with separate tracks.

To answer question 3, we develop a general formula in terms of the parameters of the problem. This can be easily done using the geometry of the time–space diagram (but would be quite difficult otherwise). Interested readers should be able to verify from said geometry that the maximum (hourly) flow of freight trains is

$$\frac{1}{H}\left\lfloor \frac{H + L(v_1^{-1} - v_2^{-1}) - (s_1 + s_2)v_1^{-1} + s_2 v_2^{-1}}{s_2 v_2^{-1}} \right\rfloor,$$

where $\lfloor \cdot \rfloor$ is the floor operator. Consideration shows that the term inside the floor operator decreases with v_1 if $L > s_1 + s_2$, and increases with v_1

if $L < s_1 + s_2$.[3] In our case, $L = 60 > s_1 + s_2 = 37.5$, so decreasing v_1 is beneficial. In fact, both the formula and the diagram reveal that when the passenger trains slow down to 90 km/h, the maximum flow of freight trains increases to 3 trains/h.

2.2. The Queueing Diagram

As we have just seen, the time–space diagram reveals in minute detail much information about the motion of groups of items along a shared path. This is surely desirable and useful to diagnose and solve problems. However, in many practical situations, it may not be possible to obtain a fine-grained diagram because of a lack of detectors. This was the case in Figure 2.4(b), where the vehicle trajectories between the stations had to be roughly interpolated. The time–space diagram is also impractical if the items in question do not move along a one-dimensional channel. Fortunately, it turns out that many practical questions can be answered without a detailed description of the vehicle trajectories. This section discusses a simple tool (the queuing diagram) that can often be used in cases like that. A more detailed description can be found in Newell (1982). The following example illustrates the idea.

Suppose a management agency will soon start to operate a new parking structure with many intersecting channels and hundreds of spaces. The agency wishes to display the number of available spaces in real time, both on the internet and on variable message signs distributed over the city. Clearly, tracking vehicle trajectories with a time–space diagram is both impractical and unnecessary for this problem. After all, the agency could just install a sensor in each parking space to detect the presence of a parked vehicle and display the number of unoccupied spaces. Or even more cheaply, it could keep track of the vehicles in the structure by recording their IDs and the times at which they enter and exit. This can be done with cameras or ticket stubs. It can be done even more simply by counting the incoming and

[3] The time–space diagram shows why. When v_1 is increased we see geometrically that the headway of the first freight train (at the start of the shared track) decreases and the backward headway of the last freight train (also at the start of the shared track) increases. These two headways are wasted time. Note from the geometry that the headway decrease corresponding to the first train is proportional to s_2 and the backward headway increase corresponding to the last train is proportional to $L - s_1$ (both with proportionality constant $1/v_1$). Thus, for $L > s_1 + s_2$ the increase in wasted time must exceed the decrease; so that the situation worsens by increasing v_1. And vice versa.

outgoing vehicles without their IDs. In this case, simple accounting reveals that if one is given an initial count of available spaces then the number of available spaces at any moment must equal the initial value minus the number of arrivals until the current time plus number of departures. Thus, for our particular problem two observation points and a simple accounting rule suffice to answer the question at hand.

Queuing theory builds on this idea. It uses a simple diagram to model the accumulation and delay of "customers" arriving at a waiting area, such as the cars arriving at our parking structure, in order to pass through specific locations (or "servers") such as our exit gate. Many transit system components can be modeled with queuing theory. In some of these, the customers are passengers and in others buses. Examples of the former include ticket booths, individual buses, bus stops and subway station platforms. Examples of the latter are traffic signals and bus stations with multiple lines.

The remainder of this section has three parts. It first introduces the basics and how the ideas relate to the time–space diagram. It then discusses how the queuing diagram can be used to interpret data, e.g. as in our parking structure, and finally, how it can be used for prediction and modeling.

2.2.1. *Basics*

We imagine there are two imaginary observers in our waiting area: an upstream observer "A," and a downstream observer "D." Observer A records all customer arrivals to the waiting area and observer D all the departures. Both start at time $t = 0$, when there is an initial accumulation of customers. Observer A keeps track of the total number of customers to have arrived, and adds this number to the initial count. We denote this sum by the function $A(t)$ and call its representation on the time-number plane the cumulative arrival curve, A. An example with zero initial accumulation is shown in Figure 2.10. Observer D keeps track of the total number of departures and does not add them to the initial count. Similarly, we label the resulting function $D(t)$, and call its representation on the time-number plane as the departure curve, D. This curve is then plotted on the same chart as curve A. Now note that if customers are not created or destroyed in the waiting area then the number in the system at any time t, denoted $Q(t)$ for "queue," must be $Q(t) = A(t) - B(t)$, i.e., it must equal the vertical separation between the A and D curves.

The diagram also neatly displays the total amount of waiting that is done over a period of time. To see this note that the area $\mathrm{Q}(t)\,dt$ enclosed between

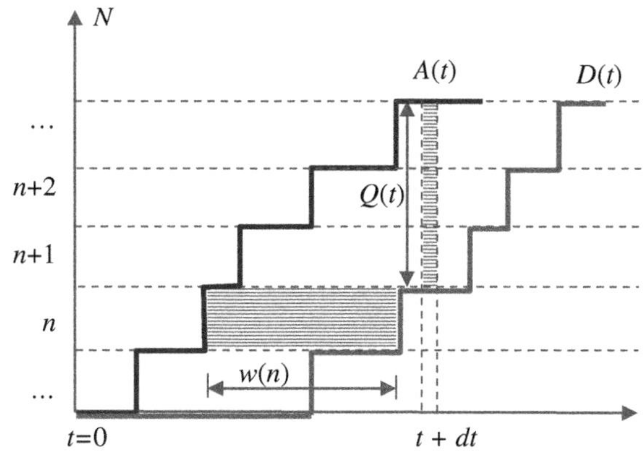

Figure 2.10. Queueing diagram: time vs. customer count.

the A and D curves and inside the thin vertical time slice $(t,\ t + dt)$, as illustrated in Figure 2.10, is the total wait done by customers during the interval $(t,\ t + dt)$. Now, by putting consecutive slices together we see that the area of a region between any two vertical lines corresponding to two inspection times and the two curves (i.e., the integral $\int_t Q(t)dt$) is the total wait done between inspections.

The queuing diagram yields further information if customers leave the waiting area in first-in-first-out (FIFO) order. This is very common. For example, it automatically happens when the waiting area is a guideway section on which passing is not possible. But it is by no means universal. Train platforms serving multiple interconnecting lines are a counterexample. The FIFO property is useful because then the nth customer to have arrived must have the same ID as the nth customer to have departed. This means that the times associated with the jumps of A and D across $N = n$ are the arrival and departure times of the same individual (customer "n"). Therefore, the waiting time for this customer, denoted $w(n)$, is the horizontal separation across ordinate n (see Figure 2.10). Mathematically, this geometrical relation can be written as $w(n) \equiv D^{-1}(n) - A^{-1}(n)$, where the superscript "-1" denotes the inverse function.

The diagram also reveals at a glance the total wait done by multiple customers. To see this, note that $w(n)$ is also given by the area of the shaded rectangle at the bottom of Figure 2.10, since both curves jump by exactly 1 customer. Hence, the total area between the $A(t)$ and $D(t)$ curves and any two horizontal lines corresponding to two different customers (i.e., the integral $\int_n w(n)dn$) is simply the total wait done by the multiple FIFO customers in between.

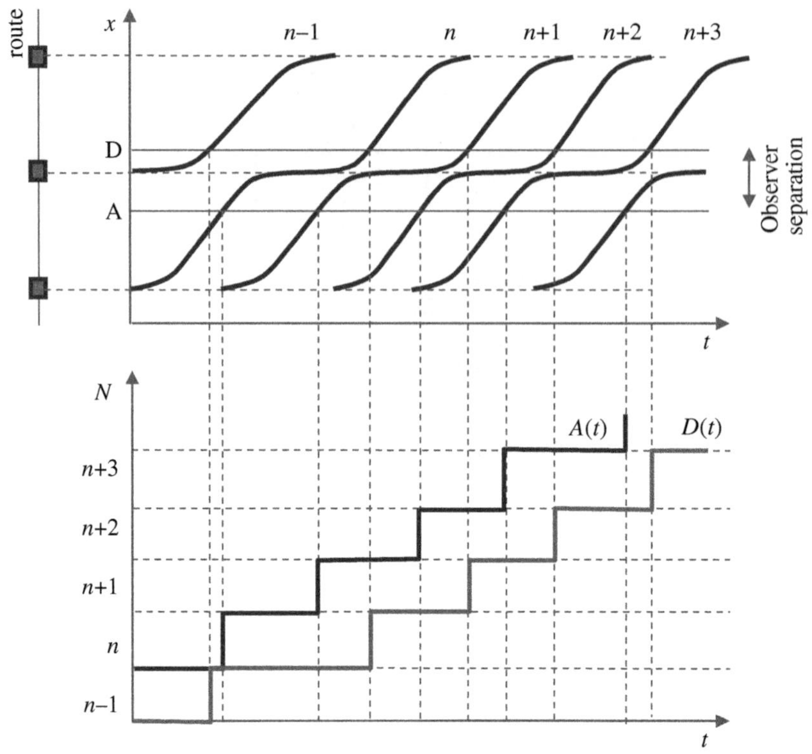

Figure 2.11. Time–space diagram vs. queueing diagram under FIFO (no passing).

Even in applications where a time–space diagram makes sense, a queuing diagram can be used to display summary data from fixed locations in a useful form. Consider Figure 2.11. On the top, there is a time-space diagram containing several bus trajectories as they navigate past three stations. Imagine we are interested in exploring the delays these vehicles suffer at the middle station, i.e., as they traverse the section from point A to point D. To do this, we can place imaginary observers at these locations and construct a queuing diagram for the intervening segment. The bottom part of the figure shows the result. Note how each curve jumps by 1 unit at the moment when a bus crosses the location of the corresponding observer on the top. By eliminating much irrelevant information for the question at hand, the bottom diagram displays the information we want more succinctly and transparently. Note how with only two curves it clearly displays not just the individual bus delays but also their statistical distribution.

Despite this neat relationship with the time–space diagram, it bears repeating that the queuing diagram can be applied to two- and

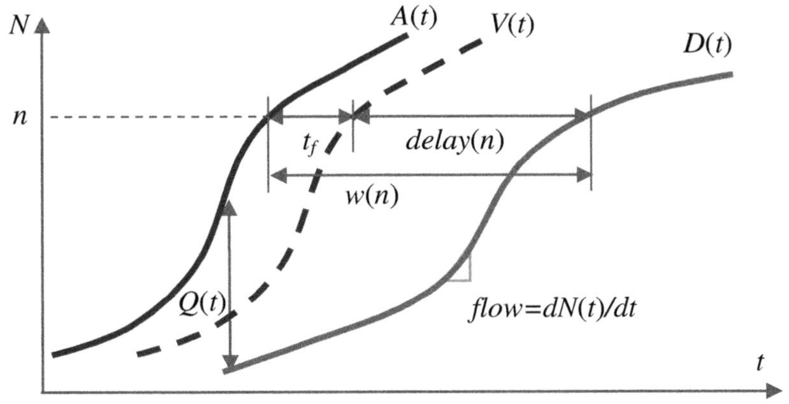

Figure 2.12. Smoothed arrival and departure curves.

three-dimensional waiting areas such as entire cities and multi-story garages, which cannot be modeled with the time–space diagram. So it is more general. In Chapter 7, for example, queuing ideas will be extensively used to model the accumulation of idle and busy taxis in two-dimensional service regions.

2.2.2. *Interpreting Data: Smoothing, Averages and Little's Formula*

Real-world applications often involve many customers and long time periods. In such cases, we normally seek statistics that summarize the behavior of the system across all customers rather than customer-by-customer details. It then makes sense to approximate the step-wise curves A and D by smooth versions such as those depicted in Figure 2.12. This is useful because it allows us to approximate areas with simple formulas and to use calculus. For example, the smoothing endows curves A and D with finite slopes, which can be interpreted as instantaneous customer arrival and departure rates, and expressed as derivatives. In the case of Figure 2.12, these derivatives first increase and then subside because the curves are intended to represent a brief burst of demand.

Note that Figure 2.12 includes three cumulative curves. This is because the time that customers spend in the system can usually be divided into two parts: a time required for traversing the waiting area without interference from other customers, and an extra time caused by the interference. Called "delay," this extra time is often used as an evaluation metric. The intermediate curve in the diagram, labeled $V(t)$, is used to sort out the delay from the total time. It is called the "virtual" arrival (or departure) curve because it expresses the imaginary number of arrivals and departures that would have been observed at the system's exit(s) had every customer been able to proceed as quickly as possible without interference from others. By construction then, the vertical separation between V and D is the number of customers that are being delayed at any given time. Thus, the area between curves V and D that is sandwiched between two vertical lines corresponding to two different instants is the total delay accrued by all the customers present in the time interval between the lines.

If the system is FIFO and all customers take the same time, t_f, to traverse the waiting area, e.g., as would occur for the transit station example of Figure 2.11, then the virtual curve yields insight about individual customers. (The subscript "f" is adopted from traffic flow theory, where the interference-free time is called the "free-flow" time.) Note that the point of V corresponding to customer n can be identified by drawing a horizontal line at ordinate n as shown in Figure 2.12, and ticking off a point t_f distance units to the right of the customer's arrival time. The horizontal separation between this point and the departure curve is the delay. Since t_f is the same for all customers the whole virtual arrival curve can be identified by shifting curve A to the right by t_f time units as if it was a rigid solid; or mathematically, by writing $V(t) = A(t - t_f)$. Finally, it should come as no surprise that the area between curves V and D that is sandwiched between two horizontal lines corresponding to two different customers is the total delay experienced by the intervening customers.

In many transit applications, the minimum traversing times are negligible. In these cases $A(t) \cong V(t)$ and there is no need to introduce the virtual curve. The rest of this section is written as if this was the case, using only curves A and D. Otherwise, the obtained results apply to waits.[4]

A simple but useful result in queuing theory, known as Little's Formula (Little, 1961), relates three time-averaged quantities of the queuing diagram: the mean number of customers, $\bar{Q}$, the average arrival rate, λ, and the average wait, $\bar{w}$. The result is

$$\bar{Q} = \lambda \bar{w}. \tag{2.2}$$

It applies exactly to observation periods that start and end with an empty system.

To see why consider Figure 2.13 and the observation period from t_0 to t_1. Because the system is empty at the beginning and the end, all the customers

[4]To obtain delay formulas, substitute V for A, delay for wait, and excess queue for queue, in the ensuing discussion.

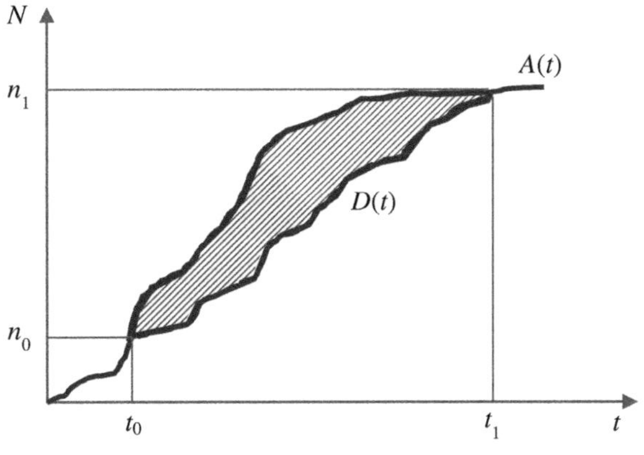

Figure 2.13. Proof for Little's formula.

from t_0 to t_1 are fully served, even if not in FIFO order. Their total wait is the (shaded) area between the $A(t)$ and $D(t)$ curves: $\int_n w(n)\,dn \equiv \int_t Q(t)\,dt$. By definition, the time-averaged queue during the observation period is the ratio of the area and the duration of the observation period. Thus, we can write: $(t_1 - t_0)\bar{Q} = \int_t Q(t)\,dt$. For similar reasons, the average wait $\bar{w}$ satisfies $(n_1 - n_0)\bar{w} = \int_n w(n) \equiv \int_t Q(t)\,dt$. Hence, the average queue and the average wait are related by $(t_1 - t_0)\bar{Q} = (n_1 - n_0)\bar{w}$. Now divide both sides by $(t_1 - t_0)$ and write: $\bar{Q} = [(n_1 - n_0)/(t_1 - t_0)](\bar{w})$. Finally, note that the quantity in brackets is, by definition, the average arrival rate λ. Thus, $\bar{Q} = \lambda\bar{w}$.

It should perhaps be intuitive that the formula also applies approximately if the initial and terminal queues are much smaller than the total number of customers served, and that it also applies to long observation periods in a steady state. Further consideration also shows that if the arrival and departure curves fluctuate randomly across days, the formula still applies to the averages across days (realizations). Books on stochastic processes explain the logic.

As a final aside, we now show that in the special case where the waiting area is one-dimensional Little's formula reduces to the traffic flow relationship (flow) = (speed) × (density), and its counterpart for transit systems, Equation (2.1). To this end, consider a bus route segment of length L where all buses travel at commercial speed $\bar{v}$. The segment is observed for a long time while in a steady state. Given these conditions, Little's formula holds. Now note the following: (i) the average time a bus spends in that segment is $\bar{w} = L/\bar{v}$; (ii) the average arrival rate λ equals the flow (frequency) seen by any stationary observer, so $\lambda =$ (flow); and (iii) the average number of buses in the system is $\bar{Q} =$ (density) $\times L$. Next, replace the three variables in Little's formula by the right sides of equalities (i), (ii), and (iii) to write: (density) $\times L =$ (flow) $\times L/\bar{v}$. Finally, cancel the L's and rearrange terms in this identity to obtain the fundamental relation: (flow) = (speed) $\times$ (density).

2.2.3. *Predicting Delay and Queue Evolution*

The queueing diagram has been described so far as a means for interpreting system data and computing performance statistics, but the diagram can also be used as a modeling tool. This subsection shows how it can be used to predict the evolution of systems with known arrivals, when the mechanism for processing customers is known; i.e., how, with a given arrival curve $\{A(t), \forall t\}$, we can determine a corresponding departure curve $\{D(t), \forall t\}$ and all the derived performance characteristics discussed in Section 2.2.2.

A queuing mechanism is a rule that allows us to determine at any given instant how many customers are served in the next few moments, given the history of the arrival and departure curves. There are many different types of servers, e.g., single servers that process customers continuously one at a time, such as turnstiles; multi-servers that process customers in parallel, such as shuttle vans at an airport; and even servers that process customers in batches such as trains and buses at stations. Here we shall describe single servers and a simple form of batch service. The ideas are similar for all server types, however, so other server types arising in the book will be examined when the need arises.

Let us now consider single servers such as turnstiles and single channel bottlenecks. These systems process customers one at a time as rapidly as possible, but without exceeding a maximum rate, which is called the service rate and denoted μ. The service rate is a property of the server and can vary with the time-of-day. We assume it is known. The servers work as follows: if there is a queue the server processes customers at the service rate until the queue vanishes since queues cannot go negative. When there is no queue the server processes customers at the arrival rate, keeping the queue at zero — provided of course that the arrival rate does not exceed μ. When this excess first occurs, the server maxes out at the service rate and a queue develops.

The dynamics of this rule are encapsulated by the following ordinary differential equation:

$$\frac{dD(t)}{dt} = \begin{cases} \mu(t), & \text{if } D(t) < A(t); \\ \min[\mu(t), dA(t)/dt], & \text{if } D(t) = A(t). \end{cases} \tag{2.3}$$

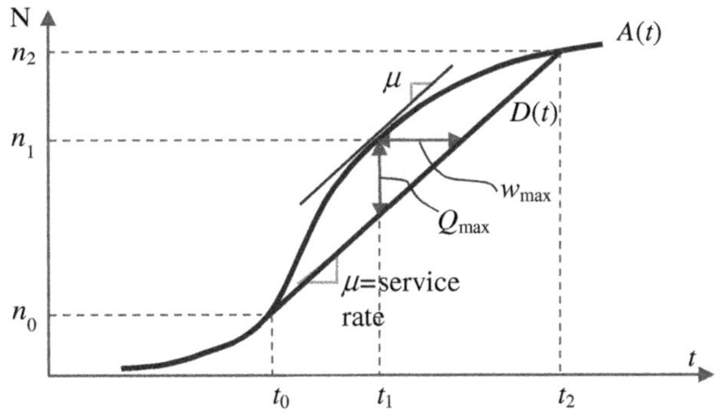

Figure 2.14. Queuing at a single bottleneck with variable arrivals.

Note the equation does not specify what to do for $D(t) > A(t)$. This is unnecessary because the condition cannot arise. Note as well that if an initial queue and the arrival process is given, the rule allows us to step through time and construct a departure curve $D(t) \leq A(t)$ for all t incrementally.

This is particularly easy when the service rate is time-independent as then a simple graphical construction can be used to draw curve D. This is now explained with the help of Figure 2.14, which depicts the queue evolution of a system of this type during an imaginary rush hour. Note how the arrival rate first increases monotonically and then decreases. The system is assumed to be initially empty, and the initial arrival rate to be lower than the service rate. To construct D, we start from the beginning when the system is unqueued and governed by the bottom part of the differential equation. We see from this equation that $dD(t) = dA(t)$, so the departure curve must track the arrival curve until $dA(t)/dt > \mu$ for the first time. This occurs immediately after $t = t_0$ in the figure. During this time the departure rate maxes out at the service rate and a queue develops. Thus, the dynamics going forward are then governed by the top part of the differential equation. The departure curve in this regime is a straight line with slope μ because the service rate is time-independent. This queued regime ends when the departure curve meets again the arrival curve at time t_2, and from then on the system is governed again by the bottom part of the differential equation. That is, curve D tracks curve A once the queue has dissipated.

The resulting diagram can be constructed quickly, even for arrival curves with multiple pulses of high demand. It can always be interpreted as explained in the previous subsections. The diagram also reveals at a glance the maximum accumulation, $Q_{\max}$. This must occur at some time t_1 where the slope of the arrival curve equals μ, as shown in the figure. If the system

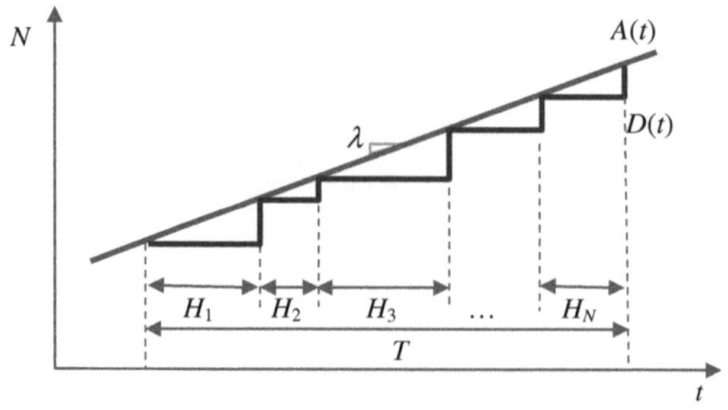

Figure 2.15. Queuing at a single bottleneck with on-off service.

is FIFO, the waiting time of a customer that arrives when the queue is $Q(t)$ is $Q(t)/\mu$. Thus, as shown by the figure, the customer that arrives when the queue is longest suffers the maximum wait: $w_{\max} = Q_{\max}/\mu$.

Servers that provide batch service are also of interest in transit applications. A prime example is a bus line processing the passengers waiting at a station. In this case, the service mechanism is defined by the paired sequences of bus arrival times and bus capacities (in terms of available passenger slots). It works as one might expect. For the most part, passengers wait, so the departure rate is zero. However, when a bus arrives the maximum number of passengers possible board the bus, subject to the capacity constraint. Thus, the departure curve jumps up by the number of passengers that board the bus, i.e., by the available capacity if the jump does not exceed the arrival curve, so some passengers are left behind, or up to the arrival curve otherwise. Clearly, if one is given the arrival curve, an initial queue, and the service mechanism, then the queuing diagram is easy to draw.

Figure 2.15 shows the result of this procedure for a special case where the customer arrival rate is constant ($dA(t)/dt = \lambda$) and the buses have plentiful capacity. The diagram is constructed for a time interval of duration T that starts and ends with an empty queue when two buses arrive; see figure. Since the arrival time of the initial bus is given, the service mechanism is characterized by the sequence of headways in the interval: $\{H_1, H_2, \ldots H_N\}$, which satisfy: $T = \sum_{i=1}^{N} H_i$. Note how every jump in the departure curve reaches up to the arrival curve since capacity is plentiful.

Clearly, the system is FIFO so individual passenger waits are horizontal separations between curves A and D. Accordingly, the diagram shows that passengers arriving immediately after a bus arrival wait for the entire duration of the ensuing headway, whereas those arriving immediately before experience no delay. As always, the total passenger delay is the area between the curves, which as the figure shows is a sum of triangular areas; one for each headway. Since the height of triangle i is λH_i, then the sum is $\sum_{i=1}^{N} \frac{1}{2}\lambda H_i^2$. And since the total number of passengers served in our interval is $\lambda T = \sum_{i=1}^{N} \lambda H_i$, we see that the average wait per passenger must be

$$\bar{w} = \sum_{i=1}^{N} \frac{1}{2}\lambda H_i^2 \Big/ \sum_{i=1}^{N} \lambda H_i = \sum_{i=1}^{N} H_i^2 \Big/ \left(2\sum_{i=1}^{N} H_i\right)$$
$$= \frac{1}{2}E[H^2]/E[H] = \frac{1}{2}(E[H] + Var[H]/E[H]), \qquad (2.4)$$

where the symbols $E[\cdot]$ and $Var[\cdot]$ stand for the sample average and sample variance operations. Equation (2.4) is a well-known expression, which we were able to unveil in a transparent way thanks to the queuing diagram.

The expression also holds for the average wait across realizations if the arrival process is random but stationary. For further information, see Newell (1982), Daganzo (1997), or many books on stochastic processes. Finally, note that the variance term in (2.4) cannot be negative. Thus, if we hold $E[H]$ constant, the average customer wait is minimized by setting $Var[H] = 0$, i.e., by ensuring that all the headways are equal. This is one of the reasons why transit service regularity and reliability are important.

2.3. Optimization

The time–space diagram and the queuing diagram are descriptive tools that help identify and predict possible problems. Throughout the book, however, we will also strive for fixing problems in the best possible way, e.g., by identifying good system designs and operating schemes. This is done with optimization tools.

The prototypical optimization problem is formulated as the maximization or minimization of a function of multiple variables, subject to constraints. The variables are called "decision" variables because they correspond to real-world decisions that address the issues at hand. They can be restricted to be integer, or allowed to be real. In this book, integer optimization is not used and therefore, this section only describes the case with real variables.

If we use vector notation for the decision variables, i.e., $\mathbf{x} \equiv (x_1, x_2, \ldots, x_n)$, the general optimization problem can be expressed as follows:

$$\begin{aligned} &\max \text{ (or min)} f(\mathbf{x}) \\ &\text{s.t. } g_j(\mathbf{x}) \leq 0, \ \forall j = 1, 2, \ldots, J \\ &\qquad h_k(\mathbf{x}) = 0, \ \forall k = 1, 2, \ldots, K. \end{aligned}$$

The function f to be optimized is called an objective function. The inequality and equality relations appended to it are the constraints. Problems of this type are called mathematical programs and are studied in the field of operations research. This section summarizes key concepts from this discipline. Section 2.3.1 examines the special case where there are no constraints and Section 2.3.2 the general case.

2.3.1. *Unconstrained Optimization*

We say that a real-valued function $f(\mathbf{x})$ has an unconstrained global maximum (or minimum) at a point $\mathbf{x}^*$ if $f(\mathbf{x}^*)$ is the maximum (minimum) value across all possible $\mathbf{x}$, i.e., if $f(\mathbf{x}) \leq f(\mathbf{x}^*)$ (or $f(\mathbf{x}) \geq f(\mathbf{x}^*)$) for all $\mathbf{x} \in \mathbb{R}^n$. Similarly, we say that $\mathbf{x}^*$ is a local maximum (or minimum) if $f(\mathbf{x}^*)$ is maximum (minimum) in some small neighborhood of $\mathbf{x}^*$, i.e., if there is a scalar $\varepsilon \geq 0$, such that $f(\mathbf{x}) \leq f(\mathbf{x}^*)$ (or $f(\mathbf{x}) \geq f(\mathbf{x}^*)$) for all $\mathbf{x}$ that satisfy $|\mathbf{x} - \mathbf{x}^*| \leq \varepsilon$. Here, $|\cdot|$ denotes the Euclidean norm, but any other norm can be used. It should be clear from these definitions that any finite global optimum must also be a local optimum, but the reverse is not true.

This book will sometimes deal with problems where $f(\mathbf{x})$ is only defined in the positive open orthant $\mathbb{R}^n_+$ and where only the solutions in this open set make physical sense. The reader can verify that the above definitions can still be used for this case if we simple replace $\mathbb{R}^n$ by $\mathbb{R}^n_+$, and furthermore, that the logic of this entire subsection continues to hold. Therefore, all the recipes about to be introduced also apply to these problems.

We are generally interested in finding global optima. Since most problems in this book will, as a rule, exhibit twice differentiable objective functions, differential calculus can be used as a powerful solution tool. This tool, however, is only good at identifying local optima. On the other hand, since global optima are local optima, we can use the tool in practice to identify many local optima (possibly all) in the hopes that the best of these is (close to) a global optimum. The tool consists in solving a set of equations that express a calculus-based necessary condition that all local optima must satisfy.

Clearly, a necessary condition for a point to be a local optimum is that it should be impossible to improve the objective function by moving away from $\mathbf{x}^*$ by a differential amount in any direction $\mathbf{d}$. Calculus tells us that the improvement rate in any such direction is the inner product of the direction vector $\mathbf{d}$ and the gradient (vector of partial derivatives) of the objective function. Thus, using matrix notation with vectors as single-row matrices, and an inverted delta for the gradient as is customary, we write this product as $\mathbf{d}[\nabla f(\mathbf{x}^*)]^{\mathrm{T}}$, where $\nabla f(\mathbf{x}) = (\frac{\partial f}{\partial x_1}, \ldots, \frac{\partial f}{\partial x_n})$. Now, since $\mathbf{d}$ is an arbitrary vector and can be reversed in sign, we see that the only way in which an improvement can be avoided for all $\mathbf{d}$ is if $\nabla f(\mathbf{x}^*) = \mathbf{0}$. These equations are the necessary conditions we just mentioned. They are also called the stationarity conditions or the first order conditions.

The conditions are not sufficient because they apply to all maxima, minima, and inflection/saddle points; and only some of these so-called "stationary points" can be local optima. Therefore, additional analysis is always needed to remove all spurious stationary points and verify that the retained points are local optima. In doing this, points at infinity must be considered carefully when functions are unbounded. An example is the one-variable function $f(x) = x\sin(x)$, which has an infinite number of local optima but is not bounded in either direction.

In some fortunate cases, additional analysis may reveal that a function has only one local optimum, which is also global. Functions of this type are called "unimodal." This occurs, for example, if the function is concave (for maximization) or convex (for minimization).[5] If we can identify a function as unimodal, then to find a global optimum we only have to look for a local optimum. This will often occur in this book.

The following illustrates the above ideas when $\mathbf{x}$ is a scalar, x. In this case, the first order condition is $f'(x^*) = 0$. Figure 2.16 shows one such function and three points satisfying the condition. Clearly, only one is a local optimum. To ensure that a point is indeed an optimum, we should check an additional "second order" condition involving the second derivative, $f''(x^*)$. Clearly, if $f''(x^*) < 0$ as occurs with the point on the left then x^* is a maximum; if $f''(x^*) > 0$ as occurs with the point in the middle it is a minimum; and if $f''(x^*) = 0$, as occurs with the point on the right it is undetermined. In this case, higher order derivatives should be examined.

---

[5]A function is said to be convex (concave) if every straight line segment connecting any two points on the function is completely above (below) the function itself.

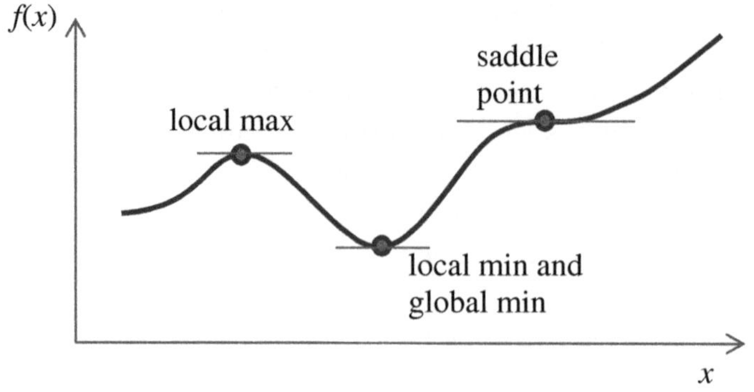

Figure 2.16. Local optimum, global optimum, and saddle point in one dimension.

Example 2.1. Considered here is a minimization problem that will arise repeatedly throughout the book. It is often referred to as the generalized economic ordering quantity (GEOQ) problem, as it originally arose from the context of inventory replenishment. We minimize the function: $f(x) = Ax^k + Bx^{-m}$ with real parameters A, B, k, $m > 0$, and $x \in \mathbb{R}_+$.

We start with the first order condition, which is $f'(x) = kAx^{k-1} - mBx^{-m-1} = 0$. It yields a single stationary point at $x^* = (Bm/Ak)^{\frac{1}{m+k}}$, and the objective value: $f(x^*) = (1 + m/k)(m/k)^{\frac{m}{m+k}} A^{\frac{m}{m+k}} B^{\frac{k}{m+k}}$. Note that both the point at infinity and the limit point at $x = 0$ yield an unbounded and positive objective value, so neither can be an unbounded minimum. Hence, the identified stationary point is the only candidate for a local minimum. Therefore, if it turns out to be a local minimum it is also a global minimum.

To check if it is a local minimum, we examine the second derivative of the objective function, which is

$$\begin{aligned} f''(x^*) &= Ak(k-1)(x^*)^{k-2} + Bm(m+1)(x^*)^{-m-2} \\ &= (m+k)(Ak)^{\frac{m+2}{m+k}}(Bm)^{\frac{k-2}{m+k}} > 0. \end{aligned}$$

This shows that x^* is indeed a local minimum, and therefore the global minimum. In the special case where $m = k = 1$ (often referred to as the economic ordering quantity (EOQ) problem), the globally optimal solution simplifies to $x^* = (B/A)^{\frac{1}{2}}$, and $f(x^*) = 2(AB)^{\frac{1}{2}}$.

For the multivariate case, the second order condition involves the matrix of second partial derivatives, which is called the Hessian:

$$\mathbf{H}(\mathbf{x}) = \begin{bmatrix} \frac{\partial^2 f(\mathbf{x})}{\partial x_1^2} & \cdots & \frac{\partial^2 f(\mathbf{x})}{\partial x_1 \partial x_n} \\ \vdots & \ddots & \vdots \\ \frac{\partial^2 f(\mathbf{x})}{\partial x_n \partial x_1} & \cdots & \frac{\partial^2 f(\mathbf{x})}{\partial x_n^2} \end{bmatrix}.$$

A second-order Taylor series expansion of the objective function around a stationary point reveals that the point is a local minimum (or maximum) if the Hessian is positive (negative) definite.[6] If the Hessian is indefinite then the stationary point is a saddle point. In the remaining cases (e.g., the Hessian being semi-definite), the stationary point cannot be classified without checking partial derivatives of a higher order.[7]

Example 2.2. Consider the bivariate function $f(x_1, x_2) = x_1 + x_2 + 8x_1^{-1}x_2^{-1}$, which is to be minimized for $x_1, x_2 > 0$. The first order conditions are $\frac{\partial f}{\partial x_1} = 1 - 8x_1^{-2}x_2^{-1} = 0$ and $\frac{\partial f}{\partial x_2} = 1 - 8x_1^{-1}x_2^{-2} = 0$. They yield a single stationary point: $x_1^* = x_2^* = 2$. To check if it is a local minimum, we derive the Hessian, $\mathbf{H}(x_1, x_2) = 8x_1^{-2}x_2^{-2}\left[\begin{smallmatrix} 2x_2/x_1 & -1 \\ -1 & 2x_1/x_2 \end{smallmatrix}\right]$, and evaluate it at $x_1^* = x_2^* = 2$. The result is $\mathbf{H}(x_1^*, x_2^*) = \frac{1}{2}\left[\begin{smallmatrix} 2 & -1 \\ -1 & 2 \end{smallmatrix}\right]$. Since the determinants of the minors are 1 and 3/2, the matrix is positive definite. Thus, our unique stationary point is a minimum. The corresponding objective value is $f^* = 6$. Now, since the point at zero and all points at infinity yield $f = \infty > 6$, we can conclude that our solution is a global optimum.

---

[6]In matrix form, the Taylor series expansion is: $f(\mathbf{x}) \approx f(\mathbf{x}^*) + 1/2(\mathbf{x}-\mathbf{x}^*)\mathbf{H}(\mathbf{x}^*)(\mathbf{x}-\mathbf{x}^*)^{\mathrm{T}}$, where we have used row vector notation, and the superscript "T" for transposition. Note that the first order term, $(\mathbf{x}-\mathbf{x}^*)[\nabla f(\mathbf{x}^*)]^{\mathrm{T}}$, has been omitted because $\nabla f(\mathbf{x}^*) = 0$. Now recall from linear algebra that a symmetric matrix $\mathbf{H}$ is said to be positive (or negative) definite if the quadratic form $\mathbf{dHd}^{\mathbf{T}}$ is positive (or negative) for every non-zero row vector $\mathbf{d}$. Thus, if the Hessian is positive (or negative) definite then $f(\mathbf{x}) \geq f(\mathbf{x}^*)$ (or $f(\mathbf{x}) \leq f(\mathbf{x}^*)$) in a small neighborhood of $\mathbf{x}^*$.

[7]The character of a matrix can be checked fairly easily. Linear algebra tells us for example that $\mathbf{H}$ is positive (negative) definite if all its eigenvalues are positive (negative); and if they are of different signs the matrix is indefinite. If the eigenvalues are non-negative (or non-positive) the matrix is positive (negative) semi-definite. The character of a matrix can also be determined by checking the determinants of its main diagonal minors. If they are all positive (or non-negative), the matrix is positive (semi)-definite. If the first minor is negative and the rest alternate in sign then the matrix is negative definite. If this pattern holds but some of the values are zero, the matrix is negative semi-definite. If neither pattern holds, and if there are no zero values, then the matrix is indefinite.

The example illustrates that multivariate optimization can be tedious. Fortunately, many of the problems in this book can be solved more expediently with a conditional decomposition approach that finds a global minimum by considering a nested sequence of univariate problems. The procedure is easiest to explain in the bivariate case where it involves two univariate problems. It is based on the fact that the global minimum operation has the nesting property: $\min_{x_1,x_2}\{f\} = \min_{x_1}\{\min_{x_2|x_1}\{f\}\}$. The first step of the procedure consists in treating one of the variables, say x_1, as a parameter and then solving the univariate inner problem, $\min_{x_2|x_1}\{f\}$. This conditional solution is a function of the parameter; thus, we write its objective value as $f^*(x_1)$ and its optimum as $x_2^*(x_1)$. The final unconditional solution is then found by solving the outer problem, i.e., minimizing $f^*(x_1)$ and plugging the resulting x_1^* into the formula for the conditional optimum.

Example 2.2 (*continued*). Let us apply the conditional decomposition approach to our function $f(x_1, x_2) = x_1 + x_2 + 8x_1^{-1}x_2^{-1}$. Note that the conditional inner problem for fixed x_1 is an EOQ optimization with respect to x_2. Therefore we see immediately from the formulas of Example 2.1 that its global minimum is: $x_2^*(x_1) = \sqrt{8/x_1}$ and $f^*(x_1) = x_1 + 2\sqrt{8/x_1}$. Now, to complete the procedure we find the global minimum of $f^*(x_1)$ with respect to x_1, and insert the result x_1^* in the above expressions. Since $f^*(x_1)$ is a GEOQ equation, we can again use the formula of Example 2.1, which yields $x_1^* = 2$. Therefore, we find in agreement with our earlier result that $x_2^* = 2$ and $f^* = 6$.

It should perhaps be intuitive that this approach can be applied with any number of variables using one nest per variable so that for example a trivariate problem can be treated as a sequence of three univariate problems. This will, in fact, occur later in the book. In some cases, however, the equations are so complicated that simple closed-form solutions cannot be obtained.

Stationary points are then identified numerically with iterative methods that gradually reduce the objective value for minimization problems, or increase it for maximization. Since a maximization problem can be converted into a minimization problem by changing the sign of its objective function we only describe the procedures for minimization problems. One of the simplest, the gradient descent (or steepest descent) method, is based on the observation that a function $f(\mathbf{x})$ decreases fastest in the direction of the negative local gradient $-\nabla f(\mathbf{x})$, i.e., when for some small scalar $\varepsilon \to 0^+$, the current solution is changed to:

$$\mathbf{x}^{new} = \mathbf{x} - \varepsilon \nabla f(\mathbf{x}). \tag{2.5}$$

Since ε is small and positive the new point should satisfy $f(\mathbf{x}^{new}) \leq f(\mathbf{x})$.[8] Thus, to implement the approach, we start with an arbitrarily chosen initial point $\mathbf{x}^{(0)}$ and construct a sequence $\{\mathbf{x}^{(0)}, \mathbf{x}^{(1)}, \mathbf{x}^{(2)}, \ldots\}$ using (2.5) recursively, always with small step sizes $\{\varepsilon^{(0)}\varepsilon^{(1)}\varepsilon^{(2)}, \ldots\}$. The recursion is

$$\mathbf{x}^{(k+1)} = \mathbf{x}^{(k)} - \varepsilon^{(k)} \nabla f(\mathbf{x}^{(k)}),\ k = 0, 1, 2, \ldots.$$

Since the steps are small, we should then have $f(\mathbf{x}^{(0)}) \geq f(\mathbf{x}^{(1)}) \geq f(\mathbf{x}^{(2)}) \geq \ldots$, so the sequence $\{\mathbf{x}^{(k)}\}$ consistently improves the objective function and will usually converge toward a local minimum.

In the steepest descent method, the value of $\varepsilon^{(k)}$ can either be fixed, or change with the iteration number k, but the perturbation is always in the direction of the negative gradient. Unfortunately, this approach often converges very slowly. For this reason, quicker-converging methods have been developed. They use the second derivatives to improve the direction of descent. The second derivatives can be either computed from their analytic expressions, estimated numerically, or inferred indirectly from the history of previous gradients. The latter strategies are collectively known as "conjugate-gradient" methods. They are considerably more efficient than the steepest descent method. In any case, the idea of all descent methods is the same: improve the objective function one step at a time until the improvements become very small or stop altogether — this usually means that a local optimum is near.

2.3.2. *Constrained Optimization*

Let us now return to the general problem. Without loss of generality, we restrict ourselves to the minimization case:

$$\begin{aligned} &\min f(\mathbf{x}) \\ &\text{s.t.}\quad g_j(\mathbf{x}) \leq 0, \forall\ j = 1, 2, \ldots, J \\ &\qquad\ \ h_k(\mathbf{x}) = 0, \forall\ k = 1, 2, \ldots, K. \end{aligned}$$

[8]To see this, enter with (2.4) into a first order expansion of f around point $\mathbf{x}$; i.e., write: $f(\mathbf{x}^{new}) \approx f(\mathbf{x}) + (\mathbf{x}^{new} - \mathbf{x})[\nabla f(\mathbf{x})]^{\mathrm{T}} = f(\mathbf{x}) - \varepsilon[\nabla f(\mathbf{x})][\nabla f(\mathbf{x})]^{\mathrm{T}}$. And since $[\nabla f(\mathbf{x})][\nabla f(\mathbf{x})]^{\mathrm{T}}$ is the sum of the squares of the elements of the gradient, the term $\varepsilon[\nabla f(\mathbf{x})][\nabla f(\mathbf{x})]^{\mathrm{T}}$ must be non-negative. Thus, $f(\mathbf{x}^{new}) \leq f(\mathbf{x})$.

No generality is lost, again, because to maximize a function we can simply minimize its negative.

The inclusion of constraints requires some new lingo and slight changes to the definitions of minima. Values of $\mathbf{x}$ satisfying the constraints are said to be "feasible." The set of all such values, $\Omega = \{\mathbf{x} \in \mathbb{R}^n | g_j(\mathbf{x}) \leq 0 \ \forall \ j = 1, 2, \ldots, J; h_k(\mathbf{x}) = 0 \ \forall \ k = 1, 2, \ldots, K\}$, is called the "feasible region." We shall say that $\mathbf{x}^*$ is a global minimum if $f(\mathbf{x}^*)$ is the lowest objective across all feasible $\mathbf{x}$, i.e., if $f(\mathbf{x}) \geq f(\mathbf{x}^*)$ for all $\mathbf{x} \in \Omega$. Similarly, we shall say that $\mathbf{x}^*$ is a local minimum if it is feasible and $f(x^*)$ is the lowest objective function value in a small feasible neighborhood of $\mathbf{x}^*$, i.e., if there is a scalar $\varepsilon \geq 0$, such that $f(\mathbf{x}) \geq f(\mathbf{x}^*)$ for all $\mathbf{x}$ that satisfy $|\mathbf{x} - \mathbf{x}^*| \leq \varepsilon$ and $\mathbf{x} \in \Omega$. As in the unconstrained case, any finite global minimum must also be a local minimum.

The inclusion of constraints also requires that the optimality conditions be modified. To this end, we say that a direction $\mathbf{d}$ from a feasible point $\mathbf{x}$ is a *feasible* direction if there is an infinitesimal arc emanating from $\mathbf{x}$ that is entirely feasible and tangent to $\mathbf{d}$. Intuitively, $\mathbf{d}$ points toward the interior of the feasible region or along its boundary. It should then be clear that if $\mathbf{x}^*$ is a local minimum, then there should be no feasible descent direction from $\mathbf{x}^*$ because otherwise the objective function could be reduced along the feasible tangent arc in the neighborhood of $\mathbf{x}^*$. This observation entitles us to define the first order condition, or stationarity condition, for a point $\mathbf{x}$ as follows: (i) $\mathbf{x}$ is feasible, and (ii) $\mathbf{x}$ has no feasible descent direction (or equivalently that every feasible direction is non-decreasing).

A natural way to find points that satisfy this condition, especially with computer programs, is an extension of the iterative descent direction search methods in Section 2.3.1. The main change is that in the present case, we must ensure at each step that only *feasible* descent directions and feasible points are considered. Below we discuss the analytical ideas underpinning the solution methods.

Consider first the case with only equality constraints. Then, the constrained problem can be changed into an unconstrained problem with the method of Lagrange multipliers from basic calculus. The method introduces a set of multipliers, $\lambda := \{\lambda_k\} \in \mathbb{R}^K$, and then defines the following augmented objective function (called the Lagrangian) with both the $\mathbf{x}$ and λ as decision variables:

$$L(\mathbf{x}, \lambda) = f(\mathbf{x}) + \sum_{k=1}^{K} \lambda_k h_k(\mathbf{x}).$$

Since this problem is unconstrained, all the analytical and/or numerical solution approaches discussed in Section 2.3.1 can be used to solve it.

It turns out that any point $\mathbf{x}$ that satisfies the first order conditions of the unconstrained Lagrangian problem, which are

$$\frac{\partial}{\partial \mathbf{x}} L(\mathbf{x}, \lambda) = \nabla f(\mathbf{x}) + \sum_{k=1}^{K} \lambda_k \nabla h_k(\mathbf{x}) = \mathbf{0}, \text{ and} \tag{2.6a}$$

$$\frac{\partial}{\partial \lambda_k} L(\mathbf{x}, \lambda) = h_k(\mathbf{x}) = 0, \ \forall\, k = 1, 2, \ldots, K, \tag{2.6b}$$

also satisfies the first order conditions of the original problem.[9] Thus, this system of equations can be used to look for local optima. It turns out, however, that the reverse is not always true. There are "pathologic" functional forms of the constraints that allow for stationary points (even local and global optima) that do not satisfy the Lagrange first order conditions. Thus, these points would be missed by solving the Lagrange equations. Fortunately, for us, these pathologies do not arise if the mathematical programs satisfy some mild regularity conditions that are met by the problems considered in this book.[10] So for our practical purposes, the Lagrange and original formulations are equivalent.

The Lagrange multiplier values that satisfy the first order conditions are also called "shadow prices." Each shadow price indicates the rate at which the optimal objective value declines if the right-hand side of its corresponding constraint is increased by a small amount, δ, i.e., if the constraint changes from $h_k(\mathbf{x}) = 0$ to $h_k(\mathbf{x}) - \delta = 0$.[11]

[9]Obviously, the point is feasible because this is what the second set of the Lagrange equations stipulates. So, it remains to prove that if the Lagrange first order conditions hold then every feasible direction from $\mathbf{x}$ is non-decreasing. To see that this is true, multiply both sides of the first equation by a feasible vector $\mathbf{d}$. Since $\mathbf{d}$ is feasible, $\mathbf{d}[\nabla h_k(\mathbf{x}^*)]^{\mathrm{T}} = 0, \forall k$, and hence all the terms involving the gradients of the constraint must vanish. Therefore, the multiplied equation reduces to $\mathbf{d}[\nabla f(\mathbf{x}^*)]^{\mathrm{T}} = -\mathbf{d} \cdot \mathbf{0}$, which is obviously zero, implying that feasible direction $\mathbf{d}$ must be non-decreasing, as claimed.

[10]The pathologies do not arise if all the constraints are linear. They do not arise either if, at the feasible points, all the constraint gradients are linearly independent and therefore non-zero. The explanation of these ideas is somewhat technical and therefore not given here.

[11]To see that this is true, consider the change in the optimum decision variable vector, $\mathbf{\Delta x}$, that results from changing the first constraint to $h_1(\mathbf{x}) - \delta = 0$. Since both the new and the changed optima must be feasible, and the $h_k(\mathbf{x})$ terms of the constraints increase by $\nabla h_k(\mathbf{x}^*)[\mathbf{\Delta x}]^{\mathrm{T}}$, it follows that $\nabla h_k(\mathbf{x}^*)[\mathbf{\Delta x}]^{\mathrm{T}}$ must equal δ for the first constraint and $\nabla h_k(\mathbf{x}^*)[\mathbf{\Delta x}]^{\mathrm{T}} = 0$ for the rest. Similarly, the change Δ in the optimum objective value

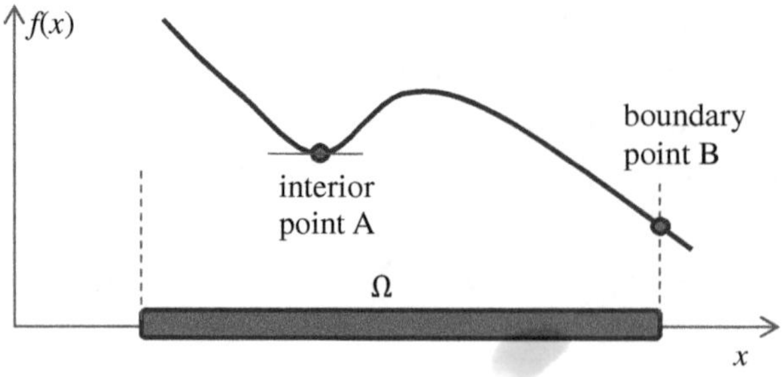

Figure 2.17. Local optimum points in a constrained set.

These ideas are now generalized for problems with inequality constraints. Inequalities are more complicated than equalities because local optimality can be achieved both when the inequalities are binding (or "active") and when they are not. Care must be exercised because if a constraint is binding at some point $\mathbf{x}$ then small perturbations in some directions may satisfy the constraint while perturbations in the opposite directions may not. And if the constraint is not binding, it will not preclude any (small) perturbations.

These ideas are illustrated with the univariate example in Figure 2.17. The feasible region Ω is an interval defined by two inequality constraints. Because the problem is univariate, at every point only two search directions are possible — to the left and to the right. Now consider point A, it lies in the interior of Ω, so no constraints are binding and therefore no directions are restricted. As a result, the first order conditions are the same as for an unconstrained problem. (Point A satisfies these conditions, as well as the second-order conditions, so it is a local minimum.) At boundary point B, however, one of the inequality constraints is binding, so it restricts the set of feasible directions. Therefore, in this case, we must only consider this limited set. (We see that point B can be feasibly perturbed to the left but not in the opposite direction; and since the only feasible direction is increasing, the point is a local minimum.)

The above illustrates that the first order conditions for a general problem must somehow account for the status (binding/non-binding) of all its inequality constraints. And indeed, the Lagrange first order conditions have been generalized for general problems by the so-called Karesh–Kuhn–Tucker

arising from $\mathbf{\Delta x}$ is $\Delta = \nabla \mathrm{f}(\mathbf{x}^*)[\mathbf{\Delta x}]^{\mathrm{T}}$. Now, multiply the first Lagrangian equation by $\mathbf{\Delta x}$ to obtain: $\nabla f(\mathbf{x}^*)[\mathbf{\Delta x}]^{\mathrm{T}} + \sum_{k=1}^{K} \lambda_k \nabla h_k(\mathbf{x}^*)[\mathbf{\Delta x}]^{\mathrm{T}} = 0$, i.e., $\Delta + \lambda_1 \delta = 0$. Clearly then, $\lambda_1 = -\Delta/\delta$ as claimed.

(KKT) conditions, which treat these constraints in a special way. They are now defined.

The KKT conditions hold at a point $\mathbf{x}^*$ if there exist constants $\mu_j \geq 0, \forall j$, and $\lambda_k \in \mathbb{R}, \forall k$, such that the following equations are satisfied:

$$\nabla f(\mathbf{x}^*) + \sum_{j=1}^{J} \mu_j \nabla g_j(\mathbf{x}^*) + \sum_{k=1}^{K} \lambda_k \nabla h_k(\mathbf{x}^*) = \mathbf{0}, \tag{2.7a}$$

$$g_j(\mathbf{x}^*) \leq 0, \forall\, j = 1, 2, \ldots, J \text{ and } h_k(\mathbf{x}^*) = 0, \forall\, k = 1, 2, \ldots, K, \tag{2.7b}$$

$$\mu_j g_j(\mathbf{x}^*) = 0, \forall j = 1, 2, \ldots, J, \tag{2.7c}$$

$$\mu_j \geq 0, \forall j = 1, 2, \ldots, J. \tag{2.7d}$$

Note that (2.7a) and (2.7b) are similar in form and function to the Lagrange conditions (2.6a) and (2.6b) of equality constrained problems. As in the equality-constrained case, the left-hand side of (2.7a) is the gradient of a Lagrangian function,

$$L(\mathbf{x}, \mu, \lambda) = f(\mathbf{x}) + \sum_{j=1}^{J} \mu_j g_j(\mathbf{x}) + \sum_{k=1}^{K} \lambda_k h_k(\mathbf{x}), \tag{2.8}$$

with respect to $\mathbf{x}$; and Equation (2.7b) restates original constraints.

What is different now is the addition of (2.7c) and (2.7d), which are used to discriminate whether a constraint is binding, as we did in the example of Figure 2.17. Equation (2.7c) is called the complementary slackness condition. It ensures that if an inequality constraint has some slack, i.e., it is not binding so $g_j(\mathbf{x}^*) < 0$, then $\mu_j = 0$. This eliminates the corresponding term from (2.7a) so the constraint has no impact on the solution, as occurred for point A of the example. If the constraint is binding, however, μ_j is allowed to take non-negative values. This allows $\nabla f(\mathbf{x}^*)$ to be non-zero and still satisfy (2.7a), as occurred for point B of the example.

This intuition is backed up by the facts. It turns out that if the KKT conditions hold at a point $\mathbf{x}^*$ then the point satisfies the first order conditions for local optimality.[12] This is why the KKT conditions are important. Unfortunately, as occurred in the equality-constrained case, the reverse is

[12] To see this, multiply both sides of (2.7a) by any feasible direction $\mathbf{d}$. Without loss of generality assume that the binding inequalities are numbered $1, 2, \ldots, J'$. Thus, we can write the result as $\mathbf{d}[\nabla f(\mathbf{x}^*)]^{\mathrm{T}} + \sum_{j=1}^{J'} \mu_j \mathbf{d}[\nabla g_j(\mathbf{x}^*)]^{\mathrm{T}} = -\mathbf{d} \cdot \mathbf{0} = 0$. (This is true because the equality constraint terms vanish, since $\mathbf{d}[\nabla h_k(\mathbf{x}^*)]^{\mathrm{T}} = 0$ for all the equalities, and the non-binding inequality terms do too because $\mu_j = 0$.) Now, since a feasible direction of a binding inequality must point in a direction where $g_j(\mathbf{x})$ does not decrease so as not

not always true. On the other hand, if the constraints of the problem satisfy the aforementioned regularity conditions, which are met by all the problems in this book, the pathology does not arise. Thus, for the purposes of the book, the KKT conditions *are* the first order conditions of the constrained problem.

The following two examples show how the KKT conditions can capture both active and inactive inequality constraints.

Example 2.3. Let us minimize the convex function $f(x) = x^2 - 4x$ subject to the linear constraint: $g(x) \equiv x - 5 \leq 0$. We can see by graphing this function in the feasibility interval that this problem has a single minimum, $x^* = 2$, where the constraint is not binding and $df(x)/dx = 0$. Let us now see how the KKT conditions can be used to identify it. Since there is no equality constraint, there is no multiplier λ. Any feasible solution has to satisfy either $x < 5$ or $x = 5$. If $x < 5$ the inequality constraint is not binding (i.e., $g(x) < 0$) and therefore the complementarity slackness condition (2.7c) implies that $\mu = 0$. Hence condition (2.7a) reduces to $df(x)/dx = 0$. This has the unique solution $x^* = 2$, which is acceptable because $x^* < 5$. To complete the analysis, we now consider the case with $x = 5$. Now $g(x) = 0$, so the constraint is binding and we require $\mu \geq 0$. Equation (2.7a) stipulates that $df(x)/dx + \mu dg(x)/dx = 0$ for $x = 5$ and some $\mu \geq 0$. However, for $x = 5$, $df(x)/dx = 6$, and $dg(x)/dx = 1$. Therefore, (2.7a) cannot hold, and hence $x = 5$ cannot be a local optimal solution.

Example 2.4. Suppose now that the same objective function $f(x) = x^2 - 4x$ is to be minimized with a tighter constraint: $g(x) \equiv x \leq 0$. Again, by graphing the function over the feasibility interval, we see that a unique local minimum exists at the upper end of the interval at $x^* = 0$. We now repeat the process and examine the KKT conditions for binding and non-binding points. In the non-binding case ($x < 0$), we have $\mu = 0$ and (2.7a) requires: $df(x)/dx = 2x - 4 = 0$. The unique solution to this equation ($x = 2$) violates the condition $x < 0$, so the solution is not valid. In the binding case, there is only one possibility ($x = 0$). Conditions (2.7c) and (2.7d) imply that $\mu \geq 0$; and for $x = 0$, (2.7a) reduces to $-4 + \mu = 0$, which is uniquely satisfied by

---

to violate (2.7b), the direction must satisfy $\mathbf{d}[\nabla g_j(\mathbf{x}^*)]^{\mathrm{T}} \leq 0$ for every binding inequality. Furthermore, since $\mu_j \geq 0$ as per (2.7d), it follows that $\sum_{j=1}^{J'} \mu_j \mathbf{d}[\nabla g_j(\mathbf{x}^*)]^{\mathrm{T}} \leq 0$ in the preceding equation. To cancel out this summation, the first term must be non-negative, i.e., $\mathbf{d}[\nabla f(\mathbf{x}^*)]^{\mathrm{T}} \geq 0$. This means that direction $\mathbf{d}$ is non-decreasing. Since $\mathbf{d}$ is an arbitrary feasible direction, the result establishes the first order stationarity condition.

$\mu^* = 4$. Since $\mu^* \geq 0$, the test solution $x = 0$ is valid, in agreement with the result found graphically.

The solution to the KKT conditions is usually obtained numerically with iterative solution algorithms. In certain situations, however, a closed-form solution can be derived analytically. This can unveil interesting properties of the solution, as illustrated by the following example.

Example 2.5. Section 2.2.3 showed that, if a steady stream of passengers arrives at a bus stop during a time interval of duration T with N bus arrivals, then the total passenger wait given by (2.4) is minimized when buses arrive with equal headways. The KKT conditions can be used to prove this statement in a different way. To do this, the question of which headways are best is reformulated as the following mathematical program:

$$\min \sum_{i=1}^{N} H_i^2$$

$$\text{s.t.} \sum_{i=1}^{N} H_i = T$$

$$-H_i \leq 0, \ \forall i = 1, 2, \ldots, N.$$

Note that the objective function is the sum of the squared headways, which is proportional to the total wait; the first constraint ensures that there are exactly N headways; and the remaining constraints that the headways are non-negative. This program has a global minimum because the objective function is bounded from below.

For this problem, the KKT conditions are

$$2H_i - \mu_i + \lambda = 0, \forall i = 1, 2, \ldots, N \tag{2.8a}$$

$$\sum_{i=1}^{N} H_i = T \tag{2.8b}$$

$$-H_i \leq 0, \forall i = 1, 2, \ldots, N \tag{2.8c}$$

$$\mu_i \geq 0, \forall i = 1, 2, \ldots, N \tag{2.8d}$$

$$\mu_i H_i = 0, \forall i = 1, 2, \ldots, N \tag{2.8e}$$

Now note that the solution, $H_i = T/N$, $\mu_i = 0$, for all i, and $\lambda = -2T/N$, satisfies the conditions. Furthermore, we now show that this is the unique solution to the KKT equations so it must be the global minimum.

Proof: To establish uniqueness it suffices to show that any solution with two different headways must violate the KKT conditions. Without loss of generality, we label the larger headway i =1, so we have $H_1 > H_2 \geq 0$. Two cases arise: either $H_1 > H_2 > 0$ or $H_1 > H_2 = 0$. We now show that the KKT conditions must be violated in both. In the former case, (2.8e) implies that $\mu_1 = \mu_2 = 0$; therefore condition (2.8a) cannot not be satisfied with a common λ for both $i = 1$ and 2; so the KKT conditions are violated. Now, if $H_1 > H_2 = 0$ then $\mu_1 = 0$; and (2.8a) implies that $\lambda = -2H_1$. This shows that $\lambda < 0$, since we are assuming that $H_1 > 0$. Now note that the left hand side of (2.8a) must be strictly negative for $i = 2$, because $\lambda < 0$ and $\mu_2 \geq 0$ (as per (2.8d)), and as we are assuming, $H_2 = 0$. Therefore, the KKT conditions are also violated in this case. □

It should be stressed that all the conditions, methods, and algorithms discussed in this subsection only produce local optima. Therefore, as in the last example, additional analysis is needed to identify a global optimum. Fortunately, for us, there are two common types of problems for which every stationary point is a global optimum, so the analysis is unnecessary: convex programs and geometric programs. Both occur often in this book. Convex programs are minimization problems where the objective function is convex and the feasible region is also convex — so any line segment connecting two points in the region lies entirely within that region. It turns out that a feasible region defined by a set of linear constraints is convex. Therefore a minimization problem with a convex objective function and a set of linear constraints such as the problem of Example 2.5 is a convex program. A geometric program is a minimization problem with only "less than or equal to" inequality constraints and non-negative decision variables; and where the objective function and all the left-hand sides of the constraints are polynomials with positive coefficients. The GEOQ problem of Example 2.1 is a geometric program. More detail on convex and geometric programming can be found in optimization books.

2.4. Dimensional Analysis

The tools explored so far produce definite answers to both descriptive and prescriptive questions. Dimensional analysis is different. It can be used with both types of questions, but only to narrow the scope of possible answers. This focusing often produces insights that are of much value by themselves and can be a basis for finding definite answers.

Dimensional analysis is based on the idea that every physical quantity has dimensions that can be expressed in some system of units. For example, acceleration has the dimension of [distance/time^2] and can be quantified by a number of measurement units such as m/s^2. It is also based on the idea that any meaningful physical relationship must hold true independent of the system of units. This simple fact implies that any such relationship must use the same system of units for all its variables, and be dimensionally consistent in the sense that any terms that need to be added, subtracted or compared must be of the same dimension. It also implies additional conditions that will be explained and exploited in the remainder of this section. Section 2.4.1 shows how to verify formulas using units, Section 2.4.2 shows the basics of dimensional analysis, and Section 2.4.3 provides some applications.

2.4.1. *Formula Verification*

Suppose we have derived a model relating multiple physical quantities $\{x_1, x_2, x_3, \ldots\}$ by means of a formula such as $x_1 = F(x_2, x_3, \ldots)$, or more generally by an implicit equation, $f(x_1, x_2, x_3, \ldots) = 0$. If the expression is to hold under all systems of units, then all the terms that are added, subtracted or compared in it must have the same units. Otherwise, non-matching terms could change disproportionately with a change of units and the equation would be violated. Similar considerations reveal that the arguments and outputs of nonlinear functions must be dimensionless. It stands to reason that one should always check for the dimensional consistency of all formulas in a model, even as they are developed. This often reveals errors in logic that can then be corrected.

To illustrate the procedure, recall the expression derived at the end of Section 2.1.3 for the maximum possible flow of freight trains along a section of single-track that is shared with pre-scheduled passenger trains. Although this is not necessary, the expression can be rewritten in implicit form as follows:

$$qH - \left\lfloor \frac{H + L(v_1^{-1} - v_2^{-1}) - (s_1 + s_2)v_1^{-1} + s_2 v_2^{-1}}{s_2 v_2^{-1}} \right\rfloor = 0, \tag{2.9}$$

where we have used "q" for the freight train flow. The units of our variables and constants were stipulated to be: freight train flow q [trains/h], headway H [h/train], length L [km], spacing s_1 [km/train], spacing s_2[km/train], speed v_1[km/h], speed v_2[km/h]. Note that $(v_1^{-1} - v_2^{-1})$ has units of [h/km] and $(s_1 + s_2)$ the unit of [km/train]. Now, to check for dimensional consistency, replace each symbol in the equation by its units and verify that compared, added and subtracted terms have the same units. In our case, we start with the argument of the floor operation and obtain:

$$([\text{h/train}] + [\text{km/train}][\text{h/km}] - [\text{km/train}][\text{h/km}]$$
$$+ [\text{km/train}][\text{h/km}])/([\text{km/train}][\text{h/km}]).$$

This expression is both dimensionally consistent and dimensionless. The latter is necessary since the expression is an argument of a nonlinear function (the floor operation), so the result raises no red flags. Since the output of the floor operation is also dimensionless, we can now write for the complete left-hand side of (2.9):

$$[\text{train/h}][\text{h/train}] - [\text{dimensionless}],$$

which is again dimensionally consistent. Although consistency is no proof that the equation is correct, it is reassuring. Otherwise, there would have been an error, which the inconsistency could help reveal.

2.4.2. *Fundamental Ideas: Reformulation with Dimensionless Groups*

Dimensional considerations can also be used to reorganize and simplify formulas. We recommend writing any dimensionally correct formula in terms of dimensionless combinations of variables, called "groups," such as the combinations $\{\pi_1 = qH, \pi_2 = L/(v_1 s_1), \text{etc.}\}$ in (2.9). For any given problem, groups can be created in multiple ways, so the reformulation in terms of groups is not unique. However, a reformulated expression is usually more compact and easy to interpret than the original. The interested reader should try this for (2.9); see homework exercise 2.9.

Perhaps it should be intuitive that the solution to any well-posed problem, meaningful formula or law of physics can always be written in terms of dimensionless groups. This idea is the centerpiece of a fundamental theorem in dimensional analysis. The theorem uses the additional concept of *mutually independent* groups. The groups in a set are said to be mutually independent if none can be obtained from the others using the operations of multiplication and raising to a power. For example, the groups qH, v_1/v_2, and $(qH)(v_2/v_1)^2$, which could be formed from the variables of (2.9), are not independent because the third group is the product of the first group and the -2 power of the second.

Theorem 2.1 (Buckingham π Theorem). *Any physical law relating p variables $\{x_1, x_2, \ldots, x_p\}$ can always be expressed as an implicit function of* *k mutually independent dimensionless groups* $\{\pi_1, \pi_2, \ldots, \pi_k\}$ *formed from the variables, i.e., as:*

$$f(\pi_1, \pi_2, \ldots, \pi_k) = 0. \tag{2.10}$$

A proof can be found in Johnson (1944).

Let us now examine the benefit of reformulating an expression in terms of groups. For maximum transparency, we shall use a simple formula. It expresses the sum of the agency and user cost of a shuttle system with a steady passenger arrival rate and even headways. The variables in the formula are passenger arrival rate λ [pax/h]; observation duration T [hours]; the number of bus dispatches N [dimensionless]; the societal value of passenger wait α [\$/pax-h]; the cost of a bus dispatch β [\$]. Then, the average passenger wait per unit time is $\lambda T/(2N)$ from Section 2.3.3, so the total user and agency cost per unit time z [\$/h] is

$$z = \alpha\lambda\frac{T}{2N} + \frac{\beta N}{T}, \text{ or in the implicit form:}$$

$$\alpha\lambda\frac{T}{2N} + \frac{\beta N}{T} - z = 0. \tag{2.11}$$

Note (2.11) has units of [\$/h] and involves six different variables. It is now rewritten using groups. The following is only one of the ways in which this can be done. First, we make (2.11) dimensionless by dividing both sides by some constant with units of [\$/h]. We choose the combination $(\alpha\beta\lambda)^{\frac{1}{2}}$. Now, since the three terms of the resulting equation are dimensionless, we define a group for each of the terms. Thus, for the second and third terms, we obtain

$$\pi_1 = N\left(\frac{\alpha\lambda}{\beta}\right)^{-\frac{1}{2}}\frac{1}{T}, \quad \pi_2 = z(\alpha\beta\lambda)^{-\frac{1}{2}}. \tag{2.12}$$

We then notice that the group for the first term is not independent, as it turns out to be the inverse of π_1 times $1/2$. Thus, we write the first term in terms of π_1 and finally write:

$$f(\pi_1, \pi_2) := \frac{1}{2}\pi_1^{-1} + \pi_1 - \pi_2 = 0. \tag{2.13}$$

This is a Buckingham π reformulation of (2.11).

Note that the reformulated expression is much simpler as it only relates two variables (groups) instead of six. So it is more transparent and can even be plotted as a single curve on the (π_1, π_2)-plane that expresses all the solutions to our problem. This was not possible with formula (2.11).

The reformulation is also useful because an examination of the groups themselves *without reference to the implicit equation* can yield significant insights. This is particularly fruitful when the groups are simple and the implicit equation is complicated — or even unknown. To derive insights from the groups alone, we exploit the fact that any changes to the original variables of a proper solution that leave the groups invariant must also be a solution.

To see how this works, let us return to our example. Although the implicit equation is simple, we shall not use it. We see from (2.12) that the group values stay the same if, for example, we double the value of N, quadruple the value of λ and double the value of z. Physically, this means that if we quadruple the demand and double the number of dispatches, then the total cost must double. What's remarkable about this statement is that it was made only on the basis of dimensional arguments, i.e., on the basis of the group structure without reference to the implicit equation.

Even for very difficult problems, it is always possible as we shall soon see to define the group structure of their solutions from the dimensions of the variables and parameters that appear in the problem statement. Therefore, dimensional analysis gives us a way of deriving insights even for problems that cannot be solved exactly. And as we show in the next subsection, it can sometimes help us find general approximations for these kinds of problems.

2.4.3. *Application*

This subsection shows how dimensional analysis can help us solve word problems, at least partially, even very hard problems. The problems in question have to be well posed in the sense that the data given should be sufficient to find the unknown answer. Otherwise, no method would work. We look for a functional relationship among the variables (i.e., the data and the unknown) named in the problem statement.

Note that Buckingham π theorem does not say how to identify the set of groups appearing in (2.10). Fortunately for us, there is a general recipe. To state it we need an important definition. We say that a set of mutually independent groups corresponding to a set of variables is "maximal" if no other group can be added to the set while preserving independence. Maximal sets can be identified in *ad hoc* ways, and also systematically as shown in Appendix 2.A.

Before the recipe is given, it is important to understand how all the groups corresponding to a set of variables are related. Appendix 2.A shows that any of these groups can be uniquely expressed as a vector in a system of dimensions; and that these vectors form a linear vector space. The appendix also shows that combinations of groups using multiplications, divisions, and power operations turn out to be linear combinations of their dimensions. It also shows that mutually independent groups correspond to independent vectors and that there is a one-to-one correspondence between maximal sets and the bases of the linear space. Now the recipe can be stated.

Corollary 2.1. *Any maximal set can be used as the argument in Buckingham's function* (2.10).

Proof: Since the dimension vectors form a linear vector space, any set of dimensions can be expressed as a linear combination of any basis. And since there is a one-to-one correspondence between groups and vectors, the previous statement implies that any set of groups can be expressed as a function of any maximal set through a change of variable that involves multiplications and power operations. In particular, if we let $\boldsymbol{\pi_B}$ be the set of groups in Buckingham π theorem and $\boldsymbol{\pi_M}$ a maximal set, it means that we can write the former as a function of the latter, i.e., as $\boldsymbol{\pi_B} = \boldsymbol{F}(\boldsymbol{\pi_M})$. Now enter with this expression in (2.10) to obtain, $f(\boldsymbol{F}(\boldsymbol{\pi_M})) = 0$, which is an implicit function of $\boldsymbol{\pi_M}$. This shows that any maximal set can be used as the argument in Buckingham π function. □

Corollary 2.1 shows that to cast a problem in group form it suffices to find a maximal set (any maximal set will do) and then use this maximal set as the argument in (2.10). This is now illustrated with two examples.

Example 2.6 (*Little's formula — stochastic case*). Let us now see what dimensional analysis can say about (2.2) in a special case involving random arrivals and waits. The word problem is "Find the average quantity of customers in a waiting area $\bar{Q}$ [quantity] if they arrive as a Poisson process with rate λ [quantity/time] and individual passengers wait for amounts of time that are independently and identically distributed negative exponential random variables with mean $\bar{w}$ [time]."[13,14]

The first step is ensuring that the problem is well posed. Note that the information given is enough to simulate the problem and derive an answer. Moreover, since infinitely long simulations should produce the same answer regardless of the initial conditions, we expect this answer to be unique. In

[13] The units of $\bar{w}$ are not wait/quantity because the numerical value of $\bar{w}$ does not change if we change the units of passenger quantity. The average wait per passenger is the same if we choose to count passengers by the dozen.

[14] The reader familiar with queuing theory will recognize the set up of our problem as an M/M/∞ queue. It is well known that Little's formula holds for this problem.

view of all this, there must be an expression linking $\bar{Q}$ to the other two variables named in the problem, so the problem is well posed.

The second step is finding a maximal group. Consideration of our three variables and two dimensions shows that only one independent group can be formed. This is the group: $\lambda\bar{w}/\bar{Q}$.

The final step is writing and analyzing Buckingham's formula: $f(\lambda\bar{w}/\bar{Q}) = 0$. If we use r_i for the ith (unknown) root of f, the formula is simply saying: $\lambda\bar{w}/\bar{Q} = r_i$ for some set of (unknown) roots. Now, since we know that the solution is unique, there can be only one root, r. Thus, we can finally write $\bar{Q} = \lambda\bar{w}/r$.

Dimensional analysis does not establish the value of r but has reduced the original question to the determination of a single constant. This is helpful because the constant can be estimated by analyzing a single instance of the problem. For example, r can be estimated by doing a long simulation with $\lambda = 1$ and exponential waits of $\bar{w} = 1$, then recording the simulated $\bar{Q}$ and finally evaluating the combination: $\lambda\bar{w}/\bar{Q} \equiv r$ (which would turn out to be close to 1) as our estimate. It would have established that the derived approximate formula holds for all instances of the original problem.

Example 2.7 (*redistribution of shared vehicles*). A transportation company provides shared vehicles at uniformly distributed pods in an urban region of area S [distance^2]. The pod density is δ [1/distance^2], where $\delta S \gg 1$. During the night, the company repositions the vehicles to even out the contents of all the pods, using an algorithm that minimizes the total distance traveled L [vehicle quantity-distance]. During the day, the net change in the number of vehicles stored at each pod is normally distributed with mean 0 and variance σ^2 [quantity^2]. The changes are mutually independent across stations and always smaller than the number of vehicles stored at the start of each day. We look for an approximate formula that expresses L as a function of S, δ, and σ^2.

The problem is well posed because from the information given one could build a simulation to answer the question for any set of parameter values.[15] And, we are justified in looking for an answer in the form of a single-valued function because one would expect similar answers from different simulation runs.

[15] This assumes that any even distribution of pods can be used in the simulation — because any even distribution should produce a similar D. This seems reasonable because the number of pods is very large.

Now we can proceed with the second step. From our four variables $\{S, \delta, \sigma, L\}$ and two dimensions $\{\text{quantity}, \text{distance}\}$ it is easy to identify two independent dimensionless groups $\{\pi_1 = \delta S,\ \pi_2 = L/(\sigma\sqrt{S})\}$. Note other groups could be formed, but in all cases, they could be expressed as a combination of these two, so they would not be independent. Thus, we conclude that the identified set is maximal. Therefore, we can now apply Corollary 2.1 and write:

$$f(\delta S, L/(\sigma\sqrt{S})) = 0.$$

Because the solution is unique, and our unknown L only appears in the second argument, it follows that this second argument must be a function of the first,[16] i.e., that we can write: for some function F: $L/(\sigma\sqrt{S}) = F(\delta S)$, or

$$L = \sigma\sqrt{S}F(\delta S).$$

Although the exact form of $F(\cdot)$ has not been determined, the result is already quite helpful. For example, it reveals that for any values of δ and S, the answer L is always proportional to σ; and that if the company wants to replicate existing service in a larger region with the same number of pods, then the distance traveled for repositioning will increase by a factor of $[\sigma S^{1/2}]_{\text{new}}/[\sigma S^{1/2}]_{\text{existing}}$.

If necessary, the function F can be identified approximately by running simulations for, say, 10 representative values of its argument. This would solve the problem in complete generality. By contrast, without dimensional analysis we would have to treat L as a function of three separate inputs, so we would need 10^3 simulations to provide answers for all possible combinations of representative input values. This is not just more laborious, the resulting table (or sets of figures) would be less intuitive.

The above examples have in common that the number of groups in their maximal sets equals the number of variables minus the number of dimensions. Although this is often true, it is not always the case. Fortunately, a simple formula exists to determine the common size of a problem's maximal sets. The formula is useful as it can tell us when to stop looking for maximal sets. Appendix 2.A presents the formula, as well as a systematic procedure for identifying these sets.

[16]For this reason, it is good practice to include the unknown in only one of the groups whenever possible.

References

Daganzo, C. F. (1985). "The uniqueness of a time-dependent equilibrium distribution of arrivals at a single bottleneck." *Transportation Science*, 19(1), 29–37.

Daganzo, C. F. (1997). *Fundamentals of Transportation and Traffic Operations*, Pergamon-Elsevier, Oxford, U.K.

Daganzo, C. F. (2005). *Logistics Systems Analysis* (4th ed.), Springer-Verlag, Berlin, Heidelberg.

Johnson, W. C. (1944). *Mathematical and Physical Principles of Engineering Analysis*, McGraw-Hill, New York.

Little, J. D. C. (1961). "A proof for the queuing formula: L = λW." *Operations Research*, 9(3), 383–387.

Mannering, F. and Kilareski, W. (1990). *Principles of Highway Engineering and Traffic Analysis* (1st ed.), John Wiley and Sons, New York, NY.

Newell, G. F. (1982). *Applications of Queueing Theory* (2nd ed.), Chapman & Hall, London.

Homework

2.1 The headways between consecutive bus arrivals at a station are independent and identically distributed random variables with mean h and variance σ^2. Suppose a stationary observer has recorded n bus arrivals (where $n \gg 1$) in an arbitrarily chosen time period of length T. The observer decides to use T/n as the estimator for the average headway. Is this estimator biased? What is the variance of this estimator? What happens if T (and n) tends to infinity?

2.2 Derive the formula for the maximum flow of freight trains on the shared track of Section 2.1.3.

2.3 Devise a two-way bus service scheme on a regional route {A–B–C; C–B–A} offering service with even but different headways on links A–B–A and on links B–C–B. The target service frequency on A–B–A is 12 buses/h, and on B–C–B, 8 buses/h, To do this, some buses will shuttle back and forth between A and B and others will travel the complete route (A–B–C–B–A). Buses can be held at B for as long as necessary to produce the requisite even departures from B in both directions. If the trip times from A–B, B–A, C–B, and B–C are each 30 min, then depict a set of bus trajectories that will meet the stated requirements, i.e., have regular headways on all legs. You need to determine how much to delay each bus at B, and if a bus just came from A, whether to turn it around at B. How many buses are required to accommodate your scheme?

2.4 (Courtesy of Gordon Newell) Suppose that two shuttle buses operate (without coordination) between the campus and downtown. One bus can carry 20 people and has a round trip time of 8 min (including loading and unloading passengers), the other can carry 25 people but has a round trip time of 10 min. There is always a queue of people waiting at the downtown bus stop, but passengers come to the campus bus stop at a steady rate of 2/min. If neither bus is idle at any time, determine:

(a) The flow of people from downtown to campus.
(b) The flow of people from campus to downtown.
(c) The average bus occupancy from campus to downtown.

If one insisted on dispatching these buses from downtown to campus at equal headways, we would only slow down the faster bus. Would you expect this strategy to reduce the average waiting time of passengers going from campus to downtown?

2.5 In a heavy morning rush, passengers arrive at a metro station at a rate of 3 pax/sec. A passenger (A) joins a 200 ft queue that contains 100 passengers in front of the entrance gate at time $t = 0$ sec. People in this queue enter the gate at a rate of 1 sec/pax. Determine the following:

(a) The time in queue for passenger A.
(b) Determine the distance that a passenger (B), who joins the queue at $t = 20$ sec, would have to walk before entering the gate. Assume the spacing between adjacent passengers in queue remains constant.

2.6 Find the optimal value of s for the following problem, where $\alpha, \beta, \gamma > 0$.

$$\begin{aligned} &\min \alpha/s + \beta s \\ &\text{s.t.} \quad 1/s \leq \gamma \\ &\qquad\quad s \geq 0. \end{aligned}$$

2.7 *(Robustness of EOQ) We discussed optimization problems of the form

$$z^* = \min\{z = Ax^n + Bx^{-m} : 0 < x \leq \infty; m, n > 0\}$$

where z is a cost, x is a design variable, and the optimum is achieved for some x^*. Show how the actual cost achieved, z_a, depends on our design, x_a ($x_a \neq x^*$). To do this use dimensionless variables $\Delta \equiv z_a/z^*$ and $\delta \equiv x_a/x^*$, find Δ as a function of δ and plot it for different values of n and m. Discuss the sensitivity of the cost as the design deviates from the optimum and the effect of the constants n and m.

Extra credit: find a simple form for the relative error $(\Delta - 1)$ in terms of $(\delta - 1)$ where $\delta \approx 1$.

2.8 Consider an entire two-dimensional plane on which customer demand for some service is homogeneously distributed with density $= \lambda$ [calls/time-area]. Identical service facilities are distributed translationally symmetrically, each serving the calls in an area of equal shape and equal size A. These areas tessellate the plane. The cost for providing service to a unit demand is proportional to the Euclidean distance between a customer and its facility with a proportionality factor, c [\$/distance]. Then, (a) use dimensional analysis to show how the average cost of service per unit demand depends on A and λ; and (b) if the amortized cost of constructing one facility is f [\$/time], derive a closed-form formula for the optimal A.

2.9 (Multiple ways to generate independent groups) Equation (2.9) shows how the maximum freight train flow q is related to passenger train headway H, shared track length L, passenger train speed v_1, freight train speed v_2, passenger train spacing s_1, and freight train spacing s_2.

(a) Write down the dimension vector of all involved variables. Find a set of groups that can be used to rewrite (2.9). Write down the structure vector of each group, and the dimension matrix. Compute the dimension vectors of these groups. Express (2.9) in terms of these dimensionless groups.

(b) Find a maximal set of dimensionless groups that can be used to simplify (2.9). You can do this either in an *ad hoc* way, or use the method described in Appendix 2.A. Repeat (a) for this maximal set of groups.

(c) Can you find another maximal set of dimensionless groups that can also simplify (2.9)? How is this set of maximal set related to that in (b)?

Appendix 2.A. How to Identify Maximal Sets

This appendix shows that the sets of dimensionless groups that can be formed from a set of variables $\{x_1, x_2, \ldots, x_p\}$ involving d independent dimensions, $\{\alpha_1, \ldots, \alpha_d\}$ form a finite dimensional linear vector space. It also shows that every point of this space is a dimensionless group (and vice versa); that the bases of the space correspond to maximal sets; and how the bases can be identified.

Definitions:

(1) A group g is a monomial of the variables with coefficient "1," characterized by a $(1 \times p)$ vector $\mathbf{e} = (\ldots, e_i, \ldots)$ listing the exponents of the variables, which we call the group's *structure vector*, i.e.,

$$g \equiv \prod_{i=1}^{p} x_i^{e_i}. \tag{2.A1}$$

(2) The dimension of a group, $D(g)$ is a monomial of the group's dimensions with coefficient "1." It is characterized by a $(1 \times d)$ vector $\mathbf{a} = (\ldots, a_j, \ldots)$ listing the exponents of the dimensions, which we call the group's *dimension vector*. In other words,

$$D(g) \equiv \prod_{j=1}^{d} \alpha_j^{a_j}. \tag{2.A2}$$

(3) Variables are elemental groups with their own dimension vectors. The dimension vector for variable, x_i, is denoted $\mathbf{m}_i = (\ldots, m_{ij}, \ldots)$, i.e.,

$$D(x_i) \equiv \prod_{j=1}^{d} \alpha_j^{m_{ij}}. \tag{2.A3}$$

The vectors for all variables, $\{\mathbf{m}_i : \forall i\}$, are given. The $(d \times p)$ matrix, $\mathbf{M} = (\ldots, \mathbf{m}_i^{\mathrm{T}}, \ldots)^{\mathrm{T}}$ summarizing these dimension vectors is called the problem's *dimension matrix*.

(4) The groups in a set are said to be mutually *independent* if none can be obtained from the others using the group operations of multiplication and rising to a power.

It should be clear from Definition 1 that our two group operations can be re-expressed in terms of their structure vectors as follows:

(I) The power operation of a group translates to a scalar multiplication of its structure vector. This can be seen from (2.A1), which yields: $g^z \equiv \prod_{i=1}^{p} x_i^{ze_i}$ for any constant z.

(II) Group multiplication translates to structure vector addition. Again, (2.A1) reveals that the structure vector of the product of two groups is the sum of their structure vectors.

Furthermore, some consideration of Definitions 1–3, and Properties I and II shows that

(III) The groups in a set are mutually independent if and only if their structure vectors are linearly independent.

(IV) The dimension vectors of all possible groups based on a set of variables form a linear vector space, $\mathcal{V}$.

Now note that the dimensions of a group are calculated from those of its variables with the formula:

$$D(g) \equiv D\left(\prod_{i=1}^{p} x_i^{e_i}\right) = \prod_{i=1}^{p} D(x_i)^{e_i}. \tag{2.A4}$$

This formula is a bit abstract but merely restates what one does. For example, consider the group $g = vt^{-1}$ that can be formed from the variables v and t. Assume the independent dimensions of this problem are [distance] and [time] and that our variables v and t have dimensions of [(distance)(time)$^{-1}$] and [time], respectively. Now apply (2.A4) and note how it reproduces what we would do: $D(g) \equiv D(v^1t^{-1}) = D(v)^1D(t)^{-1} = (\text{distance/time})^1(\text{time})^{-1} = (\text{distance})(\text{time})^{-2}$.

Now note how for this particular example the structure vector of our group is $\mathbf{e} = (1, -1)$, its dimension vector is $\mathbf{a} = (1, -2)$ and the problem's dimension matrix is $\mathbf{M} = \left[\begin{smallmatrix} 1 & -1 \\ 0 & 1 \end{smallmatrix}\right]$. Note how these three vectors/matrices satisfy the relationship: $\mathbf{a} = \mathbf{eM}$. This turns out to be generally true as we now show.

Theorem 2.2. *The dimension vector of a group with structure vector* e *is:* $\mathbf{a}=\mathbf{eM}$.

Proof: From (2.A2) and (2.A4) we can write:

$$D(g) \equiv \prod_{j=1}^{d} \alpha_j^{a_j} = \prod_{i=1}^{p} D(x_i)^{e_i}. \tag{2.A5}$$

Now substitute the right side of (2.A3) for $D(x_i)$ in this expression to obtain:

$$\prod_{j=1}^{d} \alpha_j^{a_j} = \prod_{i=1}^{p}\prod_{j=1}^{d} (\alpha_j^{m_{ij}})^{e_i} = \prod_{j=1}^{d}\prod_{i=1}^{p} \alpha_j^{e_i m_{ij}} = \prod_{j=1}^{d} \alpha_j^{\sum_{i=1}^{p} e_i m_{ij}}. \tag{2.A6}$$

Now compare the exponents of the first and last terms to note that $a_j = \sum_{i=1}^{p} e_i m_{ij}$, i.e., that : $\mathbf{a} = \mathbf{eM}$. □

This theorem is useful because it allows one to find the dimensions of a group from the group's structure vector with a simple matrix multiplication. It is also useful because it allows us to characterize all dimensionless groups (i.e., groups with a zero dimension vector) as the solutions of the homogeneous equation:

$$\mathbf{eM} = \mathbf{0}, \tag{2.A7}$$

i.e., as the left null space of $\mathbf{M}$, which we denote $\mathcal{S} \subset \mathcal{V}$.

The rank-nullity theorem of linear algebra tells us that the dimension of $\mathcal{S}$ is $p - rank\,(\mathbf{M})$. Also, recall that a basis of $\mathcal{S}$ can be found with standard methods such as Gauss–Jordan elimination. This solves our problem because the bases of $\mathcal{S}$ are (by definition) maximal sets of linearly independent vectors corresponding to dimensionless groups, which (by virtue of Property III) are maximal sets of independent groups.

Summary: All of the above reduces to the following: To find a maximal set of groups, write the problem's dimension matrix $\mathbf{M}$ and find $p - rank(\mathbf{M})$ independent solutions of (2.A7). The elements of each solution vector are the exponents of the variables in each group of the maximal set.

These ideas are now illustrated by revisiting Examples 2.6 and 2.7 of Section 2.4.

Example 2.8. This problem involved $p = 3$ variables $\{\bar{Q},\ \bar{w},\ \lambda\}$ and two dimensions [quantity], [time]. The dimensions of the variables are: [quantity] for $\bar{Q}$, [time] for $\bar{w}$ and [(quantity)(time)$^{-1}$] for λ. Thus, with these orderings for variables and dimensions, the problem's dimension matrix is

$$\mathbf{M} = \begin{bmatrix} 1 & 0 \\ 0 & 1 \\ 1 & -1 \end{bmatrix}.$$

It is easy to see that $rank(\mathbf{M}) = 2$. Thus, (2.A7) has only $3 - 2 = 1$ independent solution. The reader can verify that one such solution is $(-1,\ 1,\ 1)$. The corresponding group is $\pi = (\bar{Q})^{-1}(\bar{w})^1(\lambda)^1 = \lambda\bar{w}/\bar{Q}$. This is the maximal set identified heuristically in the example.

Example 2.9. The system involves now $p = 4$ variables $\{S,\ \delta,\ \sigma^2$, and $L\}$ and two dimensions: [quantity], [distance]. The units of the variables are [(distance)^2], [(distance)$^{-2}$], [(quantity)^2], and [quantity–distance], respectively. With this ordering, the dimension matrix of the problem is

$$\mathbf{M} = \begin{bmatrix} 0 & 2 \\ 0 & -2 \\ 2 & 0 \\ 1 & 1 \end{bmatrix}.$$

Again, it is easy to verify that $rank(\mathbf{M}) = 2$. Thus, (2.A7) has only $4 - 2 = 2$ independent solutions. Two such solutions are (1, 1, 0, 0) and $(-\frac{1}{2}, 0, -\frac{1}{2}, 1)$. They correspond to the maximal set $\{\pi_1 = \delta S, \pi_2 = L/(\sigma\sqrt{S})\}$ of the example.

Chapter 3

Planning — General Ideas

Planning a transit system consists in choosing its constituent elements and their spatiotemporal organization for the long term, often on a large scale. Typical choices include the technologies to use, the region to cover, the stop locations, the layout of the routes and their advertised frequencies or schedules. These decisions are strategic, for the long run, because they are too costly or impractical to change on a frequent basis. Planning decisions are important because they fundamentally define the accessibility and mobility delivered to transit users, as well as the infrastructure and operating costs imposed on the owner and operator.

Good transit planning requires not just sound engineering, but also knowledge of the community to be served and the demand it can generate. In particular, it depends on anticipating how many people will use the system and how their trips will vary over space and time during the planning horizon. This can be tricky because the planned systems can generate their own demand. Thus, in view of the multitude and complexity of decisions involved in planning, the topic will be covered in five chapters: the current chapter which describes some basics, and four methodological chapters (Chapters 4–7). Of course, to successfully run a transit system, shorter term, tactical and operational issues also need to be addressed. These include the following: how to manage vehicle fleets on a day-to-day basis; how to schedule drivers; and how to control vehicles in real-time to respond to disruptions. Due to their nature and their importance, these topics are not considered in the planning chapters and are instead examined in detail later in the book.

The basic ideas in the current chapter have been organized in three subsections. Section 3.1 discusses big-picture issues, and the next two technicalities: Section 3.2 shows how to build plans that take into account the diverse non-monetary consequences of a particular system design; and

Section 3.3 shows how to deal with demand while recognizing that it may depend on the system design. The chapter closes with an additional short section (Section 3.4) that describes the organization of Chapters 4–7.

3.1. Planning Decisions

This section discusses big-picture decisions; in particular, how to choose a plan's temporal horizon, the scope of its solutions, and the modeling philosophy. It will be argued that the temporal horizon should be long, the scope wide and that plans should be developed in two stages: rough first and then fine.

3.1.1. *Temporal Horizon*

It is generally accepted that the success of a transit system over its life is co-dependent with the socio-economic activities and land uses of the region it serves. Proof of this belief is the emergence of so-called transit-oriented development, or transit-oriented community initiatives, where the urban land use near transit stations is planned in conjunction with the transit system (Curtis *et al.*, 2009). In most instances, however, practical realities dictate that transit systems should be designed for a given urban structure. Although not ideal, this framework can still produce good results for society if: (i) one can roughly approximate the demand over the planning horizon and (ii) the horizon's duration is chosen to be comparable with the life of the system. Section 3.3 discusses (i). Here we focus on (ii); in particular on how the results may disappoint if the planning horizons are chosen too short.

Short horizons are often adopted because policy makers, and sometimes the general public, tend to favor simplicity, gradual change, low-cost solutions, and flexibility. In the worst case an agency may myopically modify a transit system incrementally, e.g., to address immediate changes in demand or respond to complaints, without a view to the long term. This type of shortsightedness is encouraged by the brevity of political cycles, *vis-à-vis* the lifetime of a system, since decision-making bodies are typically staffed by political appointees who need favorable outcomes during their short terms of office.

We now show with an example the dire consequences of shortsightedness, and why it should be resisted. We will see that if the demand increases gradually and indefinitely with time, then incrementally chasing optimality with short-term gains in mind can lead to a much worse fate than if one designs with foresight. The example is simple and idealized to best illustrate the point.

The setting is an imaginary city that can be served by only two modes: automobile and transit. The demand rate increases gradually over time, and we must periodically choose how best to serve it. The decisions are made at the beginning of each political cycle and the choice set is simple: invest in a subway, invest in automobile infrastructure, or do nothing. Our two investment choices have the following properties:

Automobile infrastructure: a divisible capacity with unit cost c_g per unit of capacity;

Subway: an indivisible but very large capacity with total cost c_0.

The goal is supplying at all times enough capacity to meet the current demand rate. The expected rate by the end of the ith political cycle is denoted by α_i. At each decision instant, the decision-making body must choose whether to invest a large amount of money, c_0, in digging tunnels and laying track for a subway that will have enormous capacity to handle all the demand for the foreseeable future, or to incrementally expand the street infrastructure capacity in each cycle just enough to barely meet the increased demand rate by the cycle's end.

If the decision-making body were to choose the automobile option every time, then the incremental investment cost in period i would be $c_i = c_g(\alpha_i - \alpha_{i-1})$ for every i. And if as we are assuming, c_0 is sufficiently large — so large, in fact that c_i is always smaller than c_0 — then each of these individual choices would be less costly than the subway alternative. Therefore, if cost is the main concern and the decision-making body is myopic, the automobile option will win in every period. The cumulative costs arising from this decision making process at different times are shown by the step shaped line of Figure 3.1.

The figure also shows that if the demand continues to increase indefinitely then the cumulative automobile infrastructure expenses would eventually exceed, in some cycle i^*, what the subway would have cost; and that the expenses would continue to accrue beyond this time. A tragedy is that had decisions been made with a long run horizon with more cycles than i^* then the less costly subway would have been chosen.

Of course, the real world is not as dire. Real decision-making bodies are not totally myopic and they have financial tools at their disposal to spread costs over time so they are in a position to avoid the pitfalls of the example. But to succeed they must also have a long term vision. This is an important factor. A general rule is that planning actions that have long-term

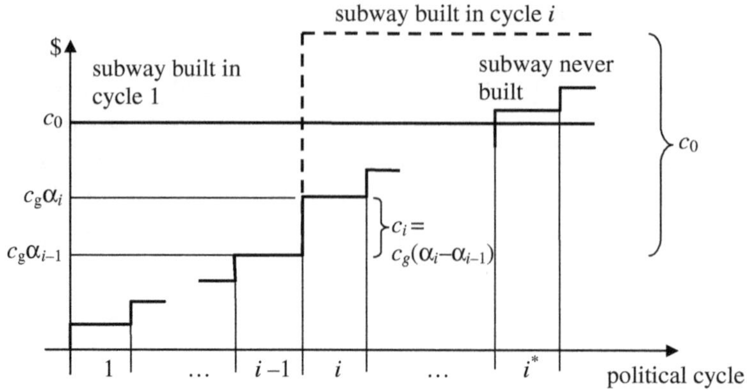

Figure 3.1. Investment options and associated costs for auto vs. subway over time.

consequences should be made with both, a long-term horizon and long term vision.

In developing this vision, it should be remembered that over long periods of time transit systems can create their own demand. This potential to shape the demand should be recognized, even exploited, when decision-makers formulate the goals of a plan. This idea is further discussed in Section 3.3.

3.1.2. *Scope of the Solution Domain*

We propose in this section that in transit planning one should always try to evaluate the broadest possible set of candidate solutions, provided of course that methods exist to compare them all. We suggest that the broadest solution domain can be evaluated by starting with a clean slate and then using a mathematical approach to narrow the set of candidates and identify a final choice.

In many practical cases, solutions to a particular urban problem are identified by looking at successfully implemented solutions to similar problems in other cities and adapting them for the current application. This is how the concept of *Bus Rapid Transit* (BRT) has gained and continues to gain acceptance. First implemented in Curitiba, Brazil, this form of bus transit with dedicated lanes and off-board fare collection has expanded all over the world. It has the advantage of providing a metro-like experience at considerably less cost. It is therefore popular in application sites where cost is of utmost importance. BRT has spread all over the world because success breeds success, and because copying what another city is doing well is a safe way of trying something different.

Although the idea of adapting solutions from other places for a particular application can and does work well, it has a drawback. If the city in question is not sufficiently similar to other cities where possible relevant solutions have been tried, then there may be no good models to copy. And even in the best of cases where many other cities are similar, the menu of possible solutions is limited by what has already been tried elsewhere. Therefore, the scope of the solution domain is narrow.

This book will show how to broaden this scope by starting with a completely clean slate, and then systematically figuring out how to assemble generic transit components (vehicle types, guideway types, etc.) to build a system that will best serve the practical situation at hand. We call this the "clean-slate" approach. The clean-slate approach enables a city to come up with innovative solutions that have not been tried anywhere. The main advantage is that by exploring a large solution space one is likely to come up with a better solution. We recognize that a systematic search over a large space can be difficult, but this barrier can be overcome as explained in Section 3.1.3. A more serious problem is that if the solution is substantially different from other existing systems then a transit agency cannot build on the experience of other localities. We believe however that the ensuing risk can be far outweighed by the potential benefit. A case in point is the innovative high-performance bus system in Barcelona, which was developed with the clean-slate approach and is attracting unexpectedly high demand (Badia *et al.*, 2017); see also Section 6.4.2.

3.1.3. *Modeling Philosophy*

Although the clean-slate and comparative approaches to planning are technically quite different, both use the procedural steps displayed on the left side of Figure 3.2. The main difference arises in the third and fourth steps: alternatives are generated and evaluated much more broadly with the clean-slate approach. Let us now review the steps one by one.

In the problem definition step, a policy making body often in conjunction with the public, defines the scope of the planning task considering the issues at hand. Ideally, these issues should be quantified with data as this will help better define the planning objectives and provide sound benchmarks for evaluation. For example, in a city where a transit system is being retrofitted, one may want quantify its population distribution, the travel demand, the transit system itself, and its current performance. The latter should include all the system's outputs and costs.

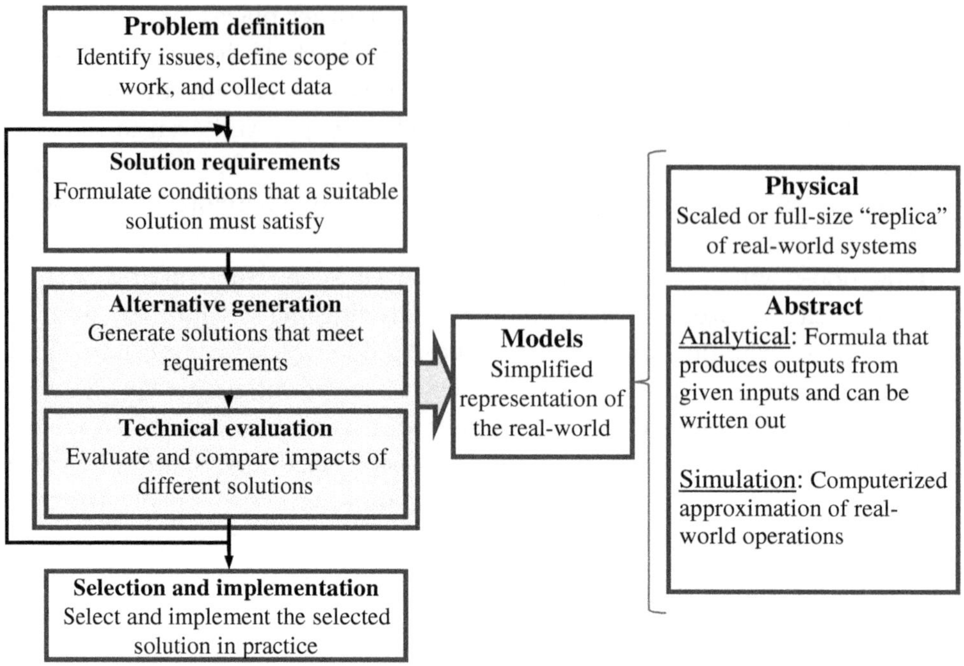

Figure 3.2. Typical planning procedure and the use of models.

In the solution requirement step, the policy setting body identifies the constraints that define what constitutes a valid solution, and sets overall objectives. An example objective is lowering the average door-to-door travel time in the service region as much as possible; and an example constraint is not exceeding a capital expenditure budget. This step may rely on political goals more than technical considerations, but it is fundamental.

The next two steps build on this information and are more technical: first one generates a menu of valid candidate solutions, and then evaluates them. In the comparative approach to planning the candidate solutions are modifications of existing systems and few in number, so each can be developed comprehensively and evaluated in detail. Solutions are usually developed in *ad hoc* ways and then systematically evaluated or validated with computer simulation models.

In the clean slate approach the candidate solutions are not enumerated. Instead, they are indirectly defined by means of mathematical equations that specify the number and type of the basic transit components used, and how they are organized to define a working system. The required constraints are introduced as equality or inequality equations that rule out invalid combinations of components. This is useful because then the feasibility region of these equations includes *all possible* feasible systems — so the solution domain is the broadest possible. To evaluate these solutions policy objectives are also formulated mathematically, as equations that express the achievement level in terms of the components used. These equations are then combined into an objective function that is to be maximized or minimized. Optimal solutions are then obtained by combining all the equations into a mathematical optimization program of the type discussed in Chapter 2 and solving it. This is usually quite a complex task, so the end result is often only approximately optimal.

In both the comparative and clean-slate approaches, the alternative generation and selection processes involve the use of models, and this is stressed in the figure by the boxes on the right. Models are abstractions of the real world that help planners and engineers avoid costly and inefficient full-scale experimentation. Physical models are scaled-down replicas of the real world, and are only used for alternative evaluation, e.g., when testing in the real world aspects of a plan such as its communications and control schemes, or when doing a full blown pilot project to demonstrate a new operational concept.

More relevant for our purposes are abstract models, which are mathematical representations of the real world that can be manipulated in an office with computers and/or pencil and paper. Models can be classified as being analytical or simulation based. An analytical model is a set of formulas that describe approximately how the system works when the formula's variables are set equal to values representing relevant features of the real-world and the system. Questions about the real system are then answered by evaluating the formula with inputs of interest. A simulation model is a computerized replica of the system with digital agents and rules of interaction that mimic those of the real world. Questions about the real system can be answered very rapidly in this case too by running the simulation with relevant inputs.

As a rule, simulation models produce better approximations than analytical models, but are more difficult to understand and optimize. Analytical formulas on the other hand can at a glance reveal relationships amongst their variables (as in dimensional analysis; see Section 2.4) and be cast as mathematical programs that allow analysts to compare an infinite number of possible solutions, and to identify the best. For these reasons, analytical models are ideally suited for the clean-slate approach. By contrast, simulation models shine when only a few options need to be compared; so they are better suited for the comparative approach.

The fifth step is the solution's final selection and implementation. This obviously involves the decision-making body and can be the final step. Sometimes, however, none of the solutions emerging from the analysis are found to be acceptable and one must go back to the drawing board. This is why the figure includes a feedback loop. This type of failure may occur if some of the stakeholders have changed their mind; or because some their original desires may have been improperly communicated, or poorly understood, when the solution requirements were formulated.

Since the focus of this book shall be the clean-slate approach, which heavily depends on the use of analytical models, some additional ideas on modeling are now introduced.

3.1.4. *The Two-Step Method: Integer vs. Continuous Variables*

Most transit planning problems are so complex that formulas can only roughly approximate their performance. Yet mathematical formulas are useful because they allow analysts to explore infinite numbers of possible solutions. To address this conundrum, we shall use a two-step modeling method that has been used in the planning of logistics systems (see Chapter 1 in Daganzo, 1991). The first step exploits the benefits of analytic formulas, and the second remedies the inaccuracies. The method is now explained in more detail.

Its first step is "sketch planning." In this stage, the real world is idealized to some extent so it can be represented with tractable formulas as a mathematical program. This allows us to explore exhaustively an enormously large domain of solution possibilities and to identify the best (see Daganzo *et al.*, 2012) for further discussion of this idea. The identified solutions are then treated as concepts, which need to be adapted to the real world considering all the details that were ignored in the idealization. This is the purpose of the second "fine-tuning" step. During fine-tuning, the chosen solution concept is adjusted in *ad hoc* ways to fit all the details of the real world, and then evaluated with detailed simulations. The step is important and time consuming but straight-forward. Therefore, Chapters 4–7 of this book will focus on the other step, sketch-planning. This activity is critically important because it is when the large space of possibilities is explored. The activity is also challenging because it hinges on the development of tractable mathematical models that are both, accurate and tractable.

It turns out that there always is a tradeoff between accuracy and tractability. If the emphasis is on accuracy, then very detailed models with much input data and many decision variables, both integer and real, would have to be used. The downside is that the resulting mathematical programs would then be so complex and difficult to solve that one cannot be sure that the solutions produced would be any good. In essence, the solution domain would be very precisely described but not efficiently explored.

Fortunately, for us, tractability improves considerably if one represents integer quantities with real-valued variables, as if the quantities were continuous, and in so doing eliminates all integer variables. For example one could treat the number of stops in a corridor as a real variable, with the understanding that the field implementation will use the closest integer to the obtained solution. With this type of "continuum approximation," the mathematical program often adopts a form that can be solved efficiently with numerical algorithms. The approach is particularly beneficial if the approximation of integer variables by real ones does not introduce large errors. Interested readers are referred to Ansari *et al.* (2017) for a review.

In view of this, all decision variables that describe counts of indivisible items such as buses, stations and routes in the mathematical programs of this book will be treated as real variables. This turns out to work particularly well when, as often occurs, the problem's solution involves integer variables that are large compared with 1. This should be intuitive, but the following example fleshes out the idea.

Example 3.1. A certain transit agency wants to determine how many vans and small buses to purchase with a given budget to maximize its passenger carrying capacity [seat-km/min]. It is known that each bus produces 22 capacity units and costs 2.1 monetary units; and that each van produces 18 units and costs 1.9. We shall solve this problem for a small budget of 2, and for a large budget of 200.

If the budget is 2 monetary units and we use x for the number of buses and y for the number of vans, the mathematical program is

$$\max\{z = 22x + 18y\}$$

$$\text{s.t. } 2.1x + 1.9y \leq 2; \text{and } x, y \geq 0$$

$$x, y \in \mathbb{Z}_+ (\text{integer valued})$$

This is an integer program. It is simple (by design) because it only has two feasible (x, y) points: (0, 0) and (0,1). So, obviously, the optimum solution is

$$x^* = 0, \quad y^* = 1, \quad \text{and} \quad z^* = 18.$$

Now let us see what happens if we use the two-step approach, treating the decision variables as continuous variables in the sketch-planning step.

The continuum formulation is obtained by dropping the last constraint from the original. Note, the problem becomes a linear program so it can be solved easily. This can be done by plotting the feasible region on the (x, y)-plane and then evaluating the objective function at its three corner points. The reader can verify that the optimum is

$$x^* = 0.952, \quad y^* = 0, \quad z^* = 20.95.$$

This solution, however, is only a concept because in reality x and y must be integers. To turn it into an implementable solution we need to perform the *ad hoc*, fine-tuning step. For this particular problem, this would consist in rounding any non-integer variables to the nearest integer that preserves feasibility. In our case, the budget constraint prevents rounding up, so the fine-tuned solution is

$$x^* = 0, \quad y^* = 0, \quad z^* = 0.$$

Note that this is much worse than the optimum solution. On the other hand, the severe error only arises because the adjustment in the fine-tuning step is large compared with the unadjusted value; much smaller errors should arise when the optimum unadjusted values turn out to be larger.

Before considering what happens for a larger budget, it will be useful to understand how the quality of a fine-tuned solution can be evaluated without knowing the true optimum. To do this, we use two facts: (i) the objective value of the first step is an upper bound to the true optimum, since the problem solved is a relaxed version of the original; and (ii) the objective value of the fine-tuned solution is a lower bound, since it is a feasible point of the original problem. Thus, the difference between the objectives of the first and second steps is an optimality gap that bounds the error in the produced solution. In the problem, we just solved, the optimality gap is 100%. It is so large because it is formed by a small difference between two small numbers.

Now, let us repeat the analysis for the larger and more realistic problem with a budget of 200. The optimum solution is obtained by solving:

$$\max\{z = 22x + 18y\}$$
$$\text{s.t. } 2.1x + 1.9y \leq 200; \text{and } x, y \geq 0$$
$$x, y \in \mathbb{Z}_+ (\text{integer valued})$$

This can be solved in a few minutes by enumerating in a 2×2 table, e.g., in a spreadsheet, all the feasible values and evaluating their objective values.

The unique optimum is

$$x^* = 95, \quad y^* = 0, \quad z^* = 2,090.$$

Now drop the integer constraint to obtain the LP and perform the sketch planning step. As before, we evaluate the objective value at the three corners of the feasible region and find

$$x^* = 95.2, \quad y^* = 0, \quad z^* = 2,095.$$

Finally, fine-tune the solution by rounding the decision variables down to the nearest integer. The result is

$$x^* = 95, \quad y^* = 0, \quad z^* = 2,090.$$

This solution happens to match the true optimum so the error is zero, but this is a coincidence that will rarely happen in large problems. A more relevant point is that the objective values of the two solutions *must* be close for budgets of that magnitude. We can see this from the optimality gap, which for a given budget B cannot exceed the ratio of the maximum possible adjustment (which is 22) to the smallest possible optimal solution. The latter is bounded from below by the objective value of the feasible solution obtained by using the maximum number of buses and no vans, i.e., by $22\lfloor B/2.1 \rfloor \approx 10B$ for $B \gg 1$. Thus, the optimality gap can be no greater than about $22/(10B) \approx 2/B$. So the larger the budget is, the more the vehicles in play and the smaller the gap. In our particular case, the gap is $0.24\% = 5/2,095 \approx 1/400$, which is indeed below our estimated upper bound, $2/B = 2/200$.

This outcome illustrates that only small errors arise in an optimization when one treats counts of large indivisible quantities as real variables. This will be done repeatedly in this book because the practical advantages of this type of simplification are considerable.

3.2. Accounting for Non-Monetary Performance Outputs

Transit planning involves tradeoffs among quantities that are measured in monetary and non-monetary units, often affecting different stakeholders. This section explores how to account for these conditions realistically in the framework of a sketch-planning optimization.

At the heart of all transit planning there is a tradeoff between agency costs and user level of service (LOS). The latter is affected by non-monetary quantities such as distance and time. To be more specific, imagine a linear city served by a single transit line, as shown by the time-space diagram

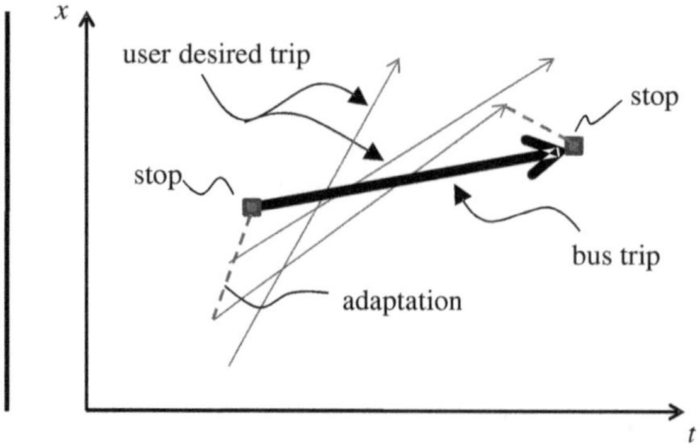

Figure 3.3. Trips in a linear city.

of Figure 3.3. The thin arrows in the diagram represent desired passenger trajectories for individual trips. The thick line is the trajectory of one bus, stopping at particular points in time at two consecutive stops. The dotted lines illustrate the gaps between what the bus can do and what one of the users wants. Consideration of the figure shows that in order to overcome these gaps this user would have to leave earlier than desired, walk some distance to access the station, and then upon arrival walk again, finally arriving later than desired. Other persons would have to adapt in similar ways, perhaps using other buses (not shown).

Consideration of the figure also reveals that this waste could be reduced by using more buses, i.e., providing more thick arrows, and by increasing their speeds — as then people would have better and more numerous travel options. Unfortunately, both of these improvements cost money to the operator, so clearly there is tradeoff involving cost to the agency, and time and distance to the users. This kind of tradeoff with different stakeholders, and outputs with mongrel units, is at the heart of all transit planning.

As Section 3.1 strongly suggested, the main goal of planning should be finding an optimal way of serving a target level of demand, which should have been previously identified in the "solution requirements" step of the planning process.[1] Efficiency is quantified by a combination of the users' level of service and the agency/societal costs. In doing this, all outputs of the transit system, regardless of who experiences them, must be expressed using a common unit for cost. This can be dollars, minutes, or whatever. And, of course, benefits should be changed into costs by changing their sign.

[1]That this is highly desirable (and reasonable) for the purposes of design is explained in Section 3.3, and Chapter 11.

To quantify the level of service component of the objective function, we define what we call a "generalized cost" or disutility, which can be defined both for individuals and groups. For example, for an individual experiencing time T and fare \$, the generalized cost could be the following:

$$\text{Generalized Cost} = \beta_T T + \beta_\$ \$.$$

The coefficients β_T and $\beta_\$$ will, of course, vary across individuals. The agency's generalized cost would itself be quantified using pertinent coefficients for the agency's different outputs. Clearly, because there are many individuals with different coefficients, and an agency with its own agenda, the choice of appropriate coefficients to represent all the stakeholders in the objective function cannot be resolved objectively. These constants must be chosen with input from the policy-making body during the "solution requirements" phase of the planning process.

To further complicate matters, transit systems also involve external benefits and costs to non-users, such as congestion reduction, energy consumption, pollution, and noise generation; and people also disagree about how these should be valued. These valuations inject another subjective element into the decision-making picture, which should also be addressed in the "solution requirements" phase.

Clearly, good planning must account for all the outputs and not just some. This can be achieved most directly if reasonable weights can be identified for all outputs, as then the generalized cost formula combining all outputs will quantify the effectiveness of our planning efforts as the policy-making body desires. Then, if we can find a mathematical function that expresses this generalized cost in terms of decision variables that characterize the plan, a "best" plan can be identified by minimizing this function without any constraints. For reasons that will soon become apparent, we call this approach the "*Lagrangian*" approach.

Alternatively, it may be easier in practice for decision-makers to identify lower or upper bounds for each (or some) of the outputs; and then set these bounds as standards for the plan. As we now show, the planning problem then takes the form of a constrained optimization problem which we shall call the "*standards*" approach. The rest of this subsection describes the standards approach in more detail and then shows how it relates to the Lagrangian approach. It turns out that the two approaches often produce the same plans.

3.2.1. *The Standards Approach*

The goal of the standards approach is choosing the cheapest plan that meets a set of standards. To illustrate, suppose for the moment that a planned transit system produces the following four outputs, and that the standards in parenthesis have been identified.

T — door-to-door time; a disbenefit (no more than T_0)
E — energy consumed; a disbenefit (no more than E_0)
M — mobility; a benefit (at least M_0)
A — accessibility; a benefit (at least A_0)

Using \$ for cost, we can write the standards optimization problem as the following mathematical program:

Mathematical Program (MP):

$$\min\{\$: T \leq T_0; E \leq E_0; M \geq M_0; A \geq A_0; \ldots\}$$

Note how each standard defines an inequality that constrains the value of the performance output to the stipulated level.

To simplify the notation and generalize the formulation we shall denote all output variables in the standards MP, both monetary and non-monetary, by $\mathbf{y}$. So, for our example, $\mathbf{y} = (\$, T, E, M, A)$. We shall also denote by $\mathbf{x}$ a vector of variables that completely characterize the design, e.g., containing all the routes and schedules; and by $\boldsymbol{\alpha}$ the system's demand, which is assumed to be given.

It stands to reason that the output vector $\mathbf{y}$ is a function of $\mathbf{x}$ and $\boldsymbol{\alpha}$, i.e., that given the design and given the demand, the system will perform in a predictable way. With this in mind, let $\mathbf{F}_m$ be the function relating $\mathbf{y}$, $\mathbf{x}$, and $\boldsymbol{\alpha}$, where the subscript "m" stands for a particular mode, i.e., so that $\mathbf{y} = \mathbf{F}_m(\mathbf{x}, \boldsymbol{\alpha})$. We assume that this output-predicting function has been identified through modeling.

We then look for a value of $\mathbf{x}$ that minimizes the \$-component of $\mathbf{y}$ without violating the constraints. The result is a best design, $\mathbf{x}^*(\boldsymbol{\alpha})$, which if implemented would yield an output vector $\mathbf{y}^* \equiv \mathbf{F}_m(\mathbf{x}^*(\boldsymbol{\alpha}), \boldsymbol{\alpha})$. The function relating $\mathbf{y}^*$ to $\boldsymbol{\alpha}$ will be denoted $\mathbf{G}_m(\boldsymbol{\alpha})$. This function represents the best performance that can be achieved by mode m when the demand is $\boldsymbol{\alpha}$. We will, in this book, compare the $\mathbf{G}_m(\boldsymbol{\alpha})$ for different modes. Let us now see how the standards and Lagrangian approaches compare.

3.2.2. *Connection between the Lagrangian and Standards Approaches*

The standards approach and the Lagrangian approach are intimately related. The reader will recall from Section 2.3.2 that if one chooses for weights in the Lagrangian of an optimization problem the optimal shadow prices then the first order optimality conditions of the unconstrained problem with the Lagrangian as its objective function match the KKT conditions of the constrained problem. Therefore, the unconstrained optimization (corresponding to our Lagrangian approach) produces the same result as the constrained optimization (corresponding to the standards approach). This equivalence gives us flexibility to choose the formulation that best suits the available information. We suggest using the Lagrangian approach if all the weights are known; choosing the standards approach if all the bounds are known; and using a hybrid if weights are known for some of the outputs and bounds for the rest.

The following simple example illustrates the above ideas more concretely, and shows in particular how the Lagrangian and standards approaches are related. We consider a very simple transit system with all user and vehicle trips emanating and ending at the same two points. The only design variable is the vehicle trip frequency. The passenger demand rate per unit time is assumed to be constant in time. Thus, for this simple example we only have two variables:

x — frequency of service (a single design variable expressed in [buses/h]);
α — demand (a single demand variable expressed in [passengers/h]).[2]

Suppose we are interested in minimizing the operating cost subject to meeting a passenger time standard. So we start with the standards approach. To this end, let us now develop formulae for the two components of $\mathbf{F}_m$. These should be expressed as totals over the lifetime of the project; or since the project is of indefinite duration and in a steady state, over a unit of time. Since the demand is fixed, we can also use average values per passenger trip by prorating the total amounts in a unit time to the (fixed) number of passenger trips that use the system during that time; and this is what we shall do. The reader can verify that the results of the example stay the same with the other two interpretations.

Consider first, the cost component. We assume that dispatching a vehicle costs c_f monetary units. This includes the vehicle's operating and depreciation cost for a round trip, including the driver wages. Since the dispatching cost and the number of passenger trips carried in a unit time

[2]Throughout this book, demand will be quantified by the number of passenger trips served by the transportation system. Its measurement unit, a passenger trip, will often be expressed by the single word "passenger" (further abbreviated as "pax") or "trip."

are $c_f x$ and α, respectively, the prorated cost per passenger trip becomes:

$$\$ = \mathbf{F}_{\mathrm{m}}{}^{\$}(x, \alpha) = c_f x/\alpha \ [\$/\text{pax}].$$

The superscript \$ refers to the component type.

Now consider the passengers' travel time, combining both the in- and out-of-vehicle times. We assume that the former is fixed at some value t. To evaluate the latter, we assume that the headways are constant and equal to $1/x$, but the schedule is not advertised. We also assume that people have appointments at the destination so they have to give themselves extra time to ensure on-time arrivals. Under these assumptions, a passenger is guaranteed to meet his/her appointment if (s)he adds a buffer of one headway to the trip time, i.e., (s)he arrives at the origin station $t+1/x$ time units before the appointment.[3] Therefore, for every passenger, $T = t + 1/x$. Now, since t is fixed and independent of x, we can omit it and consider the out-of-vehicle delay only. Thus, we use

$$T = \mathbf{F}_{\mathrm{m}}{}^{\mathrm{T}}(x, \alpha) = 1/x \ [\text{h}/\text{pax}].$$

With this modeling done, we are now ready to write the standards MP. If we choose the monetary units so $c_f = 1$, it reduces to the following:

$$(\text{MP}) \ \min\{x/\alpha : 1/x \le T_0\}.$$

To solve the problem, note that its objective function is increasing. Therefore, the problem is solved by the smallest value of x that is feasible Also note that the constraint merely defines a lower bound for x, so it must be binding at optimality. Hence, the solution is

$$x^* = 1/T_0.$$

Now, inserting this into the objective function, we find the optimum monetary cost per passenger trip:

$$\$^* \equiv \mathrm{G}_{\mathrm{m}}{}^{\$}(\alpha) = 1/(\alpha T_0).$$

Let us now compare the above with the Lagrangian approach. Recall, this involves choosing some shadow prices β for the time delay to form a

[3]Chapter 4 will discuss the passenger out-of-vehicle time and possible penalties for early arrivals in more detail.

generalized cost function that is then minimized without any constraints. For our problem, the expression is

$$\text{(GCP) } \min_x\{\$ + \beta T \equiv x/\alpha + \beta(1/x)\}.$$

Compare this with the Lagrangian of problem MP as per the definition of Chapter 2, which is

$$\text{(LP) } \min_x\{\$ + \beta(T - T_0) \equiv x/\alpha + \beta(1/x) - \beta T_0\}$$

Note that GCP and LP only differ by an additive constant, $-\beta T_0$, which can be ignored for the purposes of identifying the optimal x. Simple algebra reveals that this optimum is (for both problems):

$$x^* = \sqrt{\alpha\beta}.$$

We now show that there is a value of β that produces the standards solution, in this way, demonstrating that the two approaches are equivalent. As explained in Section 2.3.2, the constant is identified by finding the value of β that makes the MP constraint $1/x \leq T_0$ binding, i.e., by solving for β in $1/\sqrt{\alpha\beta} = T_0$. The result is $\beta = (1/T_0^2)(1/\alpha)$; and as expected, inserting it into the Lagrangian solution, we find that $x^* = 1/T_0$ and $\$^* = x^*/\alpha = 1/(\alpha T_0)$, which match the standards solution.

Note more generally that by changing the value of a weight one changes the output it multiplies (time in our example); and that the matching of solutions occurs when the resulting output equals the standard specified for it. This is true in general, showing that standards problems can be solved using a Lagrangian formulation with adequate weights. As we saw in Section 2.3.2, the correct weights, which are unique, can be found by adjusting them until the complementary slackness conditions are met, which is what we did in this example. In this book, closed-form analytical solution much like the ones we have developed will often be obtained using this Lagrangian approach.

In summary, there are two very similar approaches to obtain sketch plans, and either will work. Their key features are

(1) Standards: min $\{\$ \text{ s.t. } T \leq T_0,\ E \leq E_0, \ldots\}$
This minimizes the dollar cost subject to constraints on system outputs such as trip time and energy consumption. The approach is appropriate if the output limits are known. Usually, as shown in the example, constraints become binding when solved; i.e., $T = T_0,\ E = E_0, \ldots$

(2) Lagrangian: min $\{\$(x, \alpha) + \beta_T(T(x, \alpha)) + \beta_E(E(x, \alpha)) + \cdots\}$

This minimizes the generalized cost considering all outputs, when the unit (shadow) prices of the various outputs are given. The Lagrangian approach can also be used as a method to solve standards problems since it produces the same solutions when suitable shadow prices $(\beta_T, \beta_E, \ldots)$ are chosen. Since the Lagrangian approach is simpler than the standards approach, it will be used often in this book.

3.3. Demand Endogeneity

So far, we have assumed that the demand, $\boldsymbol{\alpha}$, is given and fixed, and we solve either a "standards" problem or a Lagrangian problem. Under these conditions, minimizing the total user and agency cost is equivalent to maximizing the agency and user benefits, or the total societal benefit if externalities are included as part of the agency cost. Critics can fairly say that fixing the demand is not realistic because the demand that materializes should depend on the system design. Therefore, this topic will be revisited in Chapter 11. Here, we show briefly that if we anticipate the correct demand and fix it at that level, then the optimum design obtained with our fixed demand methods happens to be the same as if the demand had been elastic, treated endogenously and obtained concurrently with the design. We shall show this using the standards approach formulation but the same result is obtained for the Lagrangian formulation. For more information and extensions of these ideas, see Daganzo (2012).

We assume that in addition to the demand, $\boldsymbol{\alpha}$, and the system design, $\mathbf{x}$, we look concurrently for an optimum set of prices (fares) $\mathbf{p}$ that users must pay. For convenience we also introduce in the formulation below the disutilities, $\boldsymbol{\pi}$, experienced by the various population groups arising from the prices $\mathbf{p}$ and the times $\mathbf{T}$ they experience. The standards formulation with endogenous demand is then

$$\begin{aligned}
&\max S(\boldsymbol{\pi}, \boldsymbol{\alpha}) - \$(\mathbf{x}, \boldsymbol{\alpha}) && \text{(society's benefits minus agency costs and externalities)} \\
&\text{s.t.} \quad \mathbf{T}(\mathbf{x}, \boldsymbol{\alpha}) \le \mathbf{T_o} && \text{(standards on system performance)} \\
&\qquad \boldsymbol{\alpha} = \mathbf{D}(\boldsymbol{\pi}) && \text{(endogenous demand function)} \\
&\qquad \boldsymbol{\pi} = \mathbf{G}(\mathbf{p}, \mathbf{T}) && \text{(disutility function)}
\end{aligned}$$

The objective function is the difference between the society's benefit arising from everyone's travel, S, and society's costs, \$. The latter include the agency's monetary expenditures and the non-monetary externality costs.

Both S and \$ are expressed in monetary units per day.[4] The benefit S obviously depends on the number of people traveling, $\boldsymbol{\alpha}$, and the disutility they experience, $\boldsymbol{\pi}$. The agency cost and externalities, logically, depend on how the system is designed $\mathbf{x}$ and how many people use it.

The constraints are expressed in vector form, with elements corresponding to disaggregated population subgroups of interest. The first constraint imposes standards on travel time, logically assuming that the experienced travel times are a function of the design and the demand; the second specifies the demand function with the disutilities as inputs, as is conventional; and the third how the fares and travel times translate into disutilities. Note that the vectors $\mathbf{T}$, $\mathbf{T_o}$, $\boldsymbol{\alpha}$, and $\boldsymbol{\pi}$ have the same dimension, with one element per population group. We also assume that $\mathbf{p}$ has the same dimension. The dimension of $\mathbf{x}$ is unrelated, however.

None of the assumptions in this formulation seems exceptionally harsh, so it should be fairly general. We look for the formulation's optimum ($\mathbf{x}^*$, $\boldsymbol{\alpha}^*$, $\mathbf{p}^*$, $\boldsymbol{\pi}^*$). We want to show that if $\boldsymbol{\alpha}^*$ is known then it can be treated as a constant, and the last two constraints can be ignored in the determination of $\mathbf{x}^*$. This is not obvious (or necessarily true) because the last two equations link $\boldsymbol{\alpha}$ and $\mathbf{T}$. It will be shown to be true, however, if we make two additional assumptions.

First the demand functions are assumed to be utility-invertible, i.e., there always exist a set of disutilities that will produce any $\boldsymbol{\alpha}$. This is usually the case, so this assumption is not restrictive. The assumption allows us to replace the third equation by an inverse demand model:

$$\pi = \mathbf{V}(\boldsymbol{\alpha}).$$

Second, and more restrictively, the agency in charge of the system is assumed to have enough pricing flexibility to be able to produce any desired set of disutilities $\boldsymbol{\pi}$ by manipulating the fares $\mathbf{p}$.[5] This will happen for example if the agency has the ability to apply a different price to each of the recognized population groups, e.g., by charging different transit fares for different origin-destination pairs. Mathematically, this means that the disutility function $\mathbf{G}$ is itself invertible and can be replaced by an inverse:

$$\mathbf{p} = \mathbf{P}(\boldsymbol{\pi}, \mathbf{T}).$$

[4]These items cannot be prorated to each passenger trip because the demand is endogenous.
[5]The more extended discussion in Chapter 11 will relax this assumption.

Now we can prove the result. To this end, let us rewrite the endogenous-demand formulation using the inverses as follows:

$$\begin{aligned} &\max\{S(\boldsymbol{\pi}, \boldsymbol{\alpha}) - \$(\mathbf{x}, \boldsymbol{\alpha})\} \\ &\text{s.t.}\quad \mathbf{T}(\mathbf{x}, \boldsymbol{\alpha}) \leq \mathbf{T_o} \\ &\boldsymbol{\pi} = \mathbf{V}(\boldsymbol{\alpha}) \\ &\mathbf{p} = \mathbf{P}(\boldsymbol{\pi}, \mathbf{T}) \end{aligned}$$

Note that the prices $\mathbf{p}$ only appear on the left-hand side of the last constraint, so this equation is feasible for any values of ($\mathbf{x}$, $\boldsymbol{\pi}$, $\boldsymbol{\alpha}$). Therefore, the prices and the equation can be dropped from the formulation; and we can limit ourselves to searching for ($\mathbf{x}^*$, $\boldsymbol{\pi}^*$, $\boldsymbol{\alpha}^*$) using only the first three equations. This is now done.

Conditional on $\boldsymbol{\alpha}^*$, the optimum disutility vector $\boldsymbol{\pi}^*$ is given by the third constraint and this fixes the first term of the objective function. Thus, if $\boldsymbol{\alpha}^*$ is given the optimal design $\mathbf{x}^*$ can be found ignoring both, this term and the third constraint, i.e., by solving what is left:

$$\begin{aligned} &\max\{-\$(\mathbf{x}, \boldsymbol{\alpha}^*)\} \Leftrightarrow \min\{\$(\mathbf{x}, \boldsymbol{\alpha}^*)\} \\ &\text{s.t.}\mathbf{T}(\mathbf{x}, \boldsymbol{\alpha}^*) \leq \mathbf{T_o}. \end{aligned}$$

Note, this is the standards approach *with fixed demand* as we had claimed.

If the optimum demand is not given, it can be found in two stages. First, we treat the demand as a parameter and look for functions $\$^*(\boldsymbol{\alpha})$ and $\mathbf{x}^*(\boldsymbol{\alpha})$ that return the optimal solution of the standards problem with fixed demand for every $\boldsymbol{\alpha}$ — these functions can be found either analytically or numerically. Then, we solve the mathematical program that arises after inserting $\$^*(\boldsymbol{\alpha})$ into the endogenous-demand formulation. Note after doing this that the first constraint is redundant since $\mathbf{x}^*(\boldsymbol{\alpha})$ satisfies it; and that the last one is definitional for $\mathbf{p}$. Thus, the two constraints can be dropped so we only need to solve the following:

$$\begin{aligned} &\max\{S(\boldsymbol{\pi}, \boldsymbol{\alpha}) - \$^*(\boldsymbol{\alpha})\} \\ &\text{s.t. } \boldsymbol{\pi} = \mathbf{V}(\boldsymbol{\alpha}) \end{aligned}$$

Although solving this problem is technically straight-forward, pinpointing the societal benefit function, S, is difficult; even fraught with controversy. Therefore, we do not discuss this second step any further here. The issue is more appropriately discussed in a course on economics and public policy, but we will return to it in Chapter 11. Often times, however, governing bodies will eliminate the need for this economic analysis by specifying target demands they want to serve in the future based on a societal vision. The job of transit system designers is then to find the best design that meets the target demand, using the standards approach with fixed demand — or the Lagrangian approach. To finish the job, the governing body should then take the necessary steps (e.g., price the system) to make sure that the targeted demands actually materialize.

We close this section with two additional ideas related to demand:

(i) The demand $\boldsymbol{\alpha}$, fares $\mathbf{p}$, and design $\mathbf{x}$ vectors in our formulation can cover many years. Thus, our analysis can produce designs and pricing plans that are staged over years. However, we will see later in the book that the optimum solution $\$^*(\boldsymbol{\alpha})$ and $\mathbf{x}^*(\boldsymbol{\alpha})$ are usually fairly insensitive to α so that one-stage plans can efficiently serve a broad range of multi-year demands.

(ii) Even though only rough demand targets are needed to obtain reasonable design and pricing plans, a vision for the future is fundamental to guide the process.

3.4. Building Blocks for Planning

Since this book aims to develop insights, as well as planning recipes, Chapters 4–7 will use a building-block approach, starting with simple systems and building on that knowledge to analyze more complex ones. Complexity is defined by the type of service region, as illustrated in Figure 3.4. We shall start with shuttles serving two points (a zero-dimensional region), then systems serving one-dimensional corridors, and finally complete networks serving two-dimensional regions. The following is a road map of what is to come.

(1) Chapter 4: Shuttle Systems — This chapter will assume that the population is concentrated at two points, an origin and a destination, so that there is no need for trip consolidation in space. Collective transportation, in this case, will only involve temporal consolidation as individuals adjust their departure times to match the scheduled departure of transit vehicles from the shared origin to the shared destination.

(2) Chapter 5: Corridors — This chapter will assume that the population is spread along a corridor so that all travel occurs in the one dimension along which transit service is provided. Now spatiotemporal consolidation is necessary, as individuals must travel to discrete stations where they can board transit vehicles which depart at discrete times.

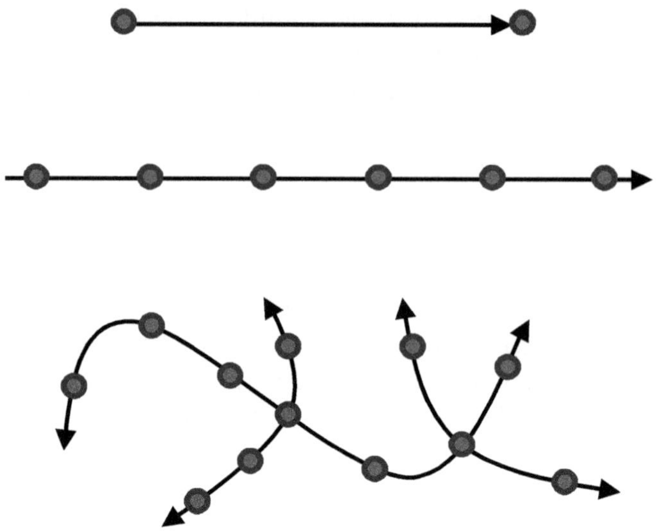

Figure 3.4. Transit systems with increasing spatial dimensions: shuttle, corridor, and network.

(3) Chapters 6 and 7: Cities — These chapters consider the most general case, where the population served is spread over a two-dimensional region. Transit service must then be designed considering the possibility of using multiple routes to cover the two-dimensional space. This adds complexity as it introduces elements of routing. Circuity considerations affect both the transit users and the vehicles that carry them. Because the topic is broad, it is covered in two chapters: Chapter 6 discusses systems with fixed routes and Chapter 7 those with flexible routes and/or schedules.

References

Ansari, S. B., M., Li, X., Ouyang, Y., and Smilowitz, K. (2018). "Advancements in continuum approximation models for logistics and transportation systems: 1996–2016." *Transportation Research Part B*, 107, 229–252.

Badia, H., Argote-Cabanero, J., and Daganzo, C. F. (2017). "How network structure can boost and shape the demand for bus transit." *Transportation Research Part A*, 103, 83–94.

Curtis, C., Renne, J., and Bertolini, L. (eds.) (2009). *Transit Oriented Development: Making It Happen*, Ashgate Publishing, Farnham, UK.

Daganzo, C. F. (1991). *Logistics Systems Analysis* (Chapter 1), Springer-Verlag, Berlin.

Daganzo, C. F. (2012). "On the design of public infrastructure systems with elastic demand." *Transportation Research Part B*, 46, 1288–1293.

Daganzo, C. F., Gayah, V. V., and Gonzales, E. J. (2012). "The potential of parsimonious models for understanding large scale transportation systems and answering big picture questions." *EURO J. of Transportation Logistics,* 1, 47–65.

Homework

3.1. *Consider a simple point-to-point shuttle serving passengers that arrive at a constant rate, α [pax/h]. The operating agency determines the vehicle dispatch frequency, $x > 0$ [bus/h], and the fare, $p \geq 0$ [\$/pax]. Each dispatch costs c_f monetary units, regardless of the number of passengers on board. The passengers' total travel time including waiting is $t+1/x$ for some constant t; and the total passenger disutility is assumed to be $\pi = p+t+1/x$ (i.e., the user's value of time is "1" [\$/h]). The demand is assumed to decline monotonically with the disutility value, and satisfy $\alpha = 0$ when $\pi \geq D$ for some constant $D > 0$, and $\alpha = \frac{1}{\pi - t}$ when $\pi < D$.

(a) Write an expression for the consumer surplus. (Recall from economics that it is the difference between the total amount that consumers are willing to pay and the total amount that they actually pay at the market price.)

(b) Formulate the social benefit function as the summation of the passengers' consumer surplus and the agency's collected fare. Express the result in closed form.

(c) Write down the demand-endogenous formulation of the design problem as in Section 3.3; derive the optimal frequency x^*, fare p^*, and the associated customer arrival rate α^*. Express all the results in analytical form.

(d) Write down the fixed-demand formulation of the design problem, assuming we have already correctly anticipated the optimum demand α^*. Can you find the resulted optimal frequency and fare? How do they compare with your answer from Part (c)?

(e) Suppose our guess of the optimal customer arrival rate in Part (d) is erroneous, such that we have actually used rate $\alpha^e = (1 \pm \varepsilon)\alpha^*$. What is the percentage error in your system design (the frequency) for $\varepsilon = 10\%, 20\%, 50\%, 100\%$? And for the fare?

3.2. *The text has discussed the downside of shortsightedness associated with short political cycles, and mentioned that it could be mitigated with financial tools that spread costs over time. With this in mind, reconsider Problem 3.1 and do the following:

(a) Assuming that the customer arrival rate in political cycle i is now perfectly inelastic (i.e. we may assume $p = 0$) and increases linearly with i (i.e., $\alpha(i) = ki$ for some $k > 0$), derive the optimal frequency that minimizes the system-wide user and agency cost in cycle i. How does the system-wide performance change when i tends to infinity?
(b) How would you change the objective function in Part (a) to incorporate a long-term vision?
(c) Can you think of existing social or political mechanisms that encourage decision-makers to take a long term view? Can you come up with new ideas?

Chapter 4

Planning — Shuttle Systems

This chapter is the first of four chapters explaining how to plan transit systems. The chapters are organized in a modular fashion, starting with simple systems and ending with the most complex. Here we shall focus on systems that connect a single origin and a single destination. The only planning decision for these types of shuttles is temporal: scheduling. The basic question is determining how the system's headways should vary with time to meet the demand. Chapters 5 and 6 introduce the notion of space, one dimension at a time. Chapter 5 examines one-dimensional systems (corridors) where in addition to schedules one must determine the spacing between stations, and Chapter 6 discusses two-dimensional systems covering entire cities, where one must also determine how many routes to have and how they should be laid out and connected. Chapter 7 is also two-dimensional and arguably the most complex. It describes flexible transit systems that operate adaptively without routes or schedules, such as taxis, jitneys, and other demand-responsive transit.

In addition to providing scheduling recipes for shuttle systems, the present chapter compares the economics of individual and collective transportation by presenting idealized models of both modes under different types of conditions. In both cases, we look first at the time-independent case where both the demand and the service conditions are invariant in time and then examine the time-dependent case. This chapter will show that individual and collective modes behave very differently. Whereas for a given guideway size the capacity of individual modes is fixed, the capacity of collective modes can be adapted by dispatching more and larger vehicles. This fundamental difference manifests itself in a most remarkable fact: whereas individual transportation modes exhibit diseconomies of scale with demand due to guideway congestion, collective transportation exhibits significant economies.

To explore these features systematically, the chapter is organized as follows: Section 4.1 considers individual modes, i.e., where people travel in single-occupant vehicles; Section 4.2 then examines the collective modes, including a scheduling recipe; and finally Section 4.3 describes what happens when individual and collective modes coexist and compete for the demand. Although all the results in these sections are derived using the Lagrangian approach of Chapter 3, the standards approach could have been used instead (see homework exercise 4.1).

4.1. Individual Transportation

This section considers the transportation guideway that connects a single origin–destination pair and allows single-occupant vehicles to travel from one end to the other. The question is how much capacity to provide. The system is examined under three demand scenarios: under steady demand in Section 4.1.1, under given but time-dependent demand in Section 4.1.2, and with adaptive demand in Section 4.1.3.

4.1.1. *Time-Independent Demand*

This subsection considers as a preliminary step the case with steady demand. Since the demand is constant, for travel to be feasible, the capacity of the guideway at its most restrictive bottleneck μ [passenger/h] must exceed or equal the demand, λ [passenger/h]. Otherwise, infinite delays would arise. The cost of providing the capacity is assumed to be proportional to the capacity, and the proportionality constant is denoted c_g (this is the dollar expenditure per unit of capacity provided per unit time, which as the reader can verify has the unit of [\$/passenger]). The operating cost of one passenger trip, including vehicle depreciation cost, is denoted c_f [\$/passenger].

Given these assumptions, the monetary cost per passenger trip is

$$\$ = \frac{c_g \mu}{\lambda} + c_f. \tag{4.1}$$

We assume that congestion is negligible when the capacity is not exceeded, so the passenger's travel delay as a function of capacity μ is

$$T = \begin{cases} 0 & \text{if } \mu \geq \lambda \\ \infty & \text{otherwise} \end{cases}. \tag{4.2}$$

The above two equations allow us to evaluate the optimum capacity that should be provided to satisfy an arbitrary standard for the delay, $T \leq T_0$. The expression for \$ indicates that we should choose the smallest feasible μ.

The expression for T indicates that μ is feasible (for all T_0) if and only if $\mu \geq \lambda$. Obviously then, the optimum capacity is $\mu^* = \lambda$. It follows that the optimum monetary cost per passenger combining the infrastructure and vehicle operating costs is given by

$$\$^* = c_g + c_f. \tag{4.3}$$

For our μ^*, we have $T^* = 0$. Thus, the optimum generalized cost combining monetary expenses and delay, $z^* = \$^* + \beta T^*$, where $\beta(\$/\text{time})$ is a given positive constant, is

$$z^* = \$^* + \beta T^* = c_g + c_f. \tag{4.4}$$

Note that both (4.3) and (4.4) are independent of λ, so there are no economies of scale in our idealization, i.e., the *total cost* per unit time accrues at a rate $\lambda(c_g + c_f)$ per unit time. Doubling the number of drivers would double the total transportation cost. We now look at the time-dependent case, both for the evening and morning rush hours. We will find that in these cases $T^* \neq 0$ and z^* is even larger, so we still do not find economies of scale.

4.1.2. *Known Time-Dependent Demand: The Evening Commute*

So far, we have assumed that demand is time-independent, so as long as capacity matches demand there is no delay. In reality, however, travel demand rises and falls over the course of a day, $[0, \tau_D]$, and delays arise. This section extends the analysis of Section 4.1.1 to a more realistic scenario that resembles the evening rush hour. The demand rate is assumed to be a known function of time, $\lambda(t)$, which can temporarily exceed the fixed guideway capacity, μ.

To simplify the exposition, we idealize the daily demand as if it had a single peak instead of two (see Figure 4.1(a)). The figure displays a hypothetical cumulative virtual demand curve V giving the number of virtual arrivals as a function of time. The figure also displays the average demand rate, $\bar{\lambda}$, and the maximum rate during the peak, λ_m. Note that $\lambda_m \geq \bar{\lambda}$, and that in a time-independent system where the demand rate does not fluctuate over the course of the day, λ_m would equal $\bar{\lambda}$.

Consider now the departure curve D for different values of μ and how the average delay $T(\mu)$ depends on μ. This function is depicted in the bottom panel of Figure 4.1(b). The function is unbounded for $\mu < \bar{\lambda}$ because such small capacities cannot meet the daily demand; thus, queues would steadily

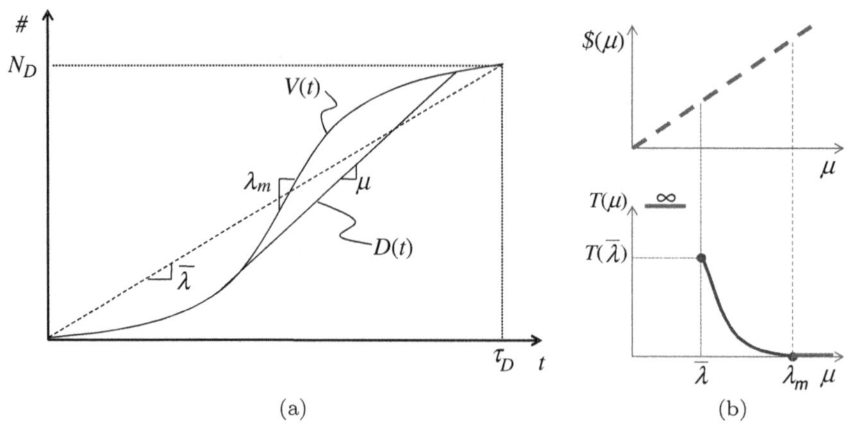

Figure 4.1. Queuing analysis of evening commute.

grow day by day. On the other hand, if $\mu \geq \lambda_m$, then queues would not develop at all, and $T(\mu) = 0$, as shown in the figure. In the intermediate range where $\mu \in [\bar{\lambda}, \lambda_m]$, transient queues of finite duration arise — this is the case depicted in part (a) of the figure. This part of the figure shows that in this intermediate range $T(\mu)$ declines gradually with μ since the area between curves V and D gets smaller as μ is increased. Accordingly, a maximum value for the delay must be achieved in this intermediate range at $\mu = \bar{\lambda}$, and the maximum value is $T(\bar{\lambda})$. The solid curve in the bottom panel of Figure 4.1(b) displays the overall pattern followed by $T(\mu)$.

The following example illustrates how to derive an explicit expression for $T(\mu)$ in a simple case. This passage can be skipped without loss of continuity.

Example 4.1. An analytical expression for $T(\mu)$. Consider a piecewise linear arrival curve with two demand periods: a peak period of duration τ_p (hours) with demand rate, λ_p, and an off-peak period with lower demand, λ_o, that lasts for the remaining $\tau_o \equiv \tau_D - \tau_p$ hours of the day. Figure 4.2 shows a prototypical cumulative demand curve with these properties. Although the curve is shown for only one day, we should imagine that it extends in both directions for many days.

Note that $\bar{\lambda} = N_D/\tau_D = \frac{1}{\tau_D}(\tau_p\lambda_p + \tau_o\lambda_o)$ and $\lambda_m = \lambda_p$. The formula we are about to derive is only valid for $\mu \in [\bar{\lambda}, \lambda_p]$. For μ in this range, the queue dissipates before the start of the next peak period and, as the figure shows, the duration of the time interval when the queue declines in size, τ, satisfies

$$\tau_p\lambda_p + \tau\lambda_o = (\tau_p + \tau)\mu, \quad \text{for } \mu \in [\bar{\lambda}, \lambda_p].$$

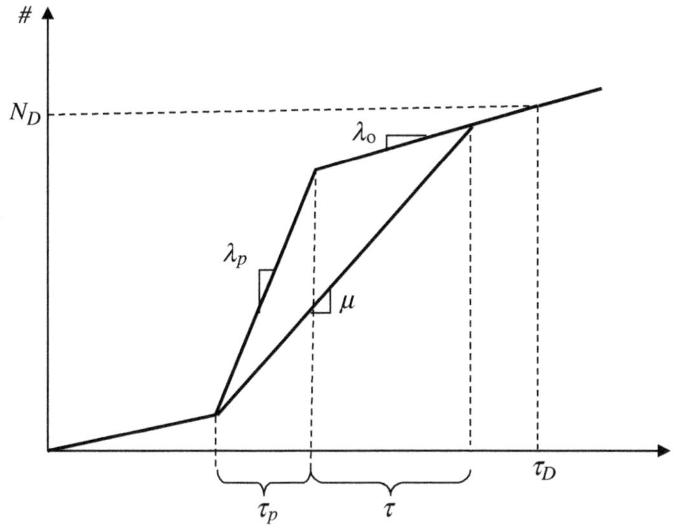

Figure 4.2. Time-dependent demand for collective transportation.

This yields $\tau = \tau_p(\lambda_p - \mu)/(\mu - \lambda_o)$, which enables us to write the area of the triangle between curves V and D as

$$\begin{aligned}\text{Area} &= \tfrac{1}{2}(\tau_p + \tau)[(\lambda_p - \mu)\tau_p] \\ &= \tfrac{1}{2}\tau_p^2(\lambda_p - \lambda_o)\tfrac{(\lambda_p-\mu)}{(\mu-\lambda_o)}, \quad \text{for } \mu \in [\bar{\lambda}, \lambda_p].\end{aligned}$$

Since this area is the total daily delay, we can finally write $T(\mu)$ as the ratio of the above expression and the total daily demand, $(\bar{\lambda}\tau_D)$, i.e.,

$$T(\mu) = \tfrac{1}{2}\tau_p^2(\lambda_p - \lambda_o)(\bar{\lambda}\tau_D)^{-1}\tfrac{(\lambda_p-\mu)}{(\mu-\lambda_o)}, \quad \text{for } \mu \in [\bar{\lambda}, \lambda_p].$$

Note how this expression decreases with μ, as anticipated.

Consider now the generalized cost, $z(\mu) = \$(\mu) + \beta T(\mu)$. Here, the expression for $\$(\mu)$ is still (4.1), with the coefficient λ replaced by $\bar{\lambda}$. This is represented in the top panel of Figure 4.1(b) by the dashed straight line. Graphically, $z(\mu)$ is the sum of the dashed and solid lines in both panels of the figure. We are interested in finding the minimum cost, which as shown by the figure must be achieved for some $\mu = \mu^* \geq \bar{\lambda}$. Thus, we can write $z^* = \min_\mu\{z(\mu) : \mu \geq \bar{\lambda}\}$. Because μ^* and z^* depend on the shape of curve V in a complicated way, we shall not look for a closed form of expression for z^*. Instead, we show that the optimum cost can be bounded from above and below in a simple way.

A lower bound is obtained by ignoring the non-negative term $\beta T(\mu)$ in the objective function of our problem, i.e., by writing $z^* \equiv \min_\mu\{z(\mu) : \mu \geq \bar{\lambda}\} \geq \min_\mu\{\$(\mu) : \mu \geq \bar{\lambda}\} = \$(\bar{\lambda}) = c_g + c_f$. Note that this is the same result as (4.3). An upper bound is provided by the objective value at any feasible μ. We choose $\mu = \bar{\lambda}$, which yields $z(\bar{\lambda}) = \$(\bar{\lambda}) + \beta T(\bar{\lambda})$. Thus, we have $z^* \leq \$(\bar{\lambda}) + \beta T(\bar{\lambda}) = c_g + c_f + \beta T(\bar{\lambda})$. The two bounds can be combined as follows:

$$\text{Time-independent cost} \equiv c_g + c_f \leq z^* \leq c_g + c_f + \beta T(\bar{\lambda}). \tag{4.5}$$

In the above, T was assumed to be the average passenger delay. If the analysis is repeated assuming that T is the delay for the worst-case passenger, then the function $T(\mu)$ would change but the lower bound would remain the same. In fact, it should be clear from the derivation that this bound must remain valid for all reasonable definitions of T. This fact establishes that, regardless of the standard chosen, a demand pattern with a rush hour can only increase cost relative to the time-independent case. So, we still do not see economies of scale.

4.1.3. *Adaptive Demand: The Morning Commute*

The problem is now further generalized by assuming that the demand curve is not just time-dependent but endogenous. As before, we consider a period (day) of duration τ_D under conditions that resemble those of the morning commute. We assume that exactly N_D people with rigid work-starting times repeat the same trip each day. Thus, the average demand rate is $\bar{\lambda} = N_D/\tau_D$.

These people are assumed to choose their virtual arrival times at the bottleneck so as to minimize a combination of waiting in the car and waiting at the destination (for work to start), but their demand curve is not given. For this reason, it is a bit more difficult to determine in this case how the commuter delay depends on the bottleneck capacity. To do this, we shall first determine the virtual arrival curve building on ideas first introduced by Vickrey (1969) and then analyze the economics as in the previous subsection.

We shall assume that the bottleneck is first-in-first-out (FIFO) and the demand is characterized by a cumulative curve of desired departure times through the bottleneck, which we call the "wish" curve, W. We assume for simplicity that $W(t)$ is Z-shaped, rising with slope λ_w from zero at some specific time until it reaches N_D (see the slanted dotted line in Figure 4.3).

Note that if the maximum slope of the wish curve, λ_w, is less than the capacity of the bottleneck, μ, then all drivers could pass through the bottleneck at their preferred times without delay, i.e., the V, D, and

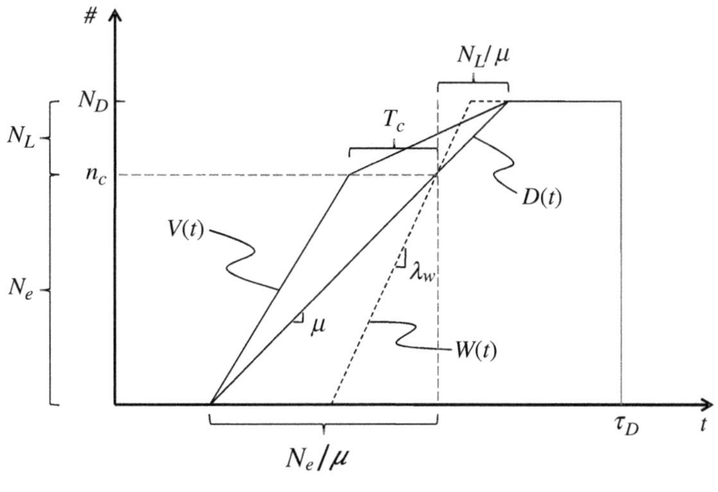

Figure 4.3. Queuing analysis under Vickrey equilibrium.

W curves would match. Thus, the average delay per commuter would be $T(\mu) = 0$ if $\mu \geq \lambda_w$. Also, as in the previous section, if $\mu < \bar{\lambda}$, then $T(\mu) = \infty$.

However, if $\mu \in [\bar{\lambda}, \lambda_w)$, then we shall find that there will be a positive and finite delay and that, as in the previous subsection, this delay decreases with μ. (It should be intuitive that there must be some delay because if the slope λ_w exceeds the capacity, then the D and W curves could not match and some drivers would have to depart the bottleneck earlier or later than they wish. These commuters would then compensate by adjusting their arrival times and creating delays.)

To see what would ensue as the result of this adjustment process, we need to know how drivers evaluate wait in queue vs. wait at the destination. We shall assume for analysis that each driver values one unit of time waiting at the destination when arriving early as the equivalent of e units of queuing delay. Likewise, the commuter values one unit of "lateness" at the destination as L units of queuing delay. The constants e and L are dimensionless and such that

$$e \leq 1 \leq L.$$

We shall assume that drivers try to minimize the sum of the delay in queue and the earliness/lateness penalty at the destination, which we call the "combined delay."

Given this setup, we look for an "equilibrium" set of arrival times, where no driver can reduce his/her combined delay by changing its own arrival time. We shall then evaluate this delay as a function of μ. As Vickrey did, we shall restrict our search to equilibria in which the order of arrivals to the bottleneck is the same as the order of wished departures.

Figure 4.3 shows an equilibrium pattern involving curves V and D for a given W. Note there is a critical driver, numbered n_c in the sequence of arrivals and departures, who experiences no earliness or lateness and whose entire cost is a time in queue, T_c. (Note how the departure curve D crosses W at the ordinate of this driver.) All drivers who arrive before n_c depart the bottleneck before their wished departure time. The number of such (early) drivers, denoted N_e, is $N_e \equiv n_c$. All drivers who arrive after the critical driver depart the bottleneck after their desired departure time. We define N_L as the count of such (late) drivers. Since there are a total of N_D drivers, the following is true:

$$N_e + N_L = N_D.$$

Appendix 4.A proves that at equilibrium, the virtual arrival curve V is piecewise linear as shown in Figure 4.3, and such that

$$T_c = \frac{N_D L e}{\mu(L+e)}; \quad N_e = \frac{L N_D}{L+e}; \quad \text{and} \quad N_L = \frac{e N_D}{L+e}. \tag{4.6}$$

This result holds independent of the value of λ_w (as long as $\lambda_w > \mu$) or more generally the shape of curve W. The reader can also verify from the geometry of the figure that the combined delay of the critical commuter, which is T_c at its current arriving position since there is no earliness or lateness, stays the same if arriving at the times when queuing begins or ends. (Note how the combined delay when the queue begins is $eN_e/\mu \equiv T_c$ and when it vanishes: $LN_L/\mu \equiv T_c$). In fact, it can be readily seen that the combined delay stays the same at all points in between, and is higher if arriving outside the queuing interval. Appendix 4.A shows that all other commuters are also content with their current costs, so the whole system is in equilibrium.

Let us now evaluate the average combined delay T across all commuters. To do this, first calculate the total delay by adding the total queuing (the area between curves V and D), the total penalty due to earliness (e times the area between curves D and W, only where $D > W$), and the total penalty due to lateness (L times the area between curves W and D, only where $D(t) < W(t)$). A simple formula for this sum can be readily obtained by combining the areas of the relevant triangles in the figure. The ratio of the result and N_D is the average combined delay per person, $T(\mu)$.

Consideration shows that $T(\mu)$ is a decreasing function of μ. This can be most easily seen in the special case where the wish curve is a step function

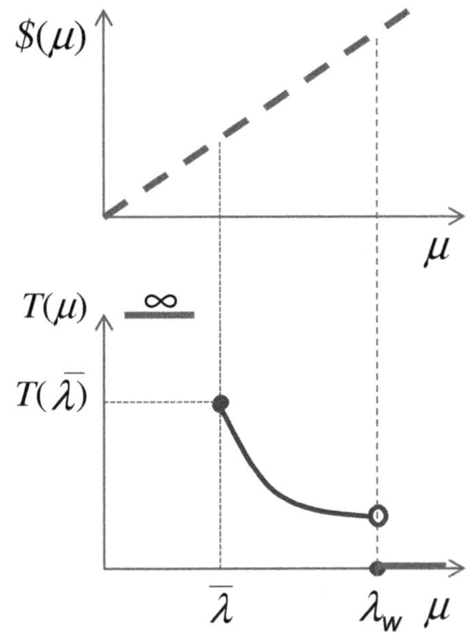

Figure 4.4. System costs for the morning commute.

($\lambda_w = \infty$) because then all commuters are identical and everyone experiences the same combined delay. Therefore we can use for $T(\mu)$ the combined delay of the critical commuter, T_c, and then use the first equality of (4.6) to write

$$T(\mu) = [N_D Le]/[\mu(L+e)] = [Le\tau_D\bar{\lambda}]/[\mu(L+e)]. \tag{4.7}$$

Note, the expression simplifies further to

$$T(\mu) \approx e\tau_D\bar{\lambda}/\mu, \quad \text{if } L >> e. \tag{4.8}$$

Note as well how both (4.7) and (4.8) vary inversely with μ.

Recall that these observations apply to the range $\mu \in [\bar{\lambda}, \lambda_w)$. The cost is infinite for smaller values of μ and zero for larger values. The complete function $T(\mu)$ is plotted on the bottom part of Figure 4.4. Compare with Figure 4.1(b) and note how $T(\mu)$ is now discontinuous at both ends of the interval, and not just the lower end. The discontinuity at the upper end is interesting. It means that with endogenous demand it is possible to increase capacity just a little when major queues exist, and the next day all queues could vanish as if by magic.

We are now ready to make an informed decision about the bottleneck capacity, recognizing how capacity affects the average combined delay. As in the previous subsection, we still want to minimize the generalized cost,

$z(\mu) = \$(\mu) + T(\mu)$, and the problem still remains finding $z^* \equiv \min_\mu\{z(\mu) : \mu \geq \bar{\lambda}\}$. Although the specific form of $T(\mu)$ has changed slightly from that of the previous subsection, the arguments about the two bounds can be repeated successfully. Thus, in this case, we also have

$$\text{Time-independent cost} \equiv c_g + c_f \leq z^* \leq c_g + c_f + \beta T(\bar{\lambda}). \tag{4.9}$$

The only thing that changes with respect to the previous section is the specific value of $T(\bar{\lambda})$, which now happens to be $Le\tau_D/(L + e)$. We continue to be worse off than in the time-independent case, and economies of scale still do not appear.

4.2. Collective Transportation

The analysis of Section 4.1 is now repeated for public transit. In the process, we shall derive scheduling recipes to optimize headways and determine the resulting generalized cost. We shall find that collective transportation exhibits economies of scale, which is encouraging.

4.2.1. *Time-Independent Demand*

Consider now a shuttle service provided on an existing guideway from a single origin to a single destination. The only planning task is determining the service headway. We assume that vehicles are large enough to carry all the passengers that show up before any departure, and define the following variables: H, headway between vehicle dispatches [h/dispatch]; c_g, prorated capital cost for infrastructure construction [\$/h]; c_f, prorated capital and operating cost for vehicles per dispatch [\$/dispatch]; and λ, demand [pax/h].

The average monetary cost per passenger trip, \$, is the ratio of the total cost per hour, which is $c_g + c_f/H$, to the number of passenger trips served in an hour, λ, i.e.,

$$\$ = \frac{c_g + c_f/H}{\lambda}. \tag{4.10}$$

The average out-of-vehicle delay T experienced by passengers in the system (ignoring the time in motion between the origin and destination, which is the same for every traveler and beyond our control) is now evaluated assuming that service is reliable. This delay is proportional to the headway in most cases. For example, if people do not have appointments at the destination and do not know the service schedule, then they would only experience a delay at the origin, and the wait would then be $T = H/2$ time units for an average passenger, or $T = H$ for the unluckiest. Of course, if passengers knew the schedule they would not wait at all.

Table 4.1. Average and worst-case waiting time per passenger trip under different scenarios.

Passenger type		Waiting time (average, worst) At origin	At destination
No appointment at the destination	Schedule unknown	$(H/2, H)$	$(0, 0)$
	Schedule known	$(0, 0)$	$(0, 0)$
Appointment at the destination	Schedule unknown	At origin and destination (H, H)	
	Schedule known	$(0, 0)$	$(H/2, H)$

The top half of Table 4.1 summarizes these results. The bottom half summarizes the cases with appointments that cannot be missed. In these cases, people must arrive early and therefore experience some delay at the destination. If they do not know the schedule, all passengers must plan for a worst-case trip, including a buffer of total duration H. And they experience this by waiting some time at the origin and the rest at the destination. Thus, the combined wait at the origin and the destination for passengers with appointments is H for *all* passengers. In contrast, if they know the schedule, they will arrive in time to catch the latest feasible shuttle and will only experience wait at the destination. This delay will depend on the synchronicity of their appointments and the schedule. Therefore, it will be $T = {}^1\!/_2 H$ for the average passenger and $T = H$ for the unluckiest (see Table 4.1).

In our evaluation, we will use the trip of an average passenger who does not know the schedule and does not have an appointment. Thus,

$$T = H/2. \tag{4.11}$$

Other scenarios including cases where people value differently the delays at the origin and at the destination would only differ in the coefficient of H; thus, the results we are about to derive would be qualitatively similar. Of particular interest is the case where people place less wait on delay at the destination, T_D, as we did in the Vickrey model. Then the amount of ordinary wait equivalent to T_D is eT_D, where $0 < e < 1$. This conversion can be applied to all the entries of Table 4.1. For example, if people have appointments and the schedule is unknown, we see that the equivalent total wait is ${}^1\!/_2\,(1+e)H$ for an average passenger and H for the unluckiest — note the former split their wait equally between the two trip ends and the latter spend it all at the origin. As we mentioned, only the coefficient of H changes.

Now combine (4.10) and (4.11), using β to convert delay into money, and write the following expression for the generalized cost $z(H)$:

$$z(H) = c_g/\lambda + (c_f/\lambda)/H + \beta H/2. \tag{4.12}$$

Note this is an economic order quantity (EOQ) equation in H. Thus, it is minimized for $H^* = (2c_f/\lambda\beta)^{1/2}$ and we have

$$z^* = c_g/\lambda + (2\beta c_f/\lambda)^{1/2}. \tag{4.13}$$

This expression shows that, unlike for individual transportation, the optimum generalized cost declines toward zero as demand increases, [1] i.e., there are strong economies of scale in collective transportation. The reason for this fortuitous outcome is that with more demand more individuals can consolidate their travel into each vehicle for a given number of dispatches. The increased vehicle occupancy lowers the average monetary cost per person. We now show that similar economies arise if we allow the demand to vary with time.

4.2.2. *Time-Dependent Demand*

Let us now see how best to schedule the system and evaluate its generalized cost if the demand varies with the time of day. As before, we shall assume for maximum transparency that there is a single rush, that vehicles have infinite capacity, and that the demand is given.[2] We also continue to assume that the vehicle operating cost is proportional to the number of dispatches, as if the costs of vehicles and drivers accrued only when used.[3]

To make the ideas more concrete, we consider a piecewise linear demand curve similar to that of Figure 4.2 (see Figure 4.5). As before, there are two demand periods in each day: a peak with demand rate, λ_p, lasting τ_p hours,

---

[1]In reality, z^* drops toward a small value greater than zero because transit vehicles have finite capacities and the cost of increasing that capacity is not zero. To verify this assertion, the reader may repeat the derivation of (4.11) assuming that the cost of a vehicle dispatch increases with the number of passengers that it needs to hold, e.g., as a linear function $c_f = c_0 + c_1(\lambda H)$, where c_1 is a constant that is assumed to be 0 in the original analysis.

[2]In our current framework, there is no congestion, i.e., every passenger can board a bus exactly when he/she wants independent of what other passengers are doing. Therefore, since passengers do not have to compete for the best buses, the assumption of a given demand captures both the evening and morning commutes.

[3]Although perhaps counterintuitive, this assumption is reasonable for both wages and vehicle costs. As we will see in Chapters 8 and 9, idle time for both drivers and vehicles can be reduced to nearly zero with good management strategies if the shuttle system is part of a larger system where drivers and vehicles can be reallocated across lines.

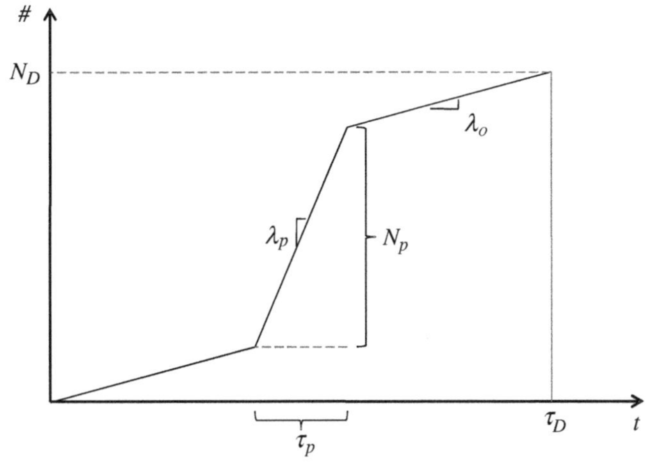

Figure 4.5. Time-dependent demand for the collective transportation.

and an off-peak with demand rate, λ_o, lasting $\tau_D - \tau_p$ hours. We denote by N_D the total number of passenger trips in a day, by N_p the number in the peak, and by $N_D - N_p$ the remainder, which occur in the off-peak hours.

Logically, we will want to use different headways in the peak and off-peak periods. Thus, we shall look for both, a peak period headway, H_p, and an off-peak headway, H_o. The goal is minimizing the generalized cost over the course of the day including all passengers and all vehicles. This can be expressed as the following optimization program with H_p and H_o as decision variables:

$$\begin{aligned}
&\min \{\text{Guideway cost} + \beta\,(\text{Total amount of waiting time}) \\
&\qquad + c_f\,(\text{Number of bus dispatches})\} \\
&\equiv \min \left\{ c_g \tau_D + \beta \left[\frac{1}{2} H_p N_p + \frac{1}{2} H_o (N_D - N_p) \right] + c_f \left(\frac{\tau_p}{H_p} + \frac{\tau_D - \tau_p}{H_o} \right) \right\}.
\end{aligned} \tag{4.14}$$

In the above, the cost for guideway infrastructure is a constant and hence can be ignored in the determination of the headways. The passengers are again assumed to wait for half of the headway at the origin on average, and not to wait at the destination. Note that the objective function is separable into two parts, each containing only one of the headways, and that each part has an EOQ form. Thus, the headways that minimize the generalized cost are

$$H_p^* = \sqrt{\frac{2c_f\tau_p}{\beta N_p}} = \sqrt{\frac{2c_f}{\beta\lambda_p}}, \tag{4.15a}$$

$$H_o^* = \sqrt{\frac{2c_f(\tau_D - \tau_p)}{\beta(N_D - N_p)}} = \sqrt{\frac{2c_f}{\beta\lambda_o}}. \tag{4.15b}$$

This is the scheduling recipe. It can be generalized for real-world applications in several ways:

(a) The day (or week) can be divided into more periods with different demand rates (e.g., peak, off-peak, weekend, night). Then, it should be clear by repeating the above derivations that a different headway can be determined for each period, using (4.15a) the average demand for each period.
(b) If a continuous demand curve is given, then one should approximate it by a piecewise linear curve with multiple parts and use recipe (a). Consideration shows that the approximation does not have to be very accurate for the result to be good (see homework exercise 4.4).
(c) If the cumulative demand curves vary stochastically from day to day, it is possible to prove that for any given schedule the average daily delay over many days is the same as if the demand had been deterministic with the same average curve (see homework exercise 4.5). Therefore, procedure (b) can be effectively applied to this case, using the average demand curve.

These scheduling techniques need to be modified if the vehicle occupancies they produce exceed on occasion the vehicle capacity. Modifications for deterministic demand can be developed with the queuing diagram ideas described in Chapter 2. They can include a constraint requiring that all buses clear the passenger queues at the origin or not (see homework exercise 4.6). Modifications for stochastic demand can be developed using the average demand as an approximation.[4]

In all cases, once the schedule has been obtained, one can evaluate the final cost. In our particular case, the generalized cost per passenger trip z^* is obtained by inserting (4.15a and b) in the objective function and dividing

[4] An exact treatment is possible with dynamic programming, but the timetables that arise are too variable for the public to remember, and the marginal benefits too small, so this is rarely done in practice.

the result by N_D, i.e.,

$$z^* = \left[c_g\tau_D + \sqrt{2\beta c_f}\left(\sqrt{\tau_p N_p} + \sqrt{(\tau_D - \tau_p)(N_D - N_p)}\right)\right] \Big/ N_D. \tag{4.16}$$

It is interesting to see how (4.16) behaves if N_D is increased without changing the character of the peak, i.e., without changing the durations of the peak and off-peak intervals, and keeping the same peak demand ratio, $r \equiv N_p/N_D$. This alteration is the same as rescaling the vertical axis of Figure 4.5. To explore the effect of the altered demand, replace N_p by rN_D in (4.16) and note how the resulting expression depends on N_D. It is easy to see that the new formula includes a term inversely proportional to N_D and a second term inversely proportional to $\sqrt{N_D}$. Or equivalently, a term inversely proportional to the average demand, which is $\bar{\lambda} = N_D/\tau_D$, and a term inversely proportional to its square root.[5] Thus, the problem continues to exhibit economies of scale.

There is an additional important difference between individual and collective transportation. It turns out that the optimum cost per day z^*N_D does not change much for a given N_D if the demand is spread differently across the whole period. Suppose for example that $\tau_D = 24\,\text{h}$ and that 30% of the trips are made in 4 of the 24 h in the day (i.e., there is quite a bit of peaking). If we use a dummy value $N_D = 10$ in the formula and use the values $c_g\tau_D = 10$ and $\sqrt{2\beta c_f} = 1$, we find that the total generalized cost for this time-dependent case is

$$10 + 1(\sqrt{4 \times 3} + \sqrt{(24-4)(10-3)}) = 25.30.$$

Using the same logic, we see that if the $N_D = 10$ trips had been spread uniformly across the entire 24 h, the generalized cost would have been $10 + (24 \times 10)^{1/2} = 25.49$.

Note the very small difference, and that peaking actually *reduces* the cost to society, which was not the case for individual modes! The relative difference is so small because with large vehicle capacities there is no congestion and we can adapt the provision of transit service to match the demand. The smallness of this difference suggests that the spatial layout of collective transportation systems can be studied as if the demand was time-independent, without having to consider the actual time-dependent schedules

[5]This is very similar to (4.13). In fact, the reader can also verify that if there is no peak ($\tau_p = N_p = 0$), then (4.16) reduces to (4.13).

that turn out to be optimal. This simplification will be used repeatedly in this book.

4.3. Bimodal Equilibrium: Congestion Abatement by Prioritizing Transit

This section analyzes the consequences of providing transportation services in a bimodal setting where some trips are made by car and the rest by transit, while recognizing that passengers can choose which mode to take. We are interested in examining how improvements to one mode affect the whole system.

To enhance transparency, we assume that the system has N_D identical users and examine the morning commute only. Although the total number of users is given, the split across modes is allowed to depend on the generalized costs of the two modes, which combine the user out-of-pocket expenses and time delays. The split should be in a Nash equilibrium, in the sense that the users should lack an incentive to change modes. This can occur in two ways: (i) if both modes are used and the generalized costs for both modes are the same; or (ii) if only one mode is used and its generalized cost is equal to or lower than that of the unused mode.[6] Case (ii) is uninteresting as it is analogous to the single-mode situations we have already considered. Therefore, we shall focus on case (i).

To solve this problem, we need to express the generalized user costs for both transit and cars as a function of the number of users. For maximum generality, we allow for fixed transit subsidies and car levies, e.g., in the form of parking charges. In recognition of this, we shall use the symbol C instead of z to express the generalized *user* cost per trip including the subsidies and charges they experience.

For car, we include a fixed component $c_f^{(\text{car})}$ that includes all monetary costs and the fixed travel time without congestion, and a variable component that depends on the number of car users and consists of the combined delay given by (4.8). Thus, the car cost function is

$$C^{(\text{car})}(N^{(\text{car})}) \equiv c_f^{(\text{car})} + \beta e \tau_D \bar{\lambda}^{(\text{car})}/\mu = c_f^{(\text{car})} + \beta e N^{(\text{car})}/\mu. \qquad (4.17)$$

For transit, we include a fixed component $c_f^{(\text{transit})}$ and a delay component given by (4.11). Consistent with the Vickrey model, now we assume the

[6]These two conditions constitute the so-called Wardrop's First Principle (Wardrop and Whitehead, 1952).

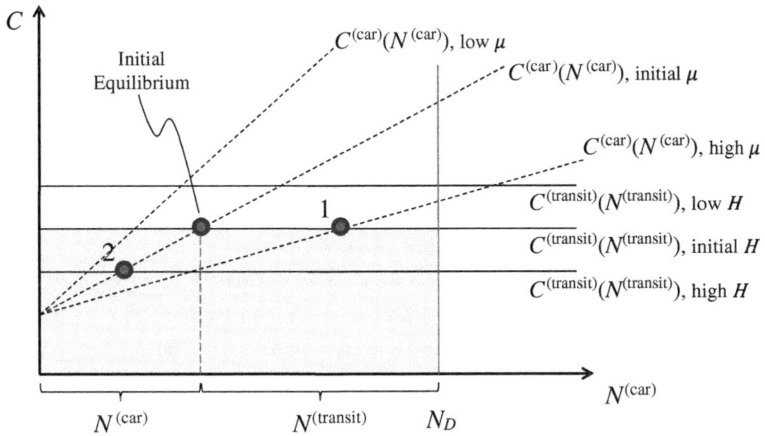

Figure 4.6. Cost comparison and demand equilibrium between auto and transit modes.

passengers have appointments at the destination and discount earliness (i.e., waiting at the destination). Thus, the transit cost function is

$$C^{(\text{transit})}(N^{(\text{transit})}) \equiv c_f^{(\text{transit})} + \beta\gamma H, \tag{4.18}$$

where $\gamma = e/2$ if the schedule is published and $(1 + e)/2$ otherwise; see discussion after equation (4.11).

Note that only the car cost function depends on the number of users.

Since we are assuming case (i), the equilibrium number of users is determined by setting the right sides of (4.17) and (4.18) equal to each other and solving for $N^{(\text{car})}$. And since the right side of (4.18) is fixed, it is the optimum cost for everyone at equilibrium.

This idea is expressed graphically in Figure 4.6. It displays two families of cost curves (for transit and car) expressing their generalized costs under different conditions. The increasing lines correspond to "car" and the horizontal lines correspond to "transit." The point at which a horizontal and a slanted line intersect marks an equilibrium. Its abscissa gives $N^{(\text{car})}$ and its ordinate the generalized cost for everyone.

Assume now that the automobile and public transit systems are initially described by the two curves labeled "initial" in the figure, and consider whether it is best to increase the street capacity (e.g., with a smarter traffic signal system), or else invest the same resources in a subsidy for transit users. The investments are assumed to be small, so both modes will continue to be used.

If we expand the street capacity, μ, the slope of the slanted line (4.17) drops and we obtain a new equilibrium point "1." The practical outcome is an increase in the number of car users without an improvement in anyone's generalized cost. Now, if we instead provide a transit subsidy, reducing the value of $c_f^{(\text{transit})}$ in (4.18), the horizontal line is lowered. As a result, the new equilibrium point becomes "2," showing that transit attracts more riders, and the cost is reduced *for everyone*.[7] The reason for this fortuitous outcome is that the riders switching to transit end up freeing up enough road space for car drivers to experience the same improvement.

In view of this, where should one spend the money? The obvious answer to this question (transit) was anticipated in Section 1.1 of this book, where it was argued that transit expenditures are not just desirable, but essential for the smooth functioning of crowded cities.

References

Daganzo, C. F. (1985). "The uniqueness of a time-dependent equilibrium distribution of arrivals at a single bottleneck." *Transportation Science*, 19(1), 29–37.

Daganzo, C. F. and Garcia, R. C. (2000). "A Pareto improving strategy for the time-dependent morning commute problem." *Transportation Science*, 34(3), 1–9.

Vickrey, W. S. (1969). "Congestion theory and transportation investment." *The American Economic Review*, 59(2), 251–260.

Wardrop, J. G. and Whitehead, J. I. (1952). "Correspondence. Some theoretical aspects of road traffic research." *ICE Proceedings: Engineering Divisions*, 1(5), 767.

Homework

4.1. In this chapter, the capacity μ of a transportation service was optimized by minimizing the generalized cost of the service, which was expressed as a weighted sum of an agency cost and user cost per unit time: $z(\mu) = \$(\mu) + \beta T(\mu)$. This approach is in the spirit of the "Lagrangian approach" discussed in Chapter 3. You are asked to repeat the analysis for the following scenarios using the "standards approach" of Chapter 3. To do this, please determine the capacity μ that minimizes the agency $\$(\mu)$ while guaranteeing the level of service, $T(\mu)$, does not exceed a standard, $T_0 > 0$. Discuss whether economies of scale exist.

---

[7] In reality, the cost of transit increases a little with the number of users because vehicles have finite capacities and may get crowded, and the guideways themselves may get crowded with transit vehicles. However, because many people must ride in each transit vehicle before these effects are felt, the slope of the transit curves in a more realistic depiction of Figure 4.6 would always be much shallower than the slope of the car curves. Thus, the qualitative effects would be much the same as those illustrated in the figure.

(a) Individual transportation under time-independent demand;
(b) Individual transportation under known time-dependent demand;
(c) Individual transportation under adaptive demand.

Can you do the same analysis for collective transportation under a single peak demand period by optimizing the headway H?

4.2. The wish curve introduced in Section 4.1.3 allows that people passing through the bottleneck have different destinations. When this happens, the trip times from the bottleneck to the final destinations should be factored in when developing the wish curve. You are asked to do this in the context of the security checkpoint of an unnamed small airport. Assume that at this airport 100 passengers want to arrive at the departure gate for a flight to Chicago by 8:00 am, and another 200 want to arrive at the gate for Dallas by 8:15 am. Suppose that all passengers must pass through a common security checkpoint and that the travel times from the security checkpoint to these two gates are exactly 10 min and 5 min, respectively. Plot the wish curve with respect to the security checkpoint.

4.3. The discussion of shuttle service under time-dependent demand in Section 4.2 focused on an average passenger. How do the results change if the focus is instead on the worst-case passenger?

4.4. *Passengers arrive at a busy shuttle station according to a time-varying rate $\lambda(t) = 600/t$ for $t \in [6{:}00, 12{:}00]$. The station is empty at 6:00 o'clock. You are given the following parameters: prorated guideway cost, $c_g = 1$ [\$/h]; dispatch cost, $c_f = 10$[\$/dispatch]; and passenger value of time, $\beta = 10$ [\$/h]. Please, do the following:

(a) Plot the cumulative demand curve V over time, and find three piecewise linear functions $\tilde{V}_i(t)$, with one, two and three linear parts ($i = 1, 2, 3$), such that in each case $\max_{t \in [6,12]} |V(t) - \tilde{V}_i(t)|$ is minimized;
(b) Find the optimum headway design $H_i(t)$ for each case, and compare the results;
(c) Discuss the impacts of approximation precision on the headway design and the associated system cost.

4.5. *At a shuttle station on a university campus, passengers arrive at a rate $\lambda_p = 50$ [passengers/h] during the evening rush, and $\lambda_p = 30$ [passengers/h] during the rest of the day (6 a.m. to 12 a.m.). The onset and duration of the rush vary across days. On half of the days, the rush lasts from 2 p.m. to 6 p.m., while on the other days, it lasts from 4 p.m.

to 7 p.m. The transit agency wants to provide the same level of service every day with a time-dependent headway. The system parameters are the following: prorated guideway cost, $c_g = 1[\$/\text{h}]$; dispatch cost, $c_f = 10[\$/\text{dispatch}]$; and passenger value of time, $\beta = 10[\$/\text{h}]$. To do this:

(a) Plot the cumulative demand curves for the two types of days; find the exact optimal headway design that will minimize the total system-wide cost in the long run;
(b) Plot the average cumulative demand curve across the two types of days; determine the optimum headway design based on this average demand profile; compare the result with that from (a);
(c) Based on the finding, discuss the impact of stochastic demand fluctuations on the headway design.

4.6. *The derivation of (4.14)–(4.15a) assumes that transit vehicles always have sufficient capacity to serve all waiting passengers. When transit vehicle capacity is limited, two strategies can be adopted.

(a) When demand surges are temporary and scattered, the transit agency may allow some passengers to be left behind in order to ensure regularity of the headway. Derive a mathematical program that minimizes the total system cost under this strategy, using a fixed headway as the decision variable and recognizing the extra delay of the passengers that are left behind.
(b) When high demand would cause oversaturation over an extended period of time under strategy (a), then it might be better to let the headways vary. In recognition of this, discuss how the multi-headway formulation and solution procedure leading to (4.14)–(4.15a) should be changed in the presence of vehicle capacity.

4.7. *Prove that the cumulative virtual arrival curve at the bottleneck, $V(t)$, under a Vickrey equilibrium is also piecewise linear for those who arrive late.

4.8. *How would you revise the recipe for the Vickrey equilibrium if the wish departure curve W is so highly nonlinear that there are multiple intersection points between curves D and W?

4.9. This problem examines the bimodal distribution of benefits during a congested morning commute. A residential suburb has $N = 15{,}000$ commuters who travel to jobs in a central business district (CBD). Their desired arrival times at the CBD are uniformly distributed from 8:00 am to 9:00 am. All car traffic destined for the CBD must cross a bridge with limited capacity $\mu = 10{,}000$ [veh/h], but road capacity upstream and downstream is sufficient to serve any amount of demand without delay. Assume that all commuters value their queuing delay at $\beta = 15$[\$/h], and that the out-of-pocket cost of driving a car is $c_d = 2$[\$/trip].

(a) Initially, the only way to get to work is by car. Using Vickrey's model of the morning commute with earliness time valued at 1/2 the rate of queuing delay ($e = 0.5$) and lateness time unacceptable ($L = \infty$), draw to scale a queuing diagram for a typical day. Show on it the maximum queuing delay, and the maximum earliness. (Note: free flow travel time can be ignored; explain why.) What are the total generalized cost and the average generalized cost per commuter? The city government is considering starting a high-speed ferry service to relieve congestion and reduce the cost of the morning commute. How many commuters (n), uniformly distributed between 8:00 am to 9:00 am, must be diverted to the ferry to eliminate the delay caused by the bridge bottleneck?

(b) Write an expression for the generalized cost per passenger in terms of the ferry's scheduled headway $h \ll 1$ [h/trip], the prorated cost per commute period for the ferry's terminal infrastructure c_f [\$/period], and the variable cost per revenue ferry trip c_v [\$/trip]. (Assume: fares reflect the ferry's costs; the ferry carries the n commuters identified in part A; a ferry trip excluding the waiting delay takes the same amount of time as a car trip exclusive of the bottleneck delay; headways are so short that people do not bother checking the schedule; and the number of seats on the ferry always exceeds the demand.)

(c) What h^* minimizes the average ferry rider's time cost if $c_f = 10{,}000$ [\$/period], $c_v = 1{,}000$ [\$/trip], and all the costs of ferry operation are passed on to ferry riders as fare? What is the average generalized cost to each ferry rider, and how does it compare to the average cost to each driver on the now uncongested bridge? Who is better off? How much total generalized cost would now be associated with the morning commute?

(d) *Could the solution of part (c) be an equilibrium solution if commuters were free to choose their transportation mode? If your answer is "no," suggest a toll and ferry fare policy that would achieve the solution we have assumed, and discuss the role of cross-subsidies.

(e) *Extra credit: Determine whether the total generalized cost could be reduced by changing n.

Appendix 4.A. Equilibrium of the Morning Commute Problem

We assume the functions $V(t), W(t), D(t)$ are invertible and use $V^{-1}(n)$, $W^{-1}(n), D^{-1}(n)$ for their inverse functions. Under "first-wish, first-in, first-out" (FWFIFO) rules which we are assuming, these functions denote the times at which a generic traveler n arrives, wishes to leave, and leaves the bottleneck. We also assume that curve W is S-shaped with a single peak (see Figure 4.A1).

We say that a system is in equilibrium if no traveler can unilaterally find an alternative arrival time with lower cost, and look for an FWFIFO equilibrium. To find a formula for an equilibrium, it will be convenient to define $V'(n)$ as the slope of the $V(t)$ function at time $t = V^{-1}(n)$, i.e., when commuter n arrives (see Figure 4.A1).

We first look at the critical driver who currently experiences a delay of T_c but no earliness/lateness penalty and see how he/she could be indifferent to arriving at other times. If he/she decides unilaterally to arrive at the same time as an arbitrary earlier commuter $n < n_c$ (see figure) and follow the same schedule, then the queueing delay is reduced to T, while the earliness penalty increases to $W^{-1}(n_c) - D^{-1}(n) = (n_c - n)/\mu$. (Note, the new earliness is *not* $W^{-1}(n) - D^{-1}(n)$). We look for an equilibrium where the critical driver is

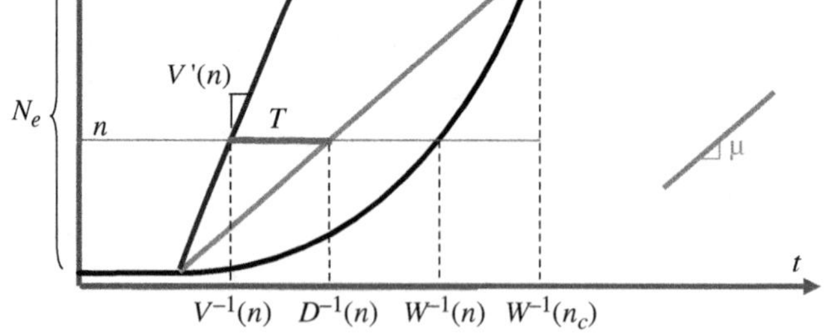

Figure 4.A1. Travel opportunities of critical commuter n_c.

indifferent after the move.[8] For this to happen, the earliness penalty cancels out the reduction in queuing time, i.e., $T_c - T = e(n_c - n)/\mu$. Now, solving for T we find

$$T = T_c - \tfrac{e}{\mu}(n_c - n), \quad \text{for } n < n_c.$$

This shows that T varies linearly with n for all $n < n_c$, implying that the corresponding branch of curve V must be linear. To see how, note that $V^{-1}(n) = D^{-1}(n) - T$. Therefore, the inverse slope of curve V, $dV^{-1}(n)/dn$, is related to T and to the inverse slope of the departure curve by $dV^{-1}(n)/dn = d[D^{-1}(n) - T]/dn = 1/\mu - e/\mu$, which is a constant. Therefore, we can write

$$V'(n) = \mu/(1-e), \quad \text{for } n < n_c. \tag{4.A1}$$

The same logic reveals that if the critical commuter decides to follow the travel schedule of a late commuter n such that $n_c < n \leq N_D$, then the corresponding branch of the V-curve is still linear, but the slope is now

$$V'(n) = \mu/(1+L), \quad \text{for } n > n_c. \tag{4.A2}$$

Since these slopes are given, curve V will be fully determined once we find the location of its inflection point; i.e., the values of $n_c \equiv N_e$ and T_c. We shall do this by forcing curve V to meet curve D at the points where queuing begins and ends. (This ensures that the critical driver has no incentive to travel outside the queuing interval.) The values can be found using the known slopes of the curves, or even more easily by ensuring that the combined delays at the two extreme points, which are N_e/μ (earliness only at the beginning of queuing) and N_L/μ (lateness only at the end), match T_c.; i.e.,

$$T_c = \frac{N_e e}{\mu} \quad \text{and} \quad T_c = \frac{N_L L}{\mu}. \tag{4.A3}$$

---

[8]Strictly speaking, the equilibrium holds for the critical driver as long as the new generalized cost is no smaller than βT_c. We anticipate, however, that we will succeed in finding an equilibrium where the critical driver is truly "indifferent."

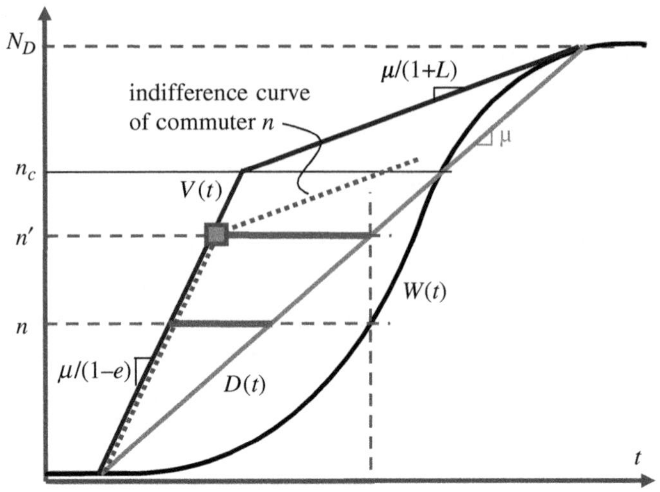

Figure 4.A2. Travel opportunities of an arbitrary commuter n.

Combining these two equalities with the relation[9]

$$N_e + N_L = N_D, \tag{4.A4}$$

we find the answer

$$T_c = N_D\left(\frac{\frac{1}{\mu}}{\frac{1}{L}+\frac{1}{e}}\right) = \frac{N_D L e}{\mu(L+e)};$$

$$N_e = \frac{LN_D}{L+e} \quad \text{and} \quad N_L = \frac{eN_D}{L+e}.$$

The V-curve we have derived is depicted in Figure 4.A2. We have demonstrated that it satisfies the equilibrium condition for the critical driver. It now remains to show that it also satisfies the equilibrium conditions for all other commuters.

To do this, consider an arbitrary early commuter $n \le n_c$ and examine how his/her cost changes if the commuter follows another commuter. To do

[9]In this chapter, we have implicitly assumed that all considered travelers experience some delay during the morning rush, and hence N_D is known. For a general $W(t)$, some "off-peak" travelers may not experience delay at all. While it continues to be true that only those passengers between the $(n_c - N_e)$th and the $(n_c + N_L)$th travelers will experience some delay, and the N_e early travelers and the N_L late travelers occupy the bottleneck for durations of N_e/μ and N_L/μ, respectively, the value of N_D is not known *a priori*. Therefore, N_D must be determined jointly with the values of n_c, T_c, N_e, and N_L. These five unknowns can be solved simultaneously by considering the relations $W^{-1}(n_c - N_e) + N_e/\mu = W^{-1}(n_c)$ and $W^{-1}(n_c) + N_L/\mu = W^{-1}(n_c + N_L)$ in addition to (4.A3) and (4.A4).

this, we denote by n' the commuter that departs exactly when n wishes to depart and mark with a square its arrival point (see Figure 4.A2). Note that $n \leq n' \leq n_c$. Note that if commuter n follows commuter n' (i.e., arrives at the square), then the commuter only experiences queuing delay. Then, the same analysis that led to (4.A1) and (4.A2) and showed that the critical commuter was indifferent across all the points of V reveals that commuter n would be indifferent to a virtual arrival curve with the same slopes as the original, *but pinned at the square* (see the slanted dotted line in the figure). Now note that curve V is never to the right of the indifference curve. Thus, it cannot produce a lower queuing time at any position. Since the commuter cannot improve, we conclude that V is an equilibrium for all $n \leq n_c$.

The same analysis reveals that V is an equilibrium for all $n \geq n_{c.}$ Therefore, V is an equilibrium curve for all commuters. Additional information can be found in Daganzo (1985) and Daganzo and Garcia (2000).

Chapter 5

Planning — Corridors

This chapter examines the organization and performance of transit systems serving corridors with a diffuse distribution of passenger origins and destinations. The analysis allows for multiple transit lines, e.g., providing express and local service, but the lines follow a common path and use a definite set of stations in the corridor.

The transit planning task is considerably more challenging for corridors than it is for shuttles because now passengers have to be consolidated, not just in time, but also in space. The analysis is also more complex because, in addition to designing a timetable, the planner must also choose the location of the stops. Waiting time is no longer the only metric affected by the design since the number and location of the stops influences both, the customers' walking/access times and the buses' commercial speed.

Given the complexity of the task at hand, it is impractical, if not impossible, to cover the full range of scenarios and possible solutions. Therefore, this chapter will introduce a generic analysis method and use it to derive results only for some basic cases. Homework problems at the end introduce additional scenarios. The hope is that an interested reader can develop both fundamental insights and the necessary skills to analyze any problem.

The chapter starts with the simplest ideas and builds toward the most complex. Section 5.1 develops preliminary insights; it uses a highly idealized model both, to identify some of the factors that enhance performance, and to explore the challenge of serving a population that is spatially spread out. Section 5.2 relaxes the assumptions in this model and examines more realistically a basic case involving a single line under homogeneous demand. This is done with the "standards" mathematical programming approach of Chapter 2, using a single standard, and also with the Lagrangian formulation.

Section 5.3 then shows how to deal with multiple standards. The final two sections generalize the ideas underpinning the basic case. Section 5.4 considers multiple lines and Section 5.5 deals with inhomogeneous problems.

5.1. Preliminary Insights: An Idealized Analysis

The new challenge introduced by corridors is that to be served in groups, passengers must be consolidated in space at discrete points where "buses" stop to collect and deliver the groups. Outside the buses, this consolidation takes place at small speeds, e.g., by walking or bicycling. Inside the buses passengers also experience limited speeds, since buses must be brought to a stop at each station. These two effects limit the door-to-door speed that transit can deliver.

In view of these limitations, and in order to set reasonable expectations before design methods are introduced, this section quantifies the door-to-door speed issue with an extremely idealized model that isolates it. The setting is an infinitely long homogeneous corridor with time-independent and uniformly distributed demand where service is provided by one or more lines. The lines themselves are also spatiotemporally homogeneous in the sense that they each have a fixed separation between stations and a fixed headway. Formulas for door-to-door speed are presented under the most favorable assumptions possible: Section 5.1.1 for single-line corridors; and Section 5.1.2 for multi-line corridors with local and express services.

5.1.1. *Limits to Door-to-Door Speed: A Single Line*

Figure 5.1 plots a section of our corridor. We assume that service is bidirectional and that passengers use the direction that aligns with their trips.[1] The line's stop spacing is s. We look for a formula that will predict both the commercial speed and the door-to-door speed for various passenger trip lengths under the most favorable assumptions possible. The assumptions are unrealistically optimistic but they will yield a formula that does not depend on demand or other parameters; so it is very general.

The following is assumed:

- Passenger origins are continuously distributed along the corridor.
- The service headways is $H \approx 0$, so passengers do not wait at stops.

[1]It shall become obvious as the results are derived that they also apply for one-directional service.

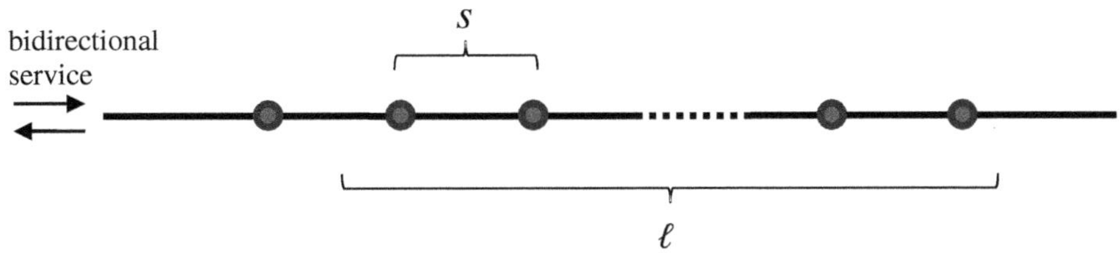

Figure 5.1. An idealized transit corridor.

- Vehicle doors open and close instantly, and passengers take no time to board and alight.
- The vehicles' top speed is $v_{max} = \infty$.

Despite these extremely favorable conditions, two limitations remain:

(1) The passenger's access speed is finite ($v_a \approx 1\,\mathrm{m/s}$ if walking).
(2) The vehicle's acceleration/deceleration rate is also finite ($a_0 \approx 1\,\mathrm{m/s^2}$ for passenger comfort).

We shall use these assumptions to derive formulas for the commercial speed, and for the door-to-door travel time. We shall evaluate the latter for the worst-case passenger rather than the average because we can then avoid specifying the spatial distribution of the origins and destinations, so the results are more general. Furthermore, the results are qualitatively similar in both cases.

Consider first the commercial speed v_v. It is the ratio of the station spacing s to the time T_s a vehicle takes to cover it when traveling as fast as possible. Since $v_{max} = \infty$, and the acceleration limit is the same as that for deceleration, T_s is achieved if the vehicle accelerates at the maximum rate after departing a station and then switches to deceleration upon reaching the midpoint between stations (see Figure 5.2). Every inter-station vehicle trajectory can then be decomposed into two symmetric parts of length $s/2$ and duration $T_s/2$. Moreover, since acceleration is constant, these quantities are related by $s/2 = {}^1\!/_2 a_0 (T_s/2)^2$ (see Chapter 2). Thus we can write:

$$T_s = 2\sqrt{\frac{s}{a_0}}. \tag{5.1}$$

This yields for the commercial speed:

$$v_v = \frac{s}{T_s} = \frac{\sqrt{s a_0}}{2}. \tag{5.2}$$

Note how v_v increases with the square root of the station spacing, and that close spacings imply slow speeds. Note as well that the equation does not

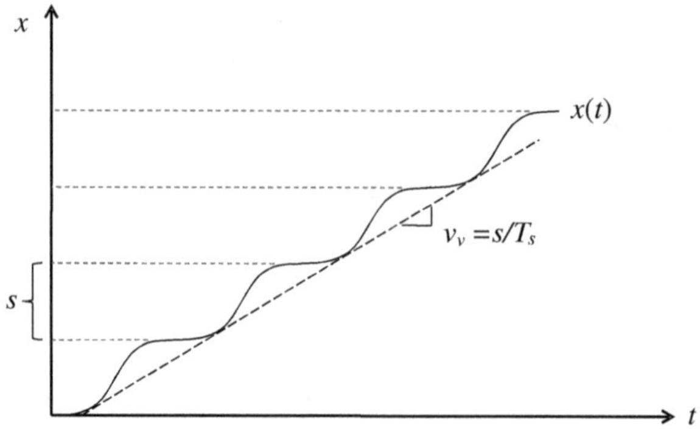

Figure 5.2. Transit vehicle trajectory showing the vehicle's acceleration–deceleration cycles.

involve anything about technology; only the human factors constant a_0. Therefore, vehicle technology cannot change this fact.

Let us now turn our attention to door-to-door speed. To do this, we assume that people walk to the nearest station and that their trip lengths are continuously distributed in an interval $[0, \ell]$. As before, we focus on the unluckiest passenger. The reader can now verify that if ℓ is not a multiple of s then for any spacing s, there always is an unlucky passenger who would have to walk a distance s and then ride for a distance $s\lceil \ell/s \rceil$.[2] As a result, the access time for this passenger is

$$T_a = \frac{s}{v_a}, \tag{5.3}$$

and the riding time is

$$T_r = \frac{s}{v_v}\left\lceil \frac{\ell}{s} \right\rceil = 2\sqrt{\frac{s}{a_0}}\left\lceil \frac{\ell}{s} \right\rceil \geq \frac{2\ell}{\sqrt{sa_0}}. \tag{5.4}$$

The second equality follows from (5.2), and the last inequality from removing the ceiling function. The reader is encouraged to verify that this expression also holds as a pure equality if ℓ is a multiple of s. Thus, (5.4) is general. It is also easy to see that the right side of (5.4) is not only a bound but also

[2] To see this, draw a picture with an unlucky trip as follows: (i) an origin displaced by an infinitesimal amount ε toward the left of a mid-point between stations, and (ii) a trip length $y = s\lfloor \ell/s \rfloor + 2\varepsilon$ where ε is infinitesimally small. Since ε is negligibly small and ℓ is not a multiple of s, it follows that $y \in [0, \ell]$, so the trip length is admissible. Now note from your drawing that the passenger backtracks at both ends of the trip, spending $\lceil \ell/s \rceil$ intervals in the vehicle, and that both the origin and destination points are half way between stations. Thus, the total access distance is s and the total riding distance $s\lceil \ell/s \rceil$.

a good approximation if $s \ll \ell$. Now combine this approximation with (5.3) to obtain the total door-to-door time:

$$T = T_a + T_r \approx \frac{s}{v_a} + \frac{2\ell}{\sqrt{s a_0}} \quad \text{for } s \ll \ell. \tag{5.5}$$

Recall from Chapter 2 that this is a GEOQ expression in s. This implies that (5.5) is globally minimized by a finite $s^* > 0$ that can be expressed in closed form using the first order conditions. The underlying tradeoff can be understood physically — if s tends toward infinity, the access time tends toward infinity, so T is unbounded; and if s tends toward zero, the riding time tends toward infinity so T is still unbounded. Clearly, there must be some optimum intermediate value. As shown in Section 2.3.1 (Example 2.1) the optimal solution of a GEOQ problem of the form $z = Ax^k + Bx^{-m}$ (with k, $m > 0$) is $x^* = (Bm/Ak)^{\frac{1}{m+k}}$ and $z^* = (1 + m/k)(m/k)^{\frac{m}{m+k}} A^{\frac{m}{m+k}} B^{\frac{k}{m+k}}$. Therefore, if we replace A, B, k and m in these expressions with the corresponding quantities of (5.5), we find the optimum of (5.5). The result is

$$s^* \approx \left(\frac{v_a^2 \ell^2}{a_0}\right)^{1/3}; \text{ and} \tag{5.6}$$

$$T^*(a_0, v_a, \ell) \approx 3\left(\frac{\ell^2}{v_a a_0}\right)^{1/3}. \tag{5.7}$$

Now, on dividing ℓ by (5.7), we obtain the maximum possible door-to-door speed for the worst-case passenger:

$$\hat{v} \approx \frac{\ell}{T^*} = \frac{1}{3}\left(\ell v_a a_0\right)^{1/3}. \tag{5.8}$$

It should be remembered that (5.6)–(5.8) are valid only if $s^* \ll \ell$, i.e., only if $\ell \gg v_a^2/a_0$. However, since realistic values of v_a^2/a_0 are comparable with 1 m, this requirement is comfortably satisfied in practice for any trip of reasonable length. Furthermore, equations (5.6)–(5.8) are quite general. For example, even though they were derived for the worst-case passenger they also hold qualitatively for the average passenger if the demand is uniform.[3]

It is informative to see which variables appear in these equations (and which do not) and how the variables that do appear influence the results.

[3]The only substantive difference in this case is that the average access time is one half of that predicted by (5.3), i.e., the same as obtained with this expression if one doubles the value of v_a. Since nothing else changes in the derivation, (5.6) and (5.7) can also be used for this case after replacing v_a by $2v_a$.

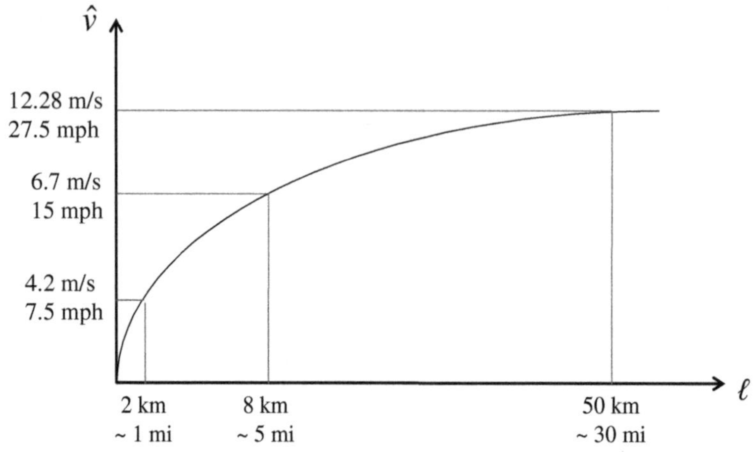

Figure 5.3. Maximum door-to-door speed along an idealized corridor.

Curiously, all the variables appearing in (5.6)–(5.8) describe the users, and none the technology. Thus, as in the case of the commercial speed, equation (5.8) imposes a hard bound on the door-to-door speed that can be achieved with a single transit line. Also note that the optimum spacing (5.6) increases with the 2/3 power of ℓ, and that the door-to-door speed (5.8) increases with the 1/3 power of both ℓ and v_a.

Let us now plug some numbers into (5.8) to quantify these observations and see how the door-to-door speed changes with the length of trips made. The result for $v_a = 1\,\text{m/s}$ and $a_0 = 1\,\text{m/s}^2$ is plotted in Figure 5.3. Note how the curve increases at a rapidly decreasing rate. The overall numerical outcome is somewhat disappointing, especially in view of the very favorable assumptions we have made.[4] The low speeds arise mostly because consolidation requires that passengers travel the "last mile" on foot. Since people walk slowly the stops must be spaced closely, and this lowers the vehicle's commercial speed. Clearly, if we could increase the access speed, so would the door-to-door speed. The next section explores how transit itself can be used to achieve this feat.[5]

[4]If we made similar favorable assumptions for individual transportation modes on uncongested guideways, their commercial speed would be close to the mode's maximum speed for all ℓ, i.e., much better than for public transit. The reason is that by being individual these modes do not require much of an access displacement: a great virtue.

[5]Although the feat can also be theoretically achieved with bicycle access, the issue is more complicated because it is not fair to assume that all passengers can ride a bicycle.

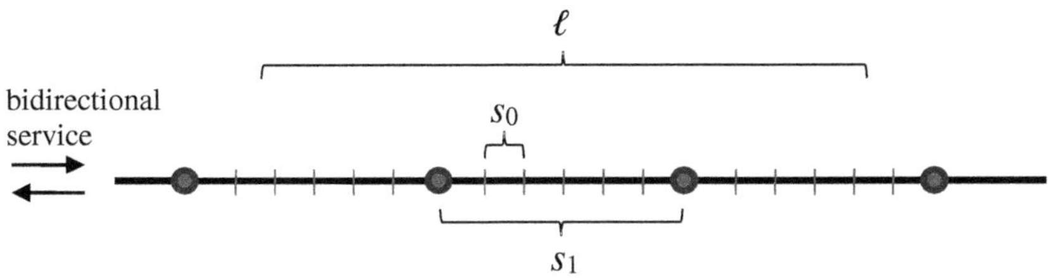

Figure 5.4. An idealized hierarchical transit corridor.

5.1.2. *The Effect of Access Speed: Usefulness of Hierarchies*

Consideration of (5.6) revealed that long trips benefit from long spacings and short trip from short spacings. Therefore, in situations where passengers take trips of widely different lengths, it makes sense to provide a local line that covers all the stations to serve the short trips, and an express line that skips stations for the long trips. And if this is done, passengers with long trips could use the local line as a feeder to the express, in effect using transit itself to increase their access speed. As a result of the faster access, they would enjoy an even higher door-to-door speed.

In view of this, let us now explore the door-to-door speeds that can be achieved by this type of hierarchical service using the same assumptions as those in the previous subsection. We shall assume that both lines are bidirectional and that passengers transfer to/from the express mode at the stations that minimize local travel, backtracking if necessary. As in that subsection, the analysis shall be done for trips of length ℓ, assuming that the lines have been optimized to best serve the unluckiest passenger with this trip length. Figure 5.4 shows the setup. Note we have two spacings to choose s_0 (local) and s_1 (express), where $s_0 \ll s_1$.

There are also two types of access speed: a speed to access the local stations, which is given and we denote v_w (the subscript is for "walking") and a speed to access the express stations, v_a, which is a derived quantity that depends on the spacings. Our solution strategy consists in calculating v_a in terms of the spacings, and then treating the express system as a single-line system, using v_a as the access speed.

So, let us now find the expression for v_a. It should be clear from the figure that the unluckiest passenger will start and end the trip at two midpoints between express stations. To do this, we shall assume for simplicity but not unreasonably that transfers take $s_0/(2v_w)$ time units, i.e., the same amount of time as walking from the origin to the local access station. With this assumption, the local inbound trip to the transfer point can be treated as a single-line trip of length $\ell = s_1/2$. By symmetry, the same can be said for the outbound egress trip toward the destination. Therefore, the optimum local spacing to serve these trips conditional on s_1, which we denote $s_0(s_1)$, is given by (5.6), with $\ell = s_1/2$. The result is

$$s_0(s_1) = \left(\frac{v_w^2 s_1^2}{4a_0}\right)^{\frac{1}{3}}. \tag{5.9}$$

Likewise, the optimum access speed, $v_a(s_1)$, is obtained by entering with $\ell = s_1/2$ into (5.8). Thus, we have

$$v_a(s_1) = \frac{1}{3}\left(\frac{s_1}{2}v_w a_0\right)^{\frac{1}{3}}. \tag{5.10}$$

Now that the access speed has been calculated, we can replace v_a in (5.5) with the right side of (5.10) to obtain the optimum hierarchical door-to-door time conditional on s_1. The result is

$$T(s_1) = \left[3 \times 2^{\frac{1}{3}}(v_w a_0)^{-\frac{1}{3}}\right] s_1^{\frac{2}{3}} + \frac{2\ell}{\sqrt{a_0}} s_1^{-\frac{1}{2}}. \tag{5.11}$$

We can now optimize (5.11) with respect to s_1. Note again the equation is a GEOQ formula, and that the usual manipulations yield:[6]

$$T^* \approx 5.3\left(\frac{\ell^4}{a_0^3 v_w}\right)^{\frac{1}{7}} = 5.3 a_0^{-\frac{3}{7}} v_w^{-\frac{1}{7}} \ell^{\frac{4}{7}}. \tag{5.12}$$

Finally, the ratio of the trip length ℓ to (5.12) yields the door-to-door speed:

$$\hat{v} \approx 0.19 a_0^{\frac{3}{7}} v_w^{\frac{1}{7}} \ell^{\frac{3}{7}}. \tag{5.13}$$

It is interesting to compare this expression with the corresponding expression for single-line systems, i.e., with (5.8), using v_w instead of v_a. Note the lower power of the walking speed. This means that the hierarchical system is better for very low walking speeds, as one might expect, and the reverse occurs for high walking speeds. Equation (5.13) also increases with a *higher* power of trip distance than (5.8). Thus, hierarchical systems are better than single-line systems for long distances, and the opposite occurs for short distances. Also, note that there must be a critical distance where the two are the same; Figure 5.5 shows the pattern qualitatively. The pattern strongly suggests that large cities where many trips are long can benefit from the superposition

[6]The reader is encouraged to use the generalized economic ordering quantity (GEOQ) formula to obtain s_1^*, and then verify using (5.9) for s_0^* and (5.6) for s^* that $s_0^* < s^* < s_1^*$.

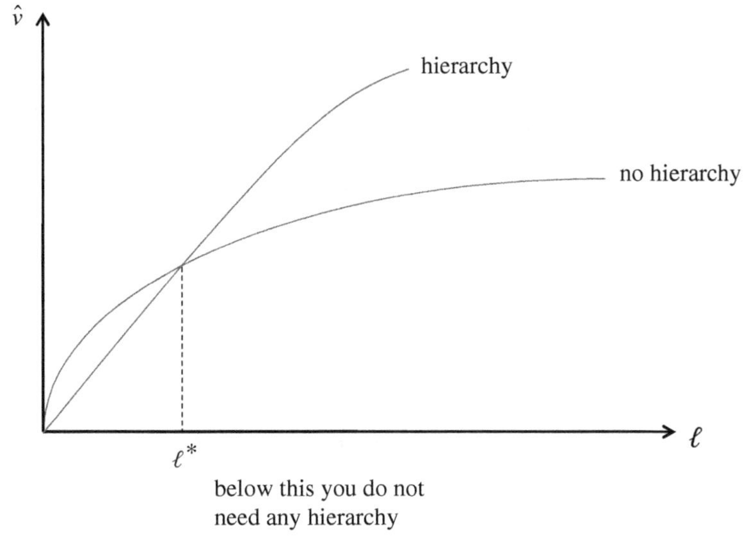

Figure 5.5. Maximum commercial speed in a corridor with and without hierarchical service.

of two transportation systems: long-distance subways or commuter trains with widely spaced stops that can provide express service, and a local network of slower buses that can serve both a feeder function for long trips and as the main mode for short trips.

There is a caveat, however. For the values of Figure 5.3 (v_w =1 m/s and a_0 =1 m/s^2), the two curves of Figure 5.5 intersect at $\ell^* = 393.7$ m and the hierarchical values quickly become much higher beyond this value. Since most transit trips exceed this critical distance, the results would seem to suggest that transit agencies should always use hierarchical systems to a great benefit. This is not true, however. Our analysis used very favorable assumptions for the purposes of deriving a *bound* (not an approximation). The assumptions are especially favorable for hierarchical service because they set the waiting time during transfers at zero, and also allow express vehicles to travel at unreasonably high speeds between far apart stations. This favors express service too much. On the other hand, the suggestion that long trips are better served with some form of hierarchical service is supported by more accurate analyses. Moreover, we often see multi-line corridors in real life — and as predicted, only when corridors are long enough to serve long trips. For all these reasons, the following sections will present the tools to analyze with considerable more realism not just single-line systems but also hierarchical systems. Single lines are taken up first.

5.2. Single Lines: Analysis Method and Solution Properties

The analysis will now allow for both spatial and temporal consolidation, combining the ideas of Section 5.1 and Chapter 4. We shall see that in addition to ℓ, two other important demand-side variables affect a corridor system's structure: the trip generation rate, λ, and the "user's value of time" β. Section 5.2.1, below, describes the assumptions and how they affect what passengers do outside the vehicle; and Section 5.2.2 presents the design method and the solution.

5.2.1. *Assumptions and Preliminary Modeling Considerations*

It will be assumed that the line is infinitely long and that the demand is uniformly generated all along the route with the same distribution of trip lengths across all origins. We will also assume that the same service is provided all along the corridor and that the "buses" have sufficient capacity to carry the generated bus loads. This symmetric setup is useful because, as in Section 5.1, it allows us to define and solve the design problem with only a few variables.

To improve realism, however, the three favorable assumptions in Section 5.1 are replaced by the following:

(1) Transit vehicles have a finite "cruising" speed $v_{\text{max}} < \infty$. This is defined as the average speed at which they would travel if they did not stop at stations. So, for example, in the case of buses operated without special lanes, $v_{\text{max}} \approx v_{\text{auto}}$. It is also assumed, quite reasonably, that stations are spaced sufficiently apart for vehicles to be able to reach their cruising speed. Then, as shown in Figure 5.6, the moving part of a vehicle's trajectory in between two stations has three phases: an acceleration phase, lasting $t_{(\text{a})}$ time units, where the trajectory is convex; a cruising phase where it is linear; and a deceleration phase lasting $t_{(\text{d})}$ time units where it is concave.

In the models of this section and later in the book, the moving part of the trajectory is approximated by the piecewise linear dotted line of the figure, as if vehicles could accelerate instantaneously. Note that the duration of the moving part of the idealized trajectory is s/v_{max} time units, i.e., it is proportional to the station spacing, and

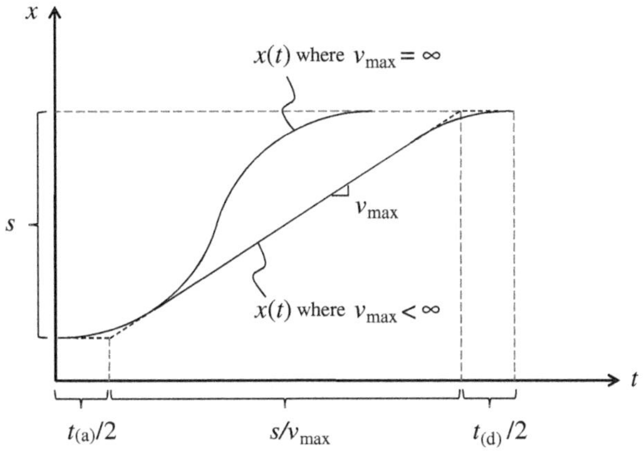

Figure 5.6. Transit vehicle trajectories between two stations: old and new assumptions.

that of its stopped part is a constant: $[t_{(a)} + t_{(d)}]/2$ time units.[7] This fixed stopped time, which represents the time lost during acceleration and deceleration, is then added to the real vehicle dwell time at each station. The idealization is useful because it eliminates the need for acceleration/deceleration parameters, and as we explain below, it allows us to express the dependence between vehicle travel time and station spacing in a simple way.

(2) Transit vehicles take some time, $t_s > 0$, to stop at stations. In addition to the the time lost due to acceleration and deceleration that we have just mentioned, $[t_{(a)} + t_{(d)}]/2$, the stop time includes an additional fixed part for opening and closing the doors, and a variable part with the doors open. The latter depends on the number and type of passengers boarding and alighting.

(3) All the buses operate with the same headway, $H > 0$. Since the corridor and the service provided is assumed to be uniform, this implies that the headway is the same everywhere along the corridor, i.e., that all bus trajectories are translationally symmetric versions of one another in the direction of time.

Our simplified treatment of moving and stopped times is useful because it allows us to express a vehicle's travel time over a number of stations as the sum of a fixed moving part that only depends on distance and a variable part

[7]Recall from Section 2.1.1 that the lengths of the two horizontal parts equal half of the corresponding durations of the acceleration and deceleration phases.

that only depends on the number of people and the number of stops made at stations, *independent of how people and stations are distributed over space.* For example, if all the stops take roughly the same time (e.g., as occurs for metros with wide doors), we can write:

$$time = (distance)/v_{\max} + (\#stops)t_s, \text{ and} \tag{5.14}$$

$$\begin{aligned} commercial\ pace^8 \equiv 1/v_v &= time/distance \\ &= 1/v_{\max} + (\#stops\ /distance)t_s. \end{aligned} \tag{5.15}$$

Furthermore, given all the above assumptions, we see that the organization of a line is fully defined by just two decision variables: the spacing between stations s and the headway H. Figure 5.7 displays some bus trajectories corresponding to a possible design (s, H). The trajectories have been represented by straight lines after smoothing out the stop times at each station, i.e., so that the slope of the trajectories is the commercial speed, v_v.

As an aside, Figure 5.7 also compares transit's accessibility with that of the automobile for a person ready to start a trip at the origin. The darker shade marks the space-time region that can be accessed by transit, and the lighter shade the extra region that the automobile allows. Note how the transit region is a subset of the automobile region. This is not always true, however. The reader can verify that if $v_v > v_{\text{auto}}$, as may occur in a congested city, then there are transit regions that cannot be reached by automobile. The reader may also find it instructive to imagine how the two accessibility domains change with s, H and the commercial speed of transit. It will become readily apparent that increasing v_v and shortening s and H is very desirable.

This type of time–space diagram is also useful to figure out in which direction a passenger should walk if (s)he knows the schedule, i.e., knows the location of the dots in the figure. To clarify this idea, Figure 5.8(a) shows two passengers and their accessible domains by walking. These domains enclose the vehicle departures that they can feasibly reach. Consideration of the figure shows that if the passengers' objective is catching the first bus possible, then the two passengers would walk in opposite directions and board the same bus.

This reasoning can also be used to identify the catchment area of each schedule point, i.e., the region of space–time where originating passengers use the point in question. Figure 5.8(b) shows as an example the catchment area for the shown schedule point under the assumption that people walk to the closest station.

---

[8] Pace is the inverse of speed.

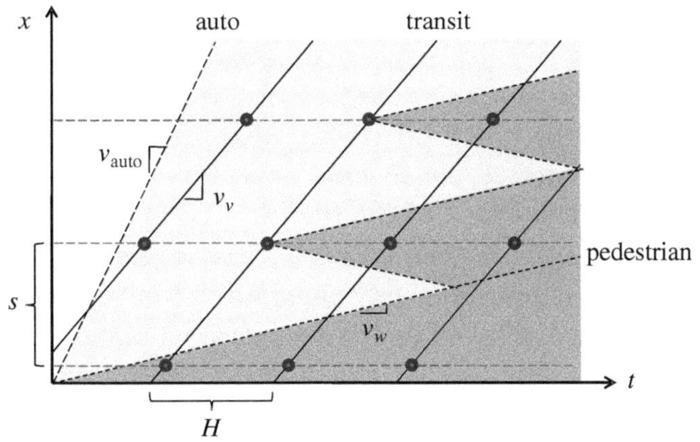

Figure 5.7. Transit trajectories and the accessibility they provide, compared with auto.

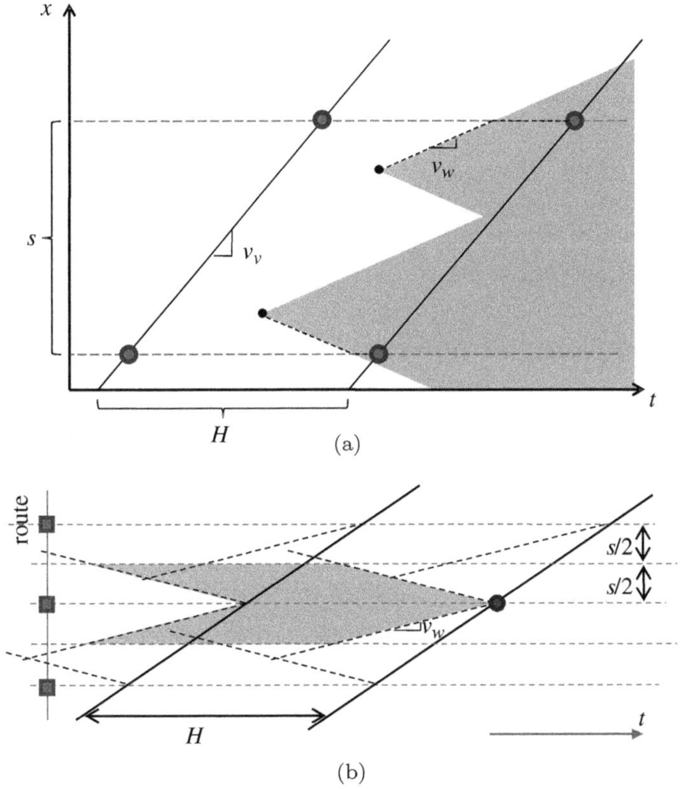

Figure 5.8. Access domains: (a) passenger perspective and (b) schedule point perspective.

Note, because the problem is translationally symmetric, this region is a tile that if repeated at every schedule point tessellates the (t, x)-plane. Also, note that this must be true regardless of the strategy that passengers use to choose a point — provided, of course, that the same strategy is used by all passengers. Therefore, in all cases, the tile has an "area" equal to sH (distance–time). The shape of the tile can also be used to calculate the average walking and waiting time across all passengers. For example, if we assume that passengers are uniformly generated at all points in space-time as if they did not know the schedule, then the average walking distance is the average vertical separation between the points in the tile and the horizontal line passing through the schedule point. In our example, this is[9]

$$\text{Average walk} = s/4. \tag{5.16}$$

We also see that under the same assumptions the maximum walk is $s/2$. A passenger's waiting time can be similarly obtained by drawing the passenger's walking trajectory to determine the point at which it reaches the station, and then evaluating the horizontal separation between this point and the schedule's point. For our example, we clearly see on considering all passengers that

$$\text{Average wait} = H/2, \tag{5.17}$$

and that the maximum wait is H. This, as one would expect, is consistent with the first row of Table 4.1 in Chapter 4.

5.2.2. *Formulation and Solution*

This subsection has two parallel goals: (i) showing methodologically how problems with realistic assumptions can be treated; and (ii) unveiling the nature of their solutions. To do this in a simple way, we shall consider a transit line with very large vehicles and quick boardings, so the vehicles never reach capacity and the boarding times t_s can be treated as fixed. We shall look for both, a cheapest design that provides a door-to-door travel time guarantee, T_0, for all trips of length ℓ, and a design that minimizes the generalized cost. We shall show that the two procedures produce the same set of outcomes and that the second is more intuitive and insightful. In both cases, the design variables are the station spacing s and the headway H.

On the demand side, given is the average passenger trip generation rate per unit distance of corridor in one direction, λ [trips/time-distance],

[9] The result is a bit different for other strategies, e.g., if passengers wish to catch the earliest train possible. However, the differences turn out to be minor if $v_v \gg v_w$.

which we call the demand density and the speed at which these passengers access the stations, v_a. Also fixed is the distribution of trip lengths. This distribution is not specified, however, because as we shall see it does not enter the formulation.

On the supply side, given are the cost rates the agency pays for vehicles and infrastructure. These rates will be used throughout the planning chapters of this book, and therefore, a bit of discussion is in order. For vehicles, the rates shall account for both their mechanical and human components. On the mechanical side, the cost includes the vehicle's depreciation, operation, and maintenance. We assume that this cost accrues only while the vehicle is moving and that it does so at a rate proportional to distance, which we denote c_m[\$/vehicle-km].[10] This rate can be estimated approximately by the ratio of (i) the sum of a vehicle's cost and its expected maintenance and operating costs over the vehicle's lifetime net of any salvage value to (ii) its expected mileage before decommissioning. On the human side, we assume that the cost for the operating crew accrues at an hourly rate that we denote c_t [\$/vehicle-h]. This rate is the ratio of (i) the sum of the vehicle operator(s) wages, benefits, and overhead and (ii) the fraction of time that a typical operator is actually engaged in the act of driving.[11]

It should be clear from the above that the cost for a crewed vehicle accrues at a different rate when moving and stopped. When stopped, the vehicle cost rate is simply c_t [\$/vehicle-h] and while moving: $c_d \equiv c_m + c_t/v_{\max}$ [\$/vehicle-km]. The second term in the last expression recognizes that drivers cruise at speed $v_{\max}$. Note that the cost of a vehicle stop (at a station) is the product of c_t and t_s. It will be convenient to denote it by $c_s \equiv c_t t_s$ [\$/vehicle stop]. The constants c_t, c_d, and c_s will be used repeatedly in our models.

The infrastructure costs include a part for the guideway and another for the stations. Both account for depreciation, maintenance, and operation. The cost of the guideway is reasonably assumed to be proportional to distance, so it is captured by a constant c_g with units [\$/time-distance], and the cost of a single station by a constant c_r with units [\$/time] which is akin to rent. The data in Chapter 1 should give an order of magnitude idea for these quantities.

---

[10]Assuming that vehicles depreciate only while moving is reasonable since typical transit vehicles are heavily used. Chapter 1 includes rough information about the various component costs. In a specific instance, more accurate numbers should be available from the operating agency.

[11]The chapters on vehicle and personnel management show that if crews are managed properly idle time is low, so the assumption is reasonable.

5.2.2.1. *The Standards Problem: Solution Methodology*

To illustrate how to obtain optimum designs we consider the standards design problem. The standards approach is appealing because it allows one to develop a design without knowing much about the demand. The idea is to minimize the agency cost, which for transit systems is fairly insensitive to the demand, subject to the level of service standards which are specified independent of demand distribution.

To illustrate the theory, we shall use as the standard a door-to-door travel time guarantee for a worst-case passenger with trip length ℓ, assuming that passengers do not have appointments and do not know the schedule. We shall write the standard as $T(\ell) \leq T_0$, where $T(\ell)$ denotes the travel time of the unluckiest passenger with trip distance ℓ. Note that both ℓ and T_0 are policy variables. The objective is minimizing the total cost per year per unit length of the corridor for the vehicles and the infrastructure. Since the demand density λ is fixed (although it may be unknown), we can instead minimize the ratio of this cost to λ, which can be more meaningfully interpreted as the cost per trip (customer served) and we denote \$ [cost/customer] or [cost/trip]. Thus, the design problem can be expressed as either of the following two optimization problems:

$$\min_{(s,H)}\{\$: T(\ell) \leq T_0\}; \text{ (agency cost per trip)} \tag{5.18a}$$

$$\min_{(s,H)}\{\lambda\$: T(\ell) \leq T_0\}.$$

$$\text{(cost per unit time and unit distance of corridor)} \tag{5.18b}$$

To proceed, we must find expressions for \$ and $T(\ell)$ in terms of the decision variables.

A formula for \$ can be obtained by considering the costs accruing in any repeating tile of our problem, such as that shown on Figure 5.9, and dividing the result by the number of customers in it, λsH. Since the tile includes a single vehicle that travels a distance s and stops once in the direction of interest, the accrued vehicle cost is $c_s + c_d s$. Furthermore, every part of the guideway and the station are in operation for H time units inside the tile. Therefore, the tile's infrastructure cost is $c_r H + c_g sH$. Therefore, the total agency cost in the tile is $c_s + c_d s + c_r H + c_g sH$.[12] Finally, divide this by λsH

[12]In order to prevent errors when doing derivations like this, we recommend checking for dimensional consistency every step of the way; see Section 2.4. Note the equation we have derived is dimensionally correct.

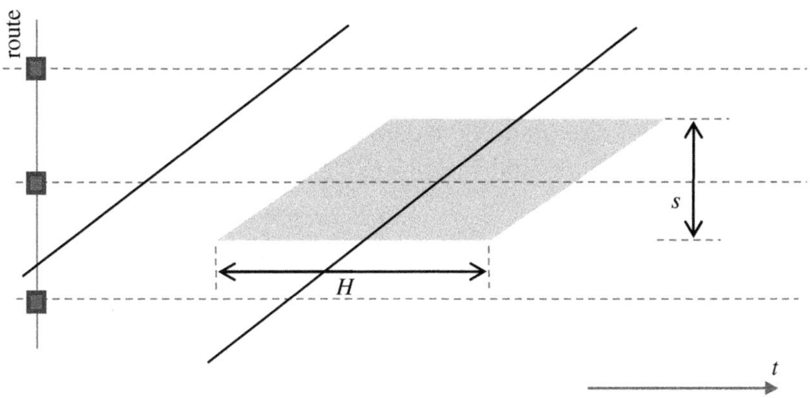

Figure 5.9. Basic unit of analysis in a time–space diagram.

to obtain the objective function of (5.18a):

$$\$ = \frac{c_s}{\lambda s H} + \frac{c_d}{\lambda H} + \frac{c_r}{\lambda s} + \frac{c_g}{\lambda}. \tag{5.19}$$

Note that the demand λ appears in this expression only as part of a multiplicative constant $1/\lambda$, and that the demand does not appear at all in the objective function of (5.18b).

The formula for the worst-case door-to-door travel time $T(\ell)$ is easier to obtain. This time includes three components: (i) the in-vehicle travel time, which is approximately given by (5.14) using ℓ for the distance and ℓ/s for the number of stops; (ii) the walking/access time, which for an unlucky passenger approximately totals s/v_a when considering both ends of the trip; and (iii) the worst-case wait H, which occurs entirely at the origin since passengers have no appointments (see Table 4.1). Thus,

$$T(\ell) \approx \frac{\ell}{v_{\max}} + t_s\frac{\ell}{s} + \frac{s}{v_a} + H. \tag{5.20}$$

Note, the only demand-side parameter appearing in this expression is the access speed, which is easy to estimate. Therefore, the solution to our optimization problem can be found without knowing much about the demand, as we had claimed.

This concludes the formulation part. If we had made different assumptions, e.g., been interested in average performance, or placed different weights on waiting, walking, and riding time, the process would have been similar. Only the final expressions (5.19) and (5.20) would differ slightly. For example, if standards have been defined for the average passenger rather than the unluckiest then the walking distance term in (5.20) would have to be changed from s/v_a to $s/(2v_a)$, and the waiting term from H to $H/2$.

Let us now solve the problem. To this end, note on examination of (5.19) and (5.20) that the inequality constraint in (5.18) must be binding at optimality. Otherwise, we could increase H a little and reduce the objective function. Therefore, we can treat the problem as if the constraint was a pure equality and use the Lagrange multiplier method.

To proceed, we choose (5.18a) and write its Lagrangian by combining (5.19) and (5.20) with the multiplier β [\$/time]. After grouping terms with the same decision variables together, and omitting the term $-\beta T_0$which is a constant (see also the discussion in Section 3.2.2), we obtain the truncated Lagrangian:

$$z = \$ + \beta T(\ell) = \frac{c_s}{\lambda s H} + \left(\frac{c_d}{\lambda H} + \beta H\right) + \left(\frac{c_r + \lambda\beta\ell t_s}{\lambda s} + \frac{\beta s}{v_a}\right) + \left(\frac{\ell\beta}{v_{\max}} + \frac{c_g}{\lambda}\right). \tag{5.21}$$

Let us now analyze (5.21). Note that its first parenthesis only includes the variable H, the second parenthesis, the variable s, and the last parenthesis is a constant.

The solution to our problem is the global minimum of this function for the value of β, denoted β^*, that yields $T^*(\ell) = T_0$. Before looking for it, we inspect the objective function to see if multiple minimums can be ruled out. In our case, (5.21) depends on the decision variables as a posinomial of the type arising in geometric programming, which as we saw in Chapter 2 must have a unique minimum. Therefore, if the parameters of the problem are given specific values, the global minimum for any β can be found with numerical search techniques such as the steepest descent method introduced in Section 2.3.1. The search should then be repeated until we succeed in finding the s and H solution that matches $T^*(\ell)$ to T_0.

Alternatively, since (5.18) only involves two decision variables, its solution can be readily identified by tabulating (5.19) in a spreadsheet, and setting the result equal to infinity when the variable combinations violate the constraint. This procedure can also be used if the problem were to include additional constraints. The value of (5.19) would then be set to infinity when *any* of the constraints is violated. An extra constraint is needed for example if the vehicle occupancy for envisioned combinations of (s, H) exceeds the vehicle's capacity, C. In that case, one may want to develop a formula for the vehicle occupancy in terms of (s, H) and exclude combinations that yield values greater than C.[13] Homework exercise 5.2 gives an example.

Equivalence and Interpretation of the Standards and Lagrangian Approaches: Note that we solved the standards problem by minimizing a Lagrangian (5.21) and that this Lagrangian is a function with units of cost in which β is the "value of time" that converts time into money. This β was a derived quantity that exists as a real number if the problem is feasible. This means that for any feasible combination of policy variables (ℓ, T_0) there is a corresponding pair (ℓ, β) that produces the same solution. Therefore, a decision-making body can specify either a length and a time-standard (ℓ, T_0) or a length and value of time (ℓ, β) with the knowledge that the space of possible solutions is exactly the same either way. The choice is just a matter of taste. For our purposes, specifying (ℓ, β) is more appealing because it yields an unconstrained optimization problem, and in some cases more easily interpreted results.

Regarding interpretation, note that the truncated Lagrangian (5.21) is a combination of average agency costs and worst-case passenger costs. In our particular example, this combination does not have a neat physical interpretation. However, if we had defined standards for the average passenger (using for ℓ the average distance across all passengers) then $T(\ell)$ would have been the average door-to-door time across all passengers; and the truncated Lagrangian $\$+\beta T(\ell)$ would have turned out to be the sum of all generalized costs to society per trip. This means that for average-case analysis the truncated Lagrangian can be interpreted as society's generalized cost. Conversely, minimizing a generalized cost function is always equivalent to solving a standards problem that constrains the door-to-door travel time of the average passenger.[14]

5.2.2.2. *Nature of the Solution*

Let us now look for insights. We assume here that the policy-making body is interested in minimizing the average-case generalized cost per trip for a given value of time, β, i.e., in evaluating performance considering all users

[13] The reader may verify as an exercise that the expected bus occupancy formula is $\lambda H\bar{\ell}$, where $\bar{\ell}$ is the average customer trip length. (*Hint*: use Little's formula of Chapter 2, and treat the bus as the "waiting area.") In real world applications, a small buffer may also be considered on account of stochastic fluctuations.

[14] The particular problem uses as the standard the value T_0 that corresponds to the chosen β.

equally. Accordingly, ℓ is now the average trip length across the population of travelers. We also assume that we are given the value of λ. One way of thinking intuitively about these three parameters is that they indicate something about the city in which the corridor lies: λ correlates with the population density; β correlates with the city's wealth; and ℓ correlates with its size. The analysis that follows shall show how the optimum design and the various components of its generalized cost depend on these demand-side parameters.

We use the same setup as before, except we now focus on the average passenger and assume that passengers do not know the schedule and have appointments at the destination. Clearly then, the agency cost \$ must stay the same so the expression for the cost per trip (5.19) still holds. To keep the formulas short, we shall focus on bus systems operated in mixed traffic, and rightfully neglect the cost of the guideway and the stations by using $c_g = c_r = 0$.

Now consider the average travel time T across all passengers. We see from (5.16) that the average access time at the origin must be $s/(4v_a)$. Therefore, the total access time considering both ends of the trip is $s/(2v_a)$. Next, since our passengers only wait at the origin, the average total wait should be $H/2$ as per (5.17). Finally, since the expression for the average in-vehicle travel time is the same as before, we see that T continues to be given by the right side of (5.20); except now the last two terms include a coefficient of ½.

Now, combining the expressions for \$ and T, we obtain our generalized cost per trip:

$$z = \$ + \beta T = \frac{c_s}{\lambda s H} + \left(\frac{c_d}{\lambda H} + \beta\frac{H}{2}\right) + \left(\frac{\beta \ell t_s}{s} + \frac{\beta s}{2v_a}\right) + \left(\frac{\ell\beta}{v_{\max}}\right). \tag{5.22}$$

The first two terms of this expression constitute the agency cost (\$) and the remaining terms show the user cost (βT). In units of time (i.e., excluding the multiplier β), the third term captures the waiting delay (WD) while the passenger is still; the fourth and fifth the delay while moving (MD); and the sixth the minimum time required to overcome distance, which we call the "line-haul" (LH) time.

Although the minimum of (5.22) does not have a neat closed form, we can find a simple expression for a lower bound, and also for an upper bound that is fairly tight. Since all the terms in (5.22) are positive, a lower bound is obtained by omitting the equation's first term, which represents the relatively small cost of stopping a vehicle, and then minimizing the truncated expression that results:

$$\left(\frac{c_d}{\lambda H}+\beta\frac{H}{2}\right)+\left(\frac{\beta\ell t_s}{s}+\frac{\beta s}{2v_a}\right)+\left(\frac{\ell\beta}{v_{\max}}\right). \tag{5.23}$$

This yields a closed form solution because the first two parenthetical terms are economic order quantity (EOQ) expressions that can be optimized separately. The result is

$$H^* \cong \left(\frac{2c_d}{\lambda\beta}\right)^{\frac{1}{2}}; \quad s^* \cong (2v_a t_s \ell)^{\frac{1}{2}}. \tag{5.24}$$

The lower bound to (5.22) is obtained by inserting this result into the truncated objective function (5.23). And, since (5.24) is a (very good) feasible solution of (5.22), as we have not changed any constraints, an upper bound is obtained by inserting (5.24) into (5.22). (Homework exercise 5.3 asks you to prove the correctness of these two assertions in a more generic setting.) The reader can verify that the result of these insertions is $z^* = \$^* + \beta T^*$, where

$$\$^* = \left(\frac{\beta c_d}{2\lambda}\right)^{\frac{1}{2}} + \left\{\begin{matrix} 0 \\ \frac{c_s}{2}\left(\frac{\beta}{\lambda c_d}\right)^{\frac{1}{2}}(v_a t_s \ell)^{-\frac{1}{2}} \end{matrix}\right\} \begin{matrix} \textit{lower bound} \\ \textit{upper bound} \end{matrix}, \quad \text{and} \tag{5.25a}$$

$$T^* = (WD)^* + (MD)^* + (LH)^* = \left(\frac{c_d}{2\lambda\beta}\right)^{\frac{1}{2}} + \left(\frac{2t_s\ell}{v_a}\right)^{\frac{1}{2}} + \frac{\ell}{v_{\max}}. \tag{5.25b}$$

Expressions (5.24) and (5.25) show at a glance how the system design and the basic components of the generalized cost depend on the demand parameters (λ, β, ℓ). They illustrate for example that, independent of trip length, headways should cut in half if the demand quadruples; and that independent of demand, the station spacing should double if the trip length quadruples.

Figure 5.10 plots all these measures of performance vs. β, and shows how the elements of T depend on the other two parameters. Recall that β is a proxy for the wealth of the city in which the corridor lies. Not surprisingly, the figure shows that wealthier cities should spend more money on transit service and experience less WD. And perhaps less obviously that the moving delay (MD) is unaffected by wealth.

Since the standards approach and the Lagrangian approach are equivalent, the figure can also be used to solve cost minimization problems like (5.18) where the standard pertains to the average passenger rather than the unluckiest. To see the connection between the two problem formulations,

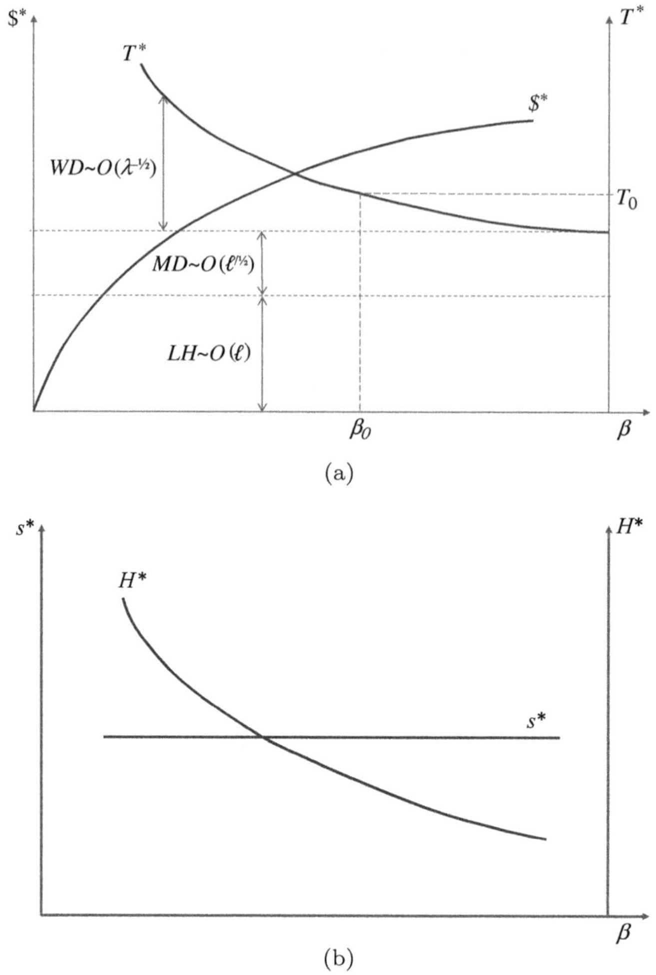

Figure 5.10. Impacts of transit passenger wealth on (a) travel time and cost and (b) headway and stop spacing design.

recall that the value of β corresponding to a given standard, T_0, is the one that produces a travel time that matches the standard, i.e., the value β_0 shown in the figure as the abscissa of curve T^* for the standard's ordinate T_0. Therefore, to solve a standards problem with the results just derived, we can simply evaluate all the measures of performance for this abscissa.

5.3. Multiple Standards

Here we consider a generalized version of (5.18) with multiple constraints, each specifying a separate standard for each individual trip length. This makes practical sense as it gives policy-making bodies some additional flexibility in stipulating what should be done. On the other hand, since specifying multiple standards is a much more complex task which is mostly based on judgment, simplifications should be considered. This will be explored in this section. We will find that that every multi-standard problem has an equivalent single-standard problem for a unique trip length that produces approximately the same design. Therefore, decision-making bodies need only specify a time standard for a single reasonable trip length, e.g., the trip length that is expected to produce the largest number of passenger-kilometers. Furthermore, since single-standard problems have an equivalent generalized cost version, as we have just seen, the decision-making body may choose instead to specify a trip length and a value of time parameter. In either case, analysis can proceed with the tools of the previous section.

The bulk of this section is devoted to developing a method to solve multi-standard problems, which is then used to establish the equivalent result just mentioned. Somewhat technical, the section can be skipped without much loss of continuity since it is only used in Section 5.4.1, and the affected passages can also be skipped.

In a multi-standard problem, the single-standard T_0 of (5.18) is replaced by an arbitrary non-decreasing function $T_0(\ell)$ such as the one of Figure 5.11, which returns a time limit for each trip length. To enforce the new standards, the single constraint in (5.18) must be replaced by the set $\{T(\ell) \le T_0(\ell), \forall \ell\}$, and the problem becomes:

$$\min_{(s,H)}\{\$: T(\ell) \le T_0(\ell), \forall \ell\}; \text{(agency cost per trip)} \quad (5.26a)$$

$$\min_{(s,H)}\{\lambda\$: T(\ell) \le T_0(\ell), \forall \ell\}.$$

$$\text{(cost per unit time and unit distance of corridor)} \quad (5.26b)$$

The theory that follows is developed assuming that the standards are applied (as in Section 5.2.2.1) to a worst-case passenger without appointments who does not know the schedule, but as the reader can easily verify, the steps and the final result are the same for other passenger types, including the average. For reasons of clarity, it will be convenient to express explicitly how the objective function and the constraints of this problem depend on the decision variables. Thus, we shall write the constraints in full, and use $\$(s, H)$ instead of $\$$ for the agency cost per trip. With this notation, (5.26a) is

$$\min_{s,H}\{\$(s,H)\} \quad (5.27a)$$

$$\text{s.t. } H + \frac{s}{v_a} + \frac{\ell}{s}t_s + \frac{\ell}{v_{\max}} \le T_0(\ell); \ \forall \ell \quad (5.27b)$$

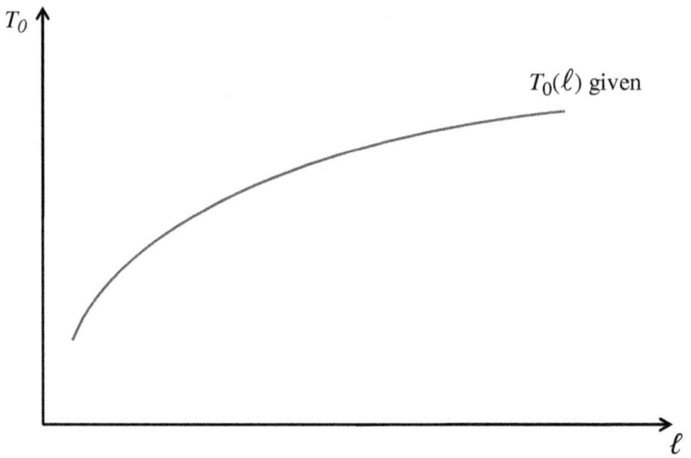

Figure 5.11. Multiple standards.

where

$$\$(s,H) = \frac{c_s}{\lambda s H} + \frac{c_d}{\lambda H} + \frac{c_r}{\lambda s} + \frac{c_g}{\lambda} \approx \frac{c_d}{\lambda H}. \tag{5.28}$$

Recall that the last approximation is particularly good for buses operated in mixed traffic.

This constrained optimization problem can be transformed into its Lagrangian and solved numerically by any nonlinear optimization algorithm. Because it includes only two decision variables, it can also be solved approximately by enumeration, e.g., by evaluating the objective function over a fine rectangular lattice of (s, H) combinations, assigning an infinite objective value to any combination that violates (5.27b), and choosing the combination with the smallest calculated objective.

However, if we adopt the approximate objective function in (5.28) then the solution can also be found exactly, both analytically and graphically. We write the problem as follows:

$$\min_{s,H} \left\{ \$ = \frac{c_d}{\lambda H} \right\} \tag{5.29a}$$

$$\text{s.t. } H + T_M(\ell|s) \leq T_0(\ell), \forall \ell, \tag{5.29b}$$

where

$$T_M(\ell|s) = \frac{s}{v_a} + \frac{\ell}{s} t_s + \frac{\ell}{v_{\max}}. \tag{5.30}$$

The function $T_M(\ell|s)$ has been introduced to stress that the left side of (5.29b) separates into a part that depends only on H and a part that depends only on s. Note that $T_M(\ell|s)$ increases linearly with ℓ for every s. The two

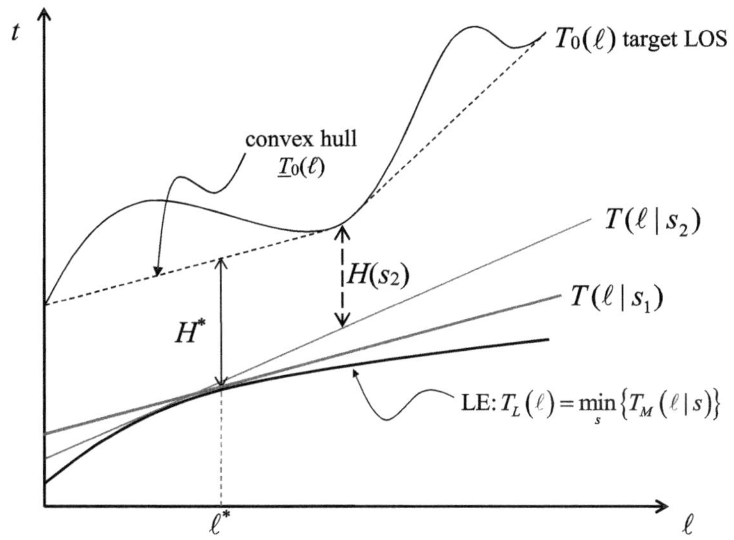

Figure 5.12. Graphical solution of the multi-standard problem.

straight lines of Figure 5.12 depict two instances of these functions. Moreover, since $T_M(\ell|s)$ expresses a passenger's moving time, these straight lines will be called "MT-lines."

Let us now solve (5.29). Since its objective function is both, strictly decreasing in H and independent of s, the optimum is achieved by the maximum value of H that satisfies (5.29b) for some s. This allows us to find the maximum with conditional decomposition in two easy steps.

First, we maximize H conditional on s and denote the result by $H(s)$. We see from (5.29b) that

$$H(s) = \min_{\forall \ell} \left[T_0(\ell) - T(\ell|s)\right]. \tag{5.31}$$

And second, we maximize $H(s)$ to find the unconditional maximum, H^*, i.e., we solve

$$H^* = \max_{s \geq 0} H(s). \tag{5.32}$$

It is now shown that the two steps can be performed graphically. Refer to Figure 5.12 and note from (5.31) that $H(s)$ is the minimum vertical distance between the travel time standard $T_0(\ell)$ and the straight MT-line corresponding to $T_M(\ell|s)$, i.e., $H(s)$ is the maximum vertical displacement of the straight line until it becomes tangent from below to the $T_0(\ell)$ curve. This is illustrated for $s = s_2$. Note as well that a displaced straight line cannot possibly touch $T_0(\ell)$ in an upward bulge; thus, the same displacement is obtained if we replace $T_0(\ell)$ by its convex hull $\underline{T}_0(\ell)$ (see Figure 5.12).[15] In other words, we can perform the first step by solving:

$$H(s) = \min_{\forall \ell} \left[\underline{T}_0(\ell) - T_M(\ell|s)\right]. \tag{5.33}$$

As the second step, we look for the s that yields the maximum displacement. We shall obtain the answer graphically, using the lower envelope (LE) of the MT-lines, which we denote $T_L(\ell)$ and define as follows (see Figure 5.12):

$$T_L(\ell) = \min_s \{T_M(\ell|s)\} = \frac{\ell}{v_{\max}} + 2\sqrt{\frac{\ell t_s}{v_a}}. \tag{5.34}$$

The last equality holds because $T_M(\ell|s)$ is an EOQ equation. Note it is concave in ℓ. The EOQ structure also implies that the equality is achieved for the spacing:

$$s(\ell) = \arg\min_s \{T_M(\ell|s)\} = \sqrt{\ell t_s v_a}. \tag{5.35}$$

We claim that H^* is the vertical displacement of the LE when it hits the convex hull for the first time (at some length ℓ^* where the two curves would be tangent to each other as shown in the figure), and that $s^* = s(\ell^*)$ is the unique optimum spacing.

To see this, let H' be the displacement of the LE. Then, note that the MT-line that is tangent to the LE at ℓ^*, i.e., corresponding to spacing $s(\ell^*)$, does not protrude above the convex hull after being displaced. (This is true because the line must also be tangent to the convex hull, which by virtue of convexity must remain above the tangent). It follows that the pair $\{H', s(\ell^*)\}$ is a feasible solution of (5.29). Finally, note that a displacement larger than H' is infeasible because then all tangents to the LE protrude above the convex hull at ℓ^*. Therefore, we can conclude that $\{H', s(\ell^*)\}$ is an optimum of (5.29), as claimed. Furthermore, because the LE is strictly concave, only the tangent at ℓ^* achieves this feat; all other spacings are infeasible. Therefore, the optimum is unique.

Summary of the LE Method: The specific recipe is the following: (i) construct curve $T_L(\ell)$ as per (5.34), as well as the convex hull of the time standard, $\underline{T}_0(\ell)$; (ii) slide $T_L(\ell)$ up until it touches (and is tangent to) the convex hull $\underline{T}_0(\ell)$; (iii) the resulting vertical displacement is the optimum headway H^*; (iv) the abscissa at the point of contact is ℓ^*; and (v) enter with ℓ^* in (5.35)

[15]The convex hull, shown by the dashed line in Figure 5.12, is the highest convex curve that can be drawn without exceeding $T_0(\ell)$.

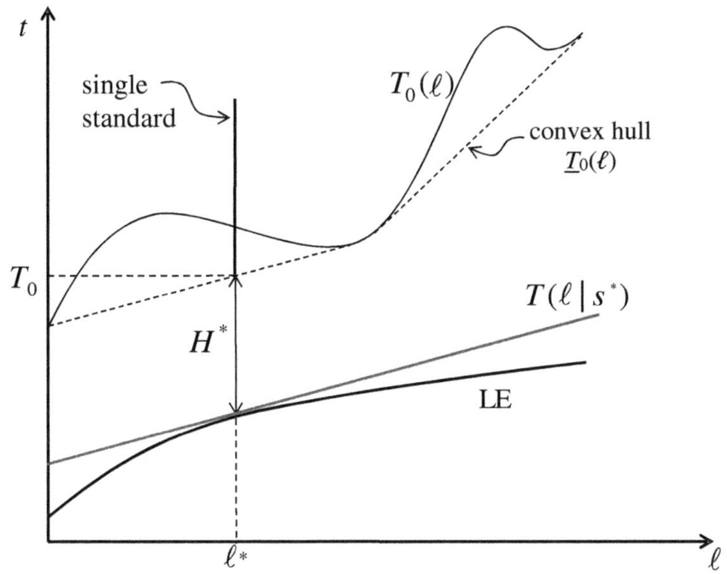

Figure 5.13. Simplification of length-based standards.

to obtain the optimum spacing, which is $s^* = \sqrt{\ell^* t_s v_a}$. The procedure can be carried out analytically in simple cases (see homework exercise 5.5).

Caveat: This recipe is approximate because we left out some terms in the original objective function (5.28). So the analysis above actually gives us a lower bound of optimal cost. To get an upper bound, we can plug H^* and s^* into (5.28). This is very similar to the discussion in Section 5.2.2.2. In cases where the station costs are comparatively small (e.g., when dealing with buses), the two bounds should be close. If this happens, the solution is near-optimum and no further analysis is necessary.

Existence of an Equivalent Single Standard: We claim that the single-standard problem defined by a limit on travel duration $T_0 \equiv T_0(\ell^*)$ for trips of length ℓ^* achieves the same optimum design. To see this, model this standard by a $T_0(\ell)$ curve that takes the value T_0 for $\ell = \ell^*$ and infinity for $\ell \neq \ell^*$. This curve is depicted in Figure 5.13 by means of a vertical dark line that reaches down to ordinate T_0.

Now apply the five step method described in the summary with this curve, and note that the LE stays the same; none of the steps change and we obtain the same H^*, ℓ^*, and s^*. Thus, the exact same solution is obtained if we use multiple or single standards, provided, of course, we know the single standard to apply.

Finally, as we pointed out at the outset, the logic leading to this result can be repeated with only trivial changes if the standards are defined for an average passenger rather than the unluckiest. Thus, the single-standard simplification also applies to standards based on averages.

5.4. Multiple Lines: Hierarchical Systems

This section extends the ideas to multi-line corridors where interline transfers are possible. Section 5.4.1 focuses on analysis methods and Section 5.4.2 on the results. We will see that the methods stay pretty much the same; the only thing that changes are the formulas that go into the objective function, the constraints, and the Lagrangian.

5.4.1. *Formulation and Analysis*

There are many ways in which multi-line services can be organized. We have already mentioned the possibility of organizing them in a hierarchical way using express and local lines that provide homogeneous service all along the corridor (see Figure 5.4). This, however, is not the only possibility. For example, one can design lines in a non-hierarchical way such that each line provides both, local service (serving all stations) in some sectors of the corridor, and express service (skipping minor stations) in the rest (see Figure 5.14). To allow for transfers, all lines would stop at all major stations. Sectors would contain the minor stations in between and would be allocated to different lines to make sure that all minor stations receive service. This is shown in the figure: the three dark lines are the transit lines; the vertical dashes intersecting these lines are the minor stations they serve; and the circles the major transfer stations. Circles connected by two vertical lines designate the same station; the vertical lines signify transfer possibilities. Note how one can travel between any two stations in this corridor using only one transfer.

The type of service in Figure 5.14 is appealing, and sometimes used for rail systems such as subways where vehicles cannot pass and hierarchical systems are infeasible. Variants, sometimes called "skip-stop service," have been used in the New York City Subway, Philadelphia SEPTA's elevated line, Chicago CTA's elevated line (until the 1990s), and the Santiago (Chile) Metro; see Giesen and Munoz (2013) for an analysis of a particular variant. The main advantage of stop-skipping is that the reduced number of stops in the express service sectors increases the average commercial speed. As a result, it makes the most sense for long corridors. This is perhaps why it

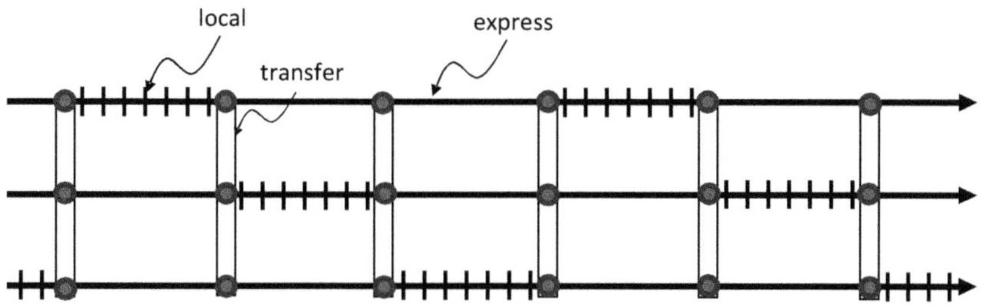

Figure 5.14. An example of inhomogeneous service lines along a corridor.

is mostly seen in large metropolitan areas, where passenger travel distances are long.

However, the system also has disadvantages. One is that there is an upper limit to the frequency that can be provided due to vehicles' inability to pass and the different average speeds at which they travel in their local and express sections. The reader can verify this phenomenon with a time–space diagram similar to the example in Section 2.1.3. A second disadvantage inherent in the lines' inhomogeneity, even if vehicles can pass, is that the headways cannot be even at every location. The reader is encouraged to quantify these two effects with formulas for the maximum frequency and for the variance of the headways measured across all stops, using as inputs the sector sizes and their internal commercial speeds (see homework exercise 5.6). Building on these results, formulas can then be developed for the average door-to-door travel time and the agency cost.

A modified version of sector-based service is also used in practice when one end of the demand is very concentrated at a few destinations (or origins), e.g., as happens during the commute hours to/from a central business district (CBD), and the other end is spatially dispersed in one or two dimensions. In these cases, it makes sense for every line to provide full local service to the segment containing the concentrated end of the demand, serving locally just another segment at the dispersed end and bypassing everything in between. This type of system eliminates transfers and increases the commercial speed. The strategy is often used by commuter buses. As in the previous case, formulas for average door-to-door travel time and agency cost in terms of the sector sizes and other design parameters can be obtained.

The analysis methods we described in Sections 5.2 and 5.3 can be applied to any type of service, including the ones we have just described. To do this, one should first choose a few design variables to identify the designs, e.g., the station spacings, the headways, and the length of the sectors in Figure 5.14; and then use physical modeling to derive formulas for the objective function and constraints of a relevant optimization problem, e.g., (5.26). For space and time reasons, we shall do this in detail only once. We have chosen the case with dispersed demand at both ends, with passing possible, because the scenario is quite common, e.g., when local buses are used as feeders, either to an express bus or a rail system. For this scenario, the express/local strategy of Figure 5.4 makes sense since it allows for regular headways and can be implemented with only two lines. Therefore, this is the focus now.

Extra wait at transfers can be eliminated completely if both lines operate with the same headway H and if service is synchronized. If service is not synchronized there is some non-zero extra wait, which depends on three particulars of the situation: whether passengers know the schedule or not; whether they have important appointments at the destination or not; and whether we are interested in the average passenger or the unluckiest. Table 5.1 summarizes the resulting extra wait for all these cases, i.e., the wait that transfers add over and above that of Table 4.1.

The results in the table should not be surprising, especially for synchronized service. For the no-appointment cases, they arise because then the added delay consists exclusively of the wait at the transfer station, which does not depend on whether passengers know the schedule or not. Obviously, it is $H/2$ for the average passenger and H for the unluckiest, as shown on the table. The results for important appointments depend on what passengers know. If they do not know the schedule, then everyone (including the average passenger) must assume a worst case in planning their trip, including one headway of wait at the origin and another at the transfer point. Thus, the transfer adds exactly H time units to both the average and unluckiest passengers, as shown in the table. If passengers know the schedule, then they would take the latest bus that gets them to their appointment on time. As a result, they would only wait at the transfer point and at the destination. And of course, the wait at the transfer station is $H/2$ on average and H in the worst case, also as shown on the table.

It turns out that unsynchronized service is much simpler and cheaper to provide, and perhaps, for this reason, it is more prevalent in practice. Figure 5.15 illustrates the issue. The figure displays the trajectories of both express buses (dark) and feeders (light) over a section of corridor between two express stations during a period slightly longer than H. The shaded portions are a repeating analysis tile. The figure assumes that service is bidirectional so bus trajectories run both up and down. Part (a) displays the unsynchronized case and part (b) displays the synchronized one.

Table 5.1. Extra waiting time per passenger due to a transfer.

Service type	Passenger type		Extra waiting time Average	Worst-case
Synchronized	All types		0	0
Unsynchronized	No appointment at the destination	Schedule unknown	$H/2$	H
		Schedule known	$H/2$	H
	Appointment at destination	Schedule unknown	H	H
		Schedule known	$H/2$	H

Part (a) is easy to understand. Since the feeder and express lines operate with the same headway the tile includes one trajectory for each line in each direction. Thus, the express buses cover s_1 distance units in the tile in each direction, as do the feeder buses. Note that the feeder buses enter and exit the tile some time away from the express buses; this separation is the extra passenger wait due to transfers that are quantified in Table 5.1. Logically we assume that the feeder trajectories continue on both sides of the tile so they can provide direct local service for short passenger trips that straddle one or two tiles. Also, note that transferring passengers can choose their transfer station and local path. For example, an inbound passenger with a destination somewhere in our tile traveling upward in the first express bus — the solid dark line on the left — could alight at either of the two stops shown in the figure and then take the first local bus that goes toward his/her destination. Outbound passengers have a similar choice. This means that all local buses both collect and distribute passengers. One final advantage of unsynchronized service is that, if desired, one can use different headways for local and express service.

Now consider part (b). Because service is synchronized, a feeder that collects passengers must be scheduled to connect with every express departure. The two light lines on the right of our figure labeled "collect" are examples. Likewise, a distributing feeder must be scheduled after each express arrival. These are exemplified by the two light lines on the left labeled "distribute." This can be achieved by restricting all feeder buses to travel back and forth between two express stations, alternatively collecting and distributing passengers. Although this is relatively simple (see Figure 5.15) it requires the insertion of unproductive dead times where the buses merely sit. In addition, our analysis tile now includes four local trips instead of two. Thus, synchronization more than doubles the cost of local service. One final minor additional disadvantage is that feeder buses can no longer provide

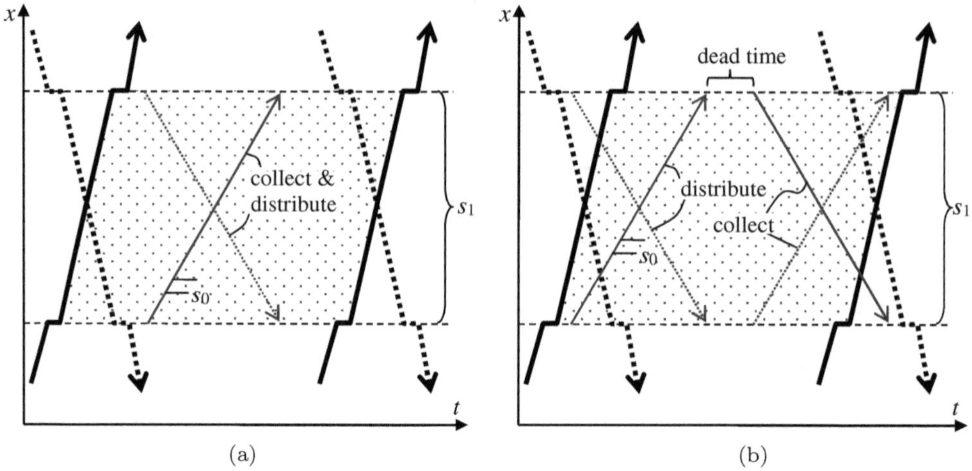

Figure 5.15. Local service to express transit: (a) unsynchronized and (b) synchronized.

direct local service for short passenger trips that straddle an express station. For all these reasons, synchronized hierarchical service is rarely seen in practice.

In the view of all this, we will assume that the service is unsynchronized and bidirectional. Accordingly, we will also assume that the demand density λ is defined for both directions combined. For simplicity, we will also assume that the same headway H is used for the two lines in both directions, and that passengers choose the express station that minimizes local travel. (The derivations turn out to be only slightly more involved if these assumptions are changed; therefore, the interested reader is encouraged to explore them in parallel as the explanation proceeds; different headways are examined in Section 5.5.1.)

Logically, we also assume that local buses stop at the express stations to facilitate transfers. Accordingly, we will force the express line spacing, s_1, to be an integer multiple of the local spacing, s_0, as in Figure 5.4. Furthermore, for generality purposes, our analysis shall use the multi-standard approach (5.26) — since the single-standard problem is a special case. And for the sake of variety, standards shall be defined for an average passenger who does not have appointments and does not know the schedule. The derivations for other passenger types are left to the reader; the results are virtually identical.

Let us now consider the agency costs. As in earlier parts of this chapter, we ignore infrastructure costs. Thus, as explained in Section 5.2.2, agency costs are approximately captured by a combination of the vehicle distance traveled at rate c_d, and the vehicle time stopped at rate c_t. Now, since four vehicles travel in the tile, a total of $4s_1$ kilometers in both directions, and since the number of person-trips generated in the tile in both directions is $\lambda H s_1$, we see that the approximate cost per trip due to distance is

$$\$_{\text{dist}} = \frac{4c_d s_1}{\lambda H s_1}. \tag{5.36}$$

To this, we must add the cost due to time stopped. In one tile, this is the product of c_t and the sum of the times at the stops, which is $2(1 + s_1/s_0)t_s$. Thus, we see that the agency cost per trip including (5.36), which we denote $\$\,(s_0, s_1, H)$, is

$$\$\,(s_0, s_1, H) = \frac{1}{\lambda s_1 H}\left\{4c_d s_1 + 2\left(1 + \frac{s_1}{s_0}\right) t_s c_t\right\}. \tag{5.37}$$

The reader may try to derive a corresponding expression assuming that the local and express headways are different ($H_0 \neq H_1$) and another expression assuming that service is synchronized as in Figure 5.15(b). Homework exercise 5.13 addresses these issues.

Let us now consider the average passenger travel time. Since service is unsynchronized and passengers have no appointments, passengers will wait on average $H/2$ time units at the transfer station and $H/2$ at the origin (see Tables 4.1 and 5.1).[16] To this, we must add the time spent moving, which we denote by $T_M(\ell|s_0, s_1)$ and will be the sum of an average access time, $s_0/(2v_w)$, a transfer penalty time Δ, and an average in-vehicle travel time. Since we are assuming that passengers always transfer at the station closest to their trip ends, the average in-vehicle travel time must equal the sum of the times it takes to travel two-fourths of the express station spacing in local vehicles, and ℓ distance units in an express vehicle. The average total moving time is therefore,

$$T_M(\ell|s_0, s_1) = \Delta + \frac{s_0}{2v_w} + \frac{s_1}{2}\left(\frac{t_s}{s_0} + \frac{1}{v_{\max}}\right) + \ell\left(\frac{t_s}{s_1} + \frac{1}{v_{\max}}\right), \quad \text{where } \ell > s_1 > s_0. \tag{5.38}$$

The notation $T_M(\ell|s_0, s_1)$ is introduced to resemble the notation of Section 5.3. (As in that section, the notation stresses that the moving time

[16] The expressions for other passenger types would only differ in the coefficient of H. If the local and express headways are different and unsynchronized then the waiting time expression is the sum of two terms, one for each headway. The synchronized case can also be expressed with a simple formula, but the result is not additive.

only depends on the spatial decision variables.) The expression on the right-hand side of (5.38) needs little explanation: the first term is the time during transfers; the second is the average walking time at both ends of the trip; the third is the average moving time in local vehicles; and the fourth is the average moving time in express vehicles. Recall that the quantities in parentheses are the commercial paces of the local and express buses.

To this, we must add the average total wait at both the origin and the transfer point, H. Thus, the average-case door-to-door time is

$$T(\ell) = H + T_M(\ell|s_0, s_1). \tag{5.39}$$

Our version of problem (5.27) can now be written as

$$\min_{s_0, s_1, H} \$\left(s_0, s_1, H\right) \geq \frac{4c_d}{\lambda H} \tag{5.40a}$$

$$\text{s.t. } H + T_M(\ell|s_0, s_1) \leq T_0(\ell), \forall \ell. \tag{5.40b}$$

Because the problem only involves three decisions variables, it can be solved approximately by enumeration on a lattice, with due consideration of the constraint, as we suggested earlier. And if we have reason to believe that the lower bound on the right-hand side of (5.40a) is a good approximation, then also by the five-step LE (lower envelope) method of Section 5.3, as we now show.

The only complication is that $T_M(\ell|s_0, s_1)$ is now a two-parameter family of straight lines. Therefore, the lower envelope found in step (i) and used in succeeding steps is now found with the following two-variable minimization instead of (5.34) and (5.35):

$$T_L(\ell) = \min_{s_0, s_1} \{T_M(\ell|s_0, s_1)\} \tag{5.41a}$$

$$[s_0(\ell), s_1(\ell)] = \arg\min_{s_0, s_1} \{T_M(\ell|s_0, s_1)\}. \tag{5.41b}$$

Because (5.38) has a simple form, the solution of (5.41) can be expressed in closed form, albeit only approximately. To find the solution, we use the two-step conditional optimization method of Example 2.2. As a first step, we minimize (5.38) with respect to s_0 conditional on s_1, and plug the result into $T_M(\ell|s_0, s_1)$. We obtain

$$s_0^* = \sqrt{s_1 t_s v_w}, \text{ and} \tag{5.42a}$$

$$T_M^*\left(\ell|s_1\right) = \Delta + \tfrac{\ell}{v_{\max}} + \sqrt{\tfrac{s_1 t_s}{v_w}} + \tfrac{s_1}{v_{\max}} + \tfrac{\ell t_s}{s_1}. \tag{5.42b}$$

We now minimize the new objective function (5.42b) with respect to s_1. Since it is a posinomial, its optimum is given by the first order conditions.

Unfortunately, these yield a quartic equation with a closed but ugly solution. Therefore, we look for an approximation. We observe that the $s_1/v_{\max}$ term is typically much smaller than $\sqrt{s_1 t_s / v_w}$, so we ignore the small term and proceed to find the optimal s_1. Note the reduced expression has a GEOQ form, such that its solution is

$$s_1^* \cong (4 t_s \ell^2 v_w)^{\frac{1}{3}}. \tag{5.43}$$

To finish, we now plug this result into (5.42a) and (5.42b) to obtain

$$s_0^* \approx (2 t_s^2 \ell v_w^2)^{\frac{1}{3}}, \text{ and} \tag{5.44}$$

$$T_L(\ell) \approx \Delta + \frac{\ell}{v_{\max}} + 3\left(\frac{t_s^2 \ell}{4 v_w}\right)^{\frac{1}{3}} + \frac{1}{v_{\max}} (4 t_s \ell^2 v_w)^{\frac{1}{3}}. \tag{5.45}$$

Note that, as expected, the lower envelope is concave. Thus, steps (ii) through (v) of the LE method in Section 5.3.1 can now be easily completed for any given $T_0(\ell)$.

The Special Case of a Single Standard: The reader can verify that when the LE method is applied to the special case of a single standard T_0 for a given trip length, ℓ, the optimum has a closed form where the spacings are given by (5.43) and (5.44), and the headway by

$$H^* = T_0 - T_L(\ell), \tag{5.46}$$

where $T_L(\ell)$ is given by (5.45). The upper and lower bounds to the objective function are then obtained by plugging these values into the objective function (5.40a), and its lower bound.

Equivalence of the Three Formulations: As occurred with a single line, every multi-standard problem can still be recast as a single-standard problem or a Lagrangian problem. Clearly, it follows from the single-standard recipe we just described that if ℓ^* and $T_0 \equiv T_0(\ell^*)$ are the binding length and time standard of the multi-standard problem, then the single standard $\{\ell = \ell^*$ and $T_0 \equiv T_0(\ell^*)\}$ achieves the same optimum. This result also holds if standards are defined for the average passenger rather than the unluckiest, as we hinted at the outset. And, as occurred before, one can always form a truncated Lagrangian that will solve the single-standard problem for both the worst and average cases. Therefore, the three formulations (multi-standard, single standard and Lagrangian) can be used interchangeably. As always, if the value of time is given, the Lagrangian formulation morphs into an unconstrained minimization of a truncated Lagrangian. For average case analysis, this truncated Lagrangian is society's generalized cost. We now use this approach to obtain some insights.

5.4.2. *Results and Insights*

If we use (5.37) for the agency cost, the objective function for our average case problem is (in units of cost per trip)

$$z = \left[\frac{4c_d}{\lambda H} + \left(\frac{2}{\lambda s_1 H} + \frac{2}{\lambda s_0 H}\right) t_s c_t\right] + \beta H$$

$$+ \beta\left[\frac{s_0}{2v_w} + \frac{s_1}{2s_0}t_s + \frac{s_1}{2v_{\max}} + \ell t_s \frac{1}{s_1} + \frac{\ell}{v_{\max}}\right]. \quad (5.47)$$

Note that the three demand-side parameters that appear in this expression (λ, β, and ℓ) can vary by orders of magnitude across cities and countries, whereas the others vary much less. Therefore, (λ, β, and ℓ) are the main determinants of optimal system structure. Here, we explore this connection.

To do this in an efficient way, we divide both sides of (5.47) by β and minimize instead the quantity z/β, so the new objective function is

$$\frac{z}{\beta} = \frac{1}{(\lambda\beta)}\left[\frac{4c_d}{H} + \left(\frac{2}{s_1 H} + \frac{2}{s_0 H}\right) t_s c_t\right] + H$$

$$+ \left[\frac{s_0}{2v_w} + \frac{s_1}{2s_0}t_s + \frac{s_1}{2v_{\max}} + \ell t_s \frac{1}{s_1} + \frac{\ell}{v_{\max}}\right]. \quad (5.48)$$

This is a valid thing to do because β is a positive constant.

An advantage of the new equation is that two of our drivers of design, λ and β, now appear together in the formula as the product $(\lambda\beta)$. Thus, the reorganization reveals that the space of optimum designs has only two degrees of freedom, and it can be fully explored by varying the parameters $(\lambda\beta)$ and ℓ.

It is interesting to assign some physical meaning to both the new objective function z/β, which has units of [time], and the parameter $\lambda\beta$, which has units of [passenger-\$/time^2-distance]. To do this for z/β, imagine that the system is not subsidized so people pay as fare their share of the agency cost, and think of β as a proxy for the average local wage. Then our objective function is the total time the average passenger has to invest in making the trip including both, the time spent traveling and the time spent working to pay for the fare. A nice feature of using time as a measure of monetary expenditures is that time is a universal currency that does not change across countries. Therefore, the results expressed in time units can be easily understood by everyone without any mental exertion for currency conversions.

The parameter $\lambda\beta$ is not as neat. It can be thought of as the "wage generation rate per unit time and distance" because λ is the trip generation rate and β is a proxy for the wage. The fact that demand and wealth can be combined into a single driver means that low-density wealthy neighborhoods in developed countries and poor dense neighborhoods in developing countries should have approximately the same system structure. Moreover, the two scenarios should also share the time-based generalized cost z/β. Isn't it nice that we can make these observations even before the system is optimized?[17]

Because (5.48) does not have a simple closed form solution we solved it numerically for a broad set of $(\lambda\beta, \ell)$ combinations and a reasonable selection of the remaining parameters. We chose the following:

$$\begin{aligned} v_w &\cong 3\,\text{km/h} \\ v_{\max} &\cong 36\,\text{km/h} \\ t_s &\cong 5 \times 10^{-3}\,\text{h} \\ c_d &\cong 1\,\$/\text{km} \\ c_s &\cong 10^{-1}\$/\text{stop} \\ c_t &\cong 20\$/\text{h} \end{aligned}$$

and explored the space of solutions for $\ell \in [3,\ 50]$ km and $\lambda\beta \in \{20, 50, 200, 1000\}\$/\text{km-h}$. Allowing for interpolation, the results exhaustively cover the space of possibilities.

Figure 5.16 displays the optimum cost results in universal units. Similar plots can be drawn for the optimum designs. To make the figure more transparent, the free-flow travel time, $\ell/v_{\max}$, which is a constant, is subtracted from the objective function. This modified objective function $(z^*/\beta - \ell/v_{\max})$ can be interpreted as a generalized form of delay combining the three-W's: walking, waiting (both in and out of the bus), and working to pay for the trip. The figure plots this generalized delay vs. ℓ for the four selected values of $\lambda\beta$. Note from the figure how the generalized delay curves are roughly parallel, slowly increasing and almost linear. And most importantly, how they decline with $\lambda\beta$.

The diagram also includes three dashed lines expressing the generalized delay for individual travel by three classes of people: wealthy, moderate (whose alternative is driving a car) and poor (who can only walk). The generalized delay is $c_d^{'}\ell/\beta$ for drivers, where we have used $c_d^{'} = 0.2\$/\text{km}$

[17]Similar ideas were introduced in Section 2.4 in connection with dimensional analysis.

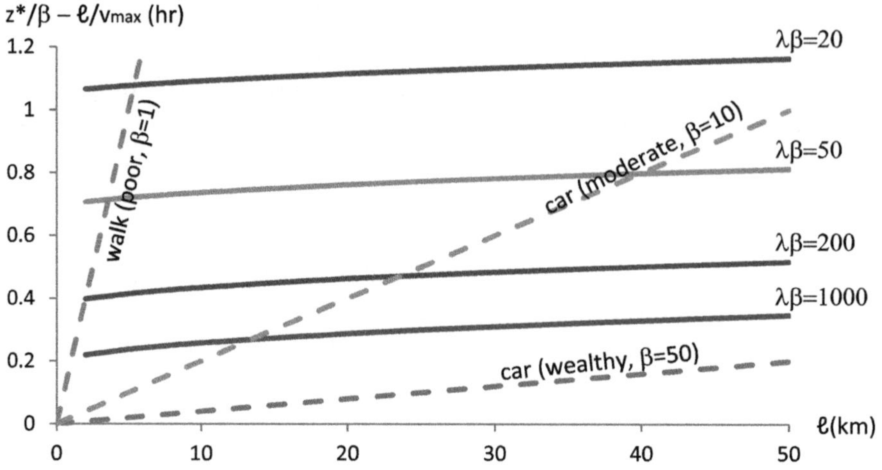

Figure 5.16. Impact of passenger wealth and demand density on transit cost and mode choice.

to reflect the lower operating cost of cars compared with buses, and $\ell[1/v_w - 1/v_{\max}] \approx 0.31\ell$ for walkers. Assuming reasonably that people choose the mode with the least generalized delay, we see from the figure that the poor walk for short trips (say comparable with a kilometer) and take transit for longer trips. Furthermore, the poorer they are and/or the lower the transit demand (both reflected by transit curves with lower $\lambda\beta$) the longer their walks. The wealthy and moderate folks also travel individually for short trips but prefer collective transportation for (sufficiently) long trips. The propensity to take transit for longer trips across all economic classes makes sense since the longer a trip is the more incentive there is for people to experience some wait and some walking in compensation for lower out-of-pocket costs — this is why people take trains rather than drive for long-distance intercity travel. By considering a vertical line for a given trip length (or indirectly, city size), we also see that transit is preferred when the demand is high. This conclusion had already been reached for shuttle systems in Chapter 4 and for idealized corridor systems earlier in this Chapter. What is different now is that the chart is quantitatively accurate for bus transit corridors.

5.5. Extensions

This section extends the ideas of Sections 5.2, 5.3, and 5.4 by showing how service should be adapted to inhomogeneous demand, with due consideration of system capacity. Section 5.5.1 explores directionally asymmetric demand; Section 5.5.2 shows more general forms of demand inhomogeneity; and Section 5.5.3 shows how to model capacity.

5.5.1. *Directional Asymmetry*

As in Section 5.4, we imagine here that buses are operated in both directions of our corridor and that the demand is different in each direction. Now, however, we allow the station spacings and the optimum headways to be different in each direction.

Four scenarios are possible, depending on whether the headways and/or the station spacings are constrained to be equal in both directions. All can be implemented in practice. Asymmetric headways can be provided by turning around only a fraction of the buses having completed service in the more frequent direction, and diverting the rest to serve other nearby lines. (Called "interlining" and examined in Chapter 8, this strategy is commonly used by transit agencies to reduce wasted bus time and the required fleet size.) Asymmetric spacings can be achieved by allowing buses to skip stations in one of the directions.

The strategy to solve problems like this is essentially the same as before and is quite general — for example, it can be applied to corridors with a single line, or with multiple (express and local) lines in one or both directions. In all cases, one should first write a generalized cost function combining the passenger, infrastructure and operating costs in both directions, expressed in terms of the decision variables for each direction. We recommend using units of total cost per unit time and unit distance rather than cost per trip because the costs of the two directions can then be added, but either approach works. In all cases, the generalized cost function turns out to be a posinomial, which can be optimized numerically or even analytically in simple cases.[18]

For illustration purposes, we now revisit Section 5.2.2.2 and briefly discuss the possibility of offering different headways and a common spacing for non-hierarchical bus systems where infrastructure costs can be neglected. We shall build on equation (5.22), which gave the cost per trip for unidirectional service. Here we shall assume that the same value of time β can be used for

[18]The headways and station spacings in the two directions should be constrained to be (sub)-multiples of each other. We recommend, however, solving the relaxed problem without these constraints, and then adjusting the optimal results a little to comply with the constraints. As was explained in Chapter 3, this adjustment increases cost very, very little. For typical problems, the optimality gap is on the order of 1% or less.

both directions, but the demand rate and the average trip length can vary across directions. Using a subscript ($i = 1, 2$) to identify the directions, we characterize the demand by pairs (λ_i, ℓ_i) for each direction. Thus, there are five demand parameters. It will also be convenient to introduce the auxiliary parameters: $\lambda \equiv \lambda_1 + \lambda_2$ for the total demand density for both directions combined, and $\ell \equiv (\lambda_1\ell_1 + \lambda_2\ell_2)/(\lambda_1 + \lambda_2)$ for the average of the two representative trip lengths. We shall continue to use a lower case z for the cost per trip, as in (5.22), and introduce a capital Z for the total generalized cost, $Z = \lambda z$.

So, let us now reformulate (5.22) considering both directions. As was mentioned, the advantage of using total cost as opposed to cost per trip for our application is that total costs can be added across directions, i.e., the bidirectional total cost is the sum of the one-directional costs ($Z = Z_1 + Z_2$) which are already known from (5.22). Specifically, the formula for Z_i is obtained by adding the subscript i to the direction-sensitive variables in (5.22) (λ_i, ℓ_i, and H_i) and multiplying both sides by λ_i. The sum of both instances of this expression is our objective function:

$$Z = \sum_i \left(\frac{c_s}{sH_i}\right) + \sum_i \left(\frac{c_d}{H_i} + \lambda_i \beta \frac{H_i}{2}\right) + \left(\frac{\beta t_s \sum\limits_i \lambda_i \ell_i}{s} + \frac{\beta s \sum\limits_i \lambda_i}{2v_a}\right) + \left(\frac{\beta \sum\limits_i \lambda_i \ell_i}{v_{\max}}\right). \tag{5.49}$$

If we now neglect the first term of this expression as we did in Section 5.2.2.2, and use the notation for total demand and average trip length, we have

$$Z \approx \sum_i \left(\frac{c_d}{H_i} + \lambda_i \beta \frac{H_i}{2}\right) + \left(\frac{\lambda \ell \beta t_s}{s} + \frac{\lambda \beta s}{2v_a}\right) + \left(\frac{\lambda \ell \beta}{v_{\max}}\right). \tag{5.50}$$

Note how this expression separates into additive parts that include only one decision variable in EOQ form: one part includes H_1, another H_2, and a third s. Note as well how the headway parts are of the same functional form as in the one-directional case. Thus, (5.24) still furnishes the optimum headways if we replace λ by λ_i. Moreover, the part of (5.50) that includes s is identical to the corresponding part for (5.22), except for the multiplicative constant λ. Therefore, the optimum spacing continues to be given by (5.23).

The above is, of course, a very simple example, but other scenarios are not much more complicated. As an exercise, the reader can duplicate these steps for hierarchical problems in which s_1 and H are allowed to vary across directions, e.g., using (5.47) as the basic model. Note in doing this that (5.47) already expresses bidirectional costs. The reader will find that this problem also decomposes neatly. Other scenarios can be similarly analyzed, as can other variants of the problem such as those in homework exercise 5.9 and mini-project 1 at the end of this chapter.

5.5.2. *Space- and Time-dependent Service*

The analysis in this chapter has assumed so far that conditions are homogeneous in space and time, at least in each direction, as this prevents the proliferation of parameters and leads to simple interpretable formulas. In reality, though, trip generation rates change with the time-of-day and time-of-week, and also spatially. Thus, any practical recipe must recognize these variations. Here we shall consider a unidirectional corridor[19] of length L, and analyze it over a period of duration W (say, a week) where demand conditions are inhomogeneous but repeat themselves approximately every week. We shall develop the ideas assuming that only one line is used but they can be easily generalized to hierarchical operations.

The main complication is in the description of demand. Whereas before we quantified the demand level with a single pair of parameters (λ, ℓ), we shall now use for maximum generality an arbitrary, time-dependent origin–destination (OD) matrix. This OD-matrix is a three-way table giving the number of trips going from a small section of the corridor to another small section in a small time slice.

We shall assume that for design purposes the corridor has been partitioned into rough sections i of length L_i that will have the same station spacing s_i. Likewise, the time has been partitioned into temporal slices j, of duration W_j that will have the same headway H_j. We further assume that $L_i \gg s_i$ and $W_j \gg H_j$, for all i, j. The sections could represent districts with different population densities, and the slices periods of time with a different character such as nights, rush periods and weekends. One such partition is shown in Figure 5.17. Ideally, the sections and slices should be designed so demand conditions are fairly uniform in each cell but this is not strictly necessary.

[19]Bidirectional service between two termini A $\leftrightarrow$ B can be equivalently represented as unidirectional service A $\rightarrow$ B $\rightarrow$ A$'$ after mirroring the segment B $\rightarrow$ A into B $\rightarrow$ A$'$, e.g., as shown in Figure 2.6 of Chapter 2.

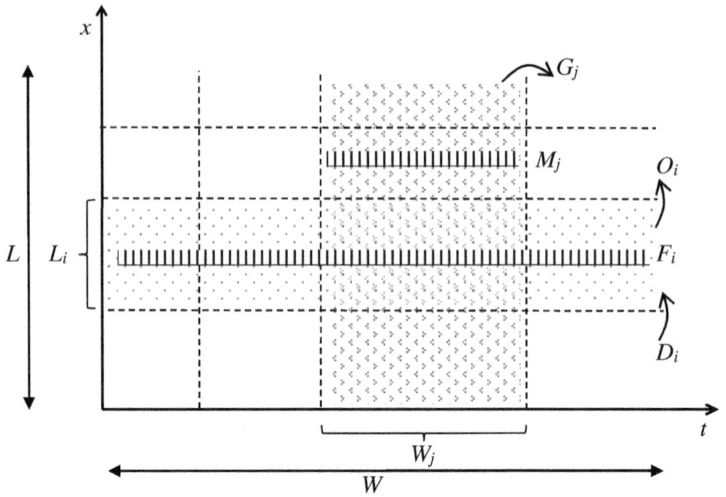

Figure 5.17. Time–space diagram for time- and space-dependent service.

Fortunately, we will soon see that the optimal solution $\{H_j^*,\ s_i^*\}$ can be obtained without full knowledge of the OD-matrix. It turns out that only the following aggregate quantities matter:

(i) the total number of trips generated throughout the corridor in each time slice j, G_j;
(ii) the number of weekly trips originating and terminating in each section i, O_i and D_i;
(iii) the average number of weekly trips that cross a random screen line in each section i, F_i.

Homework exercise 5.12 asks you to show how these aggregate quantities can be derived from any given OD-matrix. These parameters are either directly observable (if the system exists) or can be more easily estimated than a detailed OD-matrix. Furthermore, as we saw in Chapter 3, it turns out that neither the design emerging from the analysis nor its associated generalized cost is very sensitive to errors in the estimation of these parameters, i.e., the following outlined procedure is robust.

The design process is discussed for average-case analysis. It is the same as before: first, write a generalized cost function and then minimize it with respect to the decision variables. The function is understood most readily if it is expressed as a total weekly cost for the complete unidirectional corridor rather than for the average passenger. The formulas below list its various components, using the parameters just defined. An explanation follows:

$$\text{Infrastructure cost: } c_g WL + c_r W\left(\sum_i L_i/s_i\right). \tag{5.51a}$$

$$\text{Bus distance cost: } c_d L\left(\sum_j W_j/H_j\right). \tag{5.51b}$$

$$\text{Bus stopping cost: } c_s\left(\sum_j W_j/H_j\right)\left(\sum_i L_i/s_i\right). \tag{5.51c}$$

$$\text{Passenger waiting cost: } \beta\sum_j G_j H_j/2. \tag{5.51d}$$

$$\text{Passenger walking cost: } \frac{\beta}{v_w}\left(\sum_i \frac{s_i}{4}(O_i + D_i)\right). \tag{5.51e}$$

$$\text{Passenger in-vehicle moving cost: } \beta\sum_i F_i L_i/v_{\max}. \tag{5.51f}$$

$$\text{Passenger in-vehicle stopping cost: } \beta t_s\sum_i F_i(L_i/s_i). \tag{5.51g}$$

Regarding (5.51a) note that the term in parenthesis is the number of stations; also note that the guideway component is fixed, so it plays no role in the optimization. Now consider (5.51b); the term in parenthesis is the total number of headways in a week, and since the complete bus fleet travels a distance L during each headway (every bus advances to the position of the prior bus) the product of the parenthesis and L is the total distance vehicles travel in 1 week. Equation (5.51c) is also logical: the first parentheses is the number of headways, which is the number of bus stops made at any particular station, and the second is the number of stations; thus, their product is the total number of bus stops made. Equation (5.51d) should also be intuitive since the argument in the summation is the product of the number of trips made in time slice j and how much each waits, so the summation is the total wait. Formula (5.51e) should also be evident after recognizing that each trip end in section i requires $s_i/4$ distance units of walking. Formula (5.51f) holds because $F_i L_i$ is the total number of passenger-kilometers in section i and passengers move at speed $v_{\max}$; thus, the summation is the total time moving. Note, this term is a constant. Finally, (5.51g) holds because the argument in the summation is the number of passenger stops experienced by on board passengers in section i — this can be seen if we think of stations as screen lines, so an average of F_i passengers experience a stop at each station in i; thus the product of F_i and the number of stations (L_i/s_i) is the total number of passenger stops in i.

The overall generalized cost is the sum of (5.51a)–(5.51g). If we neglect from this sum the relatively small bus stopping cost term (5.51c) the result is composed of parts that are either constant or include at most one decision variable. Therefore, the problem decomposes by decision variable. In fact, each part can be seen to have an EOQ form with a closed form solution. The final expressions are easy to obtain and are not derived here for the sake of brevity. They are

$$H_j^* = \sqrt{\frac{2W_j c_d L}{\beta G_j}}; \quad \text{and} \quad s_i^* = 2\sqrt{\frac{L_i v_w (c_r W + \beta t_s F_i)}{\beta(O_i + D_i)}}, \text{ for all } i, j. \tag{5.52}$$

The reader can verify that if the OD-table is homogeneous in time and space, then (5.52) reduces to the formulas derived in earlier sections of this chapter after suitable changes of variables.

Equations (5.51) can also be used with the standards approach. For example, the sum of the first three instances could be the objective function, and the remaining formulas could be used to formulate some level of service constraints. These constraints could be defined separately for each component of travel time and/or for combinations of these times. If desired, they could also be disaggregated by slice and/or section.

5.5.3. *Capacity Considerations*

So far we have assumed in this chapter that vehicles have sufficient capacity to carry their loads. This, however, may not be the case if the demand is heavy and the service frequency is low, as then many passengers will accumulate at stations. Therefore, we now assume that all vehicles have a capacity C and derive a constraint to prevent overfilling. This is done by ensuring that the desired flow of passengers across the most critical screen line never exceeds the flow of seats. This must be true for all the time slices j.

To formulate the constraint for time slice j, we look for the screen line with the highest desired passenger flow during the time slice and for the corresponding (critical) flow, M_j; see Figure 5.17. The location of this screen line is called the "critical load point" because it is where transit vehicles will carry the most passengers. To find it we place imaginary screen lines before each individual station m and look for the largest number of screened crossings in period j across all m. To do this, let O_{mj} and D_{mj} be the number of passengers with origins and destinations at station m in period j. Note, the summations of these numbers across corresponding stations in each section yield $\{O_i\}$ and $\{D_i\}$, respectively, and they can be easily observed or roughly estimated from surveys.

Now consider the number of passengers wishing to cross screen line m in period j, which equals the number of trips that have originated before m (from all origins with $i < m$) and have not ended by m. In terms of our data, this is

$$\sum_{m'<m} (O_{m'j} - D_{m'j}). \tag{5.53}$$

This screen line count can be observed graphically on a plot of cumulative passenger count vs. m as the vertical separation between the cumulative curves of origin and destination counts at m. The critical load point in time slice j is where the two cumulative curves are maximally separated; or more formally, at the value of m that maximizes (5.53):

$$m_j^* = \arg\max_m \left\{ \sum_{m'<m} (O_{m'j} - D_{m'j}) \right\}. \tag{5.54}$$

And obviously, the critical flow M_j is the maximal separation:

$$M_j = \max_m \left\{ \sum_{m'<m} (O_{m'j} - D_{m'j}) \right\}. \tag{5.55}$$

We can now use this known quantity to evaluate the average flow of passengers across the critical load point in time slice j, which is M_j/W_j. Since this should not exceed the flow of seats, which is C/H_j, we see that a necessary capacity constraint is[20]

$$H_j \le CW_j/M_j, \quad \forall j. \tag{5.56}$$

Note this is a simple bound on the headways in each time slice. Thus, the headway optimization problem that led to the first part of (5.52) is now a constrained EOQ problem of the type analyzed in Chapter 2. We found there that the optimum is the lower of (i) the headway value of (5.52), or (ii) the right-hand side of (5.56).

[20]If demand has significant stochastic fluctuations, one should add a small buffer to this constraint. Otherwise, small transient queues will result.

The recipes in Section 5.5 can be used with minor variations to design service strategies for real corridors. When put into practice the headways and spacings may have to be changed a bit from the ideal, e.g., to make sure stations are located near intersections and headways are integer multiples of a minute. These changes, however, increase by very little the generalized cost.

In closing, note from (5.52) how the cost rate of guideway provision c_g does not affect the solution at all. This was true throughout the chapter because, for corridors, the guideway length is fixed. The following chapter deals with two-dimensional systems, where a key part of the problem is deciding how much guideway to provide.

References

Clarens, G. and Hurdle, V. (1975). "An operating strategy for a commuter bus system." *Transportation Science*, **9**, 1–20.

Clark, C. (1951). "Urban population densities." *Journal of the Royal Statistical Society: Series A*, 114(4), 490–496.

Giesen, M. F. R. and Muñoz, J. C. (2013). "Continuous approximation for skip-stop operation in rail transit." *Transportation Research Part C*, 36, 419–433.

Wirasinghe, C. S., Hurdle, V. F., and Newell, G. F. (1977). "Optimal parameters for a coordinated rail and bus transit system." *Transportation Science*, **11**, 359–374.

Homework

5.1. Consider an idealized non-hierarchical line serving an imaginary campus on the periphery of a circular lake. This line has zero headway, zero stop time, and an unbounded cruising speed. Please, optimize the design assuming that the prevailing passenger travel distance is 4 km, their access speed is 1 m/s, and for comfort reasons, the maximum bus acceleration/deceleration is $2\,\text{m/s}^2$. Now assume you can buy either a transit pass or a bike, but not both, for your travels on this campus. If you know you can ride the bike at an average speed of 5 m/s and your goal is minimizing your on-campus travel time in the long run, will you buy the transit pass or a bike? Why?

5.2. Set up a spreadsheet or a computer program to solve (5.18a) numerically, using (5.19) as the objective function and (5.20) as the level of service constraint. In addition, include in the formulation a constraint that enforces a finite vehicle capacity C. Report your optimum design (H^*, s^*) and the corresponding objective value corresponding to the following parameters: $v_a \cong 4\,\text{km/h}$, $v_{\max} \cong 36\,\text{km/h}$, $t_s \cong 10^{-2}\,\text{h}$,

$c_d \cong 1\$/\text{km}$, $c_s \cong 10^{-1}\$/\text{stop}$, $c_r \cong 10^{-3}\$/\text{h}$, $c_g \cong 2\$/\text{km-h}$, $\ell = 15\,\text{km}$, $\lambda = 20\,\text{pax/km-h}$, $C = 35$ passengers, and $T_0 \in \{0.8, 0.9, 1.0\}$ h.

5.3. Consider the following optimization problem: $z = \min_{\mathbf{x}\in\Omega}[f(\mathbf{x})]$, where $f(\mathbf{x})$ is defined on a non-empty set Ω. Suppose there exists another function $f_L(\mathbf{x})$ such that $f_L(\mathbf{x}) \leq f(\mathbf{x})$ for all $\mathbf{x} \in \Omega$. Show that a lower bound to z is $\bar{z} = \min_{\mathbf{x}\in\Omega}[f_L(\mathbf{x})]$, and an upper bound is $\hat{z} = [f(\bar{\mathbf{x}})]$, where $\bar{\mathbf{x}} = \arg\min\limits_{\mathbf{x}\in\Omega}[f_L(\mathbf{x})]$, i.e., $\bar{z} \leq z \leq \hat{z}$. Explain how the problem discussed in Section 5.2.2.2 is a special case of the above situation.

5.4. Using the assumptions of Section 5.1, show that a corridor's optimal station spacings with and without a hierarchy must satisfy, at least approximately, $s_0^* \leq s^* \leq s_1^*$.

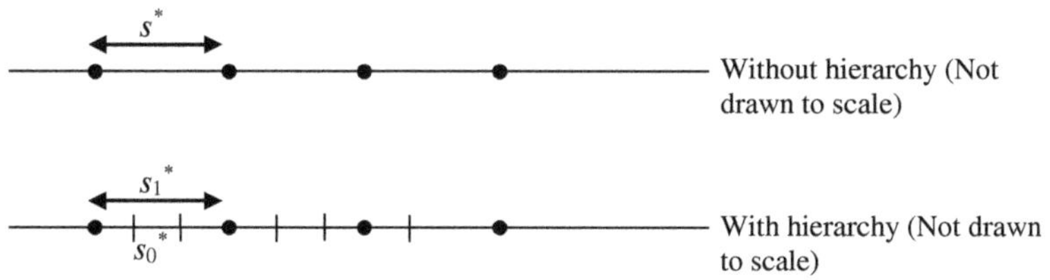

5.5. The Minimax Principle (von Neumann, 1928) says that the optimal solutions to a max–min problem, $\max_x \min_y\{f(x, y)\}$, and its min–max counterpart, $\min_y \max_x\{f(x, y)\}$, are equal in value as long as $f(x, y)$ is concave in x and convex in y. Use the Minimax Principle to derive the results on the multi-standard problem in Section 5.3, starting with the formulation expressed by equations (5.29)–(5.33).

5.6. Consider inhomogeneous service lines on a one-directional corridor as shown in Figure 5.14. There are n local stations between any two adjacent express line stations. Assume the demand is homogeneous, all passengers travel a fixed distance of ℓ, and they simply try to minimize their own total travel times. Also, assume that the express and regular services are not synchronized, and vehicles can pass each other.

(a) Formulate an optimization problem that seeks the best local stop spacing, the best headways for both the regular and express services, as well as the best value of n, so as to minimize the agency cost per trip subject to standards on the average passenger door-to-door travel time. Note that headways cannot be even at every station, so you might need to pay attention to the average passenger waiting time.

(b) Assume a set of reasonable parameter values (e.g., those in homework exercise 5.2), and solve the problem numerically. Derive a closed-form near-optimum solution based on the lower-bound method in homework exercise 5.3. Compare the numerical solution and the value of the closed-form solution. Are they close?

(c) What if vehicles can only pass each other at express line stations? A time–space diagram might be helpful.

(d) If vehicles cannot pass each other, is there a limit to the frequency that can be provided? Find the optimal design numerically, and compare the optimal performance with the result from part (b).

(e) Based on your results, discuss the performance of the skip-stop strategy of part (d) vs. that of the homogeneous hierarchical system in Section 5.4.

5.7. *Given is a linear city in the interval $x \in [-L, L]$. Commuters' residences are uniformly distributed in the interval $x \in [-L, 0]$; and their workplaces are also uniformly distributed in the interval $[0, L]$. An arterial corridor runs along the city, on which the city plans to create a transit system that runs one scheduled dispatch in the morning toward the workplaces, and one scheduled return in the afternoon. Passengers' access stops at speed v_w. The stop spacing is denoted by s. Assume $L \gg s$. The cruising speed of transit vehicles (not including dwell time at stops) is v_t. The vehicle delay per stop is τ. The operating cost to the agency (in each service direction) includes two parts: c_d per route length, and c_s per stop. The agency's budget is B [\$/day].

(a) Formulate a mathematical program that minimizes the users' average door-to-door travel time in both directions subject to a budget constraint on the agency's amortized daily expenditures.

(b) What is the lowest agency budget, B^*, that allows the user's average travel time to reach the minimum found in part (a)?

(c) Express the optimal stop spacing s^* as a function of B for $B < B^*$.

5.8. *This problem examines the design of a train line that collects commuters along a corridor to a single central destination, and its effect on transit-oriented development (TOD). Trips originate along a 10 mile radial corridor that runs from the edge of a city ($x = 0$) to its CBD ($x = 10$) where everyone works, shops, or just wants to be. The schedule of train arrivals at the city is given and such that every train carries the same number of people. We assume that schedules are published and passengers arrive just in time to avoid waiting. Also given is the cumulative number of people in each train as a function of $x, N(x)$: this is a proxy for the spatial distribution of demand. Travelers spend time to access their stations by walking, and once aboard each traveler endures an in-vehicle delay of $\tau = 2$ min for each stop experienced along the ride. The running speed of the train not including stops is $v = 40$ miles/h, and passengers access the stations by walking along the corridor at $v_a = 3$ miles/h.

(a) Assume that stops are evenly spaced with spacing S (such that $10/S$ is an integer), demand is evenly distributed with density D (i.e., $N(x) = Dx$) and passengers walk to the nearest *downstream* station. Write an exact expression for total passenger time in terms of S and D. Assume that travelers suffer τ min of stop delay at the station they board. Find the value of S that minimizes the expression you wrote. Which value of S would you choose? What would be the total passenger time?

(b) The total time for passengers using the train line can be expressed approximately by treating station spacing and demand as smooth continuous functions ($S(x)$ and $N(x)$), and integrating the time consumed in walking and stopping along the length of the corridor. Using this approach: (i) write an expression for the time consumed per unit distance at or near x in terms of $S(x)$, the cumulative demand $N(x)$ and the demand density $D(x) = dN(x)/dx$; (ii) express the total as an integral; and (iii) solve this integral for the data of part (a) and compare the result with the exact answer of part (a).

(c) The station spacing function can be chosen to minimize the total passenger time. Using the results of part (b), write an expression for the optimal station spacing, $S^*(x)$.

(d) Simplify the expression for $S^*(x)$ if the origins are uniformly distributed along the route, as in part (a), but without restricting ourselves to uniform spacings. Then, if $N(x) = 100x$, what is the total passenger time per train run into the city? What is the average travel time per passenger? Compare with part (a) and comment.

(e) If the construction and operation of the train line imposes an annual cost of c_g per mile of route operated, and if you have the freedom to give up service to some part of the city, how would your route length and location vary with c_g?

(f) Now, before the stations are constructed, the city has decided to encourage transit-TOD along the corridor. Thanks to these policies, 30% of the trips are expected to originate near the stations, without any significant access time. Then, if station spacings are redesigned recognizing the effect of TOD, find the new overall passenger travel time, the average travel time per TOD passenger, and the average time per non-TOD passenger.

(g) *Extra credit:* Densities are not typically uniform along the length of a radial corridor. Often the population density (and therefore the demand density) is roughly exponential so

$$D(x) = D_0 e^{\gamma x}.$$

If the density at the edge of the city is $D_0 = 25$ trips per mile, and the density gradient is $\gamma = 0.25$, what is the spacing $S^*(x)$? How many stations should be built along the corridor to minimize travel time? What is the average travel time per passenger in this case? To simplify the math, assume that trains depart $x = 0$ with D_0/γ passengers from outside the CBD already aboard. (For more information on population density profiles, see Clark, 1951).

5.9. In a long east–west corridor, eastbound demand is generated at a rate of λ_1 passengers per distance per time, while the westbound generation rate is λ_2. We know $\lambda_1 > \lambda_2$. All trips in both directions span a known distance ℓ. Section 5.5.1 discussed a case with the same spacing in both directions, but different headways. Solve the design problem, using similar notation, for the following three alternative scenarios:

(a) Equal spacing and equal headway in both directions.
(b) Different spacings but the same headway.
(c) Different spacings and different headways.

Suppose that interlining one bus costs c monetary units and requires τ time units of unproductive deadheading time. For the original scenario with equal spacing and different headways, please formulate an optimization problem that will yield the spacing, headways, and the fraction of deadheaded vehicles (if any) that minimize the total system-wide cost.

5.10. *The loop shown below is a transit corridor where the OD demand is uniformly distributed. You are given a certain number of vehicles to serve this route.

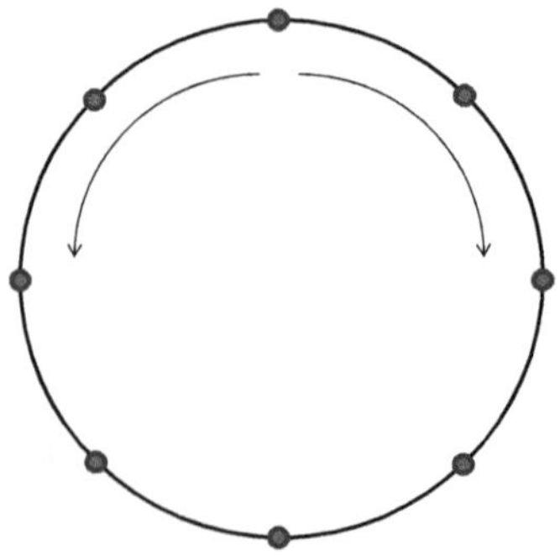

Would you want to provide bi-directional service, or is it better to allocate all vehicles to a unique direction and reduce headways? Considering the experience of an average passenger, how would the total number of available vehicles influence your choice? (*Hint*: First think in the extreme: what if you had only one bus, and what if you had an infinite number of buses. Where is the tipping point such that it becomes more efficient to operate buses in both directions?)

5.11. *Section 5.4 discussed the relationship between multiple standards and a single standard. Now, for the lower-bound formulation of the case with no transfers, show that the equivalent simple standard to the set of linear standards, $T_0(\ell) = T_0 + P_0\ell$, is

$$\ell^* = \left(\frac{t_s}{v_a}\right)\left(P_0 - \frac{1}{v_{\max}}\right)^{-2} \text{ if } P_0 > \frac{1}{v_{\max}},$$

and that

$$\left.\begin{aligned} T_0^* &= T_0 + P_0\left(\tfrac{t_s}{v_a}\right)\left(P_0 - \tfrac{1}{v_{\max}}\right)^{-2} \\ \$_0^* &= c_d\left[T_0 - \left(\tfrac{t_s}{v_a}\right)\left(P_0 - \tfrac{1}{v_{\max}}\right)^{-1}\right]^{-1} \end{aligned}\right\} \text{ if } T_0 > \left(\frac{t_s}{v_a}\right)\left(P_0 - \frac{1}{v_{\max}}\right)^{-1}.$$

Note how the solution does not involve λ or β. Then, use the Lagrangian approach to show that the shadow price that would achieve the above is

$$(\lambda\beta)^* = c_d\left(T_0 - \frac{t_s}{v_a\left(P_0 - \frac{1}{v_{\max}}\right)}\right)^{-2}.$$

Please note that standards are connected to total costs (e.g., operating cost depends on the headway), and you do not need anything else to determine this cost.

5.12. *Consider an inhomogeneous unidirectional corridor such as the one shown in Figure 5.17. A week is divided into multiple time slices. Given is a time-dependent OD-matrix $\{D_{i,i',j}\}$, where $D_{i,i',j}$ is the demand rate from section i to section i' in time slide j. Further, assume that the distribution of trip origins and destinations within each spatial section is uniform, and the distribution of trip generation time, as well as the vehicle operations, is uniform within each time slice.

(a) Derive analytical formulas for the following aggregate quantities: (i) the total number of trips generated throughout the corridor in time slice j, G_j; (ii) the number of weekly trips originating and terminating in section i, O_i and D_i; (iii) the average number of weekly trips that cross a random screen line in section i, F_i; (iv) the maximum number of trips that cross any screen line (i.e., right before each station) in the corridor in time slice j, M_j.

(b) Now suppose you do not know the OD-matrix $\{D_{i,i',j}\}$. Describe how you may conduct field measurements to obtain these aggregate quantities.

5.13. *Section 5.4.1 discussed various ways to provide multi-line service to a transit corridor and examined an optimization model for unsynchronized service with equal headways. Please formulate and solve a similar problem for each of the following scenarios.

(a) Unsynchronized service with different local and express headways ($H_0 \neq H_1$);

(b) Synchronized service as in Figure 5.15(b);

(c) Synchronized service as shown in the figure below. What are the advantages and disadvantages of this type of system?

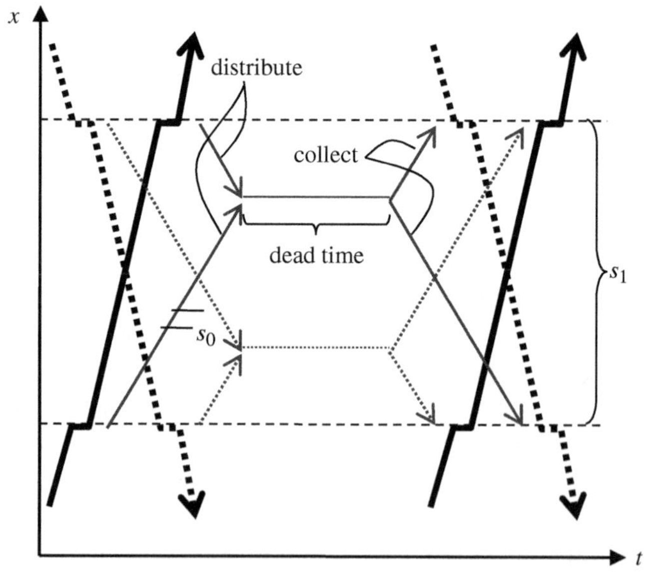

Mini Project 1. Converting a Local Bus Line into Bus Rapid Transit (BRT)[21]

The Geary transit corridor, as shown in Figure MP1.1, has one of the highest transit riderships in San Francisco. At the moment of writing, service on Geary Boulevard had high scheduled frequencies but was plagued by poor reliability due to conflicts from operating in mixed traffic, high dwell times at bus stops, excessive signal delays, and other difficulties.

In this project, you will examine the described situation on Geary Blvd. during the evening rush, as well as the benefits that can be obtained through the introduction of BRT and a hierarchy of service.

Existing Route Characteristics

The route to be improved consists of a 1.5 mile section in the CBD and 5 miles in the residential area. During a round trip, a bus stops at 84 relatively evenly spaced stations along the route including terminals at both ends. (Thus, in each direction there are 41 stations plus a terminal at the end). The round trip of a bus is 120 min. About 50% of this time is spent cruising, 25% stopped at stations (including the lost time while accelerating and decelerating the

Figure MP1.1. Geary Corridor in San Francisco.

[21]Created by Josh Pilachowsky with input from Michael Cassidy.

bus) and 25% delayed by signals. You can use a fixed loss time of 10 s per station due to acceleration and deceleration and assume that buses do not skip stops. Users take four times as long to board as to alight and we will assume that at each stop boarding and alighting movements occur in sequence. Users access to bus stops by walking at a speed of 2 mph. Muni has budgeted 40 buses for the route, which are able to provide a headway of 3 min.

Standards for Level of Service (LOS)

You will be analyzing three types of users for this project:

(1) Residential to Residential (R→ R)
(2) CBD to Residential (C→ R)
(3) Residential to CBD (R→ C)

Data

The boarding and alighting demand density (PM rush) per hour per mile for service in both directions are shown in Figure MP1.2.

Using these data, you are to perform the following tasks. Some assumptions are deliberately left for you to make, and you are expected to explain your rationale.

(a) What is the commercial and cruising speed of a bus? How much does a passenger serve increase a bus's stopped time, considering both the passengers boarding and alighting moves? What is the optimal stop spacing in the residential area for R → R passengers? What is the optimal stop spacing in the CBD and in the residential area for C → R passengers? Reconcile the two conflicting optimal values, and find the optimal stop spacings for each zone for all users. Discuss why the actual spacings on the route may differ from the calculated optima.
(b) By implementing several features of BRT, service can be improved. Converting a lane to bus service (only) will allow buses to cruise 50% faster. Transit signal priority (TSP) will cut the time spent at signals by 50%. What is the new cycle time? With these new data, what is the optimal stop spacing for each zone? Compare these values to part I and discuss the effect upgrading to BRT has on the optimal stop spacing.
(c) Service can be improved further by introducing a hierarchy of service such that local buses are available in the residential section only and express buses run the entire length of the route. For worst-case analysis

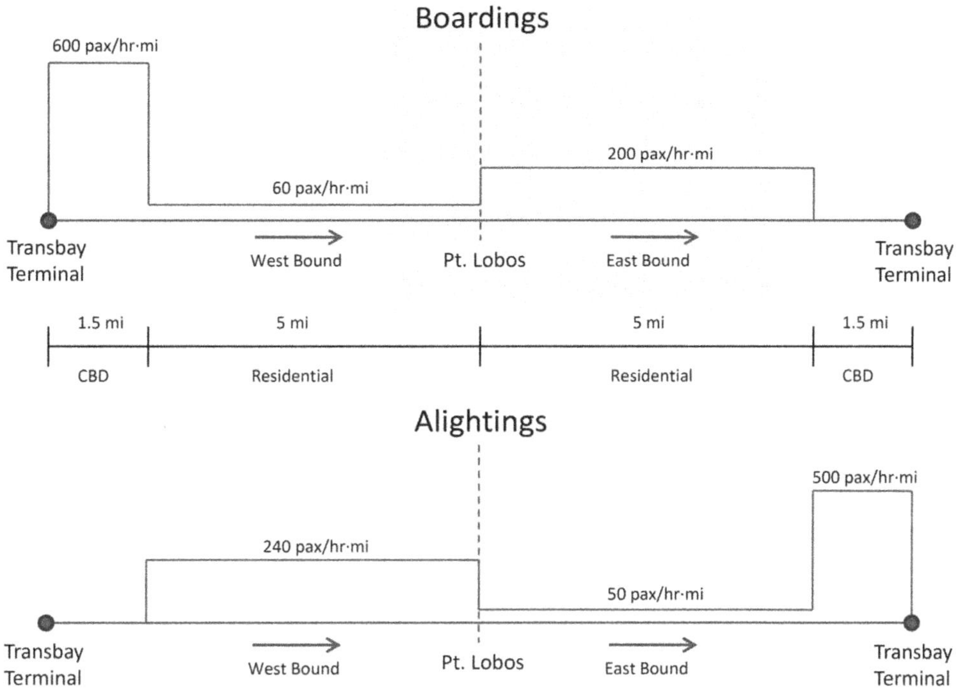

Figure MP1.2. Boarding and alighting demand along the route.

you can assume R → R users will use local service, C → R users will board an express bus and transfer to a local bus at the express stop closest to their destination, R → C users will board a local bus and transfer to an express bus at the express stop closest to their origin (Express and local schedules are not coordinated.) Determine the optimal headway and the optimal stop spacing for each zone, and for each bus type. Remember you only have 40 buses to be split between local and express service. Since this is a complex optimization problem you may solve it numerically (using Excel or a similar program).

You are to submit a formal report describing your tasks and your findings. For the discussion of part iii, you should quantify and discuss the costs of switching from your idealized solution to a real-world solution as well as your extra considerations when designing the latter. You may work in small groups of up to three people and submit a group report.

Chapter 6

Planning — Networks

This chapter examines the design of transit networks intended to cover two-dimensional regions. The types of systems analyzed in previous chapters were relatively simple. Point-to-point systems (shuttles) cover zero-dimensional regions and therefore the only design decisions are temporal — in the simplest case, the selection of a single headway, H. Corridor systems span one-dimensional regions and in the simplest cases involve only a single spatial variable (the station spacing s) in addition to the headway. The corridor problem is relatively simple because one-dimensional regions can be spatially covered with a single route of given length. As a result, the route's length and layout are not design variables. Obviously, this is no longer the case with two-dimensional regions. These regions can be covered in many different ways involving either single routes, or networks of multiple routes that allow for transfers. This flexibility introduces additional decisions of a geometric nature into the analysis, including how many routes to have, how to shape them, and how to connect them. In addition, because the layout of these routes affects the circuity of passenger travel between their access and egress stations passenger routing becomes an issue. Thus, the two-dimensional network problem is considerably more complex than anything we have seen so far. Fortunately, we will see in this chapter that despite these complexities the organization of transit systems in two-dimensional regions can still be studied with the methods we have been using.

Before delving into details, however, let us develop a bit of intuition. Take a minute to think about the transit networks in cities you have visited. Have you noticed any geometric patterns? You may have noticed that the geometry of systems serving similar communities is similar. What causes their geometrical similarities and differences? To help us think about this, Figure 6.1 gives six examples. The systems in the figure use different transit

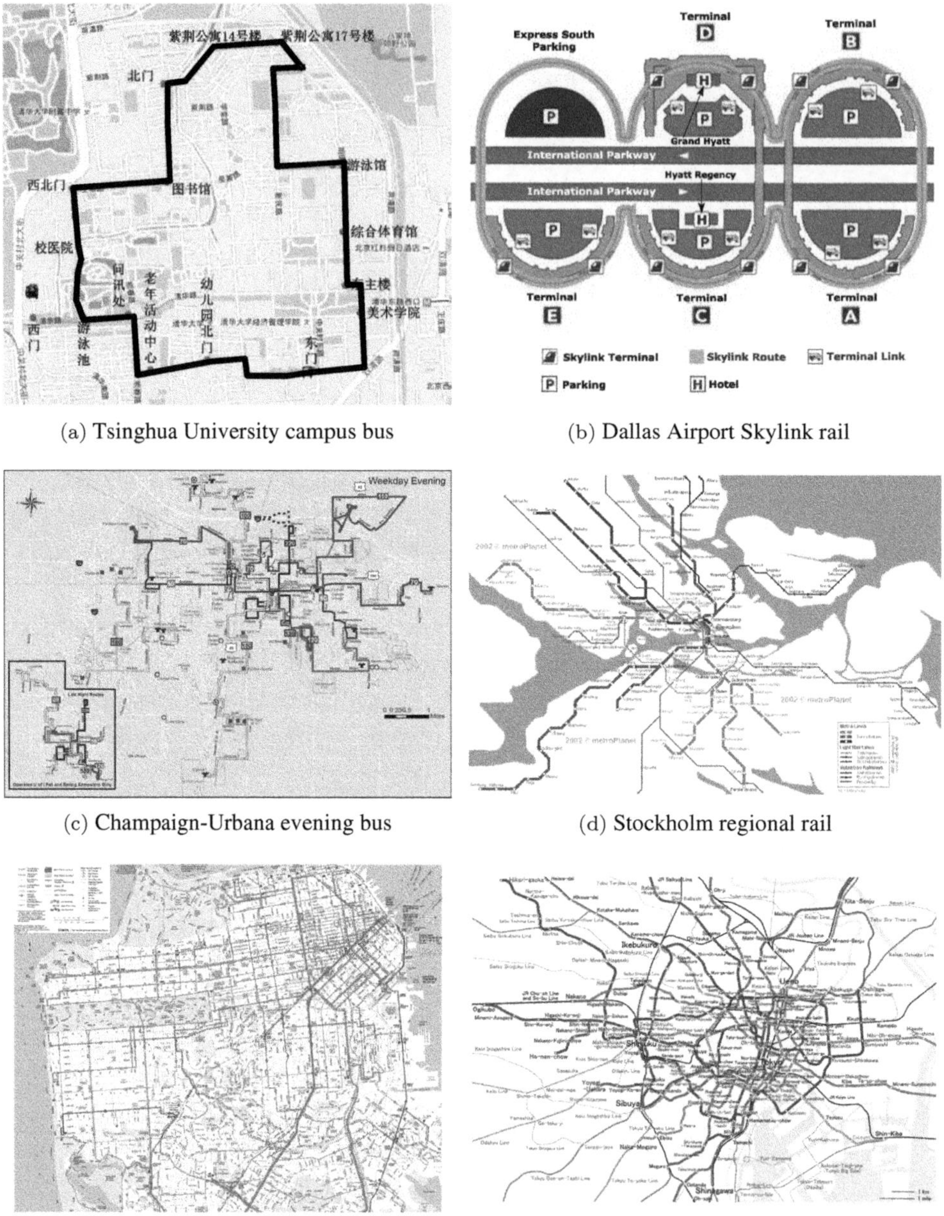

(a) Tsinghua University campus bus

(b) Dallas Airport Skylink rail

(c) Champaign-Urbana evening bus

(d) Stockholm regional rail

(e) San Francisco Muni bus

(f) Tokyo metro

Figure 6.1. Several real-world transit networks.

Sources: (a) http://t.yygongjiao.com/; (b) http://www.pinterest.com; (c) https://www.cumtd.com; (d) http://www.urbanrail.net; (e) http://www.sanfrancisco.net/bus; (f) http://tokyocityairportmetrosubwaymap.blogspot.com/; All accessed on June 1, 2016.

technologies, operate in different environments, and serve regions of vastly different sizes — from a college campus to a megacity. The figure is organized with buses operating in mixed traffic on the left column and rail systems with segregated right-of-way on the right. In each column, smaller regions are on top and larger regions on the bottom. Note the similarities and differences among these transit networks. Does the transit network in your city or town resemble any of them?

The figure reveals some patterns. Note how the top two networks are formed by single loops, the middle two consist of distorted spokes radiating from centrally located hubs, and the bottom two are distorted grids resembling spider webs. Why are the three networks in each column so different from each other despite sharing the transit technology and the type of right-of-way? To think about this, consider the left column and imagine for a second what would happen if we used a single transit loop to serve San Francisco, or built a complex grid for Tsinghua University? Can you foresee potential problems in either case? How would the transit agency and the passengers be affected?

Now you probably have recognized something: The network geometry for any given technology tends to become more complex (e.g., loop $\rightarrow$ spoke $\rightarrow$ grid) as the size of the service region grows. This definitely occurs for the bus networks in the left column, and also for the rail networks in the right. Consideration of other systems shows that the pattern is rather prevalent. You may have also noticed from the figure that technology also has an impact on geometry: San Francisco and Stockholm are cities of comparable sizes; yet, the Stockholm rail system looks like a set of spokes and the San Francisco bus system like a grid. So it appears that the more expensive technology calls for a simpler geometry. Again, the pattern turns out to be quite prevalent. Why?

In an attempt to answer these questions, this chapter starts with an idealized analysis (Section 6.1) that isolates the geometric issues introduced by two-dimensionality, and compares three different layouts: loops, hub-and-spoke, and grids. Sections 6.2 and 6.3 then use more realistic assumptions to examine grids, Section 6.2 focuses on systems without vehicle capacity issues, and Section 6.3 considers the effects of capacity. Section 6.4 then repeats the analysis for hub-and-spoke; and for a hybrid of hub-and-spoke and grid. So far the chapter will have described a "blank slate" design method where lines and stops can be laid out anywhere in the region and results can be expressed in terms of mathematical formulas that describe idealized geometries. Section 6.5 then explains how to translate these formulas into actual implementable plans for real regions, recognizing that any idealized geometry must be distorted to conform to real-world restrictions such as the location and shape of available streets. This section also reviews a case study. Finally, Section 6.6 covers additional topics that often come into play, including remedies for the "last-mile" problem, skip-stop strategies, and adaptation to time-varying demand.

6.1. Idealized Analysis

This preliminary section considers an extremely idealized model that isolates the new issues introduced by the two-dimensional character of the service region. We will see that the door-to-door speed that transit can provide is limited despite our extremely favorable conditions, which would not limit the speed in a corridor. The idealized model also explains the patterns observed in Figure 6.1.

Systems shall be compared based on average performance, considering all passengers. Such analysis will be done in two parts. Section 6.1.1 discusses the single-route case of Figures 1(a) and (b), and Section 6.1.2 discusses the multi-route cases of Figures 1(c)–(f). In both subsections, we assume that $H = 0$ and $t_s = 0$, as in the idealized corridor analysis. But now we also assume that the vehicle's acceleration/deceleration rate is $a_0 = \infty$ and its cruising speed is $v_{\max} < \infty$. These additional assumptions eliminate vehicle delay from the act of stopping. As a result, the vehicle speed is $v = v_{\max}$ at all times regardless of the station spacing. And, in a corridor, the door-to-door speed would also be $v_{\max}$, matching the performance of the automobile. Let us now see what happens in two dimensions.

6.1.1. *A Single Route: Systems without Transfers*

Consider a square region with side ϕ, and an infinitely dense grid of streets parallel to the sides. We propose serving it with a single transit line of length L that provides bi-directional service over a street path of length $L/2$. This path forms a closed loop and can weave up and down or back and forth as shown in Figure 6.2(a). No transfers are allowed. Obviously, no matter how long, a transit line cannot cover all points. Therefore, we anticipate that coverage and access become important issues in two dimensions; and consequently, that an important decision will be the length of the street path, $L/2$, and the placement of the stations.

To keep the analysis simple, and also because it is efficient, we shall assume that stations are placed on a square lattice of separation S, as shown in Figure 6.2(a). This is also consistent with Section 5.1, which placed stations on a one-dimensional regular lattice. Since no transfers are allowed,

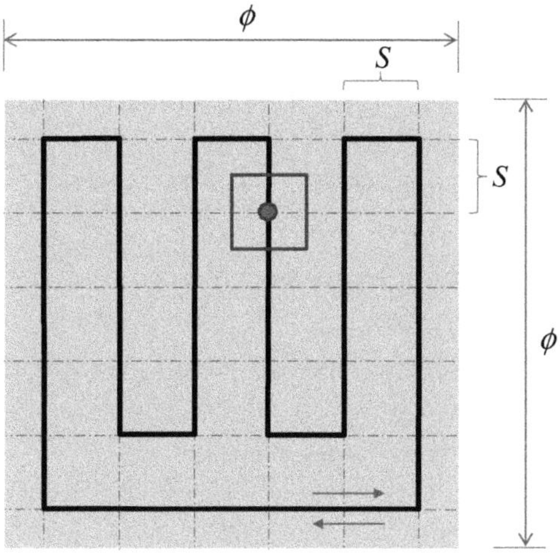

(a) Loop system (O)

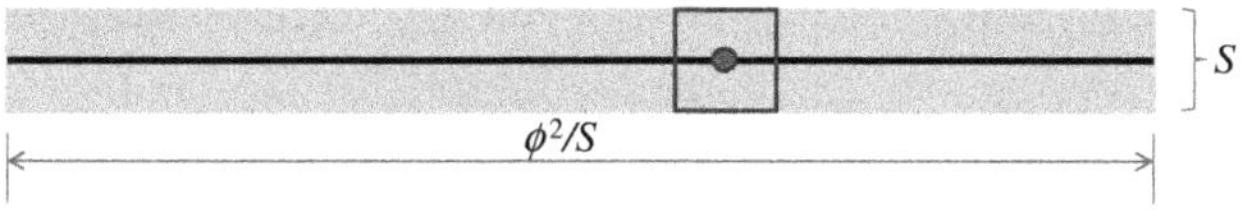

(b) Corridor analogy.

Figure 6.2. Closed loop system and an analogous corridor.

all stations must be visited by a single line, e.g., as shown in the figure. Note, the line forms a loop that sweeps back and forth across the region. Therefore, for all stations to be covered without duplication, the back and forth parts of the loop must occur in pairs. And the number of stations along each side of the region must be even, i.e., $L/2$ must be an integer multiple of 2ϕ (e.g., note, $L/2 = 6\phi$ in the figure).

Consideration of the resulting geometry also reveals that $L/2$ varies inversely with S. This can be intuitively seen if we imagine cutting out a ribbon of uniform width around the route, and then straightening it out without changing the area to form a rectangular corridor of length $L/2$ (see Figure 6.2(b)). Clearly, since the area of the rectangle is ϕ^2, it follows that $L/2 \approx \phi^2/S$. (The expression is exact if $L/2$ is an integer multiple of 2ϕ as in the figure, but we use an approximation sign to allow for exceptions.) Let us now use this relationship to evaluate the average passenger's door-to-door travel time.

To this end, we assume that passengers access and exit the system through the stations closest to their origins and destinations; and choose the direction of travel that minimizes in-vehicle time. Now, since each station has a catchment area of side S which is centered on the station (see Figure 6.2(a)) it follows that the passengers' average access distance in each direction of the grid is $S/4$ at each trip end. Thus, considering both the horizontal and vertical directions at the two trip ends, we see that the average access distance is S, and the average access time $T_a = S/v_w$. We can also express this relation in terms of route length, since $L/2 \approx \phi^2/S$. The result is $T_a \approx 2\phi^2/(Lv_w)$. The result is expressed in terms of L because route length is a proxy for the agency's costs.

Since people can choose the bus direction, their riding distance ranges from 0 to $L/4$ and must average $L/8$. Thus, the average riding time is $T_r = L/(8v_{\max})$, and the total door-to-door time:

$$T = T_a + T_r \approx \frac{2\phi^2}{Lv_w} + \frac{L}{8v_{\max}}.$$

Since this is an economic order quantity (EOQ) expression (see definition in Section 2.3) with respect to L, it is minimized for $L^* = 4\phi(v_{\max}/v_w)^{1/2}$ and the total optimum time is

$$T^* = \left(\frac{\phi^2}{v_w v_{\max}}\right)^{\frac{1}{2}} = \frac{\phi}{\sqrt{v_w v_{\max}}}. \tag{6.1}$$

This is the best that can be done without transfers. Note, if the ratio $v_{\max}/v_w$ is comparable with 10 (which is a reasonable value) then $L^* \approx 6\phi$ and $T^* \approx 3\phi/v_{\max}$. Furthermore, since the value of L correlates with agency cost, it is never in the interest of the agency to provide $L > L^*$.

Let us now express the average door-to-door travel speed as the ratio of the average distance on the street (if one were to drive) to (6.1). Dimensional analysis tells us that the average distance between random origins and destinations in a square region should be a multiple of the only distance parameter in the problem: ϕ. Intuition also tells us that the coefficient of ϕ should be somewhere between 0.5 and 1. Since the average distance between two random points in a segment is 1/3 of the segment's length (see homework exercise 6.1) it follows that in our case the average distance on the streets between a uniformly distributed origin and a uniformly distributed destination is $2\phi/3$.[1] Thus, the average door-to-door speed when the system

[1]This is true because the street distance has a horizontal and a vertical component, both equal to the separation between two random points on a side of the square.

is optimized is

$$\hat{v} \approx \frac{2\phi/3}{T^*} = \frac{2}{3}\sqrt{v_w v_{\max}}.$$

If we assume reasonable values for an urban area, say $v_w = 3$ kph and $v_{\max} = 36$ kph, we find that the resulting door-to-door speed is not attractive at all. It is about $\hat{v} \approx 6.9$ kph. This occurs despite all the favorable assumptions we have made. The underperformance arises because to achieve low access time the route needs to be very circuitous, and circuity increases in-vehicle travel time.

6.1.2. *Multi-Route Systems: The Role of Transfers*

It turns out that multi-route systems with transfers can improve matters. This happens because with transfers coverage can be provided with multiple straight routes that provide more direct passenger paths. To get an understanding of this issue, we now examine two idealized systems where passengers are allowed to transfer: a hub-and-spoke system (H) with only one transfer point; and a grid system (G) that allows for transfers at every point where lines connect. We continue to assume that stops are evenly spaced along each line, and still attempt to provide uniform spatial coverage with square catchment areas. Figure 6.3 depicts the two systems. For the grid system, we assume stops are set only at the route intersections. Note that for the same spacing between routes, the grid system requires more route-kilometers; so it should be more expensive to populate with vehicles. Another disadvantage of the grid system relative to the hub is that coordination to provide timed transfers is more difficult. On the other hand, an important advantage of a grid is that users can always choose a direct route without backtracking.

Let us now compare the performance of these two systems with each other, and with the no-transfer, single-line loop system (O) of the previous subsection for similar levels of agency cost. This is achieved by forcing the total directional service distance L to be equal for all three systems. (Recall from Chapter 5 that for any given headway H, spacing s, and commercial speed $v_{\max}$, all components of agency cost are directly proportional to L.)

We first express L as a function of S for all three cases. We shall use the subscripts "O," "H," and "G" to identify each case. We have already shown that for the "O" system $L_O = 2\phi^2/S$. Consideration of Figure 6.3 yields

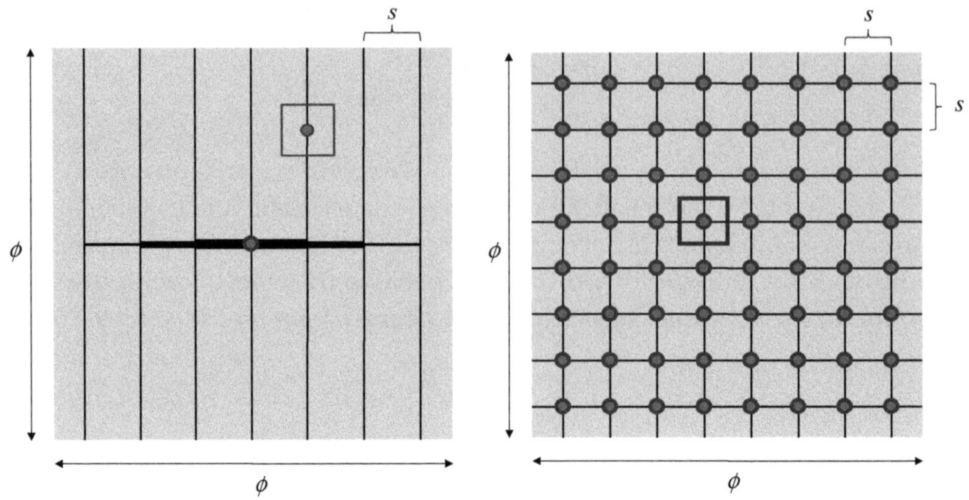

(a) Hub-and-spoke system (H)

(b) Square grid system (G)

Figure 6.3. Systems with transfers.

similar formulas for the remaining two cases. The complete results are[2]

$$L_O = \frac{2\phi^2}{S}; \quad L_H = \frac{3\phi^2}{S}; \quad L_G = \frac{4\phi^2}{S}.$$

Conversely, if we hold the three lengths equal at some value L, the corresponding spacings S will be different. They are obtained by solving for S in the above expressions, and the result is

$$S_O = \frac{2\phi^2}{L}; \quad S_H = \frac{3\phi^2}{L}; \quad S_G = \frac{4\phi^2}{L}.$$

In order to evaluate door-to-door travel time we need to specify how people navigate the system. As in Section 6.1.1, we assume that people choose the two stations that are closest to their origins and destinations, and then route themselves in the most direct way possible. With the H system, they transfer at the thick dot in the middle of Figure 6.3(a),[3] and with the G system, wherever the chosen horizontal and vertical lines intersect.

We continue to assume that origins and destinations are uniformly and independently distributed in the service region. Thus, the walking part of

[2]To get these simple expressions, it is assumed that it takes 1 spacing to turn the buses at the end of each route.

[3]Alternatively we can allow transfers wherever lines overlap. Although this reduces the average travel distance slightly — by a fraction of the difference with the other strategies — the derivation is lengthier. Therefore, the simpler scenario is used here.

travel time for the average passenger to access the nearest stops at both ends of the trip continues to be $T_a = S/v_w$ in all cases. If we now replace S in this expression by the corresponding expressions in terms of L (as given above) for each of the three systems we obtain: $T_a = 2\phi^2/Lv_w$ for "O"; $T_a = 3\phi^2/(Lv_w)$ for "H"; and $T_a = 4\phi^2/(Lv_w)$ for "G."

Regarding in-vehicle trip times, we have already seen that the average riding distance is $L/8$ for the O-system, so the average time is $T_r = L/(8v_{\max})$. For the H system, there is an inbound distance from a random point to the hub, which averages $\phi/2$ considering both directions, and an equal amount for the outbound leg. Thus, the total distance is ϕ and the time is $T_r = \phi/v_{\max}$. For the G system the distance is the sum of a horizontal component and vertical component, which we know averages $2\phi/3$. Therefore, the average riding time is $T_r = 2\phi/(3v_{\max})$.

Now, combining the average walking and riding times, we find the following totals:

$$T_O = \frac{2\phi^2}{Lv_w} + \frac{L}{8v_{\max}}; \quad T_H = \frac{3\phi^2}{Lv_w} + \frac{\phi}{v_{\max}}; \quad T_G = \frac{4\phi^2}{Lv_w} + \frac{2\phi}{3v_{\max}}.$$

If as before we now choose $v_{\max}/v_w \approx 10$ we can compare the three cases based on the dimensionless variable L/ϕ. The formulae become

$$T = \frac{\phi}{v_{\max}} \times \begin{cases} 20\left(\frac{\phi}{L}\right) + \frac{1}{8}\left(\frac{L}{\phi}\right) & (O) \\ 30\left(\frac{\phi}{L}\right) + 1 & (H) \\ 40\left(\frac{\phi}{L}\right) + \frac{2}{3} & (G) \end{cases}.$$

These expressions are depicted in Figure 6.4. The O curve has an increasing branch because as explained in Section 6.2.1, the in-vehicle travel time increases with route length. Logically, we assume that the dashed branch does not arise because an efficient agency would not waste funds to lengthen a route beyond what is optimal. The optimal strategy (O*) is illustrated by the solid horizontal line directly beneath the dashed curve.

For a region of a given size, the ratio L/ϕ on the abscissa of Figure 6.4 is proportional to the cost of the transit resources used to provide service including both infrastructure and vehicles. The graph illustrates how the desirable network structure varies with the available resources. When resources are scarce (i.e., when $L/\phi < 14$ in our example), one is better off providing single-route circulators with no transfers, whose total length is not much greater than the perimeter of their service region. As resources

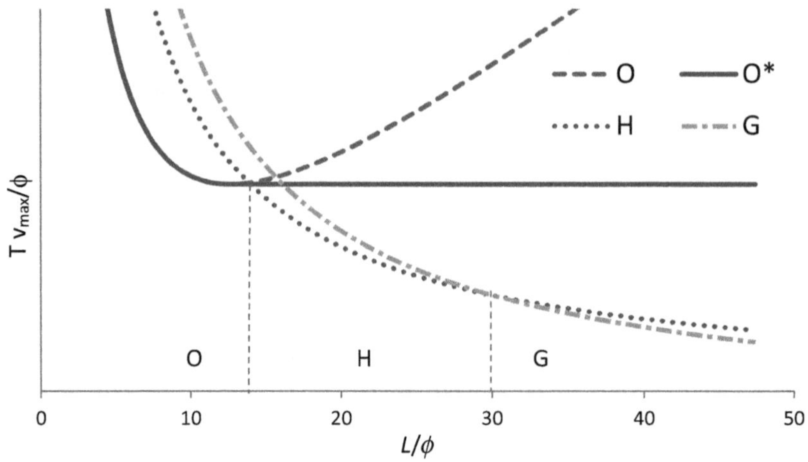

Figure 6.4. Lowest possible door-to-door travel time vs. investment level for three transit system configurations.

increase, the optimum structure morphs first into a hub-and-spoke pattern, and eventually (when $L/\phi > 30$ in our example) into a grid pattern that allows for multiple transfer points and less circuity. Figure 6.4 also illustrates that the door-to-door travel time achieved by the best system (i.e., the lower envelope of the three curves) declines with the resource level.

It is also interesting to see what these formulas say when we hold the resources invested per person constant and vary the city size. For a city of constant population density, the resources per person are proportional to the ratio $r = L/\phi^2$. Note, the more expensive the infrastructure, the less length we can provide for a fixed budget. Thus, r should be smaller for a rail system than a bus system. So let us see what happens as we increase ϕ for a fixed expenditure per capita, r. With r fixed, the system's length increases with city size as $L = r\phi^2$. Now, using this substitution in the formulas for door-to-door travel time, we find

$$T = \frac{1}{v_{\max}} \times \begin{cases} 20/r + r\phi^2/8 & (O) \\ 30/r + \phi & (H) \\ 40/r + \frac{2}{3}\phi & (G) \end{cases}$$

Figure 6.5(a) and (b) plots these expressions as a function of ϕ for both, $r = 1$ (expensive mode) and $r = 2.5$ (cheap mode). In both figures, the single-route circulator is best for small ϕ, and as ϕ increases the optimum changes to "H" and eventually "G." This is consistent with the patterns

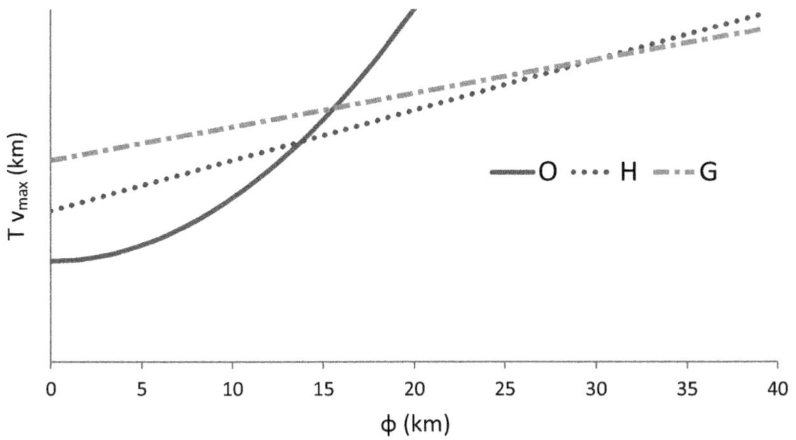

(a) r=1 (expensive mode)

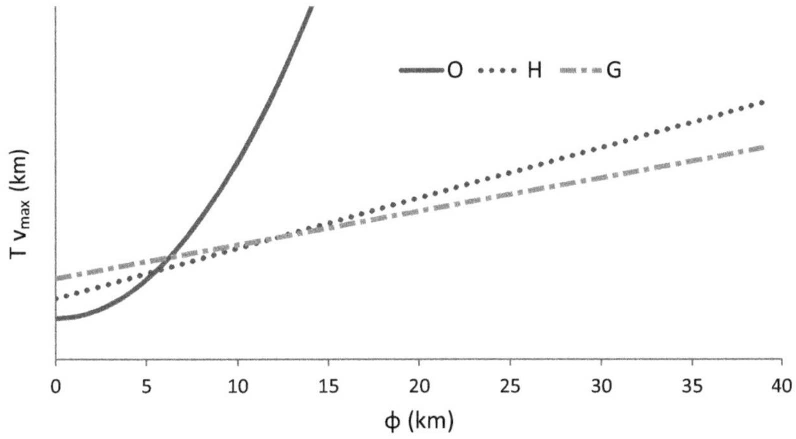

(b) r=2.5 (cheap mode)

Figure 6.5. Door-to-door travel time vs. city size for two transit modes and three system configurations.

observed in Figure 6.1. Note too that for the expensive mode of Figure 6.5(a) the transitions occur for larger ϕ. This again is consistent with the pattern of Figure 6.1 — we noted earlier that San Francisco and Stockholm have similar sizes and that while San Francisco's bus system resembles a grid, Stockholm's rail system looks like a hub-and-spoke system.

The results of this subsection also hold qualitatively if one considers more realistic models with bus stations, finite bus accelerations, significant vehicle operating costs and reasonable headways. This strongly suggests that the patterns we identified in Figure 6.1 are intrinsic to the need for covering origins and destinations dispersed in two dimensions. The results of the subsection also illustrate that to cover regions of significant size it is necessary to use networks of interconnected lines that encourage transfers, such as structures G and H.[4] The following sections explain how to design interconnected systems using realistic assumptions.

6.2. Realistic Analysis: Grids

Section 6.1 examined how best to serve dispersed demand and suggested that grids are appropriate for scenarios that have one or more of these qualities: high demand density, large area; and/or low infrastructure costs. Since these conditions are quite common this section and the next are entirely devoted to grids. The present section will assume that vehicles have enough capacity never to leave passengers stranded at a station, and Section 6.3 will examine the effects of capacity constraints. Furthermore, since it is important in practice to consider the welfare of all customers, we will use in these two sections, and throughout the rest of the chapter, an average case approach in which total generalized cost is minimized.

Since grids are especially appropriate for systems with low-infrastructure cost, the bulk of the analysis in this section will be done for bus systems. Furthermore, we will assume that service is unsynchronized (synchronizations is a practical impossibility for grids) and that passengers do not have appointments at their destinations. These assumptions are reasonable, similar to those in Chapter 5 and can be easily changed.

Here we develop analytic expressions for both, the optimal design and the resulting optimal cost of grids. To illustrate the effects of two-dimensionality, we compare these results with those for corridor systems. To be consistent with the rest of this chapter, we shall focus on a homogeneous square region, of side length ϕ, that is to be served with a homogeneous grid. For the sake of simplicity, we shall concentrate on square grids because their geometry is characterized by a single parameter: the line spacing. Rectangular grids can perhaps be slightly more efficient in some cases, but they require one more parameter, introduce directionality and add complexity to the analysis. Therefore, although analysis is possible, they are not studied here. In the following subsections, we present a generic formula for the generalized cost function of a square grid (Section 6.2.1); derive the optimal solution and

[4] For a different take on the role and need for transfers in situations with dispersed origins and destinations, see Daganzo (1987).

examine its properties under a broad array of scenarios (Section 6.2.2); and discuss some generalizations (Section 6.2.3).

6.2.1. *Derivation of the Generalized Cost Function*

The spatial structure of a square grid of transit lines is defined by both, the line spacing S and the station spacing s. Although the latter must be an integer sub-multiple of S (see Figure 6.6), in our blank slate analysis, we will only stipulate that $s \leq S$. This reasonable because we know from Chapter 3 that spacings can be adjusted *a posteriori* with little change in generalized cost. The complete spatiotemporal structure of the transit system is completed by defining a system-wide headway, H, that is used on all the lines. Again, it might be possible to use different headways on different lines, but there is no real advantage to doing this since our demand is assumed to be homogeneous in space. Therefore, the design goal is identifying optimal values of s, S, and H. This subsection formulates the optimization problem.

To start let us develop formulas for agency cost and passenger travel time (access + waiting + in-vehicle travel). As in Section 6.1, we shall assume that people choose their origin and destination stops in order to minimize their walking distance and then route themselves over the network in the most direct way possible. The trip generation density is denoted λ [passengers/time-distance^2]. Since our system is homogeneous in space and time within the confines of our square, we can base our analysis on a repeating square tile centered on a transfer station, such as the one shown in Figure 6.6, and evaluate costs during one headway. Therefore, the volume of our three-dimensional analysis unit is HS^2; and the number of people generated in it, λHS^2.

Consider now the agency cost. For simplicity of notation, we focus on bus systems with negligible infrastructure costs.[5] The agency cost for any such system is obtained by evaluating the agency cost in our three-dimensional analysis unit and dividing by its volume. Note that the unit contains $4S$ distance units of directional line and $4S/s$ stations considering both directions (see Figure 6.6). Since every portion of the network is covered by a bus exactly once during each headway, it follows that our analysis unit will encompass $4S$ distance units of bus travel and $4S/s$ bus stops. Therefore, the agency cost in the analysis unit is $4S(c_d + c_s/s)$; and the prorated amounts

[5]As we saw in Chapter 5, the general formulas for other systems are very similar but longer.

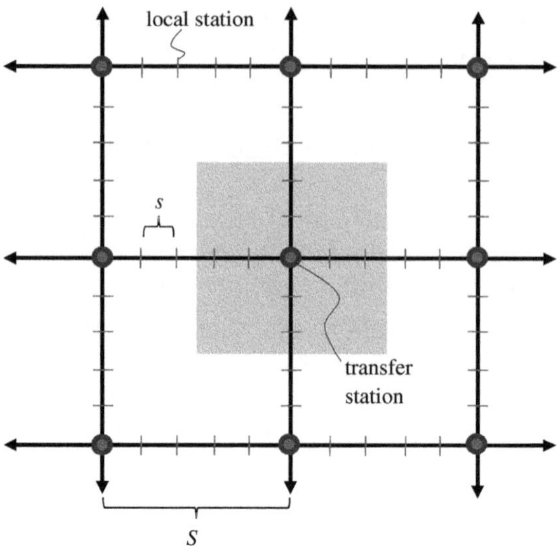

Figure 6.6. Repeating tile for a grid network.

per unit volume and per trip are

$$\$_T = \frac{4}{SH}\left(c_d + \frac{c_s}{s}\right) \text{ per unit volume [cost/time-distance}^2\text{]}, \quad (6.2a)$$

$$\$ = \frac{4}{\lambda SH}\left(c_d + \frac{c_s}{s}\right) \text{ per trip [cost/trip]}. \quad (6.2b)$$

Notice these formulas are very similar to those for a corridor; the only difference is a factor $4/S$ expressing the fact that cost depends on the number of lines.

The travel time is evaluated in parts, including the access time T_a, the waiting time T_w, and the in-vehicle riding time T_r. Because passengers have a route choice, we assume that each passenger chooses the two intersecting lines closest to his/her origin and destination. This minimizes transfers and results in the most direct routing. Every passenger has two obvious options: (i) take a horizontal line first; or (ii) a vertical line first. If they choose randomly among these options, then the average access distance in the direction perpendicular to each of the two chosen lines is $S/4$, and the distance in the parallel direction is $s/4$. Thus, the total distance at both ends of the trip is $(s + S)/2$, and the average access time is

$$T_a = \frac{1}{2}(s + S)\Big/ v_w.$$

If passengers are smart and well-informed, they can slightly reduce T_a by evaluating options (i) and (ii), and choosing the better outcome. The improvement, however, turns out to be small,[6] and has a small impact on the final results. If instead the passengers strictly aim at minimizing walking distance by choosing the closest lines at both ends of their trips, even if this requires an extra transfer, then the number of transfers necessary for the trip, the resulting T_a is similar as well.[7] Homework exercise 6.6 asks the reader to show the validity of these claims. In view of the similarity of all formulas, and recognizing as well that we do not know exactly how travelers really choose, we shall use the above expression in our evaluation.

Since passengers do not have appointments and the schedules are not synchronized, the average wait should be of the form:

$$T_w = kH/2 + \Delta, \text{ where } k = 1, 2.$$

Here Δ is a time penalty associated with making a transfer. This includes time walking plus an additional penalty for the inconvenience. The headway coefficient k is set to $k = 1$ if passengers know the schedule and $k = 2$ if they do not.

In-vehicle travel time depends on the average distance between origins and destinations, which we denote ℓ. This distance clearly depends on the distribution of origins and destinations in the square. For example, destinations could be near the origins (if the mode mainly serves short trips), uniformly distributed all over the region, or clustered around its center, e.g., as would occur in some cities during the morning rush. These different cases are represented by different values of ℓ:

$$\ell = \begin{cases} \ll \phi & \text{(nearby destinations)} \\ \frac{2}{3}\phi & \text{(uniformly distributed destinations)}. \\ \frac{1}{2}\phi & \text{(centrally clustered destinations)} \end{cases} \qquad (6.3)$$

In terms of this generic distance, the average riding time is

$$T_r = \ell\left(\frac{1}{v_{\max}} + \frac{t_s}{s}\right).$$

[6]Simple probability calculations reveal that the reduction is $S/6$ when S/s is an even integer (this turns out to be the maximum possible reduction) and $\frac{1}{6}S - \frac{1}{6}s^3/S^2$ if S/s is an odd integer. Note the reduction is zero if $s = S$.

[7]The reduction in T_a becomes $\frac{7}{60}S$ when S/s is an even integer, and $\frac{7}{60}S - \frac{1}{6}s^3/S^2$ if S/s is an odd integer (when $s = S$, the value of T_a could increase). For more insights, see Holroyd (1967).

Thus, the average-case door-to-door travel time system is

$$T = T_a + T_w + T_r = \frac{kH}{2} + \Delta + \frac{S+s}{2v_w} + \ell\left(\frac{1}{v_{\max}} + \frac{t_s}{s}\right). \tag{6.4}$$

Notice again that this is qualitatively very similar to the formula for the average door-to-door travel time in a corridor; see the corresponding terms in (5.22). The difference is in the waiting time, which now has a coefficient k and a penalty Δ (due to the transfer).

Let us now combine (6.2b) and (6.4) to write the generalized cost function, i.e., the truncated Lagrangian that is to be minimized. If as we did in Chapter 5, we neglect the cost term associated with c_s the result is

$$z(S,H,s) = \frac{4c_d}{\lambda SH} + \beta\left(\frac{kH}{2} + \Delta + \frac{S+s}{2v_w} + \frac{\ell}{v_{\max}} + \frac{\ell t_s}{s}\right) \text{[cost/trip]}. \tag{6.5}$$

To produce a design, (6.5) should be minimized subject to the constraint $s \leq S$ and possibly a second constraint, $H \geq H_{\min}$. The latter arises because stations can serve vehicles at intervals that are limited from below by the station dwell times, and by the need to maintain minimum inter-vehicle spacings for safety purposes. Typical values of $H_{\min}$ are comparable with a minute or two. The following subsection presents the general solution.

6.2.2. *Optimization and Solution Properties*

To obtain a design, (6.5) should be minimized subject to the constraints. Fortunately, it turns out that the headway constraint is almost never binding. For this reason, we shall do the analysis considering only the $s \leq S$ constraint. The headway constraint should, of course, be verified *a posteriori*. (In the rare cases where the constraint is violated we should set $H = H_{\min}$ and re-optimize the spacings; this two-variable problem turns out to be very simple.)

The proposed minimization can be done numerically because the problem only has three decision variables. Furthermore, the problem has the structure of a geometric program, which simplifies the task. Note there are two cases depending on whether the constraint $s \leq S$ is binding or not. Therefore, we now develop formulas for each of these two cases and show when each holds.

6.2.2.1. *Solution when $s \leq S$ is not Binding*

To examine this case, we solve (6.5) analytically without the constraint, and check the constraint *a posteriori*. We do this with conditional decomposition.

If the constraint turns out not to be binding then the result is optimal. The solution is found in three steps. First, hold S and H fixed to find the conditional optimum with respect to s. The result is

$$s^* = \sqrt{2\ell t_s v_w} \tag{6.6}$$

and

$$z(S, H) = \frac{4c_d}{\lambda SH} + \beta\left(\frac{kH}{2} + \Delta + \frac{S}{2v_w} + \sqrt{\frac{2\ell t_s}{v_w}} + \frac{\ell}{v_{\max}}\right).$$

Next, hold H fixed and optimize $z(S, H)$ with respect to S. The result is

$$S^* = 2\sqrt{\frac{2c_d v_w}{\lambda\beta H}},$$

and the optimum cost is

$$z^*(H) = 2\sqrt{\frac{2\beta c_d}{\lambda H v_w}} + \beta\left(\frac{kH}{2} + \Delta + \sqrt{\frac{2\ell t_s}{v_w}} + \frac{\ell}{v_{\max}}\right).$$

Finally, optimize this expression with respect to H and insert H^* into the previous expressions. The result is

$$H^* = 2\left(\frac{c_d}{\lambda\beta k^2 v_w}\right)^{\frac{1}{3}}, S^* = 2\sqrt{\frac{2c_d v_w}{\lambda\beta H^*}} = 2\left(\frac{kc_d v_w^2}{\lambda\beta}\right)^{\frac{1}{3}}, \tag{6.7}$$

$$z^* = 3\left(\frac{kc_d\beta^2}{\lambda v_w}\right)^{\frac{1}{3}} + \beta\left(\Delta + \frac{\ell}{v_{\max}} + \sqrt{\frac{2\ell t_s}{v_w}}\right). \tag{6.8}$$

The complete design is given by (6.6) and (6.7).

This solution will be valid if $s^* \leq S^*$. Entering with (6.6) and (6.7) on both sides of this inequality allows us to express this validity condition in terms of the parameters of the problem as follows:

$$\lambda\beta \leq 2\sqrt{2}kc_d v_w^{\frac{1}{2}}(t_s\ell)^{-\frac{3}{2}}. \tag{6.9}$$

It turns out however that in many practical cases the validity constraint is violated, i.e., we find that

$$\lambda\beta > 2\sqrt{2}kc_d v_w^{\frac{1}{2}}(t_s\ell)^{-\frac{3}{2}}, \tag{6.10}$$

indicating that (6.9) must be binding at optimality.

To see how common this violation is, let us evaluate the two sides of (6.10) using the common parameter values that were used for buses in Chapter 5 ($v_w \cong 3$ km/h, $v_{\max} \cong 36$ km/h, $t_s \cong 5 \times 10^{-3}$ h, $c_d \cong 1$ \$/km, $k = 2$), and assuming that $\ell = 10$ km (a medium size city) and $\beta = 20$\$/h. Making these substitutions, we see that (6.10) holds if $\lambda H > 44$ [pax/km^2-h]. In many cities worldwide, the actual bus ridership far exceeds this lower bound. Therefore, in these cases $s^* = S^*$ and one can solve the problem by setting $s = S$ from the beginning.

6.2.2.2. *Solution when $s \leq S$ is Binding and we set $s = S$*

This problem is slightly more difficult because the variables do not decompose as neatly, but it is still possible to solve it. We now develop this solution and then splice it with the non-binding case to obtain the complete solution. This complete solution will be presented in the form of two simple charts that can be used to extract the optimum solution to any problem and predict the optimum cost.

To start, we eliminate s from the formulation by rewriting (6.5) using S instead of s. This yields

$$z(S, H) = \frac{4c_d}{\lambda SH} + \beta\left(\frac{kH}{2} + \Delta + \frac{S}{v_w} + \frac{\ell}{v_{\max}} + \frac{\ell t_s}{S}\right) \text{ [cost/trip]}. \quad (6.11)$$

Note this objective function only uses two decision variables, S and H, but the expression does not decompose as neatly. As a result, its closed form solution is ugly and not easy to interpret.

To simplify matters, we shall fix some of the parameters that do not change much across scenarios at some reasonable values, and allow the rest to vary. We fix the following: $v_w \cong 3$ km/h, $v_{\max} \cong 36$ km/h, and $t_s \cong 5 \times 10^{-3}$ h. We shall also focus on the common case with $k = 2$. This leaves the parameters: λ, β, c_d, and ℓ to vary. With only four parameters, it is now possible to manipulate the formula and reduce the number of parameter combinations that affect the solution. In fact, as we show below, it is possible to write the optimum decision variables as functions of a single parameter; and to express the complete solution in chart form.

The simplification starts by dividing both sides of (6.11) by β (i.e., converting the problem to universal cost units) and then redefining the decision variables as $x \equiv S/\sqrt{\ell}$ and $y \equiv H/\sqrt{\ell}$. Using this change of variable, we find after a few algebraic manipulations that (6.11) reduces to

$$\frac{z}{\beta} = \sqrt{\ell}\left[\frac{(1/p)}{xy} + y + \frac{x}{3} + \frac{(5 \times 10^{-3})}{x}\right] + q\sqrt{\ell}, \text{ [hours/trip]}, \quad (6.12)$$

where p and q are the following constants: $p \equiv \left(\frac{\lambda\beta\ell^{3/2}}{4c_d}\right)$ and $q \equiv \left(\frac{\Delta}{\sqrt{\ell}} + \frac{\sqrt{\ell}}{v_{\max}}\right)$.

Note that (6.12) is not dimensionless and is only valid in units of [pax, km, hours]. The new parameter p is a proxy for both demand and the inverse of technology cost.

It will also be useful to rewrite (6.10) in terms of the new demand parameter. Simple manipulations reveal that it becomes:

$$p > \frac{\sqrt{2}}{2} k v_w^{\frac{1}{2}} (t_s)^{-\frac{3}{2}} = 6928 \text{ [pax-km}^{1/2}/\text{h}^2\text{]}.$$

Note now that the optimal values of the decision variables in (6.12) only depend on the expression in brackets, which only includes the parameter p. Therefore, the optimum solution (x^*, y^*) must be a function of p alone. In other words, we can write:

$$x^* = F_x(p), \quad y^* = F_y(p), \quad \frac{z^*}{\beta} \approx \sqrt{\ell}[F_z(p) + q], \tag{6.13}$$

for some functions F_x, F_y, and F_z that are to be determined. This determination was made numerically by solving the problem for different values of p.

Figure 6.7 displays the optimum design obtained with this procedure. It shows x^* and y^* as a function of p in the range $p > 6,928$ pax-km$^{1/2}$/h^2. For completeness, the figure extends this curve in the range $p \leq 6,928$ pax-km$^{1/2}$/h^2, using the solution given by (6.7), after re-expressing it in terms of p as follows:

$$x^* = \frac{S^*}{\sqrt{\ell}} = \left(\frac{2kv_w^2}{p}\right)^{\frac{1}{3}} = \left(\frac{36}{p}\right)^{\frac{1}{3}}, \quad y^* = \frac{H^*}{\sqrt{\ell}} = \left(\frac{2}{k^2 v_w p}\right)^{\frac{1}{3}} = (6p)^{-\frac{1}{3}}. \tag{6.7$'$}$$

The two curves from the figure can be used as a chart to estimate the ideal designs for grid systems without any numerical calculations. To do this in practice, first, compute $p \equiv \left(\frac{\lambda\beta\ell^{3/2}}{4c_d}\right)$. Then use the chart to read x^* and y^*; and finally, obtain $S^* = x^*\sqrt{\ell}$ and $y^* = H^*\sqrt{\ell}$. Qualitatively, note from the curves how both the spacings and the headways decline with increasing demand. However, whereas the headway tends to zero, the spacing stabilizes at a considerably higher value.

It is also interesting to examine how the optimal cost depends on p. Figure 6.8 plots F_z using a log-scale. This is the top curve of the figure. As in the case of Figure 6.7, the part for $p > 6,928$ pax-km$^{1/2}$/h^2 was obtained by solving (6.12) numerically, and the part to the left (when $p \leq 6,928$ pax-km$^{1/2}$/h^2) using (6.8). More specifically, this was done by rewriting (6.8) in the same form as (6.13), then extracting from it, the term corresponding

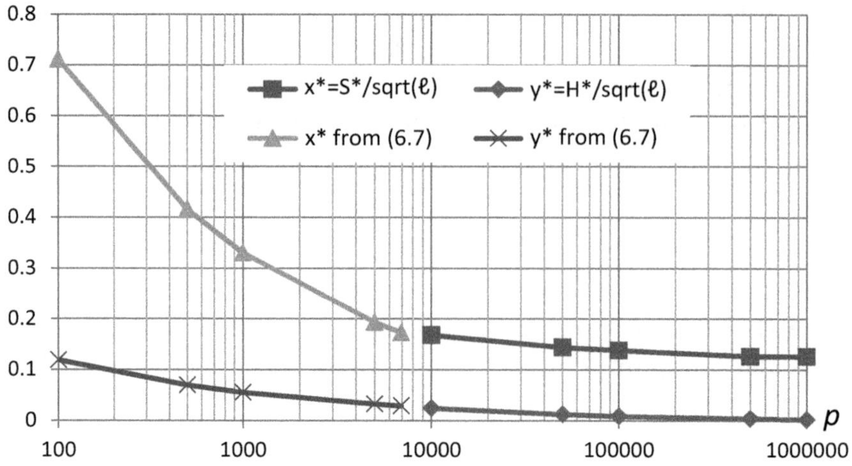

Figure 6.7. Optimal design for grid systems: rescaled line spacings x^* [$\text{km}^{1/2}$] and rescaled headways y^* [$\text{h/km}^{1/2}$] vs. rescaled demand, p [$\text{pax-km}^{1/2}/\text{h}^2$].

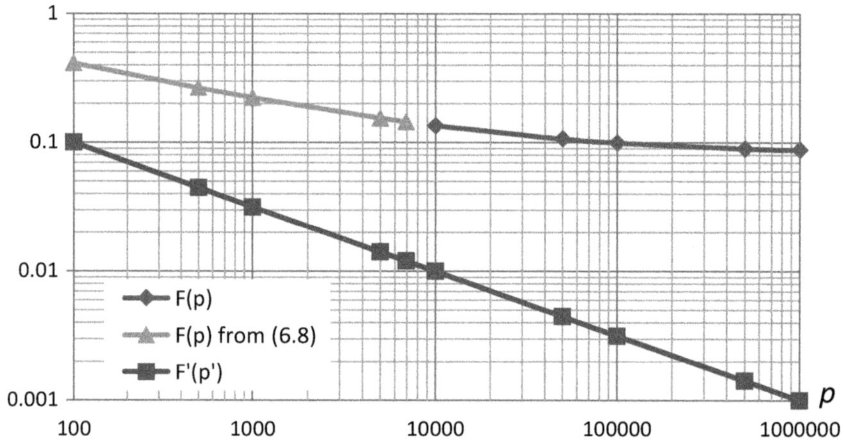

Figure 6.8. How the variable component of optimal cost F [$\text{h/h-km}^{1/2}$], depends on rescaled demand p [$\text{pax-km}^{1/2}/\text{h}^2$].

to F_z, and plotting it on the figure. Simple algebra shows that this term is $F_z(p) = 3[k/(4pv_w)]^{\frac{1}{3}} = (4.5/p)^{\frac{1}{3}}$ for $p \leq 6,928\,\text{pax-km}^{1/2}/\text{h}^2$.

As with Figure 6.7, this chart can be used to predict the optimum cost of a bus grid system without any numerical calculations. Qualitatively, and as one might expect, we see that the optimal cost curve declines, i.e., there are economies of scale. However, the curve flattens out. For comparison purposes, the figure also plots a lower curve that corresponds to corridors. Logically, the curve is lower because in corridor analysis the population is conveniently concentrated along the corridor, so there is no need for transversal travel. Note as well how the corridor curve declines more rapidly and steadily than F_z. This indicates that corridors exhibit stronger economies of scale than grids. We note as an aside that the corridor curve was obtained by rewriting (5.25) in a form similar to (6.13) and plotting the term corresponding to F_z. This similar form is

$$\frac{z_L^*}{\beta} = \left(\frac{2c_d}{\lambda\beta}\right)^{\frac{1}{2}} + 0.058\sqrt{\ell} + \frac{\ell}{36} = \sqrt{\ell}\left[F'(p') + 0.058 + \frac{\sqrt{\ell}}{36}\right], \tag{6.14}$$

where $p' \equiv {}^1\!/_2\lambda\beta\ell/c_d$ and $F'(p') \equiv (p')^{1/2}$.

6.2.3. *Generalizations*

The results of Section 6.2.2 can be generalized to scenarios where people have different trip purposes and more information about the schedule. This can be done merely by changing in suitable ways some of the coefficients in the generalized cost function. In some cases, the problems can be solved analytically and in others numerically. The task is fairly straightforward and left as an exercise (see homework exercise 6.4).

Also interesting is the inclusion of infrastructure costs. This is especially important if grids are being considered for a rail system. Again, this is achieved by adding suitable terms to the objective function. Fortunately, although this complicates its functional form it does not change the fact that the problem only has three decision variables and a favorable mathematical programming structure. Therefore, solutions for particular applications can also be obtained easily in this case (see homework exercise 6.5).

The models in this section can be further generalized by including standards for travel time. And, as occurred in Chapter 5, it still turns out that problems where different time standards are placed on trips of different lengths, considering either average passengers or unlucky passengers, can always be reduced to a problem with a single standard, which can always be expressed in a truncated Lagrangian form of the type we have just examined. Thus, the results just presented also apply to the multi-standard problem. The rest of this subsection proves that indeed, the reduction to a single standard is always possible. The discussion can be skipped without loss of continuity.

The standards approach: We show here that if we introduce different travel time standards for trips of different lengths then the general problem with three decision variables (s, S, H) still reduces to a generalized cost problem with a single representative length. Thus, the results derived so far are completely general. We formulate the multi-standard problem for an average passenger[8] using as the objective function a lower bound for the agency cost that ignores the cost of stopping. Thus, the multi-standard mathematical program is

$$\min \$ = \frac{4c_d}{SH} \tag{6.15a}$$

$$\text{s.t.} \left(\frac{kH}{2} + \frac{S}{2v_w}\right) + T(\ell|s) \leq T_0(\ell), \forall \ell, \tag{6.15b}$$

where $T(\ell|s) = \Delta + \frac{s}{2v_w} + \ell\left(\frac{1}{v_{\max}} + \frac{t_s}{s}\right)$. The main complication relative to (5.29) is that (6.15) involves an additional variable, S. The equivalence is now established.

Theorem 6.1. *There is a generalized cost problem for a single representative length that yields the same optimal decision variables as* (6.15).

Proof: Because the assertion in the theorem was already proven for problem (5.29), it suffices to show that (6.15) can be simplified to a form identical to (5.29). To begin recall that (5.29a) is monotonic decreasing in H. Therefore, (5.29) produces the same optimal decision variables if its objective function is changed to the maximization of H. Thus, our benchmark shall be instead:

$$\max\{H : H + T(\ell|s) \leq T_0(\ell),\ \forall \ell; H \geq 0\}. \tag{6.16}$$

To show that problem (6.15) reduces to (6.16), define a new variable Ω through the relation $\Omega^2 \equiv (k/v_w)SH$ and change H to $(v_w/k)\Omega^2/S$ in (6.15). Now note that the new objective function after such substitution decreases with the single variable Ω. Thus, as we did to obtain (6.16), the new objective can be changed to a maximization of Ω. With this objective, the changed version of (6.15) becomes

$$\max\left\{\Omega : \left(\frac{v_w\Omega^2}{2S} + \frac{S}{2v_w}\right) + T(\ell|s) \leq T_0(\ell),\ \forall \ell; \Omega \geq 0\right\}, \tag{6.17}$$

where $H = (v_w/k)\Omega^2/S$. Now note that variable S only shows up in the inequality constraints of (6.17), and we have the freedom to choose S. A

[8] Analysis for a worst passenger is similar; it only involves changing a few coefficients in the analysis.

value of Ω is feasible if and only if the inequality constraints are satisfied for the particular value of S that minimizes the parenthetical term, conditional on Ω. Thus, we can replace said term by its minimum and eliminate S from the formulation. Since the term is an EOQ expression, which reaches a minimum value Ω for $S(\Omega) = v_w\Omega$, it follows that (6.17) reduces to

$$\max\{\Omega : \Omega + T(\ell|s) \leq T_0(\ell),\ \forall\ell; \Omega \geq 0\}, \tag{6.18}$$

where $S^* = v_w\Omega^*$ and $H^* = v_w(\Omega^*)^2/(kS^*) = \Omega^*/k$. This completes the proof because (6.18) is identical to (6.16). □

6.3. Capacity: Grids

The previous section developed optimal designs assuming the street system was infinitely dense, and vehicles could carry any number of passengers so passengers were never left stranded at stations. In reality, however, streets able to carry buses at reasonable speeds are not infinitely closely spaced and buses have limited capacities. This has an effect on the system's capacity and limits what is possible. Therefore, this section explores the issue in detail.

We consider two types of capacity limitations: (i) the ability of the street network to carry buses, and (ii) the ability of the bus system to carry passengers. Effect (i) can be captured by enriching our "blank slate" modeling with two constraints that place a lower bound, S_{min}, on the line spacing and a lower bound, H_{min}, on the headway, i.e., we require

$$H \geq H_{\text{min}} \text{ and } S \geq S_{\text{min}}. \tag{6.19}$$

The value of S_{min} is case-dependent, perhaps on the order of 0.5 km in some American cities but it can vary widely. The value of H_{min} mostly depends on the ability of stations to process vehicles. This ability is influenced by both, the vehicles' dwell times and other operational constraints, e.g., to maintain safe and regular spacing. Generally, H_{min} will depend on the agency and in favorable cases be comparable with 5 min.

To deal with effect (i) we recommend determining H_{min} and S_{min}, and then appending (6.19) to the design problem. It is perhaps easiest to use the formulas of Section 6.2 and check that S^* and H^* satisfy (6.19) *a posteriori*. If one of the constraints is violated which is not necessarily unusual,[9] one should fix the value of the corresponding variable at the limit and re-optimize

[9]When use typical values of the parameters appearing in (6.7) and the limits, we have mentioned we find that the constraints hold if the ratio λ/c_d does not exceed values of order $O(10^1)$ in a [\$, km, h] unit system.

the design.[10] This optimization should be rather easy because the resulting problem has one decision variable less. Therefore, the procedure and formulas are not discussed here.[11]

Effect (ii) is more complicated. It was addressed for corridors in Chapter 5 by looking at the balance between the desired flow of passengers across a "critical load point" and the provided flow of bus seats. Unfortunately, with grids the critical load point is more difficult to define because passenger routings are not fixed. Therefore, the rest of this section develops and examines this constraint. As with (6.19), the new constraint should be appended to the blank-slate optimization problem. And again, it can be first checked *a posteriori*. For low demand systems typical of the U.S. the constraint is rarely binding, so the solutions developed so far should usually hold. When the constraint is violated, the problem should be re-optimized treating the constraint as a pure equality. This can be easily done numerically because as we shall see the constraint has a very favorable form.

The rest of this section is organized in two parts: it first derives the constraint and then uses it to unveil a fundamental relationship between city size and transportation overcrowding.

6.3.1. *The Passenger-Carrying Capacity Constraint*

The derivations in this section apply to both bus and rail systems. We assume that transit vehicles (buses) have a capacity to carry C passengers both standing and seated. Even though some passengers may be standing, we will refer to this number as the number of (virtual) seats. For bus systems, C depends on the type of bus, and for rail systems, on the length and width of the trains.

Note that train length is not unbounded because it cannot exceed the length of the system's boarding platforms. Thus, even for rail systems there is a finite C. Since C has an important effect on system performance, it is important to carefully consider station length when planning a rail system. If for a bus system C is found to be insufficient one can consider linking multiple buses into train-like platoons. This increases the operating cost of each "train" but also aids with the system's productivity. As with rail

---

[10]This procedure works because the log-scale contours of the objective function (a posinomial) are convex.

[11]If both constraints are binding we need to solve three problems: (i) with the headway fixed at the limit; (ii) with the spacing at the limit; and (iii) with both variables at their limits. The best feasible solution amongst these three is the sought optimum.

systems, however, the length of these platoons is limited by the "platform" length at the bus stations along the way.

We shall evaluate the effect of C for an idealized square city of the type considered in Section 6.2. As in that section, trip origins are uniformly distributed in space. The destinations, however, are assumed to be confined to a central square concentric with the city (see Figure 6.9). The sides of this square are parallel to the streets and their length is $\gamma\phi$, where $\gamma \leq 1$.

We now ask: Can a steady demand density λ be served in a city of dimensions ϕ and γ with a transit system characterized by (s, S, H, C)? The answer to this question will be an inequality that will become our capacity constraint and illustrate what is required to prevent overcrowding.

The location of the system's critical load point depends on how people choose routes. We shall continue to assume that people choose the route that minimizes walking distance. It should be intuitive that under these conditions links in the center will be more heavily used than those in the periphery, and that the critical link will lie squarely in the middle. We recognize that this is slightly pessimistic because some passengers may intentionally walk a little extra in order to avoid exposure to traveling and transferring in the central square, but fortunately, it turns out that even if this center-avoidance strategy is carried out to the extreme the results change very little (see Appendix 6.A). Therefore, our assumption is reasonable for qualitative analyses. In practice, one should be even more pessimistic and add a fractional amount to the predicted critical flow on account of stochastic fluctuations.

Formula Development: Recall that the transit grid has horizontal and vertical routes, parallel to the sides of the service region. We expect intuitively that the center of the central square must be a critical point; or more precisely that the four directional links that point into (or out of) the center will carry the maximum flow.[12] Therefore, we shall evaluate the desired flow of passengers on the center-most horizontal transit link pointing to the right. Consideration shows that these are passengers who (i) travel vertically first; (ii) have an origin in the left side of the big square; and (iii) have a destination in the narrow rectangle corresponding to the small shaded area in Figure 6.9(a). (This rectangle is the intersection of the right side of the inner square and the loci of destinations served by the horizontal

[12]The reader can verify by repeating the derivations about to be presented that this statement is true. It will be found that maximal flow is also carried by horizontal links that are vertical translations of the one in the center and vertical links that are horizontal translations. The flow on all other links is lower.

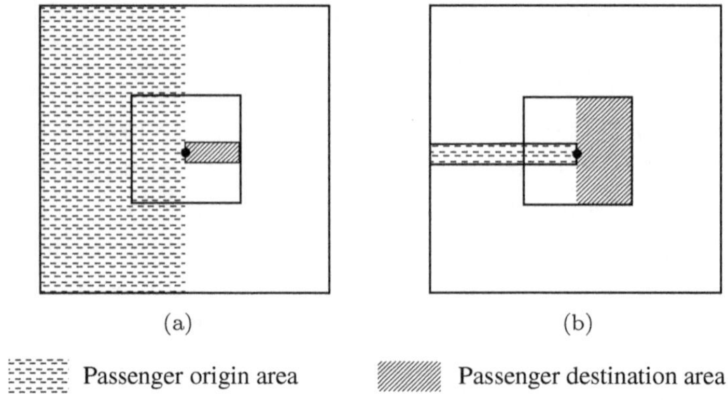

Figure 6.9. Origin and destination catchment areas of passengers using a central horizontal link: (a) traveling horizontally first; (b) traveling vertically first.

transit line passing through the center, i.e., a swath of width S straddling said transit line.) Symmetry considerations also show that there is a second group of passengers who use the same link. They (a) travel *horizontally* first; (b) have *destinations* in the *right* side of the *inner* square; and (c) have an *origin* in the *left* side of the *big* square, in the loci of *origins* for the horizontal line passing through the center (see Figure 6.9(b)).

The flow demanded by the first passenger group is $[\lambda/2][\phi^2/2][\frac{1}{2}\min\{1, S/(\gamma\phi)\}]$. The first bracket is the demand density arising from passengers satisfying (i), the second the origin catchment area (ii), and the third the fraction destinations in the destination catchment area (iii). The minimum operation is necessary to correctly express the case when S exceeds the side of the central area. The demand flow for the second group is the same. Therefore, the total demand on the critical link is twice the flow demanded by the first group, i.e.,

$$\text{Demand on critical link} = [\lambda\phi/(4\gamma)][\min\{S, \gamma\phi\}] \quad [\text{pax/h}]. \tag{6.20}$$

The flow of "seats" on the critical link is

$$\text{Flow of seats on critical link} = C/H. \tag{6.21}$$

On comparing (6.20) and (6.21), we obtain the capacity constraint:

$$\left(\frac{\lambda\phi}{4\gamma}\right) H \min\{S, \gamma\phi\} \leq C, \quad \text{where } \gamma \in [0, 1]. \tag{6.22}$$

Note, the left-hand side of this inequality is the passenger occupancy in the buses at the critical load point, which we denote O. Using this notation and assuming, of course, that $S \leq \phi$, equation (6.22) can be rewritten as:

$$O \equiv \frac{\lambda\phi HS}{4\gamma} \leq C, \quad \text{if } \gamma\phi \geq S \tag{6.23a}$$

$$O \equiv \frac{\lambda\phi^2 H}{4} \leq C, \quad \text{if } \gamma\phi \leq S \tag{6.23b}$$

Note that (6.23a) is the relevant instance for distributed demand, since in that case $\gamma = 1$ and we know that $\gamma\phi \geq S$, and likewise that (6.23b) is the relevant instance for concentrated demand, since then $\gamma \to 0$ and $\gamma\phi < S$. Finally note that neither of these constraints involves the design variable s.

6.3.2. *The Three Capacity Constraints Combined: Implications for City Size*

Here we combine the above results with (6.19) to obtain an absolute bound on the city sizes that can be served without overcrowding. The resulting formulas yield important insights at the intersection of transportation and city planning.

We assume that the central region is large enough to contain several transit links so (6.23a) applies. Since (6.23a) implies $\frac{\lambda\phi}{4\gamma} \leq \frac{C}{HS}$, and (6.19) implies $HS \geq H_{\min}S_{\min}$, we can combine the two inequalities to write

$$\phi \leq \frac{4\gamma C}{\lambda H_{\min} S_{\min}}. \tag{6.24}$$

Note how for a given transportation technology and demand pattern, the right side of (6.24) imposes a hard bound on ϕ.

It is interesting to run some numbers. Consider a fairly dense city such as Boston, Madrid, or Beijing, with a population density comparable with 5×10^3 persons/km^2, which may generate on the order of $\lambda \approx 1,000$ [pax/h-km^2] during the morning rush. Let us evaluate the feasibility of serving all this demand with transit. Assume frequent bus service is provided on every major street, so $C \approx 100$ pax, $H_{\min} \approx 0.1$ h and $S_{\min} \approx 0.25$ km (representing typical city blocks). This yields for the right side of (6.24): $1.6 \times 10^4\gamma/\lambda$ [pax/h-km]. Now, since $\lambda = 1,000$ [pax/h-km^2], this implies that $\phi \leq 16\gamma \leq 16$ km. This suggests that dense cities larger than 16 km to a side cannot be served by buses alone. And if the demand is concentrated in a central area (i.e., $\gamma \ll 1$) this critical size is even smaller.

Assume now that the city is served by an extensive subway grid with $S_{\min} \approx 1$ km, long trains with $C \approx 1,500$ pax, and fairly frequent service $H_{\min} \approx 0.1$ h. The constraint then becomes $\phi \leq 60\gamma$ km, indicating that much larger cities can be served with subways. On the other hand, note that the critical size is considerably reduced if the demand is concentrated — e.g., it is limited to only 12 km if $\gamma = 0.2$.

The reader may wonder about cars. To examine this mode, we assume that cars carry one occupant and again compare the passenger demand with the available capacity, assuming generously that the city has one bi-directional arterial street with three lanes in each direction for travel (i.e., without parking) every 0.25 km. Equation (6.20) for the demand on the critical link continues to hold if we treat streets as transit lines, and of course use 0.25 km for S. Then, for our parameter values, the passenger demand flow is $\lambda\phi S/(4\gamma) = (1{,}000\phi)/(16\gamma)$. This flow can occur only if it does not exceed the passenger carrying capacity of one direction of our arterial streets. The car carrying capacity of a three-lane signalized street is estimated to be about $Q_{\max} \approx 2{,}000$ [cars/h], which is reasonable if one considers the effects of turns, the friction of parking and allows for about 40% of green time at the traffic signals. Recall the car occupancy is $C \approx 1$ [pax/car]. Thus, the capacity limit for one of our arterial directions is $Q_{\max}C \approx 2{,}000$ [pax/h] and the constraint becomes $\phi \leq 32\gamma$ km. This is better than the bus system, but considerably less than what subways can do.

Finally, if one combines all three modes the city size limit is $\phi \leq (16 + 60 + 32)\gamma\,\text{km} = 108\gamma\,\text{km}$. Although this is a large number we also see that when the demand is concentrated, say with $\gamma = 0.2$, even all the transportation modes combined cannot meet the rush hour demand of a very large city.

On a deeper philosophical level, we may ask why such a limit arises. A simple answer is that, for a fixed population density, the number of transportation resources a city can deploy (e.g., total route length) is proportional to its area, which increases with the square of the city's diameter. In contrast, the demand for transportation, quantified by the total number of kilometers the city's population wishes to travel on any given day is proportional to the third power of the city's diameter — since it is the product of the population, which is itself proportional to the area, and the average trip length which is proportional to the city's diameter. Clearly thus, there must be a limit to this diameter. This speaks to the need of limiting the size of megacities and decentralizing the distribution of their destinations.

6.4. Other Network Structures

Although grids are very common and can perform rather well for scenarios with both distributed and concentrated destinations, other network structures can perform even better in some situations. It should be intuitive for example that hub-and-spoke systems are particularly well suited for scenarios with highly concentrated demand, as the hub can then be located near all the destinations thereby eliminating the need for transfers. Not surprisingly, these types of systems are often used for commuter bus and rail systems. The systems also make sense for distributed demand when infrastructure costs are high, demand is low and the service region is small, as hinted in Section 6.1.

This section will show how to design non-grid systems and how their optimal configurations perform compared with grids. The section can be skipped without loss of continuity. Section 6.4.1 covers pure hub-and-spoke systems, and Section 6.4.2 hybrid systems that combine the best features of the grid and hub-and-spoke concepts.

6.4.1. *Hub-and-Spoke Networks*

We assume as in Sections 6.2 and 6.3 that the underlying street network is a grid. Therefore, at most, four straight spokes can emanate from a hub. In one of the simplest configurations, every route starts from the hub, follows a spoke and then peels off to become a branch that serves local stations (see Figure 6.10). Note, the spokes do not have to contain stations because the stations on the branches can already cover the whole region. This simple layout is sometimes called "trunk" and "branch." This section focuses on the trunk-and-branch concept because its simplicity allows us to illustrate the methodology and practical issues with a minimum of derivations. To further simplify matters we shall consider a system with only two trunks, rather than four.

Trunk-and-branch is not the only possible form of a hub-and-spoke system, however. It is possible for example to break each branch into subsections that are covered by separate bus lines that would only stop at the stations in their branch; this idea is similar to the commuter bus system examined in Clarens and Hurdle (1975). It is also possible to allow buses to stop along the trunk so passengers can transfer among branches before reaching the hub. This idea was explored in Newell (1979). It is also possible to do away with the straight trunk idea completely and allow routes to weave and branch anywhere in the service region. This concept will be explored in Section 6.4.2. In the interest of brevity these ideas are only covered as exercises (see homework exercise 6.7).

The three subsections that follow develop the ideas. Section 6.4.1.1 derives the generalized cost function and a capacity formula; Section 6.4.1.2 uses them to compare the hub-and-spoke and grid concepts; and Section 6.4.1.3

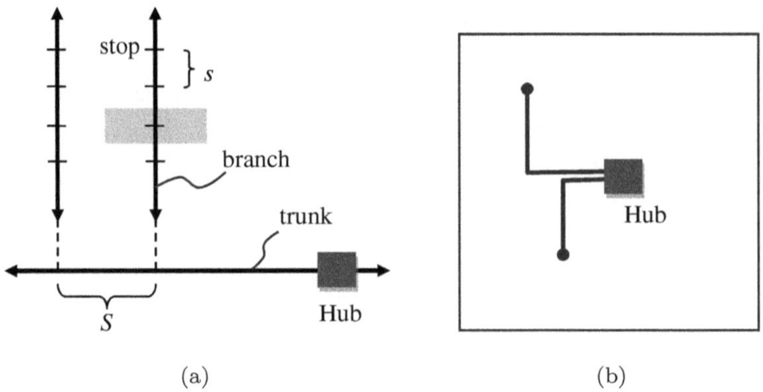

Figure 6.10. Schematic hub-and-spoke network: (a) network layout and (b) typical trip trajectory.

obtains and describes the optimal designs. The methodology is as important as the results.

6.4.1.1. *Derivations*

We study a system with trunks in the horizontal direction only, as in Figure 6.10, and continue to focus on an average passenger. As before we assume that the service region is a square with side length ϕ and that the routes are unsynchronized but operated with a common headway. Again, passengers are assumed not to have appointments, be unaware of the schedule and to walk to/from the stop nearest to their trip ends. Unlike with grids, however, they must now transfer at the hub — unless of course, their origin and destination stops happen to be on the same branch, which occurs with low probability when $\phi \gg S$.

To begin, we now develop formulas for agency cost and passenger travel time, including its components: access, waiting, and in-vehicle riding time. And again to keep the formulas short, we do this for bus systems where the infrastructure cost can be neglected. This turns out not to be too difficult because our trunk-and-branch network is characterized by the same decision variables as grids: stop spacing, s; route spacing, S; and headway, H.

Consider first the agency cost. It will be evaluated on a per trip basis and divided into two parts: the local cost of covering the branches and the line-haul cost of covering the trunk connectors. For the branches, we consider the events in a single tile during a single headway. This spatiotemporal box generates λsSH people. In every box, buses make two stops (one in each direction) and travel $2s$ distance units, entirely covering the lines inside the tile. Therefore, the bus cost in the unit of analysis is $2(c_s + c_d s)$. When this total is divided by the number of people generated in it, we obtain the local cost per trip: $2(c_s + c_d s)/\lambda sSH$.

The unit of analysis for the line haul cost is the product of a headway interval and the complete service region. We assume the hub is located at the center but formulas can also be developed if the hub is located elsewhere (see homework exercise 6.8). To do this, denote the average length of a connector by χ. It should be intuitive that $\chi \approx \phi/4$ if $S \ll \phi$ since the shortest connector is about $S/2$ km long and the longest about $(\phi - S)/2$ km.[13] Now, since the number of trunk connectors (considering both the up and down sides and inbound and outbound directions toward the hub) is $4\phi/S$, it follows that the total connector length is $4\chi\phi/S = \phi^2/S$. This is the line-haul distance traveled by all buses during each headway. Therefore, the line-haul cost per headway is $c_d\phi^2/S$. Consequently, since the region generates $\lambda\phi^2 H$ passengers per headway, we see that the line-haul cost per-trip is $c_d/(\lambda HS)$.

Finally, we must consider the cost of stopping at the hub. If we reasonably assume that a vehicle traversing the hub experiences cost c_s twice, once on the inbound stop and then again on the outbound stop, we see that the cost per headway due to stops at the hub must be the product of c_s and the number of inbound and outbound vehicle stops; i.e., the number of directional connectors. Since this number is $4\phi/S$, the stopping cost at the hub is $4c_s\phi/S$ and the prorated cost per trip is $4c_s/(\lambda\phi SH)$.

We can now add the agency cost components listed at the end of the previous three paragraphs to obtain the total agency cost per trip. After rearranging terms, we find

$$\$ = \frac{2}{\lambda SH}\left(1.5c_d + \frac{c_s}{s} + \frac{2c_s}{\phi}\right) \approx \frac{1}{\lambda SH}\left(3c_d + \frac{2c_s}{s}\right) \quad \text{[cost per trip]}. \tag{6.25}$$

The last term in the middle member, $2c_s/\phi$, has been neglected because it is small compared with the previous term (since $\phi \gg s$). Notice that (6.25) is very similar to the formula for a grid network (6.2b), except that the coefficients "3" and "2" of the two terms in (6.25) are substantially smaller than the corresponding coefficients of (6.2b), which are "4" for both terms. This occurs because service is reduced from double to single coverage, which more than compensates for the extra length of the connectors.

[13] The exact formula when ϕ/S is an integer is: $\chi \approx \phi/4$ if ϕ/S is even, and $\chi \approx \phi[1 - (\phi/S)^{-2}]/4$ if ϕ/S is odd. This can be verified by drawing a picture and counting the lengths of the connectors, which form an arithmetic series.

To complete the cost accounting we now turn our attention to passengers, considering an average case. We assume that the lines are unsynchronized at the hub, and do this for three scenarios corresponding to the three cases considered for grids (see equation (6.3)): (i) distributed destinations near the origins (short trips); (ii) distributed destinations all over the square (long trips); and (iii) concentrated destinations at the hub (long trips).

Note that scenario (iii) allows passengers to complete their trips without a transfer, using a single bus, but the other two scenarios do not. If we neglect the possibility of destinations on the same line, scenarios (ii) and (iii) require exactly two bus trips and one transfer. This is slightly pessimistic but will serve the purpose of keeping things simple. We shall use $n = 1$ or 2 for the required number of bus trip in a scenario.

Now note that the average waiting and transfer for a complete trip is $T_w = H + \Delta$ for scenarios (i) and (ii), and $T_w = H/2$ for scenario (iii). And using the indicator variable n we can combine these two expressions into the following: $T_w = -\Delta + n(H/2 + \Delta)$.

The average access time is independent of the scenario. Consideration of Figure 6.10(a) shows that it is $T_a = (S + s)/(2v_w)$.

Riding time also depends on the scenario. Note that the average passenger leg on a bus (including the local plus line-haul components) is $\phi/2$ km long, and that on average the passenger is delayed by stops only on one half of that leg. Thus, the total time for the average leg is $\phi/(2v_{\max}) + \phi t_s/(4s)$. Since the number of legs is n, we have $T_r = n\ [\phi/(2v_{\max}) + \phi t_s/(4s)]$.

Combining these expressions, we can write the total door-to-door travel time for all three scenarios as

$$T = \frac{S+s}{2v_w} - \Delta + n\left(\frac{H}{2} + \Delta + \frac{\phi}{2v_{\max}} + \frac{\phi t_s}{4s}\right). \tag{6.26}$$

To find the best design, we can minimize (6.25) subject to a level of service standard $T \leq T_0$. Or equivalently, we can minimize the generalized cost per trip $z = \$ + \beta T$ for some suitable β. If, as before, we neglect the relatively small cost associated with bus stops (not just at the hub, but everywhere) then the terms including c_s in (6.25) are eliminated and the general cost formula becomes

$$z(S, H, s) \approx \frac{3c_d}{\lambda SH} + \beta\left[\frac{S+s}{2v_w} - \Delta + n\left(\frac{H}{2} + \Delta + \frac{\phi}{2v_{\max}} + \frac{\phi t_s}{4s}\right)\right]$$
$$[\text{cost/trip}]. \tag{6.27}$$

We can also derive a formula for capacity. The number of passengers in a bus while traveling in line-haul is the product of the demand rate, λ, the bus catchment area $S\phi/2$ and the headway H. Thus the maximum bus occupancy is $\lambda HS\phi/2$ and the bus capacity constraint is

$$\lambda HS\phi/2 \leq C. \tag{6.28}$$

Also important is the flow of buses and passengers that enter the hub in one direction, since too large a flow can be problematic. The directional flow of buses is the ratio of the number of inbound lines coming from one side of the hub, which is ϕ/S, to the arrival headway H:

$$\text{Directional bus flow at the hub} = \phi/(SH). \tag{6.29}$$

The directional flow of passengers is the product of the left side of (6.28) and (6.29):

$$\text{Directional passenger flow at the hub} = \lambda\phi^2/2. \tag{6.30}$$

As one might expect, this is one half of the city's trip generation per unit time — since exactly one half of the traveling public enters the hub in one of the two directions.

6.4.1.2. *Performance Compared with Grids*

Section 6.1 suggested that hub-and-spoke systems should be better than grids for smaller cities with low demand, and when operating costs are high. Here we compare (6.27–6.30) with the corresponding equations for grids in order to determine a bit more precisely the best application domain of each system. We look at capacity/feasibility issues first, and cost comparisons second.

Feasibility and Capacity: To look at bus capacity, we compare (6.28) with the corresponding equation for grids, which is (6.23a) with $\gamma = 1$. Note that the left hand side of (6.28) is exactly double that of (6.23a) as a direct result of single coverage; so the hub-and-spoke system requires larger vehicles.

With grids, the flow of buses on a link is $1/H$ and there usually are no capacity issues with the guideway. But with the hub-and-spoke system, the guideway capacity along the trunk matters a great deal. This is because the flows of the different lines are superimposed on the trunk, reaching a maximum at the hub. As (6.29) shows, this maximum is ϕ/S times bigger than that for grids. If the city is sufficiently large, the resulting flow can approach the guideway capacity, Q_G [veh/h], so a constraint of the form, $\phi/(SH) \leq Q_G$ should be added to the problem. We write it as

$$SH \geq \phi/Q_G \equiv (SH)_{\min}. \tag{6.31}$$

It should be clear from (6.31) that hub systems, unlike grids, cannot provide an extensive mobility solution for large cities. Hub-and-spoke systems need to play a more specialized role. The particulars can be understood by looking at the cost structure.

Cost considerations: To best illustrate the issues we compare the cost structure of both systems assuming that both are set up to deliver the same spatial coverage, i.e., that the values of $(s,\ S)$ are the same for both systems.[14] The cost equations to be compared are (6.5) with $k = 2$ for grids; and (6.27) for hub-and-spoke systems. Equation (6.5) with $k = 2$ is repeated here for the reader's convenience:

$$z(S, H, s) = \frac{4c_d}{\lambda SH} + \beta\left(\frac{S+s}{2v_w} + \Delta + H + \frac{\ell}{v_{\max}} + \frac{\ell t_s}{s}\right).$$

[cost/trip for grids]

To start, let us consider the case with concentrated destinations ($n = 1$ and $\ell = \phi/2$). For this scenario, all the coefficients of (6.27) are either equal or smaller than the corresponding coefficients of (6.5). Thus, (6.27) dominates (6.5): if the main purpose of a system is to serve concentrated demand, a hub-and-spoke system is the answer. This should be intuitive since the hub-and-spoke configuration saves money for the agency, and for concentrated destinations allows passengers to travel without any circuity, zero transfers, and fewer stops.

The case with distributed destinations is more nuanced. In this case, the agency cost continues to be lower for the hub-and-spoke system by a difference of $\frac{c_d}{\lambda SH}$ but, with $n = 2$, the passenger cost is now higher for the hub system. This increase in passenger cost is $\beta[(\phi-\ell)/v_{\max}+(1/2\phi-\ell)t_s/s]$. If trips are short ($\ell \ll \phi$) then this increase is comparable with $\beta\phi/v_{\max}$ (because usually $t_s/(2s)$ is small compared with $1/v_{\max}$). As a result, the hub-and-spoke concept will win if the ratio of the two main changes exceeds one, i.e., if $\frac{c_d v_{\max}}{\lambda\beta SH\phi} > 1$. If trips are long and distributed across the city, then $\ell = 2\phi/3$ and the increase can be seen to be $(\beta\phi/3)(1/v_{\max} - t_s/(2s))$. Again, $t_s/(2s)$ is small compared with $1/v_{\max}$ so the cost increase is approximately $\beta\phi/(3v_{\max}) > 0$. Thus, the condition for hub-and-spoke system to win with long trips can be reduced to $\frac{c_d v_{\max}}{\lambda\beta SH\phi} > 1/3$. For bus systems, the constant on the left hand side of these inequalities is typically much smaller than 1. Therefore, for buses, the grid system will usually outperform trunk-and-branch in the case of distributed demand.

[14]This assumption is reasonable and simplifies the derivations. More tedious calculations reveal that the qualitative conclusions do not change almost at all if we allow the tuple $(s,\ S)$ to take different (optimal) values for each system.

Although our derivations did not include infrastructure costs, we can roughly envision what should happen for rail systems by increasing c_d by three orders of magnitude. In this case, and as we found in Section 6.1, hub-and-spoke systems can win; particularly if the service region is small and the demand and/or value of time are low. The bar is about three times higher, however, if passenger trips are short compared with the city's diameter.

6.4.1.3. *Optimal Design Formulas and Discussion*

To obtain a design, we should minimize equation (6.27) subject to the three feasibility constraints imposed by the stipulated lower bounds on line spacings and headways (6.19), and the trunk capacity (6.28). The constraint $s \leq S$ no longer needs to be considered. The problem can be easily solved numerically since it continues to be a geometric program with only three decision variables.

Note that (6.27) has the same functional form as (6.5) except for the values of some coefficients. Therefore, equations (6.6) and (6.7), which gave the unconstrained minimum of equation (6.5), can still be used if we adjust their coefficients accordingly. The analytical result for our trunk-and-branch system is found to be

$$s^* = \sqrt{\frac{1}{2} n\phi t_s v_w}, \; H^* = \left(\frac{6c_d}{n^2\lambda\beta v_w}\right)^{1/3}, \; S^* = \left(\frac{6nc_d v_w^2}{\lambda\beta}\right)^{1/3}, \tag{6.32a}$$

$$z^* = 3\left(\frac{3nc_d\beta^2}{4\lambda v_w}\right)^{1/3} + \beta\left((n-1)\Delta + \frac{n\phi}{2v_{\max}} + \sqrt{\frac{n\phi t_s}{2v_w}}\right). \tag{6.32b}$$

As in the case of grids, if only one of the constraints is binding we can set the constrained variable at the limit and solve the resulting unconstrained problem with one less variable. If more constraints are binding, it may be best to solve the constrained problem numerically.

We stress that (6.32) is only given as an example to illustrate how formulas can be derived. Real problems may require modifications, but the analysis method is the same. For example, one may want to approximate a real region by a rectangle instead of a square — or another less regular shape; or space the lines in a rectangular pattern rather than a square;[15] or place

[15] An interesting possibility is placing stations on a square lattice oriented at 45° with the streets, where stations are offset on neighboring lines such that $s = 2S$. Another is considering four trunk lines. These possibilities are discussed as part of homework exercise 6.9.

constraints on the maximum waiting time, walking times and overall travel time; or place different weights on these components of travel time; etc. In all cases, one should derive expressions for the components of generalized cost: agency costs, walking time, transfer time, waiting time, and riding time by modeling the situation and then formulate a mathematical program with a meaningful objective function and suitable constraints.

Finally, an important decision in planning a hub-and-spoke system is where to locate the hub. It should be intuitive that if one has a choice the terminal should be centrally located. This can be confirmed by repeating our analysis for non-centrally located hubs; say a distance κ from the center of the region in the horizontal direction.[16] This analysis reveals that the optimum cost increases very little with κ when $\kappa \ll \phi$. Thus, planners enjoy considerable flexibility in choosing a hub location, allowing them to select favorable locations, e.g., near major trip generators/attractors, or where land is available.

6.4.2. *Hybrid Networks*

Considered in this section is a design concept that can improve on the grid and hub-and-spoke concepts in cases of distributed demand, and can alleviate the capacity issues of the hub-and-spoke concept in cases of concentrated demand. As proposed in Daganzo (2010), this hybrid concept will be described generally and evaluated for the (more difficult) case of distributed demand.

Recall that a hub-and-spoke network covers each tile using only one bidirectional route. As a result, it provides all passengers with a single routing option with no choice of transfer location. This requires some detouring. In contrast, a grid network covers each tile with two bidirectional routes in two orthogonal directions. This is more expensive but allows transfers to occur wherever lines intersect, thus eliminating passenger detours and improving the level of service. The hybrid concept combines these ideas by using double coverage in the center of the region and single coverage in its periphery. Because the derivations are somewhat lengthy,

[16]Only two things change: the distance of the connectors, which instead of being $\chi \approx \phi/4$ is now an increasing function of $\kappa, \chi(\phi, \kappa)$; and the average line-haul distance traveled by passengers, which instead of being $\phi/2$ now becomes another increasing function of κ, $\eta(\kappa)$. Because the objective function increases with κ for any feasible combination of parameter, it follows that $\kappa^* = 0$. Furthermore, the functions χ and η have minima at $\kappa = 0$, so they increase very slowly.

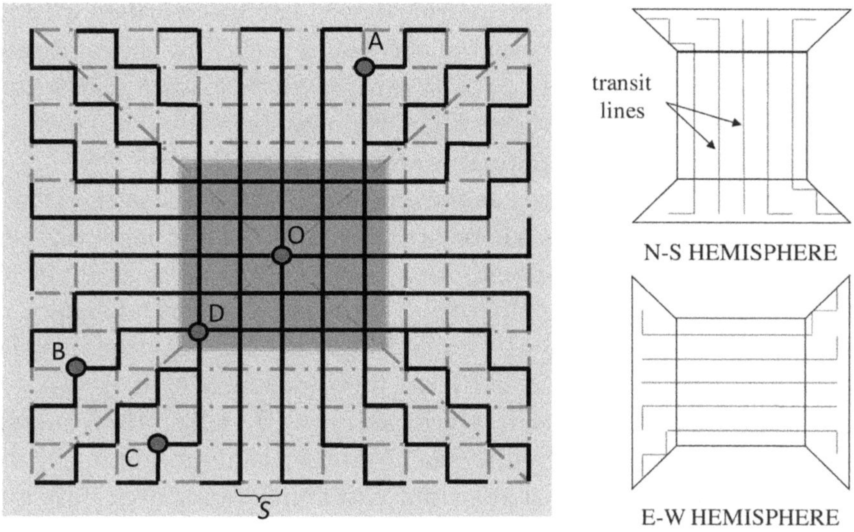

Figure 6.11. A hybrid network structure in a square city (adapted from Daganzo, 2010).

they will be done in a different way, leveraging dimensional analysis ideas and leaning (only very slightly) on Daganzo (2010).

We will assume as in that reference that the goal is to provide the best and most uniform service everywhere along the city and that to do this, stations are placed on a square lattice that is to be covered with the network of lines. This assumption forces the line spacing and the station spacing to be equal. Thus, it reduces the number of decision variables by one and helps with the derivations. Due to the simplification, the results can be slightly suboptimal under average case analysis, but only in some cases. On the other hand, the results in this section shall account for the cost of infrastructure, and will therefore apply to rail systems.

Figure 6.11 illustrates the hybrid concept for a square service region with a dense street grid. Double coverage is provided with a grid of transit lines in the highlighted central square. In this system, all lines run from one edge of the city to the opposite edge and back (N–S–N or E–W–E). Some lines run vertically through the central square and emerge at the north and south ends, and the rest run horizontally and emerge at the east and west ends. The horizontal and vertical lines crisscross in the center but not in the periphery. Passengers are assumed to take the most direct travel path possible. Note, this requires at most one transfer for most trips — two transfers are required only in the relatively rare cases when both the origin and the destination are either in the same quadrant of the periphery or else in opposite quadrants.

A homogeneous version of the hybrid concept can be described with just a few parameters. Homogeneous spatial access is achieved by setting the same stop spacing and line spacing $s = S \ll \phi$ everywhere. This requires lines to branch in the periphery (see Figure 6.11). Homogeneous temporal coverage is only possible in the central square. It is achieved by setting the same headway H for all the lines within it, even as they emerge into the periphery. However, buses must alternate service where branches split so the headway should double after each branching point. As a result, headways in the periphery are lower at the outer edge of the region than near the central square.

To complete the description of our network we must also set the side of the central square. We denote it by $\alpha\phi$ [km], where $\alpha \in [0, 1]$ is a design variable. Note that $\alpha = 1$ yields a grid and $\alpha = 0$ yields a hub-and-spoke network. The latter is slightly different from the trunk-and-branch concept analyzed in Section 6.4.1 as it does not use trunks with overlapping lines and does not provide uniform headways.

Hybrid systems are appealing because for intermediate values of α they allow a high proportion of origin destination pairs to be served with little backtracking. Thus, they offer quasi-grid service, but at a substantially lower cost. Quantitative results in Daganzo (2010) show that indeed, hybrid systems invariably reduce generalized cost and that the optimum size of the core increases with demand and declines with the cost of technology.

The rest of this subsection is about hybrid system design. As before this is done by solving a mathematical program. Therefore, we now develop expressions for the program's objective (generalized cost) function and for its constraints. Because the system is now more complicated, and also to illustrate a different methodology, these formulas will be derived with dimensional analysis.

Let us start with the generalized cost function and in particular with its components on the passenger side: access, riding and waiting time. Note that the access time depends only on the dimensions of the repeating tile around a station, and $s = S$, so it is the same as before:

$$T_a = {}^1\!/\!_2(s + S)/v_w = S/v_w. \tag{6.33a}$$

Now let us consider the parameters and variables that might affect riding time. Note that the riding time T_r [hours] is the ratio of the average riding distance, D_r [km] and the commercial speed v_v. The riding distance should only depend on the geometry of the lines, i.e., on the parameters ϕ [km] and α [dimensionless]. (Recall, the line spacing S is assumed to be small compared with ϕ so it should not affect the result.) Since at most two independent dimensionless groups can be formed with the constants pertaining to riding distance (e.g., α and the ratio D_r/ϕ) it follows that $\{\alpha, D_r/\phi\}$ is a maximal set. Therefore, Corollary 2.1 and Buckingham's π Theorem 2.1 imply that the formula for the distance must be of the form: $D_r/\phi = F_R(\alpha)$, where F_R is a function that is to be determined. So, omitting the argument of the function for notation simplicity, we have $D_r = \phi F_R$, and therefore the formula for riding time is

$$T_r = (\phi/v_v)F_R. \tag{6.33b}$$

Now consider waiting time, including the time for transfers. For any regional configuration (i.e., a fixed ϕ and α), the average across origin-destination pairs of both, the mean passenger wait at the inbound stations and the mean number of transfers should not depend on λ since passengers do not interact. And since $S \ll \phi$, these averages should also be independent of S. Therefore, the average waiting time T_h [h] and the average number of transfers E_T [dimensionless] can only depend on H [hours], ϕ [km] and α [dimensionless]. Obviously, the parameter ϕ cannot be part of a dimensionless group as it is the only one involving the distance dimension, and hence at most two independent groups can be formed for either average. We choose the sets $\{\alpha, T_h/H\}$ for T_h and $\{\alpha, E_T\}$ for E_T. With the maximum number of groups, these are maximal sets. Therefore, Theorem 2.1 and Corollary 2.1 again imply that $T_h = HF_H(\alpha)$ and $E_T = F_T(\alpha)$. Now, since the total waiting time including moving time while transferring is $T_w = T_h + \Delta E_T$, it follows that

$$T_w = HF_H + F_T. \tag{6.33c}$$

Thus, the complete door-to-door travel time is, again omitting the argument "α":

$$T = S/v_w + (\phi/v_v)F_R + HF_H + \Delta F_T. \tag{6.34}$$

Expressions for F_R, F_W, and F_T in terms of α can be obtained with geometric probability considerations. This is a bit tedious but it is shown in Daganzo (2010) that they are

$$F_R = (12 - 7\alpha + 5\alpha^3 - 3\alpha^5 + \alpha^7)/12, \tag{6.35a}$$

$$F_H = (2 + \alpha^3)/(3\alpha) + (1 - \alpha^2)^2/4, \tag{6.35b}$$

$$F_T = 1 + {}^1\!/\!_2(1 - \alpha^2)^2. \tag{6.35c}$$

This result is consistent with our earlier results for grids, i.e., equation (6.4). To see this note that for $\alpha = 1$ (grids), (6.35) yields $F_R = 2/3$ and $F_H = F_T = 1$. Thus, $T = S/v_w + 2\phi/(3v_v) + H + \Delta$. Now, since the commercial pace is $1/v_v \equiv 1/v_{\max} + t_s/s$, and since the average door-to-door distance for our region is $\ell = 2\phi/3$, we can rewrite T as:

$T = S/v_w + \ell(1/v_{\max} + t_s/s) + H + \Delta$, which precisely matches (6.4) if we use in it $k = 2$ and $s = S$.

Let us now consider the agency cost. Formulas for the agency costs will depend on the total length of one-way guideway infrastructure L [km]; the number of stations N [dimensionless] and number of transfer stations X [dimensionless]; and the distance traveled by transit vehicles in a unit of time, V [km/h], which influences the vehicle operating cost per unit time in the region C_v [\$/h]. The infrastructure length and the given "rent" per unit length c_g determine the guideway "rent" as per $C_g = c_g L$ [\$/h]. The number of stations, the known station rent cost c_r and a premium c_x for the more expensive transfer stations determine the total station rent as per $C_s = (c_r N + c_x X)$ [\$/h]. The combined bus distance per unit time V and the vehicle operating cost per unit distance c_v (which must include the cost of stopping and is therefore defined as $c_v = c_d + c_t[1/v_v - 1/v_{\max}]$) determine the total vehicle operating cost as per $C_v = c_v V$ [\$/h]. Thus, the total agency cost in the region per unit time is $c_g L + c_r N + c_x X + c_v V$. Since the region generates $\lambda\phi^2$ passengers per unit time, the cost per trip (in universal units of time) is

$$\$/\beta = (c_g L + c_r N + c_x X + c_v V)/(\lambda\phi^2\beta). \quad \text{[hours per trip]} \qquad (6.36)$$

Since N is the number of basic tiles and X is the number of transfer stations in the center square we have: $N = \phi^2/(sS)$ and $X = \alpha^2\phi^2/S^2$. Expressions for L and V can again be obtained with dimensional analysis. Note that L is independent of s and H and therefore it is a function of ϕ, S and α. The Buckingham π Theorem of dimensional analysis and Corollary 2.1 again tell us that the expression for L must be of the form $L = \phi G_L(S/\phi, \alpha)$. Now a scaling argument comes into play. We note from the geometry of the problem that if S is halved for *any* fixed α then L should double. The only way this can happen is if the bivariate function G_L factors out as $G_L = (\phi/S)F_L(\alpha)$. It then follows that $L = (1/S)F_L$. Finally, consider V, which is determined by ϕ, S, α, H and v_v. Considering that we have six variables and two dimensions we see that a maximal set of groups is $\{V/v_v,\ S/\phi,\ Hv_v/\phi, \alpha\}$. Therefore, the solution must be of the form: $V/v_v = G_V(S/\phi,\ Hv_v/\phi, \alpha)$. And again scaling arguments can be used. We see that if either S or H is halved, one at a time, while holding all other variables constant (including either H or$S$) then V must double. The only way this can happen is if the trivariate function G_V factors out as: $G_V = (\phi/S)(\phi/(Hv_v))F_V(\alpha)$. It then follows that $V = [\phi^2/(SH)]F_V$. Now inserting all these expressions into (6.36), we obtain

$$\frac{\$}{\beta} = \frac{1}{\lambda\beta}\left(\frac{c_g}{S}F_L + \frac{c_r}{sS} + \frac{c_x}{S^2}\alpha^2 + \frac{c_v}{SH}F_V\right). \quad \text{[hours per trip]}$$

It is not difficult to show from geometric probability considerations (see homework exercise 6.11 and/or Daganzo, 2010) that $F_L = 2(1 + \alpha^2)$ and $F_V = 2(3\alpha - \alpha^2)$. Thus, the final expression is

$$\frac{\$}{\beta} = \frac{1}{\lambda\beta}\left(\frac{2c_g}{S}(1+\alpha^2) + \frac{c_r}{sS} + \frac{c_x}{S^2}\alpha^2 + \frac{2c_v}{SH}(3\alpha - \alpha^2)\right). \quad \text{[hours per trip]} \tag{6.37}$$

Again the result matches previous formulas for square grids. In particular, note how if we set $\alpha = 1$, $c_g = c_r = c_x = 0$ and $c_v = c_d$ as per the assumptions of Section 6.2.1 then (6.37) yields $\$ = 4c_d/(\lambda SH)$ in agreement with the first member of (6.5).

Equations (6.34) and (6.37) can now be added to obtain our generalized cost objective function.

$$\frac{z}{\beta} = \frac{1}{\lambda\beta}\left(\frac{2c_g}{S}(1+\alpha^2) + \frac{c_r}{sS} + \frac{c_x}{S^2}\alpha^2 + \frac{2c_v}{SH}(3\alpha - \alpha^2)\right)$$
$$+ \frac{S}{v_w} + \frac{\phi}{v_v}F_R + HF_H + \Delta F_T. \tag{6.38}$$

This expression should be complemented with lower bounds for the spacing and the headway (6.19) and a passenger carrying capacity constraint akin to (6.23). To be realistic the latter should be evaluated with the average demand density for the rush hour, which we denote Λ rather than the average demand for the day λ, which is smaller. This is particularly important in the case of rail as it helps determine the necessary length of the train station platforms and the stations themselves, which are extremely expensive.

We formulate the passenger carrying capacity assuming (reasonably) that the critical load point is in or around the central square where the headways are H. In view of this, we can simply stipulate that the flux of seats provided by the system across every cross-section in the central area, which is C/SH [seats/time-distance], must exceed the desired flux of passengers across the worst (critical) cross-section. To write this condition let us introduce the variable φ [pax/time-distance] for the desired passenger flux across the critical cross-section. Now, considering how passengers choose routes we see that φ should only depend on the peak demand Λ, and the geometry of the region $\{\phi, \alpha\}$. Thus, we are looking for a formula involving four variables $\{\varphi, \alpha, \Lambda, \phi\}$ and three dimensions {passenger quantity, time, distance}, and therefore, we can form at most two independent groups. One such (maximal) set is $\{\alpha, \Lambda\phi/\varphi\}$. This implies that the solution must be of a form $\Lambda\phi/\varphi = F_C(\alpha)$, where F_C is a function to be determined. Since our capacity considerations dictate that $\varphi \leq C/SH$ it follows that $(\Lambda\phi SH)/F_C(\alpha) \leq C$. Note, the left hand side of this inequality is the vehicles' passenger occupancy at the critical load point, O. Thus, we can write our capacity constraint as

$$O \equiv \frac{\Lambda\phi SH}{F_C(\alpha)} \leq C. \tag{6.39}$$

Note the resemblance with (6.23a). Daganzo (2010) derives an approximate expression for F_C and points out that the critical link always touches the perimeter of the central square: the link is either in the periphery pointing to the perimeter or lies entirely on the perimeter. The latter occurs in most cases, unless α is very small, which explains why circulator lines around city centers tend to be heavily used. The formula for F_C is

$$F_C(\alpha) = \max\left\{\frac{(1-\alpha^2)}{2\alpha}, \frac{(3+2\alpha^2-3\alpha^4)}{8\alpha} + \frac{\phi}{S}\frac{(1-\alpha^2)^2}{32}\right\}. \tag{6.40}$$

Note how (6.39)–(6.40) and (6.23a) are the same if $\alpha = \gamma = 1$.

This completes the derivations. Although the formulas look complicated, the minimization of (6.38) subject (6.39)–(6.40) and (6.19) only involves three decision variables. Furthermore, the conditional problem with α fixed is a geometric program that can be readily solved. Thus, the optimization step of the design process can be readily completed for any city or region.

Generic insights can be developed by solving the design problem for a few representative city types and technologies. Table 6.1 presents some results for the scenarios studied in Daganzo (2010) which compare the performance of the Bus, BRT and Metro technologies.[17] Constraints (6.19) were not used. The problem's parameters are summarized in Table 6.2.

Note how the size of the central district is very large for buses across the four city types. This is consistent with our earlier findings comparing hub-and-spoke vs. trunk-and-branch at the end of Section 6.4.1.3. Note as well how, for a given city type and in agreement with our earlier findings, the size of the central square declines as technology becomes more expensive. Finally, note that bus dominates subways in the table. This is reasonable for the scenarios studied, but *would not occur for higher demand levels and cities with narrow streets where constraints* (6.19) *would come into play.*

[17]The results are slightly different as those in Daganzo (2010) because of round off error and because that reference assumes that dwell time increases by 1 s for every boarding passenger in the case of bus and BRT, whereas we assumed that the dwell time is fixed. Section 6.6.3, explains how to incorporate such an assumption in the analysis.

Table 6.1. Design and performance of hybrid transit networks in four representative cities.

	Tech.	α	s(km)	H (min)	O (pax)	$\$/\beta$ (min)	T (min)	*Total* (min)
Benchmark city	Bus	0.93	0.47	5.6	76	6.2	43.9	50.0
($\lambda = 20,000$ pax/h,	BRT	0.79	0.54	4.4	113	8.9	38.5	47.4
$\phi = 10$ km)	Metro	0.37	0.97	2.9	312	21.8	53.3	75.1
Dense city	Bus	0.97	0.40	3.1	120	3.3	40.4	43.7
($\lambda = 80{,}000$ pax/h,	BRT	0.91	0.42	2.7	146	4.2	34.4	38.6
$\phi = 10$ km)	Metro	0.69	0.66	3.0	490	11.1	41.9	53.0
Sprawl city	Bus	0.93	0.70	8.9	93	10.4	72.6	83.1
($\lambda = 20{,}000$ pax/h,	BRT	0.75	0.86	6.2	150	17.5	62.5	80.0
$\phi = 20$ km)	Metro	0.26	1.70	3.1	399	42.6	88.6	131.2
Big city	Bus	0.99	0.58	4.8	120	5.8	66.6	72.4
($\lambda = 80,000$ pax/h,	BRT	0.93	0.64	4.1	150	8.6	54.7	63.4
$\phi = 20$ km)	Metro	0.54	1.09	3.8	721	21.6	66.8	88.4

Source: Daganzo (2010).

Table 6.2. Input parameters.

Modes	v_w (km/h)	β (\$/h)	v_{max} (km/h)	t_s (min)	Δ (min)	c_g (\$/km-h)	c_v (\$/km)	$c_r = c_x$ (\$/stop)	C (pax)
Bus	2	20	25	0.5	0.9	9	3.6	0.01	120
BRT	2	20	40	0.5	0.9	90	3	0.01	150
Metro	2	20	60	0.75	3	900	8	0.03	1000

Source: Daganzo (2010).

6.5. Designing Real Systems

The chapter so far has discussed how to optimize and evaluate different transit system concepts in a "big picture" kind of way. Formulas were given for the design parameters that characterize these systems and for the optimal generalized costs that the systems can achieve under different scenarios. The purpose of this section is showing how these theoretical ideas and methods can be translated into plans for real systems. Section 6.5.1 explains how to do it, and Section 6.5.2 presents a real-world case study and implementation. The ideas also apply to corridors.

6.5.1. *The Design Procedure*

The proposed procedure has two phases: a sketch planning phase I where the system is treated in a big-picture way as in the previous sections and chapters; and an adaptation phase II where a detailed plan is finalized using the sketch plan as a guide. The procedure can be followed whether designing a transit system from scratch or retrofitting an existing system. It is broken below into smaller steps (1–4 for phase I) and (5–6 for phase II):

(1) *Define the scenario's parameters, including its demand and supply sides*: Obviously, the first step is to understand the scenario, i.e., what the city wants, has and can provide. The demand-side should be investigated in consultation with the city's decision-making body. As explained in Chapter 3, forecasts that extrapolate current use should not be used to establish a target demand level because they are highly dependent on how good the baseline system is. Instead, targets should be based on the body's vision, recognizing that a good system will generate its own demand. This vision should help us set reasonable values for the target trip generation rate, the distribution of destinations and the passenger's value of time that should be used for design. The decision-making body should also tell the designer if there is a budget and/or standards to be met. Cost parameters should also be obtained at this stage, in consultation with the city's planning body and the transit agency.

(2) *Identify a concept*: The next step consists in adopting a design concept appropriate for the area of interest. This could be a grid, a hub-and-spoke system, a hybrid network, a superposition of different patterns (e.g., hub-and-spoke and grid, or a hierarchy of grids) or something else. The patterns can be even allowed to change in space, e.g., as shown in Figure 6.12. Part of the pattern could be preexisting and fixed, and only the rest open to design. For example, the central grid in the left pane of Figure 6.12 might be a fixed part that is operated with buses, onto which we want to superimpose a yet to be designed BRT grid for longer distance trips; and the same thing could be said of the right pane. The most important thing is that the concept should do what the city wants.

(3) *Define decision variables*: Next, one should decide how to characterize a specific design form among all that are possible within the concept using a few decision variables. For the three concepts examined in Sections 6.2–6.4 we used the stop spacing s, the line spacing S, and the headway H, but different and/or additional variables may be needed if the design is different or inhomogeneous in space.

(4) *Develop an optimal sketch plan*: Given the scenario, the concept and its decision variables, it is now possible to formulate a mathematical optimization program that aligns its outcome with what the city wants.

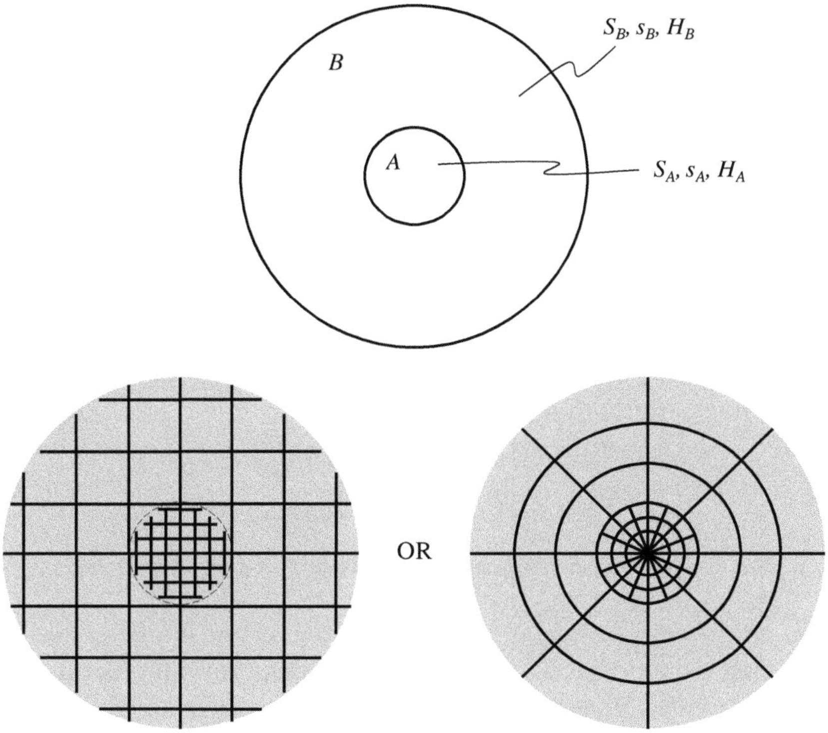

Figure 6.12. Families of design concepts — some examples.

The optimal decision variables define a specific instance of the concept, which is our optimal sketch plan.

In most cases, the mathematical program will be a generalized cost minimization problem subject to constraints. But other possibilities are not excluded. The designer's task is to obtain *approximate* formulas that express the objective function and the constraints of this problem in terms of the decision variables, much as we did in Sections 6.2–6.4. Of course, if the selected concept is covered in those sections, then there is no need for modeling and the formulas from the section can be used. Otherwise, we should develop the formulas from scratch and solve the problem numerically to obtain the optimal decision variables that apply to our scenario. The same modeling tricks can be used. Remember that high accuracy is not necessary since all we are doing is developing a sketch plan, and we have seen in Chapter 3 that our optimization methods are robust to errors in the coefficients of our formulas.

Just to be concrete, we now show the mathematical program for designing a system that is allowed to be inhomogeneous in space and time using the standards approach. Assume that the service region consists of two sub-regions ($r = A,\ B$) that are allowed to have different designs $\{(s_r,\ S_r)\colon r = A,\ B\}$, e.g., as in Figure 6.12, and that the day (or week) is partitioned into different periods such as rush hour, off-peak and night ($j = 1, 2, \ldots$) that could use different headways $\{\ldots,\ H_{rj},\ \ldots\}$. Then, using T for the door-to-door travel time, with subscripts that indicate the origin and destination regions, we can write the standards design problem as

$$\min \$ = \sum_{\substack{r = A, B \\ j = 1, 2, \ldots}} \$_{rj}(s_r, S_r, H_{rj})$$

$$\text{s.t. } T_{AA}(s_A, S_A, H_{Aj}) \le T_{0AAj}, j = 1, 2, \ldots$$

$$T_{BB}(s_B, S_B, H_{Bj}) \le T_{0BBj}, j = 1, 2, \ldots$$

$$T_{AB}(s_A, S_A, s_B, S_B, H_{Aj}, H_{Bj}) \le T_{0ABj}, j = 1, 2, \ldots$$

Note, additional constraints could be added to put limits on access time or waiting time if that is what the city wants. Anything is theoretically possible, really. However, if the problem is constrained too much a feasible solution may not exist.

If the time horizon is long, a system may be deployed in phases. The whole project should then be optimized by including in the mathematical program decision variables for each phase. Some of these variables, like the headways, can be changed easily so they can be treated independently across phases, much as in the mathematical program above. However, line locations (and station locations in the case of rail systems) cannot be changed easily after construction. Therefore, they should be constrained across phases, e.g., an existing spacing between two lines can only be split into multiple new spacings by inserting more lines in between. Obviously, if the number of lines and/or stations increases by only a small percentage from one phase to the next (i.e., only a small number of existing spacings will be split) then their spacings cannot remain even in each sub-region throughout all the phases. This should be recognized in the formulas of the mathematical program. Homework exercise 6.12 explores a way to treat this inhomogeneity while simplifying the optimization process.

Fortunately, the mathematical programs that arise from these efforts usually have a geometric programming form that can be optimized numerically. Of course, Lagrangian and conditional decomposition may further facilitate the process; especially in cases where some of the subproblems admit analytical solutions.

(5) *Adapt to the real world*: The objective here is to produce a feasible plan that will conform as much as possible to the ideal of the sketch plan. To do this the sketch plan of step 4 can be drawn and overlaid over a region's map to get an idea of how much it should be modified to fit within the available streets and for the stations to be at feasible/desirable locations. One should then use judgment to distort the lines and stations of the sketch plan so the new design conforms to all the real-world constraints. These distortions will of course change the agency costs and passenger metrics but only a little. Remember for example that infrastructure and vehicle operating costs are proportional to the length of the lines, and not to their specific shapes. Likewise, station costs depend on the number of stations more than on their locations. Passenger travel time should also change little: it may increase due to the line displacements, but may decrease if stations are placed near major origins and destinations. On the whole, the sum of these two effects should not be very large. Thus, the actual generalized cost of the system should be close to the prediction from the sketch plan.

As the reader can see, this adaptation step is an art more than a science but the process can be carried out fairly easily. Mini-project 2 at the end of this chapter gives the reader an opportunity to practice by developing a design for a transit system in downtown Chicago.

(6) *Re-estimate costs and benefits*: Of course, the final and very important step in all this is to verify with a detailed analysis that the system will operate as predicted. This can be done with simulations and continued with real-world observations once the system has been deployed.

6.5.2. *Real-World Case Study*

The purpose of this section is two-fold: (i) to show how the six-step design method works in practice; and (ii) to illustrate the real-world benefits of systematic design.

6.5.2.1. *The Design Method Applied*

The design procedure of Section 6.5.1 with a modified version of the hybrid concept in Section 6.4.3 was used to design a high-performance, high-frequency bus system for the city of Barcelona. The goal was to provide a complete mobility solution for the city that would leave no pockets uncovered. Key elements of this process are now presented. A more complete description supplied by the design team can be found in Estrada *et al.* (2011).

Prior to 2011, Barcelona's bus network consisted of 63 direct-service lines, which served a population of 2 million people in a region of 100 km^2. This system is for the most part still in place. As Figure 6.13 shows, the system is like a plate of very long and twisted spaghetti. This occurred because in order to avoid transfers routes were created in an ad hoc manner to provide direct service along important corridors. In addition, because the local population likes having stops near their residences, the agency succumbed to local pressure and the system grew to include many stops — this is common everywhere. Because the network was long, routes could not be populated with enough buses so the average headway was somewhat long (13 min); and because there were many stops, the average bus commercial speed was low (11.9 km/h) — which is bad for the agency and bad for the passengers.

Encouraged by predictions made with the method of Section 6.4 by a Catalan think-tank, the city decided to plan and build a new high-performance bus system that would function as a BRT-light network and was called the *Nova Xarxa* (New Network in Catalan). A variant of the hybrid network in Section 6.4.3 was chosen as the design concept; and steps 1–6 of the procedure just discussed were followed.

Figure 6.14(a) shows the concept. Because Barcelona is elongated in shape, the system is customized for rectangles, i.e., such that the two shape-defining decision variables (the size of the central grid d, and line spacing S) can differ for the vertical and horizontal directions. Headway and stop spacing are assumed to be identical throughout the system, however.

The design problem was expressed as a mathematical program, similar to (6.38), that minimized the sum of agency and user costs subject to a capacity constraint and two other constraints limiting the number of lines and the number of simultaneously operated buses. Closed-form formulae were derived for all the elements of this program, as generalizations of (6.33)–(6.37), and optimum solutions obtained numerically. Details can be found in Estrada *et al.* (2011).

With the optimum sketch plan in hand (i.e., at the conclusion of step 4), and in collaboration with the transit agency, the designers adapted it to the available streets, displacing lines and stops a little in order to hit the major demand generators and attractors. The adopted system includes 28 lines: 8 horizontal, 17 vertical and 3 diagonal. The latter three were not part of the sketch plan but were included anyway to serve three major corridors.

Ultimately, the *Nova Xarxa* will replace most of the old bus network — and only 20 secondary routes of the old network will be retained. The *Nova*

Xarxa was activated in stages. The first occurred in September 2012, when a few lines were opened and a few redundant lines of the old network removed. As of early 2015, 13 new lines had been opened. By December 2015, two more portions of the Nova Xarxa were opened, and two more sets of old redundant lines had been eliminated. The system should be completed in late 2018 or 2019. Figure 6.14(b) displays the system map as of November 2017.

6.5.2.2. *Real-world Benefits of Systematic Design*

This section compares the *Nova Xarxa* with the old bus network in order to highlight the benefits of systematic design. It is based on a follow-up study carried out midway through the implementation (Badia *et al.*, 2017).

The *Nova Xarxa* will eventually use 573 buses that will provide an average headway of 6.18 min, similar across all lines. In contrast, the old bus network was served by 761 buses that provided an average headway of 12.30 min. Thus, the *Nova Xarxa* will use considerably fewer buses, which is good for the city's finances, while delivering nearly twice the service frequency of the old network.

From a passenger's perspective, the *Nova Xarxa* is also comparatively very appealing because (i) it covers the whole service area with easy transfers and non-circuitous routings; (ii) it is easy to understand since it is a grid of numbered horizontal and vertical lines; and (iii) it delivers very high frequency with higher commercial speed. The only disadvantage, which pales by comparison, is that it requires longer access times due to its lower station density.

Empirical data from the first three deployment phases (from 2012 to 2015) have confirmed this appeal. Badia *et al.* (2017) find that as of 2015 the *Nova Xarxa*, despite being unfinished, already attracted more demand than the old network. And quite remarkably, at the start of each new phase, the demand jumped more than what was expected in proportion to the number of lines opened for service. This indicates that people use the *Nova Xarxa* as a network involving transfers, and not as a superposition of direct lines, presumably because of the network's desirable properties (i), (ii), and (iii). While conventional bus networks in developed big cities exhibit transfer percentages ranging from 1.5% to 16%, the corresponding percentage for the *Nova Xarxa* at the end of 2015 was approximately 26%, reaching a maximum of 57% for certain lines. According to Badia *et al.* (2017), these numbers should further increase as the *Nova Xarxa* is completed and passengers have even more opportunities for transferring.

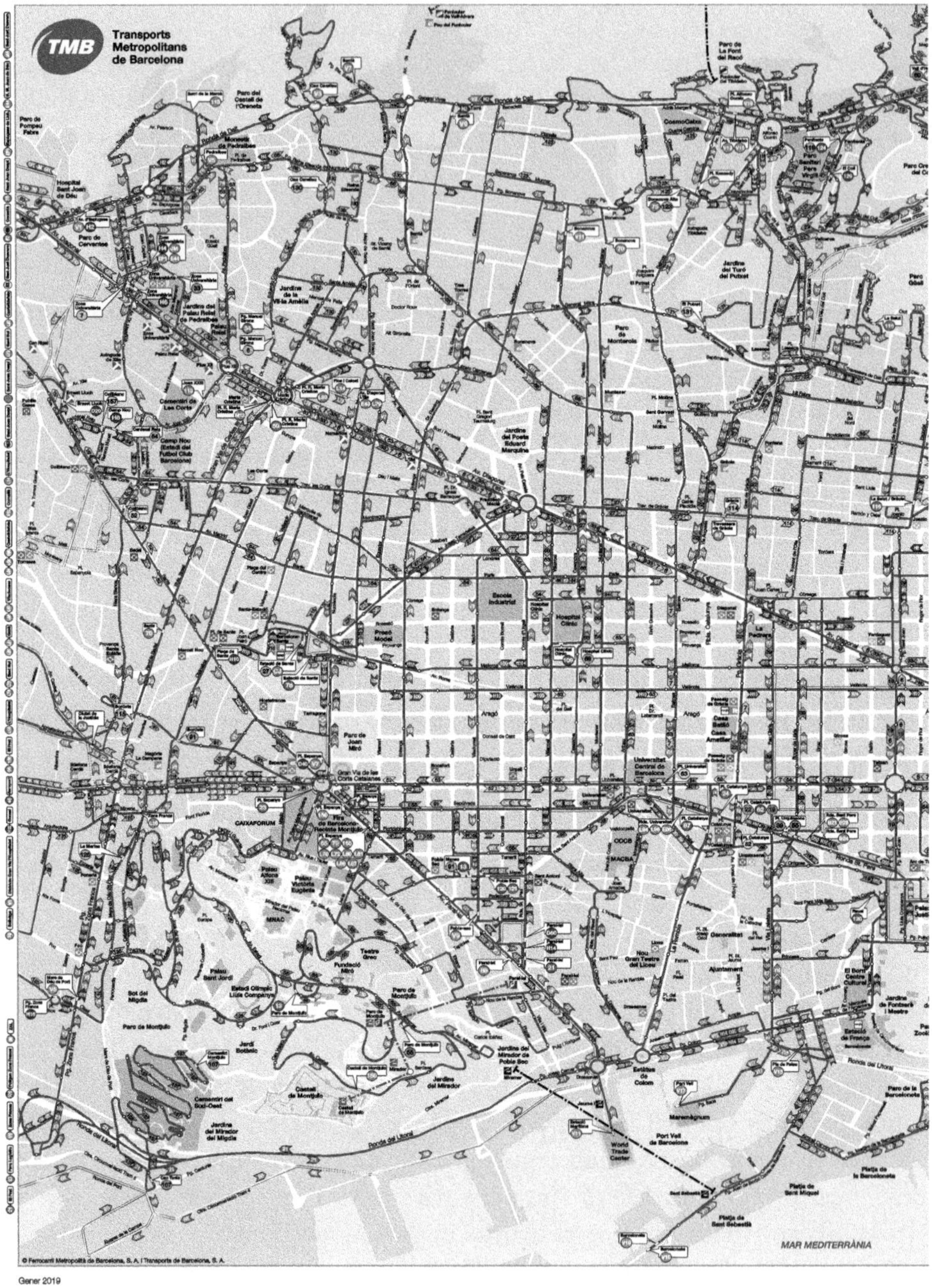

Figure 6.13. Maps of the old bus network.
Source: http://www.tmb.cat/.

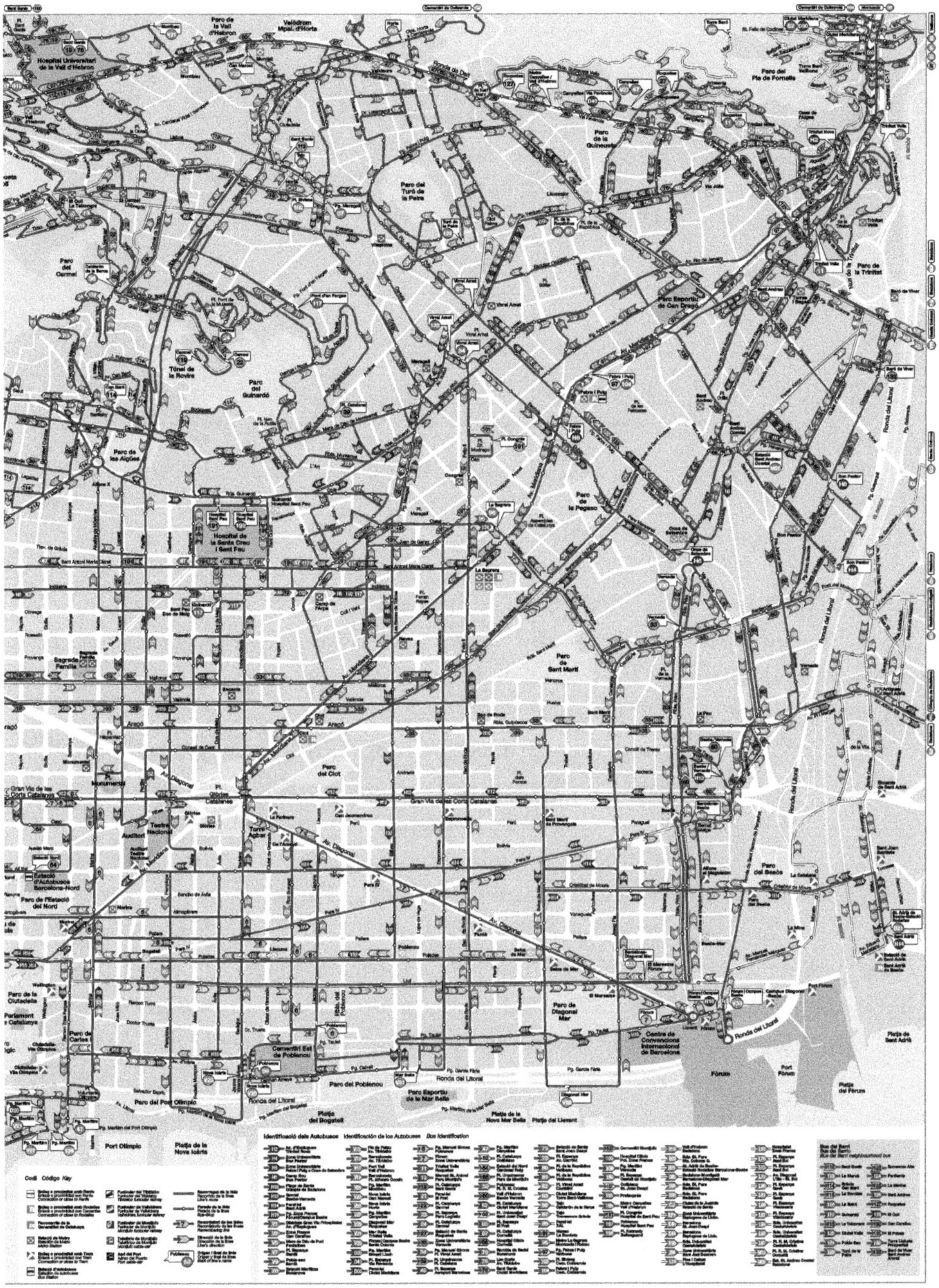

Figure 6.13. *(Continued)*

(a)

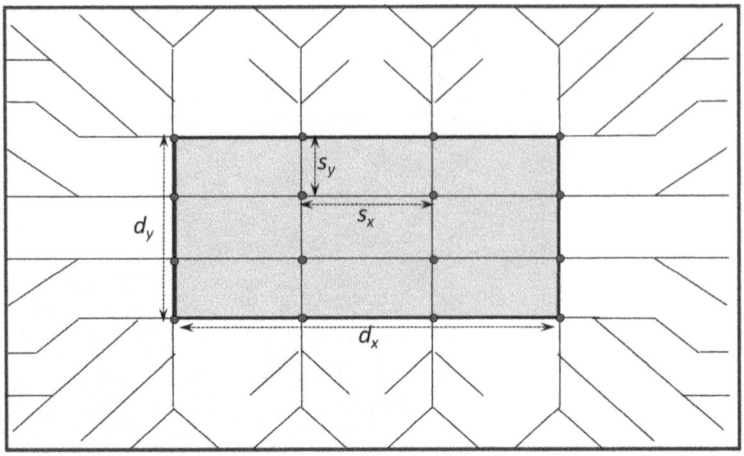

Figure 6.14. The *Nova Xarxa*. (a) Schematic network structure; (b) Actual network layout design.

Source: (a) Estrade *et al.* (2011). (b) http://www.tmb.cat/, accessed in February, 2018.

Curiously, some of the lines with the highest ridership turned out not link major origins and destinations and appear to be used as connectors. This came as a total surprise to the transit agency folks and to the designers. The anecdote offers two important planning lessons. First, it indicates that the conventional planning practice of opening lines along corridors of existing demand should be questioned, since some of Barcelona's most successful lines might not have been opened if the practice had been followed. Second, it speaks to the need of evaluating transit as a complete system and not as a superposition of independent lines. In other words, it is very important to have a grand vision and to develop a master plan that fulfills it. This master plan (like Barcelona's *Nova Xarxa*) should then function as the framework under which smaller projects are evaluated. Barcelona's success story also suggests that master plans can be successfully developed with the six-step method and the sketch planning recipes proposed in this chapter.

6.6. Additional Topics

This section addresses some important issues that often come up in planning, but are less central to the overall understanding. These issues can be addressed with small tweaks to what has already been presented. Therefore, the discussion will be brief. Section 6.6.1 examines possible network shapes when the transit lines are not constrained to follow a preexisting street

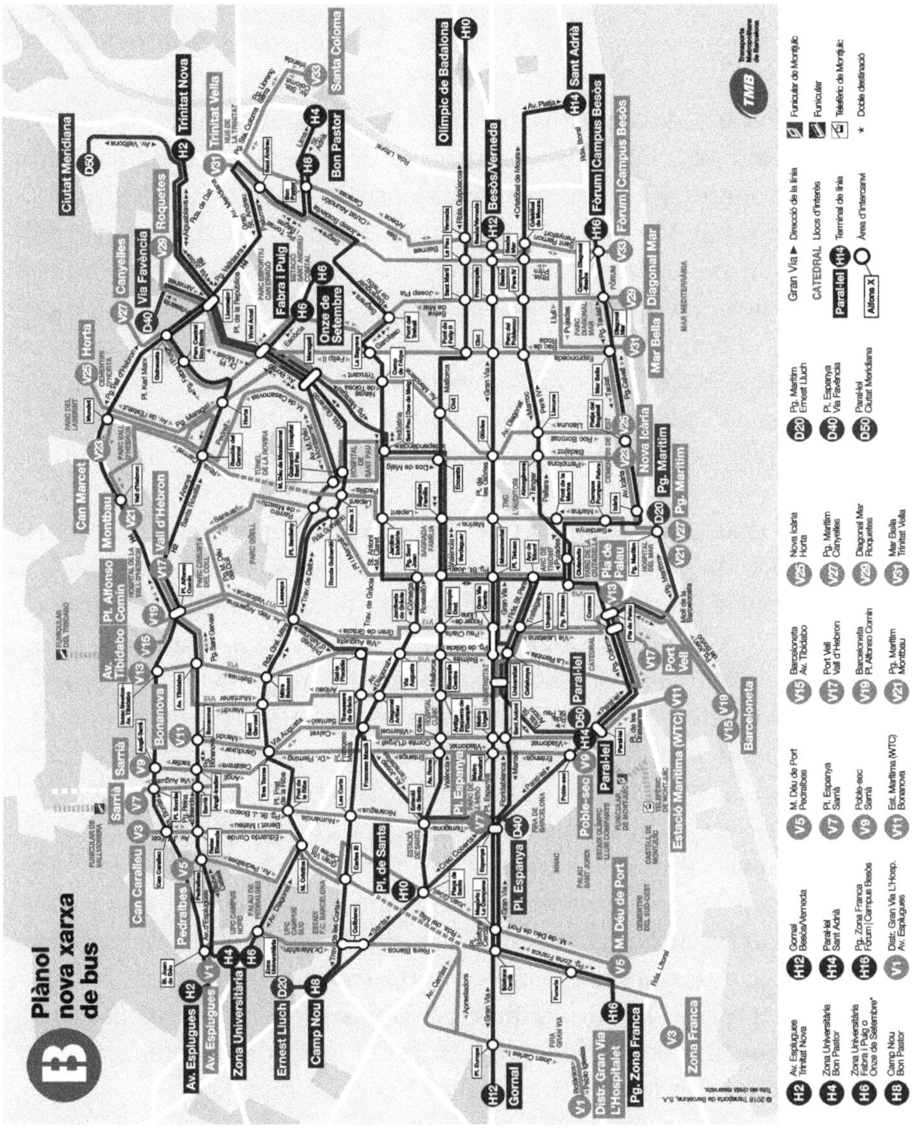

(b)

Figure 6.14. *(Continued)*

system. Section 6.6.2 discusses the synergies that arise in connection with the last mile problem when a transit network is combined with other transportation systems. And Section 6.6.3 analyzes the effect that passenger demand has on bus dwell times, and how to deal with it.

6.6.1. *Networks without Street Constraints*

The design formulas and sketch plans discussed so far assume that the transit system is to be laid on top of a square grid of streets. These idealized designs are then adapted to the real-world street system by deforming the lines and displacing the stations. Although assuming a square grid in the sketch planning phase seems reasonable in many cases, there can be instances where the assumption may be unnecessarily constraining, e.g., if planning deep underground subways or bus systems for new towns. Could we then do better? We will answer this question in two parts: first for "one-to-many" systems of concentrated destinations or their symmetric "many-to-one" counterparts, and then for systems of distributed demand.

We have seen that if the demand is concentrated in a small walkable area, then a hub-and-spoke system rooted in the area dominated grids. This occurred because the hub-and-spoke system offers more direct service than grids with less expensive single coverage. It should be obvious that this comparative advantage does not disappear if routes are not required to follow a grid. Therefore, a hub-and-spoke system is still the best way of serving concentrated demand. However, since the underground infrastructure is expensive, it makes sense to use a branching structure such as the one in Figure 6.11 (with a very small shaded central region) as this structure allows lines to share tunnels near the hub and to operate above ground farther away. This can often be seen in practice, e.g., the Muni rail system in San Francisco. Furthermore, although stations may still form a rectangular lattice since most cities have rectangular blocks, routes can now branch at oblique angles and cover the stations more directly. For a given set of stops, this can save a significant fraction of the infrastructure length and vehicle travel time — perhaps on the order of 10–30% (see Figure 6.15(a)).

Although street-independent hub-and-spoke systems can be modeled from scratch, e.g., with new formulas for the infrastructure length and route length in each tile, it may sometimes be more expedient just to reduce by a small amount the relevant coefficients of the generalized cost functions we have presented in Section 6.4.3 for a small α. Mini-project 2 at the end of the chapter uses these ideas to design a metro system for downtown Chicago.

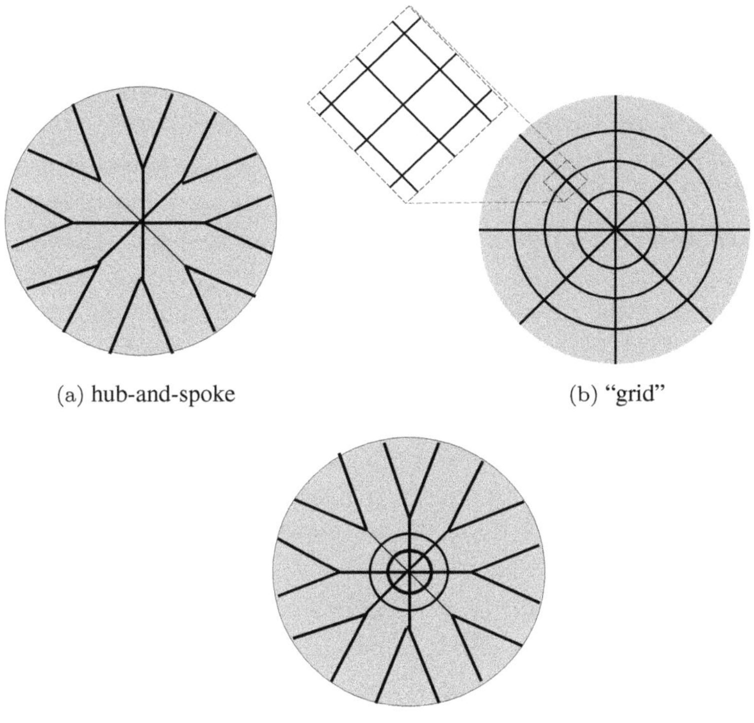

(a) hub-and-spoke (b) "grid" (c) hybrid

Figure 6.15. Representative ring-radial networks.

We have also seen that if the demand is distributed then either a grid system or a hybrid system that provides double coverage in a central region is better than hub-and-spoke. Without the constraint of streets, this double coverage can now be provided more efficiently with a "ring-and-radial" structure such as the one in Figure 6.15(b). Note that the structure contains a hub-and-spoke sub-structure (the set of radii) that can efficiently serve concentrated demand. In theory, thus, the ring-and-radial system can double-up and serve both types of demand.

For distributed demand, the ring-and-radial concept does not add significantly to agency cost and is better for passengers than grids. To evaluate the effect on the agency note that, like grids, the ring-and-radial system still exhibits two orthogonal bidirectional lines in the tiles around each transfer station (except for a few around the very center); see Figure 6.15(b). Ring-radial systems can also be configured as hybrid networks in which single coverage is provided in the periphery, simply by deleting all the peripheral rings; see Figure 6.15(c). Thus, if we control for the number of transfer stations, the agency cost expression for a ring-and-radial system should be similar in structure to that of grids, either full or hybrid.

The beneficial effect on passengers arises because the ring-radial structure allows for more direct routing. Consideration shows that the optimum routing is as follows. To minimize distance, a passenger must use the radial line passing through the trip end that is most distant from the center and then choose from two possible connecting options: (i) the ring line closest to the other trip end, transferring at the point of intersection; or else (ii) the radial line closest to the destination, transferring at the center. A comparison of the two resulting distances shows that option (ii) is faster if the angle of the two radial lines is between 2 and π radians. The availability of option (ii), which is very direct, works to reduce circuity. Geometric probability calculations show that ring-radial routing decreases average passenger distance traveled by about 10% compared with grids.

Unfortunately, all is not rosy. The reduction in distance arises mostly because $(2-\pi)/\pi$ of the passenger trips are routed through the center. This high volume of transfers and transit vehicle traffic is not physically feasible for medium and large sized cities, especially under the moderate to high-demand levels necessary to justify a rail system. Perhaps for this reason — a single hub — ring-radial systems tend to be confined to small towns. Therefore, they are not observed much in real life, *and are not recommended as mobility solutions for large cities.* Therefore, they are not analyzed here; especially since results are available elsewhere (see e.g., Badia *et al.*, 2014; Chen *et al.*, 2015). As in Section 6.4.3, these references consider extensions that allow for double coverage in a central region and single coverage in its periphery.

6.6.2. *Transit Synergies and the Last Mile Problem*

Consideration of optimal cost formulas for different network structures, such as equation (6.8) for grids, shows that network performance improves markedly with an increased walking speed, v_w. This speaks to the importance of paying attention to the last mile problem — improving access to transit stations.

Individual transportation modes such as one-way shared bicycles, shared e-bikes and shared cars can affordably increase v_w. Clearly, these transportation modes are synergistic with transit, and they will be covered in Chapter 7. The increased access speed they provide allows the designer to increase the station spacing. This, in turn, increases the commercial speed, which reduces agency cost and passenger riding time. And with higher commercial speeds

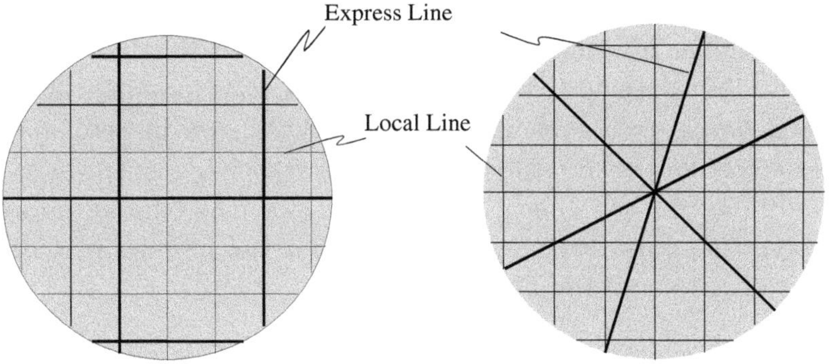

(a) grid + grid (b) hub-and-spoke + grid

Figure 6.16. Examples of hierarchical networks.

the agency can provide shorter headways. Thus, both the agency and the passengers win with a faster access speed.

The last mile problem can also be remediated with another transit system. This idea was already explored for corridors in Section 5.4, by studying the performance improvements attained by combining local and express buses on the same line. We saw that by providing a frequent local line with short station spacings it was possible to increase the spacings of the express line and in this way reduce the overall door-to-door travel time for long passenger trips. In two dimensions, we could do the same. Instead of two lines we would superimpose two networks with different line and station spacings, and the results should be similar: improved generalized costs if the system is intended to serve long trips. The analysis procedure is fairly straight-forward and case-dependent so it is not presented here. Appendix 6.B explores the case of two hierarchical grids depicted in Figure 6.16(a).

In the case of networks, the synergies are stronger because we can mix networks of different types, as shown in Figure 6.16(b). The combination in this part of the figure provides a hub-and-spoke (rail) system that could serve commuter trips with concentrated many-to-one demand, and a grid (bus) system for dispersed local and long distance trips. This combination is often seen in real life, e.g., in Chicago, Illinois, and many other cities that combine a good bus system with commuter rail. The two systems are synergistic because the grid can also function as a feeder to the rail system if it is sufficiently frequent to be appealing. This allows the rail system to be designed with longer stations spacings, which as we have seen offers many benefits.

Ideally, the two systems should be designed together, but this may not always be possible. In either case, we should start with a sketch plan that characterizes the system with as few descriptors as possible. The overall analysis and design processes are similar to those seen in Sections 6.4 and 6.5. Mini Project 2 asks the reader to design a transit system such as the one in Figure 6.16(b) for the Chicago area. Note how the express network can be described with only three decision variables: the number of radial lines, the number of rings and the headway.

6.6.3. *Endogenous Dwell Times and Flexible Stopping*

It has been assumed throughout the chapter that the average station dwell time was an exogenous parameter, t_s. While this is reasonable for subway systems, where t_s is fairly reproducible, it is less so for buses because the actual delay a bus suffers at a particular station depends quite significantly on the number of boarding and alighting moves. And since the average number of moves is influenced by the decision variables s, S and H we see that in these cases t_s should be modeled as a function $t_s = G(s, S, H)$ instead of a constant.

It will be helpful to address this complication after knowing how G behaves. Therefore, we first derive an approximate expression for G. To do this, we reasonably assume that buses skip stations when nobody wishes to board or alight — this is common practice as otherwise buses suffer unnecessary delays, which is bad for passengers and bad for the agency. We also assume that each station stop adds τ time units to the travel time, and that each passenger move, whether boarding or alighting, adds τ' units on the margin.[18] Under these assumptions, the average bus delay per station is

$$t_s = \text{E(number of pax moves at a station)}\tau' + \Pr\{\text{stopping at a station}\}\tau.$$

Now note on passenger conservation grounds that the expected number of moves during a bus stop must equal the average amount of demand generated and attracted in the catchment area of a station during one headway which is $2\lambda HsS$. And if we assume that the actual number of moves is a Poisson

[18]This constant is the inverse of the maximum possible flow of passengers across the threshold between the station and the transit vehicle. For vehicles with multiple and/or wide doors that allow passengers to board and alight on different channels, τ' is *not* the interval between successive passengers in one channel. For example if a train has 10 double doors allowing two passengers to use each door at a time and each passenger move takes 1 s, then the maximum flow is 20 pax/s and $\tau' = 1/20$ s.

random variable with mean $2\lambda HsS$ then the probability of stopping is $(1 - e^{-2\lambda HsS})$. Thus, the formula for t_s is

$$t_s = G(s, S, H) \equiv 2\lambda HsS\,\tau' + (1 - e^{-2\lambda HsS})\tau. \tag{6.41}$$

Note, for high-demand systems (with $\lambda HsS \gg 1$) stops are rarely skipped and (6.41) reduces to

$$t_s = G(s, S, H) \approx 2\lambda HsS\tau' + \tau \quad \text{(for high demand, } \lambda HsS \gg 1\text{)}. \tag{6.42a}$$

For systems such as subways and BRTs where boardings through multiple doors are quicker and decelerations/accelerations take longer, the first term of (6.42a) is much smaller than the second and we can effectively assume that $t_s \approx \tau$. Thus, the complication goes away and we can treat t_s as a constant.

At the other extreme, if the demand is so low that $\lambda HsS \ll 1$, the formula reduces to

$$t_s = G(s, S, H) \approx 2\lambda HsS(\tau' + \tau) \text{ (for low demand, } \lambda HsS \ll 1\text{)}. \tag{6.42b}$$

Now the complication cannot be assumed away. The most direct way to incorporate (6.41) or (6.42a) into the optimization is by substituting the variable t_s appearing in the formulation of the problem (e.g., in Equation (6.5) for grids) by $G(s, S, H)$ and optimize the reformulated problem.

However, because τ and τ' are usually very small compared with the other time constants appearing in the objective function (e.g., s/v_w, S/v_w, H) we find that for typical parameter values G varies much less with the decision variables than other parts of the objective function. This means that the problem can still be approximately optimized treating t_s as if it was constant. In this case, however, its value, t_s^*, is unknown. Fortunately, it can be found indirectly (and the design problem solved very easily) if closed form formulas such as those of Section 6.4 are available.

To do this, simply make an initial guess for t_s assuming a sensible value for $2\lambda HsS$ and use the available formulas to obtain tentative values for s^*, S^*, H^*, and z^*. Now insert these values in (6.41) to obtain t_s^*. The obtained values will usually be accurate enough. To verify this, one can iterate once more with the refined value of t_s and if desired adopt the new slightly changed values. Remember, however, that ultimate accuracy is not the goal since the outcome of the optimization is only a sketch plan.

We conclude with a note on operational flexibility when the demand is very low, as in (6.42b). When this occurs the probability of stopping is also low and buses will often go for several stations without stopping. Often, buses will bypass many stations. This is good because it speeds them up. An even greater benefit can be achieved if buses have the flexibility to take shortcuts when bypassing stations, e.g., using an information system to allow passengers to request stops in real time, both at the stations and onboard. This kind of improvement suggests that for cases with really low demand one would want to completely do away with the routes and the schedules and serve passengers on a demand responsive basis. This is actually done in practice in many different ways. We call these types of systems flexible transit. They are the subject of the next chapter.

References

Badia, H., Estrada, M., and Robuste, F. (2014). "Competitive transit network design in cities with radial street patterns." *Transportation Research Part B*, (59), 161–181.

Badia, H., Argote-Cabanero, J., and Daganzo, C. (2017). "How network structure can boost and shape the demand for bus transit." *Transportation Research Part A*, 103, 83–94.

Chen, H., Gu, W., Cassidy, M. J., and Daganzo, C. F. (2015). "Optimal transit service atop ring-radial and grid street networks: a continuum approximation design method and comparisons." *Transportation Research Part B*, 81, 755–774.

Clarens, G. C. and Hurdle, V. (1975). "An operating strategy for a commuter bus system." *Transportation Science*, 9(1), 1–20.

Daganzo, C. F. (1987). "The break-bulk role of terminals in many-to-many logistics networks." *Operations Research*, 35, 543–555.

Daganzo, C. F. (2010). "Structure of competitive transit networks." *Transportation Research Part B*, 44(4), 434–446.

Estrada, M., Roca-Riu, M., Badia, H., Robusté, F., and Daganzo, C. F. (2011). "Design and implementation of efficient transit networks: Procedure, case study and validity test." *Transportation Research Part A*, 45(9), 935–950.

Holroyd, E. M. (1967). "The optimum bus service: a theoretical model for a large uniform urban area," in *Vehicular Traffic Science* (L.C. Edie, Herman, R., and R. Rothery (eds.), Proc. 3*rd* ISTTT, Elsevier, pp. 308–328.

Newell, G. F. (1979). "Some issues relating to the optimal design of bus route." *Transportation Science*, 13(1), 20–35.

Homework

6.1. You are an independent taxi driver (and the only driver) who provides service to passengers in a linear region spanning the interval [0, 1].

(a) Assume there is a subway station at the origin and you only serve trips that start at the station. Prove that if the passenger destinations are uniformly distributed, the expected travel distance required to serve one passenger including the empty backhaul is 1.

(b) Assume now that there is no subway station and you decided to serve passengers on a FIFO basis. Prove that if their origins and destinations are both uniformly and independently distributed in [0, 1], the expected distance per passenger is 2/3.

6.2. Consider the situation in the diagram below, which includes a terminal and a square service region for a feeder bus. The terminal is half way up the square's west side. The bus needs to pick up one last passenger before returning to the terminal. Both the bus and the passenger are independently and randomly located in the square. All points in the square are equally likely. Streets form a dense grid parallel to the square's sides in the x- and y-directions. Under these conditions, what is the expected distance the bus must travel before reaching the terminal? What would the formula be if the square's side was ϕ instead of 1? (*Hint*: do not repeat the derivations; use dimensional analysis)

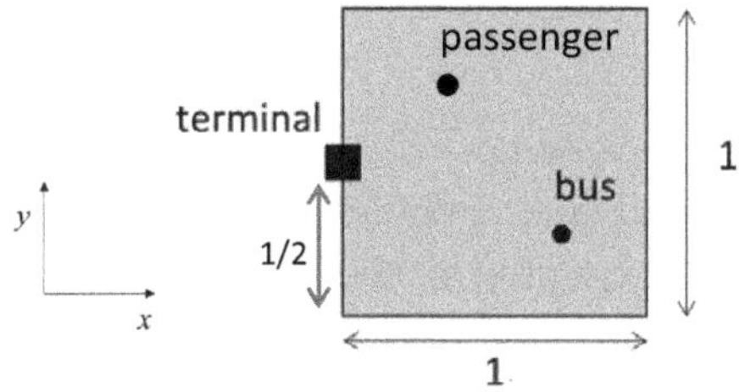

6.3. Suppose passenger origins and destinations in a square city of side ϕ are uniformly and independently distributed. The transit network forms a regular grid parallel to the sides of the square with constant line spacing S. The headway is H everywhere. Passengers use the one-transfer routing scheme described in the text and used in Section 6.3. Derive a formula for the passenger flow on a link that is (x, y) distance units away from the city center.

6.4. In the analysis of grid networks, equation (6.5) was derived for an average-case passenger who neither has an appointment at the destination nor knows the schedule, when transit service is not synchronized. Discuss how the coefficients in (6.5) vary across the 16 possible situations allowed by the following four binary choices: (i) average or worst-case passenger; (ii) passengers with or without appointments; (iii) passengers aware or unaware of the schedule; and (iv) synchronized or unsynchronized service. Optionally, do the same for hub-and-spoke networks and equation (6.27).

6.5. Equations (6.5) and (6.27) and the ensuing results only apply to ordinary bus service because they ignore infrastructure costs. Please derive more general versions that include infrastructure costs. Discuss how these additional costs impact the comparison between grid networks and hub-and-spoke networks. (*Hint*: Recall from (5.19) that infrastructure costs can be embodied in a coefficient c_g with units of [\$/km-h] and that, unlike operating costs, infrastructure costs do not depend directly on the temporal scheduling decisions.)

6.6. *Consider a transit grid network such as the one in Figure 6.6. The service region is a square with a dense grid of streets. The side of the square is an integer multiple of the line spacing S; and the station spacing s is itself an integer sub-multiple of S. Furthermore, as shown in the figure, lines intersect at a station. It was mentioned in the text that the average passenger access distance depends slightly on the routing choice. Therefore, assuming that passenger origins are uniformly and independently distributed in the region, do the following:

(a) Suppose every passenger chooses the nearest stations at both ends of the trip, even if such a choice requires additional transfers. Then, prove that the expected access time is

$$T_a = \begin{cases} (\frac{1}{2}s + \frac{1}{3}S)/v_w, & S/s \text{ even;} \\ (\frac{1}{2}s + \frac{1}{3}S + \frac{1}{6}s^3/S^2)/v_w, & S/s \text{ odd.} \end{cases}$$

(*Hint*: you may want to sketch the catchment areas of each station when S/s is odd and even, respectively. Derive the expected vertical vs. horizontal distances.)

(b) Suppose every passenger chooses the pair of stations at both trip ends that minimize the combined access distance, but conditional on making at most one transfer. Prove that the expected optimal access distance is now

$$T_a = \begin{cases} (\frac{1}{2}s + \frac{23}{60}S)/v_w, & S/s \text{ even;} \\ (\frac{1}{2}s + \frac{23}{60}S + \frac{1}{6}s^3/S^2)/v_w, & S/s \text{ odd.} \end{cases}$$

(*Hint*: The figure below illustrates every passenger's two basic options. The passenger can travel horizontal first and transfer at station 1; or travel vertically first and transfer at station 2. Note, if the passenger transfers at station 1, the total access distance is $\min\{|OA|, |OB|\} + \min\{|ED|, |FD|\}$; and something similar happens at station 2.)

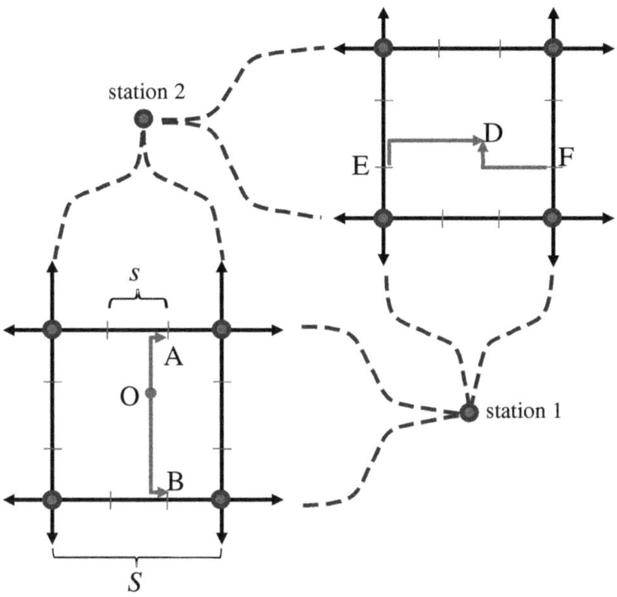

6.7. * A hub-and-spoke network can be configured in many different ways. The following figure shows two possibilities: (a) a trunk-and-branch system where transfers are allowed only along four trunks (anywhere a branch joins the trunk); and (b) a system where spokes branch out anywhere in the service region while maintaining constant line spacing. Derive the counterparts of (6.27) for these two systems using similar parameters and variables, and discuss their pros and cons.

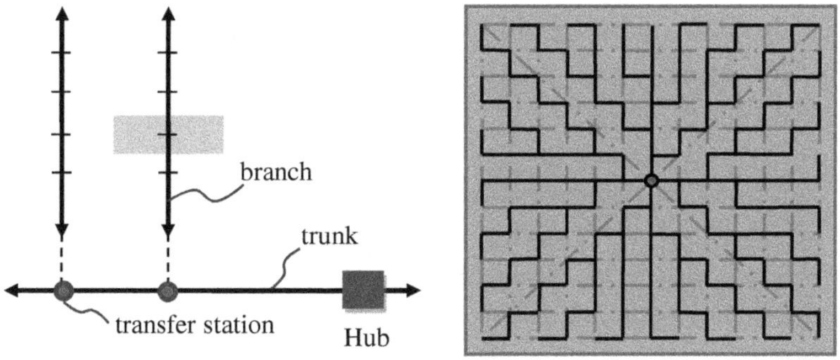

6.8. For a trunk-and-branch network, show that if the hub is $\kappa\phi$ distance units away from the center of a square region of side ϕ, either in the vertical or horizontal direction, and $0 \leq \kappa \leq 1/2$, then the average connector length along the trunk becomes $(1/4 + \kappa^2)\phi$.

6.9. * This problem deals with design of a scheduled feeder bus system for a metro station. A metro system with headway, H [h] has a steady demand such that the number of passengers embarking (and disembarking) a train at each station is $N/2$. The streets form a dense square grid in the rectangular feeder area of the station (dimensions $2L \times W$, see the figure below).

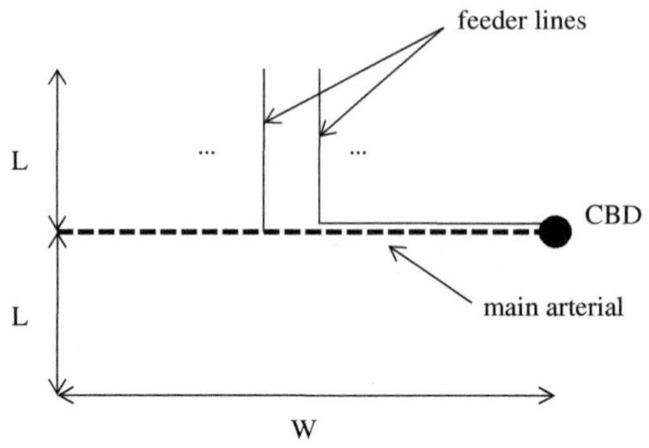

We look for a set of feeder bus lines with headway H (to provide timed transfers) and the location of their stops. Bus lines travel perpendicularly to the rail line with stop spacing $s \ll L$, W, and then along it without making any stops (see figure). The bus line spacing is also s (so that stops can form a square lattice in the service area if desired). Walking speed is much less than bus speed ($v_w \ll v_b$). The time for a stop is t_s.

(a) Draw a pattern for the stops that will minimize the maximum distance walked, and show the catchment areas for each stop. Write a formula for the maximum walking time for any passenger.
(b) Write a formula for the maximum travel time, T, of any outbound traveling passenger, assuming that they know the service headway and do not have to wait for the next dispatch.
(c) Write a formula for the number of bus-h traveled (you can find bus-km as a stepping stone) to transport the passengers.
(d) If the cost per bus-hr is c_t (\$/bus-hr), write a mathematical program for s^* that will minimize that fare required, F^*, to avoid subsidies while achieving a total time service standard, T_0, for the worst case passenger. Prove that s^* is the largest spacing that would meet the time requirement.
(e) Give the values of s^* and F^* that would apply to the distribution of travelers to and from downtown San Francisco ($L = 3$ mi; W $= 7$ miles; $N = 1000$ pax/dispatch; $H = 8\,\text{min}$; $v = 15\,\text{mph}$; $v_w = 3\,\text{mph}$; $c_t = 30\,\$/\text{h}$; and $t_s = 0.01\,\text{h}$) for different maximum travel time design standards, T_0. What is the minimum feasible standard, T_0?

(f) Extra Credit 1: Assume that the feeder lines provide express service and only make stops on a length $l \ll L$, $l > s$. Then repeat part E.

(g) Extra Credit 2: Repeat part e if we allow the line spacing to be different from the stop spacing (without the extra parameter of part f).

6.10. A transit agency serves a square/diamond-shaped town with side $\sqrt{2}R$ (see diagram below). The CBD is located at the center. The city streets on which people walk form a dense grid parallel to the x and y directions. In the evening, commuters travel from the CBD to their homes, which are uniformly distributed across the town. The passenger demand is steady during the rush, and is characterized by a density λ [pax/h-km^2]. The agency decides to design an elevated metro network with diamond-shaped ring routes and four crossing radii. Bidirectional train service with headway H is provided on all the routes. The spacing between any two adjacent rings, as well as that between adjacent stations is equal to s (with $s \ll R$) (see the figure below). The stations form a square lattice slanted at 45°. Transfers may occur anywhere routes intersect. The train dwell time is $t_s > 0$ at non-transfer stations and $\Delta > t_s$ at transfer stations. The walking speed is much less than the train speed, i.e., $v_w \ll v_t$. For this system, please do the following:

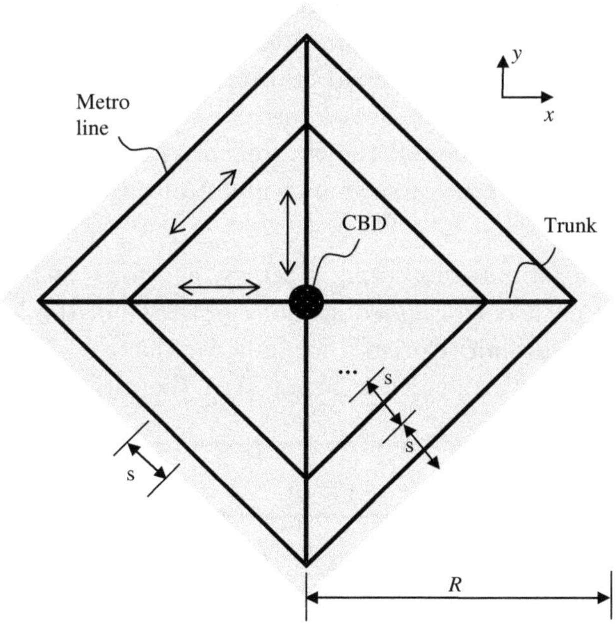

(a) Determine the number of commuters, N, who would need service during one headway.

(b) Sketch the door-to-door travel path of a worst-case commuter — i.e., with the longest travel time. Clearly, mark the walking and riding parts of the commuter's travel path.

From now on, simplify the notation by using the approximation: $R \approx R - \sqrt{2}s/2$. (This is justified since $s \ll R$.)

(c) Write a formula for the door-to-door travel time, T, of a worst-case commuter, assuming that they know the service headway and do not have to wait for the next dispatch.

(d) Write a formula for the total number of directional train-hours traveled to transport the N passengers who appear in one headway.

6.11. Consider the hybrid network sketched in Figure 6.11.

(a) Prove based on geometrical grounds the last step toward (6.37), that $F_L = 2(1+\alpha^2)$ and $F_V = 2(3\alpha - \alpha^2)$.

(b) Analyze both, the number of transfers and the riding time needed for different origin and destination locations. Compare the results with those of a hub-and-spoke network, i.e., for $\alpha = 0$. Show that two transfers are required for $(1-\alpha^2)^2/2$ of the total demand and that about $3/4$ of the total passengers experience lower riding times in the hybrid network.

6.12. *Over time, uniformly distributed travel demand along a corridor grows. Yet major stations cannot be changed easily especially for rail based modes. Suppose we add stations incrementally, one at a time, to a unit length segment of the corridor, while trying to minimize the average access distance. If the stations are fixed once they are placed, subsequent placement of stations will not always give us the minimum access cost.

(a) Consider a naïve bisecting strategy, as shown in the figure below, where each newly added station splits one of the longest existing spacings in half. Derive a formula for the average access distance within the unit length segment after the mth addition.

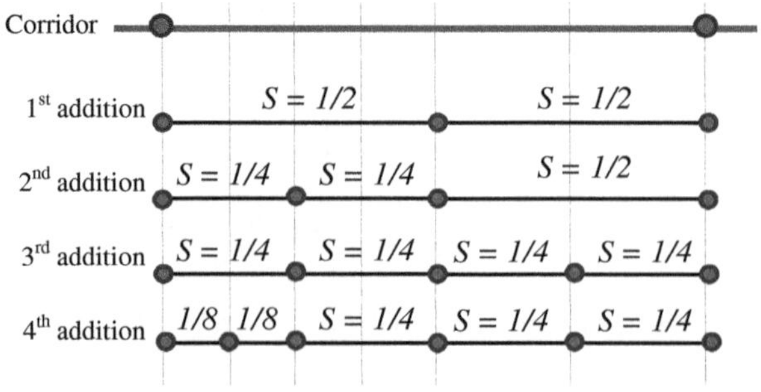

(b) Consider an asymmetric strategy, as shown in the figure below, where the mth added station splits the longest existing spacing into two uneven spacings of length ratio $\log_2(\frac{2m+1}{2m})/\log_2(\frac{2m+2}{2m+1})$. Derive a formula for the set of spacings after the mth expansion, and compute the average access distance.

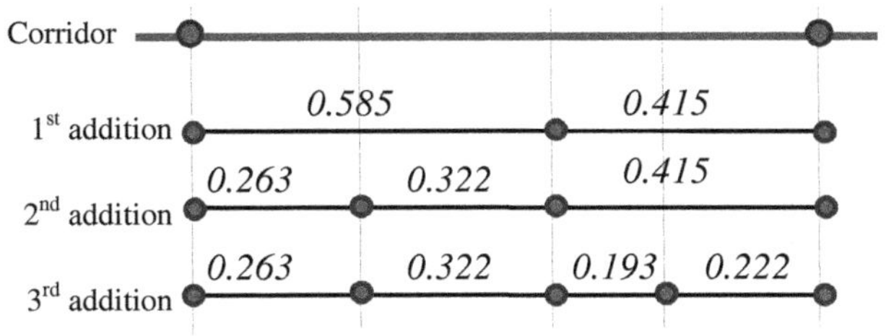

(c) For a range of m values, compare the average access distances under the asymmetric approach and the symmetric naïve approach. Which is better? Compare also with the benchmark case where the stations could be easily demolished and re-placed optimally every time the system is expanded. Discuss your finding.

(d) Repeat the analysis if the objective is minimizing the worst-case access distance.

Mini Project 2. Designing a Transit Network[19]

The Chicago Transit Authority currently serves the city of Chicago and the surrounding area with rail and bus transit. In this project you will create your own transit design for the Chicago area, combining grid bus service with a particular type of radial rail service.

Region Characteristics

The area of design is bounded by a quarter-circle of radius 15 mi northwest of the downtown area and a quarter-circle of radius 10 mi southwest of the downtown, as shown in Figure MP2.1. Streets cover the region with an infinitely dense grid.

Vehicle Characteristics

The rail service operates independent of the street grid at a cruising speed of 30 mph and has a fixed lost time per stop (including loss time due to

[19] Created by Josh Pilachowsky with input from Michael Cassidy.

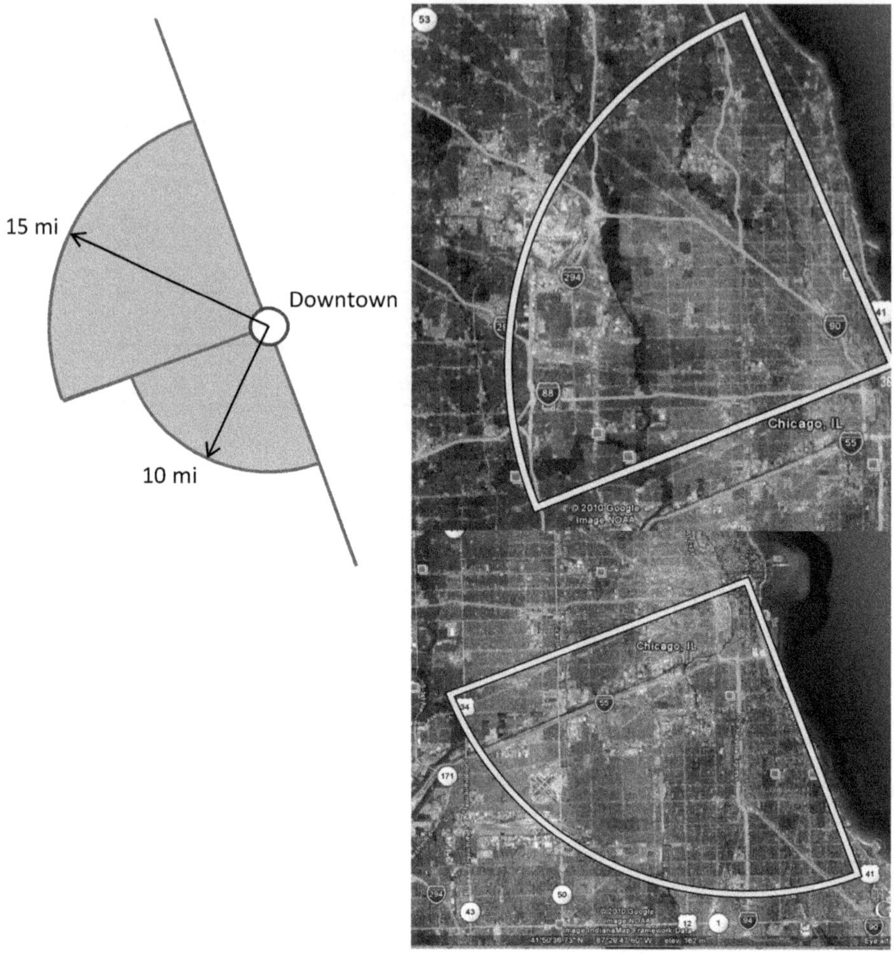

Figure MP2.1. Chicago downtown area.

acceleration and deceleration and passenger loading) of 1 min per stop. The bus service operates on the grid at a cruising speed of 15 mph (accounting for traffic and signal delays), has a loss time (due to acceleration and deceleration) of 10 s per stop, plus 4 s per boarding passenger, and negligible time per alighting passenger. Users access the system by walking at a speed of 2 mph.

Data

The transit agency values user time at \$20/h. Bus and rail service have operating costs of \$3/bus-mile and \$10/train-mile, respectively, while rail

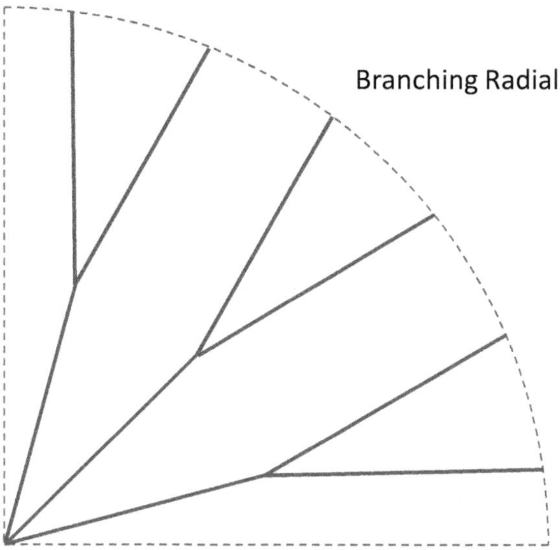

Figure MP2.2. Rail transit network layout.

infrastructure has a cost of \$650/mile-h. During peak hours, trips to/from the downtown region are generated uniformly at a rate of 360 pax/mi^2-h and trips with origins and destinations outside the downtown region are generated uniformly at a rate of 60 pax/mi^2-h.

Using these data, you are to perform the following tasks. For each part analyze a branching-radial design for the rail service as shown in Figure MP2.2, combined with a grid system for bus service.

(a) Write an expression for the hourly agency cost for this system.
(b) Define the different trip profiles for the following trip types: NW region to downtown, SW region to downtown, NW region to SW region, and SW region to NW region (e.g., a trip profile from NW region to downtown might include Walk → Bus → Rail → Walk). For each trip type and trip profile, formulate the worst case travel time in terms of the system parameters and decision variables listed in Table MP2.1, or explain why that trip profile would not produce a reasonable worst-case trip.
(c) For the following set of standards, optimize the bus and rail system with respect to the reasonable trip profiles from part B (i.e., minimize generalized cost subject to standards). To do this, first show how the agency cost can be optimized with respect to a single profile in part B analytically (assuming that this particular profile is the

Table MP2.1. Decision variable list.

Bus (grid)	S_b		Route spacing
	s_b		Stop spacing
	H_b		Headway
Branching-Radial	$S_{\text{rad,NW}}$	$S_{\text{rad,SW}}$	Max route spacing
	$s_{\text{rad,NW}}$	$s_{\text{rad,SW}}$	Stop spacing
	$H_{\text{rad,NW}}$	$H_{\text{rad,SW}}$	Headway on the periphery

binding constraint). Then optimize the agency cost with respect to all constraints numerically using Excel or a similar solver program. Identify the worst-case passenger profile for the optimal design.

STANDARDS

- NW → Downtown – 2 h
- SW → Downtown – 1.5 h
- NW → SW, SW→ NW – 3.5 h

(d) Using a street map of Chicago (a coarse map is provided to show the regions we will be analyzing), lay out the bus and rail lines using the optimal spacing values determined in part (ii) as a guideline. You should keep in mind that the existing infrastructure provides constraints that must be considered (for example, bus lines should only be placed on existing city streets and you should consider a minimum of 10 key locations in Chicago that should be connected to the transit service). Comment on the differences between the continuum approximation solution and the more realistic "on-ground" system with respect to the worst case passenger.

You are to submit a formal report describing the tasks and your findings. Extensive calculations should be placed in an appendix (and referred to in the report). You may work in small groups of up to three people and submit a group report.

Appendix 6.A. A Lower Bound for the Maximum Link Flow

This appendix revisits the derivation of (6.20)–(6.23), optimistically assuming that every passenger chooses a transit route that minimizes the distance traveled in the central square. Even more optimistically, we further assume that that the flow on the critical link equals the average flow across all links in the square. We shall see that these optimistic results are similar in form to (6.23), and only about 30–50% lower despite the very drastic assumptions.

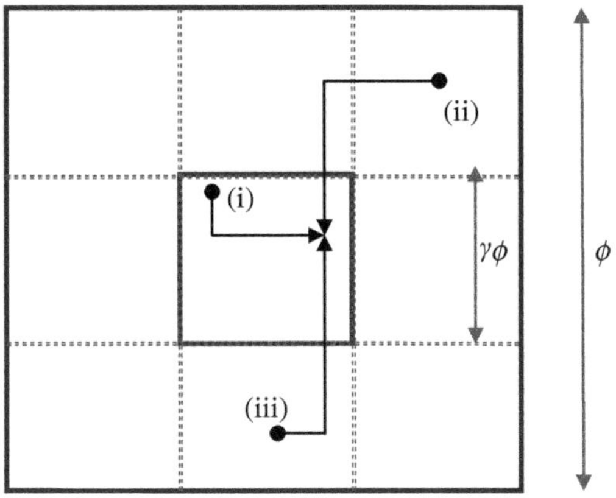

Figure 6.A1. Optimistic route choices that minimize travel in the central area.

We evaluate the average link flow (which is also the critical flow given our assumptions) by calculating the total number of vehicle-km that people want to travel in the central area per hour under our center-averse routing assumption; then prorating these vehicle-km equally to all the kilometers in the central square; and finally dividing this result by the length of a link. Of course, the derived formulas only make sense if the side of the central area is reasonably larger than the line spacing so the central area can contain several links.

To evaluate the average travel distance inside the central area we consider three classes of people depending on the location of their trip origins (see Figure 6.A1): (i) in the center square; (ii) in the four corners of the peripheral area; and (iii) in the rest of the peripheral area. People of type (i) will complete the whole trip in the central area and their total travel distance in it will be $2(\gamma\phi)/3$ on average. People of type (ii) will choose to enter the central area from the direction that minimizes the distance within it; therefore, this distance averages $(\gamma\phi)/3$ because the smaller of two independent and identically distributed uniform random variables follow a triangular distribution. People of type (iii) will transfer outside the central square and then enter it on their second leg; thus, their average travel distance in the central square is $(\gamma\phi)/2$. Figure 6.9 illustrates the travel paths of these passengers. Finally, since a fraction γ^2 of the people are of type (i), $2\gamma(1-\gamma)$ are of type (ii), and $(1-\gamma)^2$ are of type (iii), respectively, we can combine the conditional averages to obtain the unconditional average. The result is $\frac{2\gamma\phi}{3}\gamma^2 + \frac{\gamma\phi}{3}(2\gamma - 2\gamma^2) + \frac{\gamma\phi}{2}(1-\gamma)^2 = [\frac{1}{2} + \frac{\gamma^2}{2} - \frac{\gamma}{3}]\gamma\phi$ [km].

Therefore, the number of passenger kilometers of travel demanded in the central square per unit time is $[\lambda\phi^2]\ [\frac{1}{2} + \frac{\gamma^2}{2} - \frac{\gamma}{3}]\gamma\phi$. These are to be prorated equally to the number of directional kilometers in the central area. Recall from the definition of tiles in Section 6.2 that each tile has $4S$ kilometers of directional lines and that there are $(\gamma\phi/S)^2$ tiles. Therefore, the total number of line-kilometers is $(2\gamma\phi)^2/S$, and the prorated demand per link:

$$\text{Demand on critical link} = \lambda\phi S\left[\frac{1}{8\gamma}+\frac{\gamma}{8}-\frac{1}{12}\right], \quad \text{where } \gamma \geq S/\phi. \tag{6.A1}$$

This can now be compared with (6.21) to obtain the (optimistic) capacity constraint:

$$\left[\frac{1}{8\gamma}+\frac{\gamma}{8}-\frac{1}{12}\right]\lambda\phi HS \leq C, \quad \text{where } \gamma \geq S/\phi. \tag{6.A2}$$

Note that the coefficient $[\frac{1}{8\gamma}+\frac{\gamma}{8}-\frac{1}{12}]$ is strictly decreasing with respect to $\gamma \in (0,1]$. In the extreme cases (6.A2) reduces to:

$$\frac{1}{6}\lambda\phi HS \leq C, \qquad \text{if } \gamma = 1 \text{(distributed demand)} \tag{6.A3a}$$

$$\left[\frac{\phi}{8S}+\frac{S}{8\phi}-\frac{1}{12}\right]\lambda\phi HS \leq C, \qquad \text{if } \gamma = S/\phi \text{(concentrated demand).} \tag{6.A3b}$$

Comparing the lower bound of the link flow (6.A2) and the upper bound (6.23a), assuming large γ, we can easily verify that they are similar in form and different only by approximately a factor of 2. Therefore, either formula can be used for qualitative analyses. In practice, however, one should use the upper bound (6.23a), and even add an extra buffer on account of stochastic fluctuations.

Appendix 6.B. Hierarchical Grid Networks

This appendix analyzes the performance of two hierarchical grids as depicted in Figure 6.16(a). For simplicity, we only consider the extreme case when (6.9) holds, such that $s \leq S$ is not binding. Recall from Section 6.2 that formulas (6.6) and (6.7) produce the optimal design for a simple grid network when passengers travel ℓ distance units. The formulas are

$$s^* = \sqrt{2\ell t_s v_w},\ H^* = 2\left(\frac{c_d}{\lambda\beta k^2 v_w}\right)^{\frac{1}{3}},\ S^* = 2\left(\frac{k c_d v_w^2}{\lambda\beta}\right)^{\frac{1}{3}}.$$

Combined, they yield the following door-to-door travel pace:

$$\frac{1}{v^*(\ell)} = \frac{T^*}{\ell} = \left(\frac{kH}{2} + \Delta + \frac{S+s}{2v_w}\right)\frac{1}{\ell} + \frac{1}{v_{\max}} + \frac{t_s}{s}$$
$$= \frac{\Delta}{\ell} + \frac{2}{\ell}\left(\frac{k c_d}{\lambda\beta v_w}\right)^{\frac{1}{3}} + \sqrt{\frac{t_s}{2\ell v_w}} + \frac{1}{v_{\max}}.$$

The above results can be applied directly to the local line system (line spacing S_0, stop spacing s_0, headway H_0), in which the average traveler travels a distance of $\ell = (S_1 + s_1)/4$ to reach the express line system (line spacing S_1, stop spacing s_1, headway H_1). If this is done we find:

$$s_0^* = \sqrt{(S_1 + s_1)t_s v_w/2},\ H_0^* = 2\left(\frac{c_d}{\lambda\beta k^2 v_w}\right)^{\frac{1}{3}},\ S_0^* = 2\left(\frac{k c_d v_w^2}{\lambda\beta}\right)^{\frac{1}{3}},$$

and the door-to-door speed in the local system, or access speed of the "express" system, becomes

$$v_a = \left\{\frac{4}{S_1 + s_1}\left[\Delta + 2\left(\frac{k c_d}{\lambda\beta v_w}\right)^{\frac{1}{3}}\right] + \sqrt{\frac{2t_s}{(S_1 + s_1)v_w}} + \frac{1}{v_{\max}}\right\}^{-1}.$$

Plugging this expression into the objective function for the express line system, as in (6.5), we can minimize the combination of agency cost (with a larger cost coefficient $C_d > c_d$) and average passenger travel time in the Lagrangian form. The resulting problem is

$$\min\ Z = \frac{4C_d}{\lambda S_1 H_1} + \beta\left(\frac{kH_1}{2} + 3\Delta + 4\left(\frac{k c_d}{\lambda\beta v_w}\right)^{\frac{1}{3}}\right.$$
$$\left. + \sqrt{\frac{(S_1 + s_1)t_s}{2v_w}} + \frac{(S_1 + s_1)}{2v_{\max}} + \frac{\ell}{v_{\max}} + \frac{\ell t_s}{s_1}\right). \qquad (6.A4)$$

Typically, the last term $\sqrt{(S_1 + s_1)t_s/(2v_w)}$ is dominated by most other terms, e.g., $\frac{\ell t_s}{s_1}$. If we omit this last term in the optimization process (but keep it in the formula), the above optimization problem can be solved analytically. The readers are encouraged to verify that the following optimal solution is obtained:

$$s_1^* = \sqrt{2\ell t_s v_{\max}},\ H_1^* = 2\left(\frac{C_d}{\lambda\beta k^2 v_{\max}}\right)^{\frac{1}{3}},\ S_1^* = 2\left(\frac{kC_d v_{\max}^2}{\lambda\beta}\right)^{\frac{1}{3}},\ \text{and}$$

$$\frac{Z^*}{\beta} = 3\left(\frac{C_d k}{\lambda\beta v_{\max}}\right)^{\frac{1}{3}} + \sqrt{\frac{2\ell t_s}{v_w}} + \left(3\Delta + 4\left(\frac{kc_d}{\lambda\beta v_w}\right)^{\frac{1}{3}} + \frac{\ell}{v_{\max}}\right)$$
$$+ \sqrt{\left(\left(\frac{kC_d v_{\max}^2}{\lambda\beta v_w^3}\right)^{\frac{1}{3}} + \sqrt{\frac{\ell t_s v_{\max}}{2v_w}}\right) t_s}.$$

Analyses for alternative cases, e.g., $s = S$, can be conducted as in Section 6.2.

Chapter 7

Planning — Flexible Transit

We saw in the last chapter that conventional forms of collective transportation with fixed access points and routes have economies of scale, unlike the more flexible individual transportation. As a result, the rigid collective modes can deliver better service than individual transportation modes, but only if the demand rate is sufficiently high. Since there must be a range of intermediate demand where the two modes perform similarly it is fair to ask whether it is possible to make collective transportation more flexible so it can better compete with individual modes for intermediate levels of demand. These are the issues examined in this chapter.

To do this in an organized way, it is useful to classify all public transportation concepts depending on how people share rides and vehicles. The advent of internet-based technologies has facilitated new ways in which this can be done, and these new services will also be considered. At one end of the spectrum is the form of collective conventional transit with fixed routes and schedules that was discussed in earlier chapters. Travelers using these modes share vehicles in large groups at the same time. At the other end of the spectrum, we find conventional private travel, which involves exclusive use of a privately owned vehicle, usually an automobile with a single occupant. As shown in Figure 7.1, there is a host of possibilities in between these two extremes, which we call "flexible transit."

As the figure suggests, flexible transit can be a derivative of individual automobile travel or resemble collective conventional transit. In the first category, termed "individual public transportation" (IPT) and represented by the second box from the left, vehicles and rides are shared over the course of a day in an *ad-hoc* way, a few people at a time. Vehicle occupancy is not much higher than that for private travel and the travel experience is also similar. Examples are taxi and car-sharing, where different passengers use

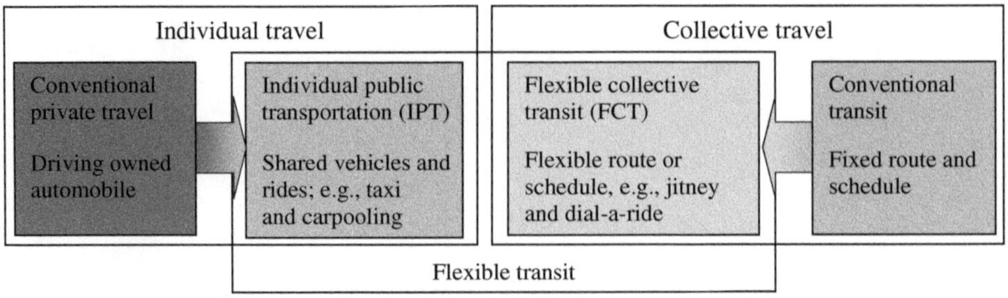

Figure 7.1. Collective vs. individual travel and flexible transit.

the same vehicle at different times, and carpooling, where the vehicle is shared simultaneously without requiring significant passenger detours.

In the second category, termed "flexible collective transit," vehicles carry more passengers at the same time, using routes and/or schedules that are adaptive. To achieve this kind of grouping, passengers may be significantly detoured on the way to their destinations. Jitneys and dial-a-ride services are examples in this category. Below is a more complete list.

Individual Public Transportation

Taxi: A chauffeur takes passengers directly from their origin to their destination without sharing a ride. This includes crowd-sourced services provided by transportation network companies (TNCs) such as Uber, Lyft, and Didi Chuxing. An example of this type of service is "Uber X."

Robo-taxi: Same as taxi, only using a driverless vehicle.[1] As of this writing tech and auto companies are fairly far along developing these types of vehicles and services.

Vehicle-sharing: Users pick up an available vehicle at a nearby location, and then return it when finished. The concept is used for cars, bicycles, scooters, and e-bikes. In some systems, the vehicles must be picked up and returned at pre-determined locations that we call "pods."

Carpooling: A driver modifies a trip slightly to pick up and deliver a passenger.

Ride-matching (RM): A taxi passenger shares a ride with another passenger with the *guarantee* that neither passenger is detoured significantly from their

[1]We call it robo-taxi because the taxi in this case is essentially a robot. The Oxford dictionary defines a robot as "a machine capable of carrying out a complex series of actions automatically, especially one programmable by a computer."

shortest paths. (To the authors' knowledge, this form of guarantee is not currently being provided by any company.)

Personal Rapid Transit (PRT) and Dual-Mode Transit (DMT): Low occupancy vehicles travel along specialized guideways carrying separate passengers. Only small scale PRT systems are operational, e.g., in Morgantown (West Virginia, USA) and at Heathrow International Airport (London, UK). The "Hyperloop" and the "TriTrack" electric car are conceptual examples of the PRT and DMT concepts, respectively.

Flexible Collective Transportation (FCT)

Demand responsive transportation (DRT): Numerous users share a vehicle that offers door-to-door service with an adaptive route and no schedule. Passengers may endure significant detours. Both, *Dial-a-Ride* paratransit services, and the ride sharing services being offered by companies such as Uber, Lyft, and Didi Chuxing (e.g., "Uber Pool") are examples. We shall call the latter *Shared Taxi*. Dial-a-ride differs from shared taxi in that the former emphasizes low cost and the latter short waits.[2] Shared taxi differs from ride matching in that the former allows for significant detours. DRT systems can offer "many-to-many" service connecting dispersed origins and destinations, and also "many-to-one" service connecting a service region to a single point. *Airport Shuttle* systems are an example of the latter.

Jitney: A transit system which dispatches vehicles with an adaptive schedule from one or more termini over published routes. Dispatches are usually made whenever the vehicles reach a pre-determined occupancy level.

Flexible-route Transit: A hybrid of DRT and conventional transit that allows for "soft" routes with variable detours, and possibly transfers. Used for low-demand small-area systems such as college campuses and resorts.

Hybrid trains with transfer on-the-fly (not in existence yet, and not covered in the chapter): This is a potential generalization of DRT. Since the concept does not exist, it requires a lengthier explanation than previously introduced ideas. It is introduced here only to illustrate the kinds of imaginary systems that can be analyzed with the analysis methods presented in this book — and this chapter in particular. Trains with hybrid cars would serve a long

[2] Another distinction which is less important from a physics perspective, but very important economically, is the vehicle ownership. Whereas dial-a-ride vehicles are typically owned by an operating agency, shared taxi vehicles are not. Instead, they are owned by the drivers, who operate under the guidance of a TNC.

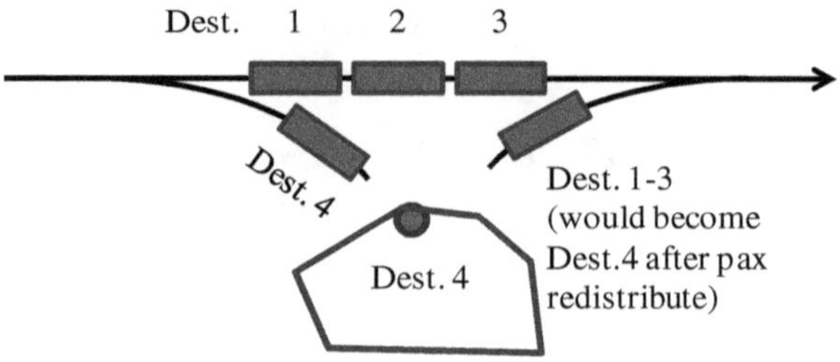

Figure 7.2. Transit with transfer on-the-fly.

corridor with widely separated cities or service regions. These cars would be equipped with both steel and rubber-tired wheels to be able to run both on rail and on local streets. Cars destined for a city would decouple from the train well before reaching the region and then travel on rail spurs to the local streets where they would act as buses. Conversely, hybrid cars would collect passengers in the city and use on-ramp spurs to meet the train on the downstream side of the region. Figure 7.2 displays a snapshot of a system with four destinations when the train is serving destination 4. The numbers near each car denote the destinations of the passengers currently in it. Transfer is on-the-fly because each passenger arriving to a moving train in a particular car would have to get up and walk inside the train to a car designated for the passenger's destination. For this to be possible, the train travel time between successive cities would have to be sufficiently long. Therefore, this idea only makes sense for long distance travel.

This chapter will examine and compare all the commonly used forms of flexible transit. We can advance that the main advantage of automobile IPT is that cars are shared, reducing both the number of vehicles required to serve a given demand level and the required parking infrastructure. Obviously good for society, these efficiencies also benefit users since their parking and garaging needs would decline considerably — it is estimated that privately owned automobiles are parked more than 90% of the time (Shoup, 2011). And the benefits may be magnified in the future when autonomous vehicles become widely adopted. The disadvantage, of course, is the inconvenience of waiting and sharing. The main promise of FCT is that it can perhaps improve the level-of-service of conventional transit, while remaining inexpensive thanks to the sharing of rides.

In order to explore these ideas and compare the various forms of flexible transit systematically, the chapter has been organized as described in Table 7.1. It classifies the covered modes according to whether the vehicles are shared with or without a chauffeur, and the level of sharing. The table

Table 7.1. A taxonomy of flexible transit services.

Usual number of passengers sharing a ride	Vehicle Type: Chauffeured	Vehicle Type: Non-chauffeured
1	*Section 7.1. Taxi* Conventional: (Yellow cab) Crowd-sourced: (Uber X, Lyft, Didi Express) Autonomous: (Waymo–Phoenix, AZ)	*Section 7.2. Vehicle-sharing* Two-way: car rentals (AVIS) One-way: podless (Car2Go, ReachNow, JUMP) One-way: pod-based (Zipcar, Maven)
2^-	*Section 7.3.1. RM* Spontaneous Reservation: (Didi Chuxing, Uber's "Scheduled Rides," Lyft–San Francisco)	*Section 7.3.2. Carpooling* Spontaneous: (Casual carpooling) Reservation: (Scoot)
2^+	*Section 7.4. DRT* Dial-a-Ride Shared Taxi: (Uber Pool, Lyft Line) *Section 7.5. Jitneys* (Chariot)	—

displays the section in which each mode is covered, the common name of the system in question and (in parenthesis) some current examples as of Year 2018. Section 7.6 concludes the chapter with a look to the future of transit in cities.

7.1. Chauffeured Vehicle Sharing: Radio-Taxis and Robo-Taxis

We consider here how a fleet of chauffeured vehicles that can be hailed remotely, which we call "*taxis,*" can serve randomly occurring trips. Taxi systems are for the most part owned and operated by for-profit companies. However, since they operate in a public space they are often regulated and given incentives (or disincentives) to operate for the public good — Chapter 11 shows how this can be done. Accordingly, both this section and the next, which also deals with for-profit operations, develop generic supply-side, physics formulas that can be used both by an operator to maximize profit and by a public agency to establish what can be expected if the mode is regulated.

In a taxi system, passenger trips are characterized by an origin, a destination and a time of occurrence. And they are covered by the taxis, one trip at a time. The chauffeurs in these systems can be dedicated employees as in conventional taxi systems, crowd-sourced part-time workers as in Lyft, Uber, and Didi Chuxing, or robots as very likely will occur in the future with self-driving taxis. The type of driver matters little from a modeling perspective since its character only enters the models through the vehicle's cost parameters.[3] Thus, the analysis about to be presented applies to all types of chauffeurs.

The main advantages of taxi vs. traveling with an owned car are the elimination of parking and the freeing up of capital tied up in owning an automobile. The main disadvantages are the cost of the "chauffeur" and the cost of waiting. However, in a hypothetical future where only self-driving cars are used, the cost of the chauffeur-robot would no longer be an issue since both a taxi and an owned car would have the same type of chauffeur. Thus the first disadvantage disappears. Furthermore, our models predict, and reality confirms, that if the demand for the service is sufficiently large then waiting should be minimal — crowd-sourced taxi services already achieve low waits in many urban areas. Therefore, it is easy to imagine a future where robo-taxis have replaced the owned automobiles for individual urban travel.

In view of this, it is important to understand the physics of taxi travel. This understanding should help operators design better systems and regulators better interventions. The key policy variable for these types of systems is fleet size. Conventional taxi companies control it directly since they can decide how many drivers to hire. Crowd-sourced companies control it indirectly through the wages they pay. Cities can affect it through regulations, tolls and other fees. Fleet size is a key decision variable because it affects both the user level-of-service and the operator's productivity. Fleet size also affects indirectly the number of taxis that are idle at any given moment, and through the deleterious effect of these taxis on parking and traffic, also non-users. In view of the importance of fleet size, a simple queuing model is now used to quantify its effects. For the sake of transparency, the model is highly idealized. It assumes that taxis are randomly and evenly distributed over space at all times, and ignores stochastic fluctuations. As a result, the model predictions are slightly optimistic.

---

[3]In reality though, drivers of taxi systems operated by TNC's are more difficult to control than robots. For example, TNC drivers require surge-pricing incentives to reposition themselves to areas of need. They may sometimes reject service to individual passengers on the basis of destination, or even race. None of this will happen with robots.

7.1.1. *The Physics of Taxi Service: A Queuing Model*

We consider a region of area R in which we provide radio-taxi service to customers with a steady demand density λ and expected trip length ℓ (see Figure 7.3). Let us assume that we provide enough taxis, m, so every call immediately gets assigned a taxi. This assumption is reasonable if this form of transportation is to be massively used.

If not enough taxis are provided, then passengers have to wait in queue for an assigned taxi. This type of undesirable service can be analyzed using an approach similar to the one used here. The model, however, turns out to be a special case of the demand responsive transit model in Section 7.5. For this reason, the discussion of this system is postponed to that section.

Taxis can be in one of three states: (i) idle, (ii) assigned and cruising toward a customer, or (iii) carrying a customer toward the destination. While the numbers of taxis in these states, as well as the times they stay in such states, fluctuate randomly over time, it will be sufficient for our purposes to use simple deterministic queuing models based on expected values that approximately capture the relationships of interest.[4] We will denote the expected number of taxis in each of these states by n_i (idle), n_a (assigned), and n_s (serving/carrying). Obviously, $n_i + n_a + n_s = m$. We can imagine that all the taxis in each state are contained in three separate reservoirs; and that taxis flow from one reservoir to the next through their life cycle.

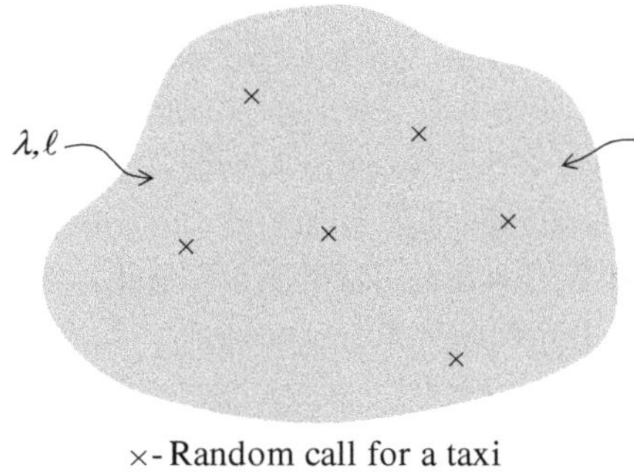

Figure 7.3. Taxi service.

[4]The idea somewhat resembles the so-called "mean value analysis" in queuing theory. Experience and theory shows that this deterministic approximation is quite accurate for our problem when the number of taxis is large.

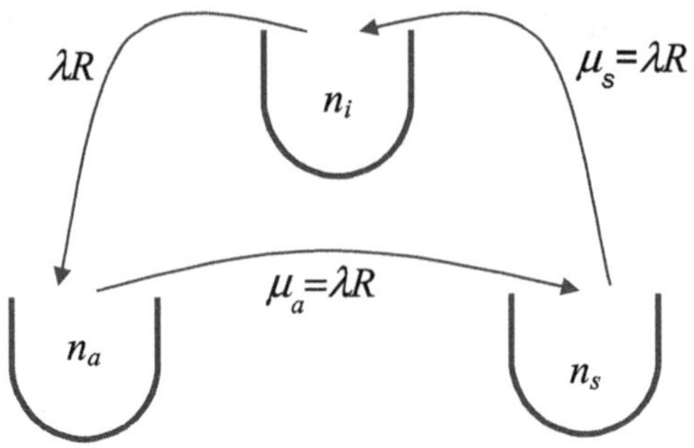

Figure 7.4. State evolution of taxis.

Figure 7.4 shows this idea, including notation for the rates at which taxis switch from one state to another.[5]

In the steady state, $\mu_a = \mu_s = \lambda R$. Therefore, only the variables n_i,n_a, and n_s expressing the expected numbers of taxis in each state need to be determined. They should depend on the demand parameters λ and R, and on the vehicle performance characteristics. For simplicity of notation we neglect the time that taxis spend stopped picking up and delivering passengers, which is small compared with average travel time, so only the vehicle cruising speed, v, will enter the results. It will also be convenient to define T_a and T_s as the expected times a taxi spends in the first and second bins. Note that T_a is also the expected passenger waiting time.

In the spirit of the standards approach, we now derive expressions for n_a, n_s, n_i, and $m \equiv n_i + n_a + n_s$ in terms of T_a, i.e., as if T_a was a rigid standard that must be met as a pure equality. The output of these functions can then be combined with suitable coefficients to define the agency cost.[6] Thus, for any given level-of-service, this cost can be predicted. We will then use the derived cost function to obtain optimal operating rules and costs using both the standards approach (with standards defined as inequalities) and the Lagrangian framework. The results will enable us

[5]A similar figure could be drawn for the undesirable case in which taxis are always engaged with a passenger, and passengers have to wait for taxis rather than the opposite. In this case the figure should track passengers rather than taxis. The first bin would contain the number of passengers waiting for an assignment n_w, the second the number assigned passengers, n_a and the third the number in service, n_s. The equations linking these state variables would be slightly different than those about to be introduced; see homework exercise 7.4 and Section 7.5 for answers to this problem.

[6]For a taxi company with salaried drivers the cost is roughly proportional to m. For a company with crowdsourced chauffeurs it should be proportional to n_s.

to compare the different forms of taxi service with both private cars and collective transportation. The derivations follow.

The expression for n_a is obtained by applying Little's formula from Chapter 2 to the first reservoir. Since the expected time a taxi spends in the first reservoir is T_a (by definition) and the average inflow and outflow of taxis into the reservoir is λR, the formula predicts that the expected number of taxis in that reservoir must be

$$n_a = \lambda R T_a. \tag{7.1}$$

Our second expression is for n_s. It is also obtained from Little's formula, now applied to the second reservoir. In this case, the expected customer service time is $T_s \approx \ell/v$ and the arrival rate is still λR. Therefore,

$$n_s = \lambda R \frac{\ell}{v}. \tag{7.2}$$

The expression for n_i is a little more complicated. It is obtained by recognizing that the mean distance from a request to the closest idle taxi, which is proportional to T_a, should decline with the number of idle taxis. Therefore, there should be an inverse relation where n_i declines with T_a.

To find it we will assume that idle taxis are uniformly distributed in the region at all times. This is an approximation, of course, but not too unreasonable if passenger origins and destinations are random. With this in mind, we now define d_{n_i} as the distance from a random request to the closest idle taxi when there are n_i idle taxis; and then look for an expression for the expectation $E[d_{n_i}] \equiv T_a v$ in terms of n_i.

This is done in two steps: first deriving the complementary cumulative distribution function of d_{n_i}, $\Pr\{d_{n_i} \geq x\}$; and then integrating it to obtain the expectation. This distribution is derived by considering the interior of the equidistance contours from a request, i.e., disks containing only those points that satisfy $d_{n_i} \leq x$.[7] These disks are relevant because $d_{n_i} \geq x$ if and only if no taxi is in the disk corresponding to x. Therefore, $\Pr\{d_{n_i} \geq x\}$ is the probability that no taxi is in the disk. Now, if the n_i taxis are randomly and independently distributed in the region as we are assuming, the probability that one specific taxi is not in the disk is the ratio of the size of the region that is not in the disk to that of the whole region. And, since taxi locations are independent, the probability of no taxi in the disk is the n_i^{th} power of

[7]For the Euclidean metric the disk is a circle. For the square grid metric the disk is a square angled at 45° to the grid.

this ratio. Thus, we have

$$\Pr\{d_{n_i} \geq x\} \equiv \Pr\{\text{zero taxis in the disk}\}$$
$$= \begin{cases} \left(1 - \pi x^2/R\right)^{n_i} & \text{(Euclidean metric)} \\ \left(1 - 2x^2/R\right)^{n_i} & \text{(grid metric)} \end{cases}.$$

The second step is calculating the expectation corresponding to this distribution function. We find through integration (see Appendix 7.A) that

$$E[d_{n_i}] \cong \kappa\sqrt{\frac{R}{n_i}},$$

where $\kappa = 0.5$ for the Euclidean metric, and $\kappa = 0.63$ for the square grid metric.

Finally, since $E[d_{n_i}] = T_a v$, we have $T_a v \cong \kappa\sqrt{\frac{R}{n_i}}$ and our inverse relation is

$$n_i \approx \frac{\kappa^2 R}{v^2 T_a^2}. \tag{7.3}$$

Now, combining (7.1), (7.2), and (7.3), we find

$$m(T_a) = n_i + n_a + n_s \approx \frac{\kappa^2 R}{(vT_a)^2} + \lambda R\left(T_a + \frac{\ell}{v}\right). \tag{7.4}$$

Note we have treated the numbers of vehicles as real variables, and we have assumed that all passenger requests get assigned to an idle vehicle immediately. Therefore, for this formula to make sense its three terms should be much larger than 1, i.e., we should check that,

$$\frac{\kappa^2 R}{(vT_a)^2}, \quad \lambda R T_a, \quad \lambda R\frac{\ell}{v} \gg 1. \tag{7.5}$$

Caveat: In reality, the number of taxis in each bin changes randomly over time due to fluctuations in demand and travel times. Stochastic queuing theory suggests that the standard deviation of the changes in the bin contents should be similar across bins and comparable with the square root of (7.3), and that the predictions of the model should be rather accurate when the bins rarely become empty — otherwise the model is optimistic. It follows from these comments that empty bins should be rare if the ratios of the expected values of the bin contents (7.1), (7.2), and (7.3) to the square root of (7.3) are several times larger than 1. And since (7.1), (7.2), and (7.3) are proportional to R, we see that these ratios decline with the square root of R. Thus, the model should be particularly accurate for large service regions. With this caveat in mind, let us now see what the model says about policy.

7.1.2. *The Operator's View: System Optimization via Standards*

Before we start, note that for a given level of service a taxi system's demand should be a monotonically decreasing function of its price. Thus, for a given level of service we can always find a unique price that generates any given demand. Therefore, if both the demand and the level of service are given, the revenue is fixed. As a result, profit is maximized simply by minimizing cost, i.e., by solving a standards problem of the type we have seen in previous chapters. This is an important problem because an operator can use its solution to explore the bigger picture, including pricing strategies.[8] So, let us now turn our attention to the standards problem.

A proxy for operating cost is the size of the vehicle fleet, m; and a proxy for the user's level-of-service is the average waiting time T_a. Recognizing that these two variables are related, as per (7.4), we formulate the standards problem using T_a as the decision variable, i.e., as

$$\min_{T_a} \{m(T_a) : T_a \leq T_0 \text{ and } (7.5)\}. \tag{7.6}$$

It turns out (7.5) is usually not binding for most problems, so we ignore it to begin with and check it *a posteriori*. Then, the solution of (7.6) has a closed form. To find it note that (7.4) is a generalized economic ordering quantity (GEOQ) expression whose unconstrained minimum is $T_a^u \cong (2\kappa^2)^{1/3}\lambda^{-1/3}v^{-2/3}$ and $m^{u*} \equiv m(T_a^u) \cong 1.9\kappa^{2/3}\lambda^{2/3}v^{-2/3}R + \lambda R\ell/v$. Thus, if $T_0 > T_a^u$ our constraint is not binding, and therefore: $T_a^* = T_a^u$, and $m^* = m^{u*}$. If on the other hand $T_0 \leq T_a^u$ then the constraint is binding and the constrained solution is: $T_a^* = T_0, m^* = m(T_0)$. Figure 7.5 displays the complete optimum fleet size function, which we denote by $m^*(T_0)$. This assumes, of course, that (7.5) is satisfied.

The value $m^{u*} \cong 1.9\kappa^{2/3}\lambda^{2/3}v^{-2/3}R + \lambda R\ell/v$ bounding the curve $m^*(T_0)$ from below is the minimum fleet size required to be able to keep up with the demand. Smaller fleets cannot maintain a steady state in which all passengers are assigned a vehicle immediately. The figure also shows that any fleet size that exceeds this minimum can achieve a standard equal to or better than: $T_a^u \cong (2\kappa^2)^{1/3}\lambda^{-1/3}v^{-2/3}$; and that the larger the fleet the better the level-of-service that can be achieved.

[8]As explained in Chapter 11, by exploring the solutions of these standards problem for different prices and levels of service, an operator can find the optimum design and pricing solution.

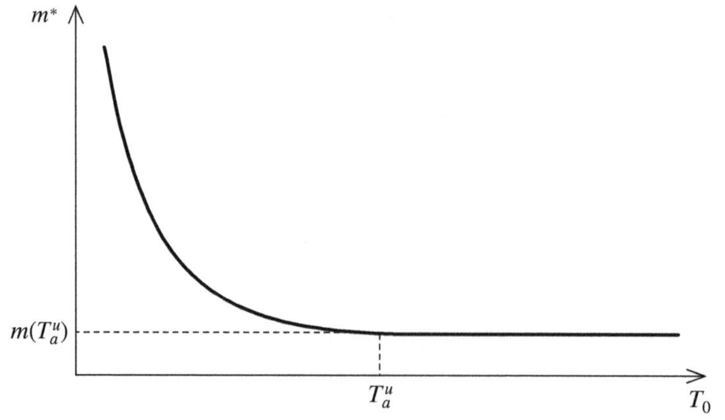

Figure 7.5. Minimum fleet size required to achieve a target level-of-service T_0.

Also interesting is to consider how the operator cost per trip served relates to demand. This cost is captured by the ratio: m^*/λ. The expressions we have developed for the two branches of $m^*(T_0)$ show that both are proportional to $\lambda^{2/3}$. Therefore, the ratio m^*/λ must be proportional to $\lambda^{-1/3}$; i.e., the operator's cost per trip served declines with demand. These economies of scale suggest that market forces favor large enterprises and that there may be a need for regulation.

Recall that the above formulas are valid only if we find *a posteriori* that (7.5) is satisfied. Fortunately, this always happens in realistic situations where either λ or R is moderately large. Otherwise, one should solve (7.6) considering (7.5) and solve it again, numerically if necessary. The following example illustrates a typical scenario.

Example 7.1. Consider a city under the Euclidean metric (i.e., $\kappa = 0.5$) and with the following parameters:

$$\left.\begin{aligned} T_0 &= 0.08\,\text{h} \\ \lambda &= 1\,\text{pax/km}^2\text{-h} \\ R &= 400\,\text{km}^2 \\ v &= 20\,\text{km/h} \\ \ell &= 10\,\text{km} \end{aligned}\right\} \Rightarrow \begin{aligned} n_a &= 32 \\ n_i &= 40 \\ n_s &= 200 \\ \hline m &= 272 \end{aligned}$$

Since all the variables satisfy (7.5), the solution can be accepted. Furthermore, the solution should be quite accurate despite our caveat since $\sqrt{n_i} \ll n_i, n_a, n_s$.

7.1.3. *Society's View: Comparison with Private Autos and Collective Transit*

Let us now see how conventional taxi compares with other modes, assuming that thanks to regulatory policies the taxi company operates for the good of society. We can do this, as in previous chapters, with the truncated Lagrangian approach, i.e., by combining the agency and passenger costs into a single generalized cost function.

In what follows we shall use $\gamma \equiv c_d/v$ for the cost per unit time of operating one taxi and β for the value of time. The operating agency side is captured by the function $m^*(T_0)$ of Figure 7.5 and the passenger side by the sum of the waiting and riding times. Thus, we formulate the minimization problem as

$$\min_{T_0} \left\{ \lambda R\beta \left[T_0 + \frac{\ell}{v} \right] + \gamma m^*(T_0) \right\}.$$

In order to derive closed form formulas, we first simplify this problem. Clearly $T_0^* \le T_a^u$ since the objective function increases with T_0 for $T_0 > T_a^u$. Thus, we can add the constraint $T_0 \le T_a^u$ to the above problem without changing the result. This is useful because, as Figure 7.5 illustrates, $m^*(T_0) = m\,(T_0)$ in the feasible range. Thus, we make this substitution and add the constraint to the above. In addition, we divide both sides by $\lambda\beta R$ to express the result in more intuitive universal units of time-equivalents per passenger carried. The simplified version after these manipulations is

$$\frac{z}{\beta} = \min_{T_0} \left\{ \left[T_0 + \frac{\ell}{v} \right] + \frac{\gamma}{\beta} \left[\frac{1}{\lambda(2vT_0)^2} + \left(T_0 + \frac{\ell}{v} \right) \right] \right\}; \text{s.t. } T_0 \le T_a^u.$$

Consideration now shows that the minimum is actually achieved for $T_0 < T_a^u$ so that the constraint is not needed. So the problem is a generalized, unconstrained GEOQ in T_0, which admits a closed form solution. This ends the mathematical simplifications.

We know from Chapter 2 that the analytic solution of this GEOQ is

$$T_0^* = \left(\frac{1}{2} \frac{\gamma}{\beta + \gamma} \right)^{1/3} \lambda^{-1/3} v^{-2/3} \quad \text{[h]} \tag{7.7}$$

and

$$\frac{z^*}{\beta} = \left(1 + \frac{\gamma}{\beta} \right) \frac{\ell}{v} + \left(\frac{27}{16} \frac{\gamma}{\beta} \left(1 + \frac{\gamma}{\beta} \right)^2 \right)^{1/3} \lambda^{-1/3} v^{-2/3} \quad \text{[h]}. \tag{7.8}$$

By varying the value of γ, the formulas represent different forms of taxi service. For example, when the taxi drivers are human we expect $\gamma > \beta$ since the cost of the vehicle includes not just its mechanical cost but also the wages of the driver, which should be comparable with β. By contrast, in the case of robo-taxis where there are no humans we expect $\gamma/\beta \ll 1$, i.e., γ is just the same as for an owned robo-car.

For robo-taxis, the first term on the right hand side of (7.8) is the cost of overcoming distance, which is the same as the total cost of driving an owned car. The second term is the extra cost associated with passenger waiting and non-productive robo-car time — while idling and traveling to a waiting passenger. This extra cost should be compared to the cost of the extra activities required by private auto travel: garaging, parking, looking for parking, and walking to/from the parking spaces. Note that the extra costs of private auto are not affected significantly by demand, but those of robo-taxis, which include the cost of empty taxis, decline toward zero as λ increases. *This is why robo-taxis could one day become the most prevalent way to travel individually in densely populated cities.*

For countries of moderate to high wealth and the taxis of today with humans at the wheel, we expect the mechanical cost of driving $(\gamma - \beta)$ to be small compared with the driver wages β so $\gamma/\beta \approx 1$. In this particular case, the equations for taxi reduce to

$$T_0^* \approx 0.63\lambda^{-1/3}v^{-2/3} \quad \text{[h], and} \tag{7.9}$$

$$\frac{z^*}{\beta} - \frac{\ell}{v} \approx \frac{\ell}{v} + 1.9\lambda^{-1/3}v^{-2/3} \quad \text{[h].} \tag{7.10}$$

Note that the left side of equation (7.10) expresses the extra cost for taxi beyond the cost of an automobile trip since this cost is $(1 + \frac{\gamma-\beta}{\beta})\frac{\ell}{v} \approx \frac{\ell}{v}$. We see that even if demand is very high, the right side does not tend toward zero; and it is bounded from below by quite a large value: ℓ/v. *Thus, human-operated taxis are only a viable option for private travel when auto driving involves considerable extra costs,* e.g., *for parking.*

Taxi can also be compared with collective transportation. We saw in earlier chapters that the generalized cost of collective transportation can come arbitrarily close to ℓ/v as the demand increases toward infinity. Thus, as (7.10) reveals, collective transit can easily outperform human-driven taxis if the demand is moderately high. However, as (7.8) also reveals, the competition from robo-taxis with their lower operating costs should be much fiercer. It should be even fiercer if ways are found to efficiently share rides. This will be explored in Section 7.3. Before this is done, however, we should complete the discussion of vehicle sharing by examining the case where vehicles are not chauffeured.

7.2. Non-Chauffeured Vehicle Sharing: One-Way Systems

We consider here how to share vehicles such as motorized scooters, bicycles, e-bikes and ordinary cars that do not come equipped with a chauffeur. The chauffeurless nature of the vehicles changes the problem considerably because, without a chauffeur, idle vehicles cannot be sent to waiting customers. Instead, they must be parked and customers must walk to the parking locations. This is clearly less desirable. For this reason, we conjecture that in a hypothetical future where robo-cars are the dominant way of traveling by car, the park-and-walk approach would be unnecessarily inconvenient. Therefore, the technology we are about to study would then persist only for those vehicles that cannot be robotized such as bicycles and e-bikes.

There are two types of systems. Some park-and-walk systems allow idle vehicles to be stored anywhere that is legal (e.g., like the Mobike and Ofo systems in China, or the Bird and JUMP systems in the U.S.) and others only at specific locations that we call "pods" (e.g., Zipcar, Divvy in Chicago, Velib in Paris, Bicing in Barcelona, etc.). At least, for now, the latter are more prevalent. The choice of technology involves a cost trade-off. In park-anywhere systems, every vehicle must have all the telematics to be tracked by the operator, be found by the customers and execute all necessary business transactions. In pod-based systems, vehicles can be dumber with some of the telematics situated in the pods. The location and number of these pods is then a central design issue.

Pod-based systems come in two flavors: two-way and one-way. In a two-way system, customers are required to return their vehicle to the pod from which it was borrowed. The "two-way" name is used because each borrowing episode involves a two-way trip from and to the pod. This is how conventional car rental companies typically work. Two-way systems are useful for long term rentals that involve multiple trips but not for quick individual trips. One-way systems, which allow a return to occur anywhere, have been developed to serve the latter. They are likely to be more useful as a mode for urban travel and for this reason, shall be our focus here.

Bicycle and e-bike systems are already commonly used as access modes for subway and regional train systems in metropolitan areas, with pods conveniently located at the stations. In one-way systems, additional pods are distributed throughout the stations' catchment areas. One way systems, however, can also be used as a stand-alone mode which can be used on its own for many-to-many travel. This type of application is examined below as it is more broadly applicable and its analysis considerably more complex.

The Achilles heel of one-way many-to-many vehicle sharing systems is that if the distribution of demand is imbalanced in the sense that some neighborhoods of the service region have systematically more trip destinations than origins (and the reverse) then the operator is forced to redistribute the empty vehicles periodically. This can be a very considerable expense. The problem is particularly serious for bicycle systems in hilly cities because people prefer to ride downhill.

It should be clear that a one-way vehicle sharing system requires consideration of three design variables that affect cost and performance: the fleet size; the redistribution strategy; and the number of pods if used. As in other parts of this book, we will only examine a simple scenario with uniform and steady demand. This is a favorable scenario as this type of demand minimizes the need for redistribution but it will serve to illustrate the methodology and yield insights. Other scenarios can be analyzed by adding systematic terms to the redistributing costs but there is no room here to explore that in detail. And in order to obtain analytic formulae, we will also use, as in other sections of this chapter, a slightly optimistic queuing model that does not track vehicles over space.

7.2.1. *Podless, Park-Anywhere Systems*

From a user's perspective, this is the better system as it greatly reduces walking especially at the destination end of a trip. In the system, we consider customers should be able to view all idle vehicles online and make a reservation at the push of a button. When an idle vehicle is reserved it is immediately assigned to the customer and removed from the list of available vehicles. A customer transaction is then initiated by the operator. The customer can then walk to the vehicle, use it, and eventually return it. At that time the vehicle rejoins the pool of available vehicles and the transaction is consummated.

To analyze this system we assume for now that the demand is perfectly balanced so there is no need for systematic redistribution, and we use the approach, scenario, and notation of Section 7.1. In both cases a vehicle cycles through the same three states: idle, assigned, and in-service. And in both cases, customers are either getting closer to an assigned vehicle or being transported by it. The only difference is that in Section 7.1 the distance between an assigned vehicle and its customer shrank with time at a rate v, and now is reduced at a rate $v_w \ll v$. Therefore, the model's basic building blocks embodied in (7.1)–(7.3) still apply if we simply replace v by v_w in (7.3).

The reader can easily verify by repeating the derivations following (7.3) that all the formulas continue to apply if we replace the variable v (only when it represents the speed of empty cars) by v_w. For example, the coordinates of the critical point in Figure 7.5 are now $T_a^u \cong (2\kappa^2)^{1/3}\lambda_w^{-1/3}v^{-2/3}$ and $m^{u*} \cong 1.9\kappa^{2/3}\lambda^{2/3}v_w^{-2/3}R{+}\lambda R\ell/v$. The optimum level-of-service is still given by (7.7), after substituting v_w for v. And the optimal generalized cost of a well-regulated operation (7.8) becomes

$$\frac{z^*}{\beta} = \left(1 + \frac{\gamma}{\beta}\right)\frac{\ell}{v} + \left(\frac{27}{16}\frac{\gamma}{\beta}\left(1 + \frac{\gamma}{\beta}\right)^2\right)^{1/3} \lambda^{-1/3}v_w^{-2/3} \quad \text{hour.} \tag{7.11}$$

Note from (7.11) that there continue to be economies of scale. And that if the demand is very large the system performance is comparable to that of the automobile. This could allow multi-car families to eliminate one or more of their cars and reduce the size of their home garages, which is not an insignificant benefit.

By using appropriate values for γ and the access speed (v or v_w) in these formulas we can also compare how optimally configured human-driven taxi service does *vis-à-vis* one-way vehicle sharing. Note the latter has a lower value for both the access speed and γ since it has no chauffeur. The structure of (7.8) and (7.11) reveals at a glance that, with these lower values, car-sharing should win for large λ, and taxi for small λ. Homework exercise 7.5 asks the reader to design optimum systems of both types for a few scenarios and to compare them.

The analysis is optimistic because it assumed balanced demand. Real systems will have systematic imbalances, and this should add to the cost of operation. These imbalances can be remedied not just with redistribution strategies but also with pricing incentives that would reward users for returning their vehicles in regions where they are needed. If systematic imbalances are important and known, the cost associated with redistribution can be modeled as a min-cost network flow problem in the theory of linear programming.

7.2.2. *Pod-based Systems*

These systems are more common because they are less demanding from a technological point of view — most of the intelligence can reside in the pods, and the system can be designed for use without reservations simply by ensuring that the pods are always stocked with at least one available vehicle. Therefore, unlike podless systems, this form of operation requires some form of redistribution. A key decision variable is now the number of pods because as we shall see it strongly influences the required fleet size and the user access time.

We initially use the balanced scenario of Section 7.2.1, with demand so steady and uniform that there is no need for redistribution. The only additional variable is the number of pods, p. We assume that the region is covered by a square grid of streets and that, for maximum efficiency, the pods are placed on a square lattice at 45° with the streets. Simple probability calculations then show that the average access distance from a passenger to the nearest pod is $\sqrt{2R}/(3\sqrt{p})$. Thus, the average customer access time varies inversely with the square root of the number of pods:

$$T_a = \frac{\sqrt{2}}{3}\frac{\sqrt{R}}{v_w\sqrt{p}}. \tag{7.12}$$

Since the demand is balanced and steady we now assume that the number of vehicles in each pod fluctuates very mildly, say within plus or minus 1 vehicle. If we then require all pods to have at least 1 vehicle at all times to ensure that all requests are fulfilled, the average number of vehicles in all pods will be about $2p$. And, since the number of vehicles in circulation is $n_s = \lambda R(\ell/v)$ [as per Little's formula; see (7.2)] the required fleet size is

$$m = 2p + \lambda R(\ell/v). \tag{7.13}$$

From an operator's perspective, cost should be approximately captured by a linear combination of p and m. The standards problem would then be a minimization of this linear combination subject to both, (7.13) and to $T_a \leq T_0$ with T_a given by (7.12).

Form society's perspective, we can still use the generalized cost approach with (7.12) and (7.13) as basic elements. This is now explored, assuming for simplicity that pods are so cheap that their cost can be ignored. To this end note that (7.13), unlike (7.12), increases with p. Thus, the generalized cost function should exhibit a trade-off involving the number of pods. Expressed in universal time units on a per-trip basis, this function is obtained by multiplying (7.13) by $\gamma/(\lambda R\beta)$ and adding to it the user trip time. Since this time is the sum of the access time at both ends of the trip, $2T_a$, and the average riding time, the generalized cost formula is

$$\frac{z}{\beta} = \frac{\ell}{v} + \frac{2\sqrt{2}}{3}\frac{\sqrt{R}}{v_w\sqrt{p}} + \frac{\gamma}{\lambda\beta R}\left(2p + \lambda R\frac{\ell}{v}\right). \tag{7.14}$$

This is a GEOQ equation with an optimal solution:

$$p^* = R\left(\frac{\lambda\beta}{3\sqrt{2}\gamma v_w}\right)^{2/3}, \text{ and} \tag{7.15}$$

$$\frac{z^*}{\beta} = \left(1 + \frac{\gamma}{\beta}\right)\frac{\ell}{v} + \left(\frac{12\gamma}{\beta}\right)^{1/3} \lambda^{-1/3} v_w^{-2/3}. \tag{7.16}$$

This "cost" expression has the same functional form as (7.11), which should not be surprising since both services are very similar. The coefficients of their second terms are different because the distribution of empty vehicles (and their number) is different. And as one would expect, the coefficient in (7.11) turns out to be smaller since we expect $\gamma/\beta \ll 1$ (as there is no chauffeur), confirming that the podless system is more efficient.

7.2.2.1. *Stochastic Effects and Redistribution Costs*

The model can be made a little more realistic by incorporating stochasticity. This requires the inclusion of "safety stocks" at the pods and regular vehicle repositioning. To estimate the associated costs we assume that the demand comes from a spatiotemporal process with independent increments with an index of dispersion (i.e., variance-to-mean ratio) I [vehicles].[9] We also assume that pods are periodically inspected in intervals of duration h [hours], and that they are perfectly rebalanced by repositioning vehicles immediately afterwards.

Let us consider the safety stocks first. The variance of vehicle pickups at a pod during an inspection interval is the mean demand times the index of dispersion, i.e., $(\lambda Rh/p)I$[vehicles^2]. Since demand is balanced and steady, this is also the variance of the vehicle returns. Therefore, since pickups and returns are assumed to be mutually independent, the variance of the surplus (i.e., returns minus pickups) is $2\lambda RhI/p$. This surplus can be positive or negative and should vary with high probability in a range of plus or minus two standard deviations. We can ensure that a pod will contain at least 1 idle vehicle at all times if its initial stock equals: $1 + 2(2\lambda RhI/p)^{1/2s}$. Since there are p pods, the system-wide number of idle vehicles including the safety stocks is:

$$n_i = p + 2(2\lambda RhIp)^{1/2}. \tag{7.17}$$

[9]For a Poisson process, $I = 1$.

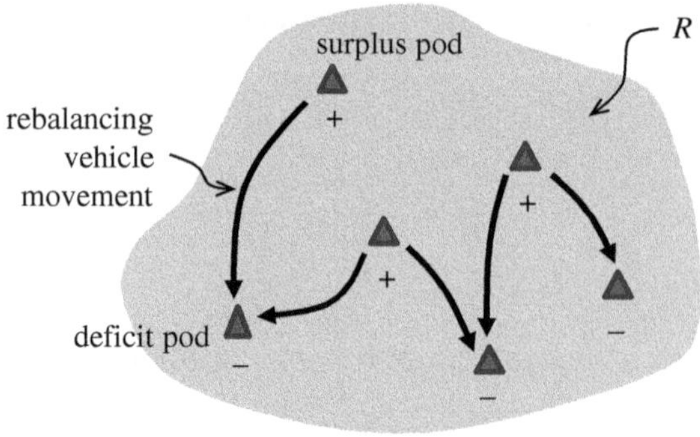

Figure 7.6. Rebalancing vehicles across pods.

Repositioning costs depend on the strategy used. The method is different for car and bicycle systems. The former are moved one at a time by employee-drivers and the latter in groups using specialized vehicles. Here we show how the cost of repositioning cars can be estimated with the help of dimensional analysis. We assume that cost is proportional to the total number of vehicle kilometers of repositioning distance.

We assume that in each repositioning event the operator solves a balanced transportation problem, a well-known network flow problem, where the number of cars to be transported from/to each pod is the pod's surplus/deficit during the previous interval (see Figure 7.6). The expected repositioning distance in one event will be denoted by D [veh-km]. The only parameters that enter in the determination of this distance are the standard deviation of the demand surplus, which we momentarily denote by σ [vehicles]; the area of the region R [km^2]; and the number of pods p [dimensionless]. (This is true because if we are given these parameters we do not need any additional information to produce simulations that would answer our question.) Obviously, only two independent groups can be formed with these constants. Therefore, the binary set $\{D/(\sigma R^{1/2}); p\}$ is a maximal set of independent groups. It follows from dimensional analysis that the solution to our problem must be of the form: $D/(\sigma R^{1/2}) = f(p)$, where f is a function to be determined; or alternatively, $D = \sigma R^{1/2} f(p)$. Simulations in Daganzo and Smilowitz (2004) show that this function increases slowly with p and that a good approximation in the range from 10^1 to 10^4 is $f(p) \approx 0.8p^{1/2}$. Therefore, we use $D \approx 0.8\sigma(pR)^{1/2}$.[10]

[10]The coefficient in the reference was 0.62 instead of 0.8. Our coefficient has been increased by a factor of $1.28 \equiv 0.8/0.62$ because the reference used a Euclidean metric but we are using a grid — and the factor 1.28 is the average circuity of random trips on a grid.

To complete the analysis, recall that the variance of the demand surplus at a pod is $(2\lambda IRh/p)$, i.e., $\sigma = (2\lambda IRh/p)^{1/2}$. Therefore, $D \approx 1.13(\lambda Ih)^{1/2}R$ [veh-km]. On dividing this result by the number of trips made in one inspection interval, λRh, we obtain the repositioning distance per customer trip: $1.13(I/\lambda h)^{1/2}$. To convert this into cost in universal units the quantity should be multiplied by $\gamma/(v\beta)$ and we have

$$\text{Expected repositioning cost per passenger trip} \approx 1.13[\gamma/(v\beta)](I/\lambda h)^{1/2}. \tag{7.18}$$

Curiously, this approximation is independent of the number of pods. This suggests that if the number of pods is large in a real-life application, changing their number should not influence the repositioning cost significantly.

We can now obtain the generalized cost function including stochastic fluctuations by combining the repositioning cost (7.18) with twice the access distance in (7.12), the idle fleet size (7.17) and the in-use fleet size $\lambda R\ell/v$, all multiplied by appropriate coefficients. The result in universal units is

$$\frac{z}{\beta} \approx \frac{\ell}{v}\left(1+\frac{\gamma}{\beta}\right) + \frac{2\sqrt{2R}}{3v_w}\left(\frac{1}{p}\right)^{1/2} + \left(\frac{\gamma}{\lambda R\beta}\right)\left(p + 2\,(2\lambda RI)^{1/2}\,(hp)^{1/2}\right)$$
$$+1.13\left(\frac{\gamma I^{1/2}}{\lambda^{1/2}\beta v}\right)\left(\frac{1}{h}\right)^{1/2}. \tag{7.19}$$

Although this equation seems complicated, we see that our cost components depend on h and p in a simple manner.

- Fleet costs $\sim (hp)^{1/2}$ (ignoring the term "p" which is usually comparatively small).
- Vehicle repositioning costs $\sim h^{-1/2}$.
- User access costs $\sim p^{-1/2}$.

These dependencies imply that an optimum design as defined by h^* and p^* can be easily found. In fact, if the small term $(\frac{\gamma}{\lambda R\beta})p$ is ignored in the optimization, the solution has a closed form solution that can be obtained by conditional decomposition. Note that the number of pods p must be at least 1. A few algebraic manipulations reveal that when $R(\frac{\lambda\beta^2 v}{I\gamma^2 v_w^2})^{2/3} \geq 2$ the optimum cost is

$$\frac{z^*}{\beta} \approx \frac{\ell}{v}\left(1+\frac{\gamma}{\beta}\right) + \left[0.5\left(\frac{\beta v^2}{\gamma I^2 v_w^2}\right)^{1/3} + \left(\frac{81.4\gamma^2 I v_w}{\beta^2 v}\right)^{1/3}\right]\lambda^{-1/3}v_w^{-2/3}. \tag{7.20}$$

This expression cannot be readily compared with (7.11) and (7.16) because the last term involves more parameters such as $v_w I/v$. But it can be shown[11] that the expression in the bracket has a lower bound of $(63.4\gamma/\beta)^{1/3}$, meaning that the system with pods generally yields higher costs.

In reality, the costs of pod-based systems should be even higher if one also considers the infrastructure cost of the pods, which was neglected in the above. On the other hand, a pod-based system may be helpful with electric vehicles because pods can double up as charging stations. In any case, equation (7.20) still exhibits economies of scale. Thus, just like driverless taxis, car-sharing without chauffeurs can improve on human-driven taxis and on the private automobile for moderate to high demand. For such levels of demand, however, we may be able to do even better by sharing rides and not just cars. This is explored in the next section.

7.3. Ride-Matching and Carpooling[12]

By ride-matching (RM), we mean an enhancement to taxi service that enables taxi riders with similar origins, destinations and desired travel times to share a ride with the guarantee of a tiny detour at most. Therefore, riders still experience taxi-like service. Although ride-matching with this type of guarantee appears not to be used, it seems to us that taxi companies could add it to their menus.[13] By carpooling (CP) we mean a system that enables car drivers to accept an extra rider with a compatible travel time, and an origin and destination that do not require a detour, without the participation of a third party chauffeur.

In this section, we shall focus on the potential of both trip matching approaches to consolidate travel on a large scale. As in other sections of this chapter, we will use a simple and slightly optimistic model. Rather than an optimization, we will look for the percent of participating passenger trips that can be matched for different scenarios. A key difference between RM and CP is that in the former all participants are alike, but in the latter, there are two classes: participants in the first class provide the car and driver, and the rest merely look for a ride as passengers. This asymmetry implies that

---

[11]To see this, note that the expression in the bracket is a GEOQ function of $v_w I/v$, so it has a lower bound.

[12]This section can be skipped without loss of continuity.

[13]In our experience, services like Uber Pool and Lyft Line are not ride-matching as they sometimes detour passengers considerably.

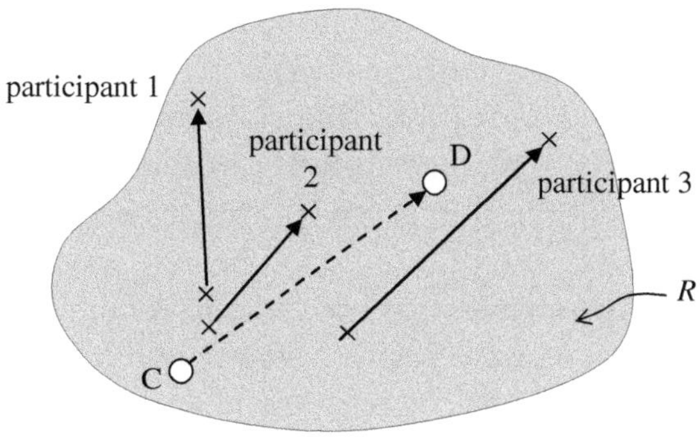

Figure 7.7. Participants in RM or CP.

the matching possibilities are slightly different for the two systems. To see this, refer to Figure 7.7, which depicts a service region, assumed to have an underlying square grid of streets. Now consider the participant depicted by the dashed arrow. This participant is currently at point C, already in a car traveling toward destination D. The arrow is dashed to signify that the participant has some routing flexibility. At that moment one of three participants (with origin–destination vectors shown in the figure by thin solid arrows) calls to request a ride. Note that under CP only participant 2 can be served without any detour, but under RM either 2 or 3 can. This illustrates that RM participants are more likely to find matches than CP participants. This difference justifies treating the two services in separate subsections. We start with ride-matching.

7.3.1. *Ride-Matching with and without Reservations*

A ride-matching system can be spontaneous (SRM) or based on reservations (RRM). In a spontaneous system, riders would declare whether they are willing to share a ride at the moment of booking. In reservation-based systems riders would declare their intentions well ahead of time, e.g., the day before. In both cases, the taxi company matches as many trips as possible and serves the rest individually. We reasonably assume that every match involves only two riders.

This subsection develops a model that can roughly predict the fraction r of participants that are successfully matched for both system types. We will find that the percent of participants matched is larger with reservations and that the relative advantage of RRM over SRM increases with city size. It will be assumed that the underlying taxi fleet is large enough to guarantee quick pickups, as in Section 7.1. And, in order to keep the analysis simple, we will use slightly optimistic approximations, designed to be most accurate when the demand for ride-matching is small.

We shall assume in our derivations that rides must be matched temporally within a tolerance of $\tau \ll 1$ [h], and perfectly in space, i.e., so that neither of the riders matched is detoured in a way that increases their trip length. No detouring is desirable to attract demand. The restriction is relaxed in Section 7.4, which describes shared taxi operations. As in previous sections, we consider a service region of area R with a steady and uniform trip generation density λ [rides/h-km^2] in which vehicles travel at speed v. However, in order to better highlight the issues, we will not specify an average trip length. We will instead assume that destinations are either concentrated at the center (concentrated demand) or uniformly and independently distributed over the square (distributed demand). This has the added benefit of eliminating an additional parameter from the problem formulation.

The problem formulation is quite parsimonious because only four parameters $\{\lambda, R, v, \tau\}$ are required to specify a scenario for a specific application, i.e., a specific city and network, a demand type (concentrated or distributed), and a service type (RRM or SRM). Therefore, including the percentage we are seeking, r, the analysis of all possible scenarios for an application involves five variables and two dimensions (time and distance). Consideration of these variables shows that a maximal set of independent dimensionless groups is $\{r, \lambda R\tau, \tau v/R^{1/2}\}$. It will be convenient to abbreviate the last two groups as

$$\pi_1 \equiv \lambda R\tau;\ \pi_2 \equiv \tau v/R^{1/2} \tag{7.21}$$

Note that parameter π_1 is a dimensionless measure of demand, expressed as the number of ride requests in the whole region during the tolerance time; and parameter π_2 is a dimensionless measure of the tolerance. Since π_2 is the ratio of the tolerance time to the time it takes to cross the region we expect $\pi_2 \ll 1$ in most applications. We will restrict our analysis to this case.

With the above in mind, dimensional analysis tells us that the answer to our question for any given application must be of the form:

$$r = F(\pi_1, \pi_2), \tag{7.22}$$

where F depends on the application. An approximate form of this function is derived next.

Before describing the logic, some terminology is in order. A solo rider looking for a match shall be called a "seeker." Now note that every match involves two participants, one who starts the trip first and a second who joins the ride later. The former shall be called the "provider" (of the ride) and the latter the "taker." Obviously, the probability that a random participant is in one of these two categories must be equal to $r/2$. We denote this probability, $q \equiv r/2$. Note, it is a proxy for the fraction of cars that could be taken off the road.

The logic is centered on finding an expression $n(\pi_1, \pi_2)$ for the expected number of valid service calls (from potential takers) a "seeker" encounters during his/her trip.[14] By valid we mean a call that satisfies two conditions: (i) the seeker can reach it temporally in the allotted τ time units, and (ii) if both trips are combined neither requires a detour. Note that valid encounters should occur independently as a Poisson process. Therefore, the total number of such encounters is a Poisson random variable with expectation $n = n(\pi_1, \pi_2)$. Thus, assuming optimistically that none of the valid calls are served by other seekers, a seeker's probability of success in finding at least one taker during his/her trip is $p \approx 1 - \exp\{-n\}$. Next note that $q = (1 - q)p$ because the probability that a particular participant is a successful provider, q, is the product of the probability that it fails to be a taker $(1 - q)$ and then succeeds in finding one along the trip, p. Thus, $q \approx (1 - q)(1 - \exp\{-n\})$, and solving for q we find

$$q \approx \frac{1 - e^{-n}}{2 - e^{-n}} \quad \text{or} \quad r \approx \frac{1 - e^{-n}}{1 - \frac{1}{2}e^{-n}}. \tag{7.23}$$

These expressions are most accurate if $n \ll 1$. In this case, they further reduce to

$$q \approx n/(1 + n) \quad \text{or} \quad r \approx 2n/(1 + n), \quad \text{if } n \ll 1. \tag{7.24}$$

Appendix 7.B shows that if the region is a square with a dense perpendicular grid of streets and the trip destinations are distributed then the expected number of valid encounters is

$$n \approx \pi_1\pi_2/36 \quad \text{(for SRM)} \quad \text{and} \quad n \approx \pi_1/72 \quad \text{(for RRM.)} \tag{7.25}$$

And, if the destinations are concentrated at a central point then

$$n \approx \pi_1\pi_2/8 \quad \text{(for SRM)} \quad \text{and} \quad n \approx \pi_1/16 \quad \text{(for RRM).} \tag{7.26}$$

Similar formulas should apply if taxis are allowed to detour passengers slightly and/or the network resembles a deformed grid.

---

[14] In doing this, we assume that taxis are never detoured to collect a second passenger when they are on the way to pick up the first; i.e., that a taker can be matched only with a participant that is already riding (a seeker). This is a sensible simplification, especially in the case of SRM, as it makes waiting times more reliable.

Example 7.2. Consider a city with the following hypothetical values for SRM service: $\lambda \approx 1.5\,\text{pax/h-km}^2; R \approx 12 \times 12 = 144\,\text{km}^2$; $\tau \approx 0.1\,\text{h}$; and $v \approx 20\,\text{km/h}$ under moderate congestion. This yields $\pi_1 = 21.6$ and $\pi_2 = 0.17$. Therefore, $n \approx 0.1$ and $r \approx 0.18$ for SRM.

It is interesting to examine how advance reservations affect the number of matches when RM is offered in cities of different sizes. Note from (7.25) and (7.26) that reservations increase the number of valid encounters by a factor of $(2\pi_2)^{-1}$; and from (7.24) that the corresponding ratio for the number of matches remains close to $(2\pi_2)^{-1}$ if π_1 is small. Otherwise, as per the nonlinearity of (7.23), the ratio is smaller. In either case, the improvement ratio varies inversely with π_2 *ceteris paribus*. It thus follows that for any given reachability distance τv the relative benefit of advance reservations increases with a city's diameter.

7.3.2. *Carpooling*

The main difference between ride-matching and carpooling (CP) is that carpooling participants are exogenously divided into two groups: a group whose participants offer rides and a group that can only take rides. Participants in the first group are chauffeurs and stay with their cars for the complete ride. If they fail to find a taker they ride solo. If a participant from the second group fails to find a match the issue is trickier. In a reservation-based system (RCP) said participant could learn about the failure with enough advance notice to make alternative arrangements and for example, drive solo. But in a spontaneous system (SCP) the participant would have to find another way to travel without any lead time. This is why spontaneous systems are mostly seen during the morning commute where the demand is concentrated (so the probability of finding a match is higher), and near busy transit stations where a travel alternative is readily available. The phenomenon of "casual carpooling" is a case in point. This section will derive expressions for both RCP and SCP, even though the spontaneous system with dispersed demand is rather hypothetical. Companies such as Scoop and Waze Pool offer reservation-based matching services.

As compared with ridesharing, the carpooling problem requires one additional parameter. This is necessary to describe the fraction of demand in each group. The parameter shall be the fraction of participants, f, that are takers. Thus, the demand density of takers is denoted as λf and that of providers is $\lambda(1-f)$. Both are known. As before, we still look for the fraction of participants that succeed in finding a ride, r. However, since every match involves one participant from each group and the number of participants in each group is different the success rates for takers and providers will be different in general. They are r/f and $r/(1-f)$, respectively.

The analysis logic parallels that of ride sharing. The main difference is that each participant has only one role either as a taker or provider. As before, we shall look for the expected number of valid encounters n found by a provider (from the provider's group). Instead of (7.23), the basic balance equation is now

$$r/(1-f) \approx 1 - e^{-n}, \tag{7.27}$$

or for small n:

$$r \approx (1-f)n \quad \text{if} \quad n \ll 1. \tag{7.28}$$

This merely expresses that the success rate of the providers is the probability of at least one successful encounter. As in the previous section, it ignores competition from other providers. Therefore, the expression continues to be an upper bound. It is particularly tight when the fraction of takers, f, is low.

Appendix 7.B derives approximate expressions for n assuming the service region is a square with a dense square grid of streets. For distributed trip destinations the results are

$$n \approx f\pi_1\pi_2/72 \quad \text{(for SCP)} \quad \text{and} \quad n \approx f\pi_1/144 \quad \text{(for RCP)}. \tag{7.29}$$

If the trip destinations are concentrated the results are

$$n \approx f\pi_1\pi_2/8 \quad \text{(for SCP)} \quad \text{and} \quad n \approx f\pi_1/16 \quad \text{(for RCP)}. \tag{7.30}$$

In combination with (7.27) and (7.28), equations (7.29) and (7.30) produce the solution for r. For example, when $n \ll 1$, the expressions for reservation-based systems reduce to

$$r \approx f(1-f)\pi_1/144 \quad \text{(dispersed RCP)} \quad \text{and}$$
$$r \approx f(1-f)\pi_1/16 \quad \text{(concentrated RCP)}. \tag{7.31}$$

Note how the matching rates are maximized in all cases for $f = 0.5$, i.e., when the number of providers and takers is roughly the same, as one might expect.

Example 7.3. We examine three cases of concentrated RCP (with $f = 0.25$, 0.5 and 0.75) using the data of the RS example. We still have $\pi_1 = 21.6$. Therefore, according to (7.30) $n \approx 0.34$, 0.68 and 1.01 for these three values of f; and we find from (7.31) that the fractions of participants matched are $r \approx 0.25$, 0.34 and 0.25. The numbers are a bit lower if we use the more accurate approximation (7.28): $r \approx 0.21$, 0.25 and 0.16. A comparison with the RS example reveals that the two sets of rates are comparable. This may be surprising given that the CP example uses a much more favorable scenario with reservations and concentrated destinations. The reason is that all CP participants play single roles as either providers or takers. In contrast, all participants in RS systems first try as takers, and if unsuccessful have another go as providers. This dual role allows for more successful matches.

It bears repeating that the results in this section are optimistic because they ignore the competition that seekers face from other seekers.[15] These interactions are negligible for "low traffic," i.e., when π_1 is small, as then the probability for two seekers to be nearby is low. For example, computer simulations of SRM service reveal that if $\pi_1 = 15$ and $\pi_2 = 0.15$, then the simulated matching rate is about 8.6%, but (7.23) and (7.25) predict $r \approx$ 11%. Although this error gap should increase with demand, refinements are not introduced because in typical cases the demand is low.

7.4. Demand Responsive Transit

We define DRT as a form of flexible transit that offers door-to-door service by grouping passengers in larger numbers than ride-matching services and allowing substantial detours. DRT's emphasis is not on matching the level-of-service of the automobile with quick responses and direct travel, but in reducing the monetary cost of service. Although detours alone allow passengers traveling at similar times to be grouped into vehicles, the size of such a group can be further increased if passengers are also asked to wait at their origins until more matches can be found. Obviously, this kind of temporal aggregation can further reduce cost.

In view of this, we further classify DRT systems depending on whether temporal aggregation is used or not. *Dial-a-Ride* (DAR) paratransit services, where cost reduction is a primary concern, use it. (DAR service is often provided by public transit agencies as a supplement to conventional transit with very small vehicle fleets; especially to serve the disabled population in low density areas.) *Shared Taxi* (ST) services such as Uber Pool and Lyft Line where passengers can be lost to competitors if asked to wait do not. (ST services are usually provided with very large vehicle fleets.)

This section will describe a simple model of DAR (Section 7.4.1) and introduce results for ST without a detailed derivation (Section 7.4.2). The

[15]The appendix also made slightly optimistic assumptions.

interested reader can find it in Daganzo and Ouyang (2018); the derivation is similar but more involved. As in previous sections, both models are slightly optimistic. They use aspatial bins to track vehicles, use deterministic queuing approximations, and make other slightly favorable assumptions. The upside of these simplifications, however, is that it is then possible to compare systematically and transparently how the various modes we have discussed, including the private automobile and conventional transit, fit in the urban transportation ecosystem. This comparison is provided in Section 7.4.3.

7.4.1. *Dial-a-Ride*

Like ride-matching taxi services, DAR can be spontaneous or reservation-based. As occurred with RRM and SRM, the performance of the matching algorithms should be more efficient when passenger trips are known beforehand, but not hugely so. Therefore, in this section, we shall only attempt to give a rough idea of what DAR service can do, regardless of service type.

To encompass both spontaneous and reservation-based systems, we will use a very simple, albeit suboptimal, matching algorithm that can be used with either type of service. The ideas are based on Daganzo (1978).[16] Although real systems can probably perform a little better than what we will predict, the simplicity of the algorithm allows us to derive simple formulas that reveal at a glance what is possible. We shall ask the same two basic questions we asked for taxis, again focusing on the case with dispersed demand: (i) what is the fleet size needed for a given level-of-service? And (ii) how does DAR compare to other modes?

As before, we analyze the steady state of a region of area R with trip generation density λ where origins and destinations are uniformly distributed; see Figure 7.8. To achieve cost efficiency, the bus fleet size m is assumed to be more limited than that in the case of taxis, so in the steady state buses carry multiple passengers and are always busy. As a result, DAR passengers do not receive assignments as early as they want — i.e., immediately upon calling in the spontaneous case. Instead, they must wait at their origins (or "homes").

The figure shows buses by means of boxes and passengers by means of dots. Passengers in the system are divided into two groups: (i) those waiting at home and not yet assigned to a vehicle, represented by the dotted arrows

---

[16]Extensions of these ideas have been recently used to model rider-sharing markets and pricing mechanisms (e.g., Zha *et al.*, 2017; 2018).

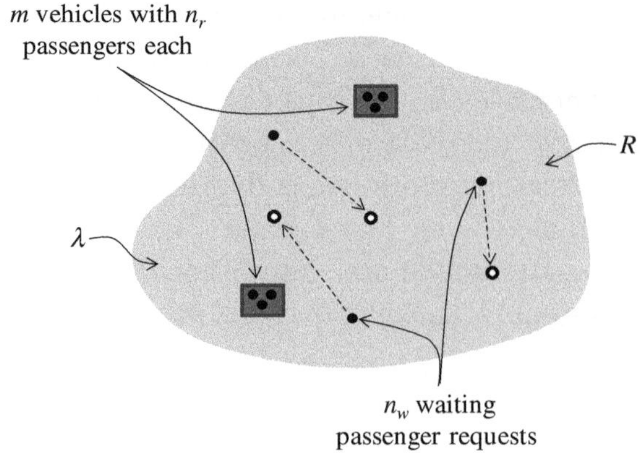

Figure 7.8. DAR service.

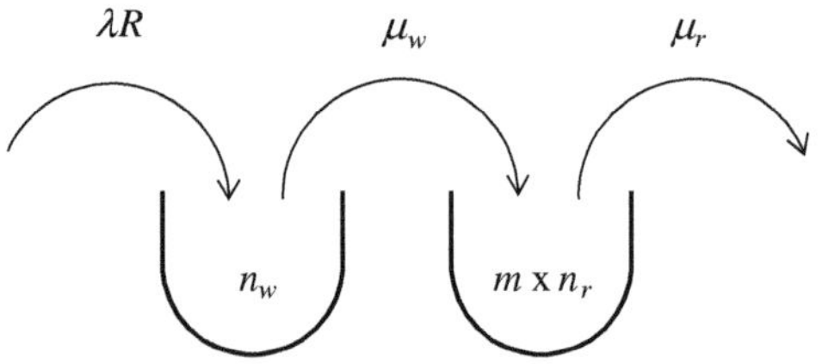

Figure 7.9. State evolution of passengers in a DAR system.

of the figure; and (ii) those already assigned to a vehicle or riding in it, represented by the dots inside the boxes. Note that all passengers assigned to a bus, even if not yet physically inside, will be called occupants, said to be "riding" and counted toward the bus's occupancy. As the figure shows, the total number of unassigned passengers waiting at home in the steady state will be denoted by n_w, and the average number of occupants in a bus, n_r.

As in the case of taxis, we use an aggregate model without keeping track of the detailed vehicle and passenger positions. However, the state of the system is now defined by the expected number of passengers in each group instead of the expected number of vehicles. The dynamics are still modeled with a deterministic approach that tracks steady state flows across space-less bins; see Figure 7.9. The symbols λR, μ_w, and μ_r represent these flows. And since these flows must be balanced, we have $\mu_w = \mu_r = \lambda R$.

The matching algorithm is defined by the following three rules that bus drivers follow:

- After achieving a desired passenger load buses alternate between pickups and drop-offs.

- After each pick-up, a bus chooses to drop-off the onboard passenger whose destination is closest to the current bus location.
- After each drop-off, a bus chooses to pick up the unassigned waiting customer who is closest to the current bus location. Said customer is immediately assigned to the bus, considered to be an occupant and included in n_r.

Note, with this algorithm the buses' passenger load is a decision variable that can be optimized. If we choose $n_r = 1$, the system operates as a taxi system with few taxis and people waiting at home. These conditions were not examined in Section 7.1 because they were deemed to be undesirable for the taxi-using population. If we choose $n_r>1$ buses can process passenger deliveries more rapidly and the system becomes more productive (cheaper) but people end up spending more time riding. The analytic model about to be developed will explore this trade-off. It will produce an expression that predicts the expected number of people waiting at home n_w as a function of the system design variables m and n_r.

7.4.1.1. *DAR Physics*

To obtain the desired expression we shall use the balance equation matching the demand rate and the rate at which the bus fleet processes passengers $\lambda R = \mu_r$. Note that this rate is the ratio of the fleet size to the average time between one bus's successive passenger deliveries. Also note that the latter is the sum of the average times needed for a passenger pickup and a passenger delivery, which are denoted by $\bar{t}_p$ and $\bar{t}_d$, respectively. Therefore, our analysis will be based on the identity:

$$\lambda R = \frac{m}{(\bar{t}_p + \bar{t}_d)}. \tag{7.32}$$

We will express $\bar{t}_p$ and $\bar{t}_d$ in terms of our decision variables, and use (7.32) to solve for n_w.

To obtain these expressions, recall from Appendix 7.A that the expected distance between a random point and the closest of n random points is: $E[d_n] \cong \kappa\sqrt{\frac{R}{n}}$. Therefore, since bus drivers choose the closest of $n = n_w$ points when considering who to pick up next, and the closest of $n = n_r$ points when considering drop-offs, our sought averages are simply:

$$\bar{t}_p = \frac{\kappa}{v}\sqrt{\frac{R}{n_w}}, \quad \bar{t}_d = \frac{\kappa}{v}\sqrt{\frac{R}{n_r}}. \tag{7.33}$$

Now, entering with these expressions into (7.32) we find

$$\lambda R = m \bigg/ \left[\frac{\kappa}{v}\sqrt{\frac{R}{n_w}} + \frac{\kappa}{v}\sqrt{\frac{R}{n_r}}\right]. \tag{7.34}$$

We can now solve for n_w in (7.34) to find the expression we were after. In doing this, we must recognize that (7.34) does not have a solution if n_r is very small. Keeping this in mind, a few algebraic manipulations reveal that

$$n_w = (mv/\lambda\kappa R^{3/2} - 1/\sqrt{n_r})^{-2} \quad \text{where} \quad n_r \geq (\lambda\kappa R^{3/2}/mv)^2. \tag{7.35}$$

The inequality condition in (7.35) is the condition required for a solution's existence. If the condition is violated, the fleet size is insufficient to serve the demand, and the number of waiting passengers would grow without bound, i.e., $n_w = \infty$.

The average passenger waiting and riding times can also be written using Little's formula. Applied to the first bin of Figure 7.9, we have

$$\text{average wait} = n_w/(\lambda R), \tag{7.36}$$

which can be evaluated using (7.35). And applied to the second bin we have

$$\text{average riding time} = mn_r/(\lambda R). \tag{7.37}$$

It should be stressed that all these formulas are only rough approximations because in real systems the distribution of waiting passengers and buses may not be perfectly uniform and independent through the course of a day. Nevertheless, the formulas give an accurate idea of cause and effect with regards to the decision variables, the problem's parameters, and the desired outcomes.

7.4.1.2. *Designing the System*

We are now in a position to tackle the design question from an agency's perspective and, as before, use the standards approach. A proxy for the agency cost is m. Level of service shall be quantified by the product of a factor χ and the average time passengers spend in the system, T. This factor makes sense if one wishes to capture the anxiety passengers suffer due to their variable service times. (Note, the factor should be smaller for reservation-based systems because in this case passengers can be told ahead of time their approximate pickup and delivery times.) With this in mind, we see that the standards problem consists in choosing a decision set $\{m, n_r\}$ that solves: $\min\{m : \chi T \leq T_{max}\}$. Of course, we can redefine the standard in terms of actual time, as $T_0 \equiv T_{max}/\chi$, and rewrite the problem more elegantly as $\min\{m : T \leq T_0\}$.

To complete the formulation, we need to express T in terms of the decision variables. To this end, it will be convenient to let p denote the number of people in the system and write $n_w(n_r, m)$ for the function that extends (7.35) to all non-negative arguments — using the convention that n_w takes on an infinite value if the qualifying inequality in (7.35) is violated. We can then write T using Little's formula as $T = p/(\lambda R)$. And since p is the combined number of people in the buses and at home we have $p = mn_r + n_w(n_r, m)$. Thus, $T = [mn_r + n_w(n_r, m)]/(\lambda R)$, which is the sought expression.

The standards problem can now be written as:

$$\min_{m,n_r}\{m : mn_r + n_w(m, n_r) \leq \lambda R T_0;\ m, n_r \geq 1\}, \tag{7.38}$$

where $n_w(n_r, m)$ is the extension of (7.35). Although this problem does not have a particularly favorable form, it can be easily solved numerically since it only has two decision variables.

7.4.1.3. *Society's View*

Let us now answer the second question and evaluate the performance of a generic DAR system that has been optimally configured considering both the user and agency costs. To do this, we first write the generalized cost function. As before, we assume that vehicles cost γ monetary units per unit time and society values passenger time at a rate of β monetary units per unit time. This rate should account for the high multiplier χ that passengers use in evaluating their own travel time. With this convention, the generalized cost function is

$$z = \$ + \beta T = \frac{1}{\lambda R}\left(\gamma m + \beta[mn_r + n_w(n_r, m)]\right). \quad [\$/trip] \tag{7.39}$$

This expression is now minimized with the conditional decomposition approach, conditioning on m. Note that with m fixed, the minimum of (7.39) is obtained by minimizing $[mn_r + n_w(n_r, m)]$; or equivalently the user time T.

Appendix 7.C derives the result, which is

$$T^* = \frac{\theta(m)}{m}\frac{c^2}{\lambda R}, \quad \text{where} \quad c = \frac{\kappa\lambda\sqrt{R}R}{v} \quad \text{and} \quad \theta(m) \equiv (1 + (m)^{-1/3})^3. \tag{7.40}$$

The constant c is a dimensionless proxy for demand. It is the number of requests that the system receives in the time it takes a bus to travel across the region. The function $\theta(m)$ defined by the last identity decreases with m

and tends toward 1 as $m \to \infty$. Therefore, T^* declines with m. This means, not surprisingly, that passenger cost declines with fleet size. This answers question (i) posed at the outset of this section.

To answer question (ii) and compare DAR with other modes, we must solve the outer (fleet sizing) problem. After inserting (7.40) into (7.39), this outer problem can be expressed as

$$\min_m \left[z = \frac{1}{\lambda R} \min_m \left(\gamma m + \beta \frac{\theta(m) c^2}{m} \right) \right].$$

A simple approximate formula for the solution can be obtained when c is large because then m^* must also be large and as result, we can use $\theta(m) \approx 1$ in the objective function. This function then becomes an EOQ expression that yields

$$m^* = \left(\frac{\beta c^2}{\gamma} \right)^{1/2} \sim (\beta/\gamma)^{1/2} c, \quad \text{for} \quad m^* \gg 1,$$

and

$$\$^* = \beta T^* = \frac{\sqrt{\gamma \beta} c}{\lambda R} = \kappa \frac{\sqrt{\gamma \beta R}}{v}, \quad \text{for} \quad m^* \gg 1. \tag{7.41}$$

Therefore, for large c (and large m^*) the total cost in units of time is

$$\frac{z^*}{\beta} = \frac{\$^*}{\beta} + T^* \approx \frac{2\sqrt{\gamma/\beta}}{\lambda R c} = 2\kappa \frac{\sqrt{\gamma R/\beta}}{v}. \tag{7.42}$$

Note that λ does not appear in the last expression. Thus, DAR has no economies of scale for large c and λ, much like the automobile.

A more detailed comparison with auto is instructive. Note that the time-denominated cost of an average automobile trip (excluding parking) is $\ell(\beta + \gamma')/(\beta v)$, where ℓ is the length of an average trip and γ' the mechanical cost of driving the car per unit time. Now, for a square city with a grid network and random origins and destinations $\ell = \frac{2}{3}\sqrt{R}$. Thus, the car cost formula per trip is $\frac{2}{3}\sqrt{R}(\beta + \gamma')/(\beta v)$. To compare apples with apples, let us use $\gamma \approx \gamma' + \beta$ for the DAR mileage cost in (7.42) — since γ must include both the cost of the vehicle and its driver. We then see in comparing the two cost expressions that they are functionally very similar. However, while the car expression varies linearly with $(\beta + \gamma')/\beta$, the DAR expression is proportional to the square root. This means (logically) that DAR outperforms the automobile for populations with low values of time, even if we neglect the cost of parking.

Moreover, the same should occur for taxis. This should be clear since as we already saw in Section 7.1.3 the automobile outperforms human-driven taxis if parking costs are neglected.

7.4.2. *Shared Taxi*

By shared taxi, we mean the form of ridesharing currently being offered by taxi companies such as Uber and Lyft. This transportation mode differs from DAR in the algorithms it uses to route the vehicles and assign vehicles to calls, and on the smaller number of seats of the vehicles in the fleet.

Because taxis operate in a competitive environment, a primary concern for the operating companies is ensuring that calls that are received are not lost to the competition due to long waits. For this reason, operators do not have the luxury of keeping a buffer of unassigned calls, as with DAR, and therefore must use $n_w = 0$. A direct result of the competitive environment is that some vehicles may be idle or with single occupants. This was also assumed in Section 7.1 for individual taxi service.

The assignment and routing rules of an algorithm define its performance. These rules are company secrets, but reasonable facsimiles can be imagined and analyzed systematically. This has been done in Daganzo and Ouyang (2018), which shows that *Shared Taxi* fills a niche in the transportation ecosystem somewhere between *DAR* and *Taxi*. The reference analyzes the same demand scenario of Sections 7.1 and 7.4.1, i.e., steady and uniformly distributed demand over a homogeneous network. As in those sections the taxis are assumed to be identical and the analysis is slightly optimistic. Because the derivations in the reference are similar to those of previous sections but involve numerous bins and more development steps, only the basic logic, and the main results are discussed here.

As we saw earlier, an operating scenario with fleet size m is characterized by the parameters $\{m, \lambda, R, v\}$, by the character of the network, and by the algorithm itself. An important property of our imagined algorithms is a maximal vehicle occupancy, which influences the level of sharing. For consistency with Section 7.4.1, we denote it by n_r.[17] This is a control variable, which cannot exceed the vehicle capacity.

Before presenting the results, note that at most two independent dimensionless constants can be defined from the four parameters defining a scenario; e.g., m and $\pi \equiv \lambda R^{3/2} v^{-1}$. The latter[18] can be interpreted as a dimensionless measure of demand expressing the average number of calls that arrive during the (approximate) time it takes for a vehicle to cross the service region. Dimensional analysis tells us that any dimensionless measure

[17]Whereas the DAR algorithm forced each vehicle's occupancy to alternate between n_r and $n_r - 1$, the ST algorithms allow the occupancy to range anywhere in $[0, n_r]$.

[18]This dimensionless constant is essentially the ratio of π_1 to π_2 defined in Section 7.3.

of performance, such as the ratio of the door-to-door travel times by shared taxi and that by private auto, must be an implicit function of the two dimensionless groups and our control variable. For example, if we use π_T to denote this ratio, we know that there must be a four-argument function F_T such that $F_T(\pi_T, m, n_r, \pi) = 0$. Daganzo and Ouyang (2018) show how to derive this function.

Figure 7.10(a), adapted from that reference, displays the result for the special case of a square grid of streets with $\pi = 100$ and $n_r = 2$. This demand level is intended to represent small, low-density semi-rural communities typified by Napa (California, USA). The two dark curves in the figure labeled "shared taxi" express the relation between π_T and m, i.e., the curve $F_T(\pi_T,\ m,\ 2,\ 100) = 0$, for to two different algorithms. They are interesting because they show qualitatively how the level-of-service of shared taxi (expressed inversely by the indicator, π_T) depends on an excellent indicator of societal and operator cost (the fleet size, m). The algorithm leading to the curve on the right assumes that assignments cannot be made to taxis with passengers on board, and the one leading to the curve on the left allows them. Preventing such assignments is desirable from a level-of-service standpoint since passengers can then be told at the time they board the approximate vehicle route and time of arrival. Note that both curves bend at a point where the fleet size is a minimum. This is the critical fleet size for the algorithm in question. Smaller fleet sizes cannot serve the demand.

7.4.3. *Comparison of Urban Transportation Modes*

Figure 7.10(a) also plots as point of comparison the corresponding curves for individual taxi ($n_r = 1$), DAR ($n_r = 2, 3, 5$) and conventional transit. The latter corresponds to a grid with fixed routes and stops, similar cruising speed as DRT, and optimized with the methods of Chapter 6.[19]

To read these curves, pick a normalized travel time π_T and mentally draw a horizontal line across the corresponding ordinate. The curve that intersects this line closest to the ordinate axis corresponds to the mode that delivers the chosen level-of-service with the fewest vehicles, i.e., most efficiently and cheaply. The collection of all these points is contained in the "Pareto frontier"

[19]The commercial speed could be higher or lower than that of DRT depending on whether the vehicles enjoy an exclusive right-of-way or not. More details of the model set-up are available in Daganzo and Ouyang (2018).

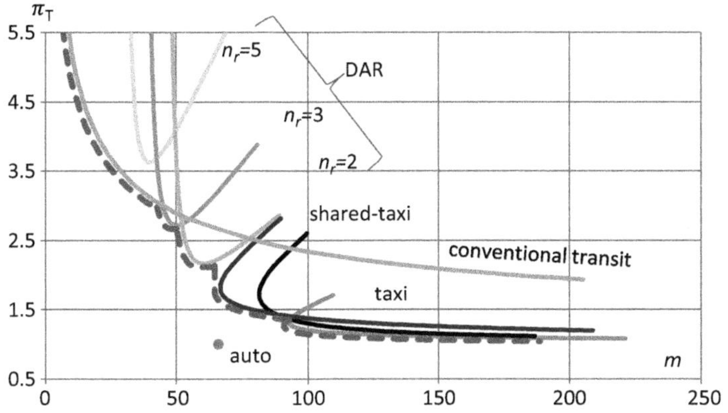

(a) $\pi = 100$ (small, low-density community)

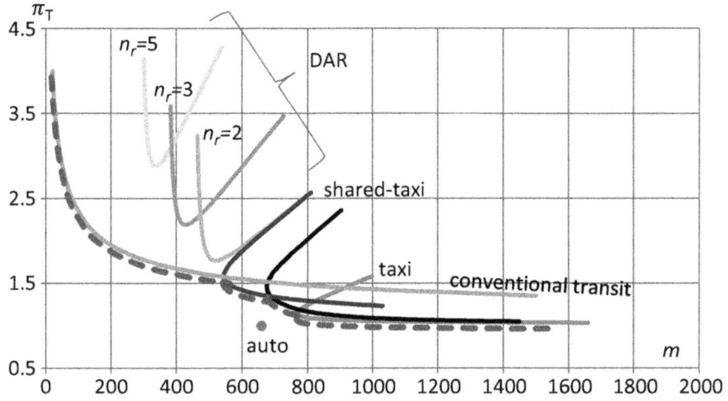

(b) $\pi = 1{,}000$ (medium size, semi-dense town)

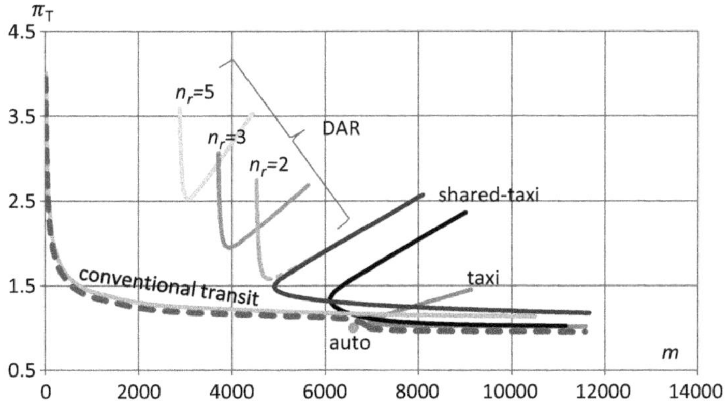

(c) $\pi = 10{,}000$ (large, dense city)

Figure 7.10. The urban transportation ecosystem: Normalized door-to-door travel time, π_T, vs. fleet size, m, for different transportation modes and three demand levels, π.

of all the transit modes, i.e., the set of points whose cost–time coordinates are not improved by any other point on any curve. Consideration of this property shows that the frontier is the highest non-increasing curve that does not protrude above any of the modal curves. For our diagram, the frontier is the thick dashed curve bordering the bottoms of the transit curves.[20] The Pareto frontier neatly demarcates the door-to-door times and societal costs that public transportation modes can deliver when used alone.[21]

Also worth noting is the part of the frontier that overlaps with a particular modal curve. Any such part is the niche of the corresponding mode in the transit ecosystem. Note from the niches in the figure how *Conventional Transit* is positioned to serve the low-cost/low-LOS section of the market, *DRT* (including *Shared Taxi*) the mid-cost/mid-LOS and *Taxi*, the high end.

The figure also includes a solid dot intended to position on the same plane travel by private automobile. In this case, the fleet size is estimated by the number of cars in motion, assuming that parking is plentiful at both ends of a trip. This is optimistic but not unreasonable for the small, low-density communities that the figure is intended to represent. Note how for this scenario *Shared Taxi* and *Taxi* are dominated by the automobile. This strongly suggests that for small, low-density communities *Shared Taxi* mainly serves travelers who do not have access to their private cars.

Before we examine how the picture changes for other city types, it is interesting is to examine the feasibility domains of the various modes. Part (a) of the figure reveals that while conventional transit can serve our $\pi = 100$ demand with a fleet of any size (because vehicle capacity can be adjusted as necessary), the remaining modes cannot — they become infeasible if the fleet size is too small. To see this, note how the *DAR* curves increase asymptotically toward infinity for positive values of m and how the (*Shared*) *Taxi* curves bend back at their left ends. The abscissas at these

[20]Note, the curve has a small step-shaped jump between the shared taxi and DAR curves. This occurs because the poorly performing branches of the modal curves cannot be on the Pareto frontier.

[21]The curves in the figure are based on proxies and therefore only intended to give a qualitative picture. Quantitative comparisons of costs and times would require that our proxy for cost (the fleet size, m) be converted into an actual cost. This can be done by rescaling the axis for each separate mode using an appropriate conversion factor. We know for example that the cost of operating a bus (including a human driver) is about twice the cost of driving a car. Therefore a more accurate comparison between shared taxi and conventional (bus) transit could be made by rescaling the cost axis for of the latter by a factor of two while leaving the former alone.

bends are the critical fleet sizes. As the figure shows, they decrease with the level of sharing: the critical size is highest for *Taxi*, intermediate for *Shared Taxi*, and lowest for *DAR*. This ordering should not be surprising since fewer vehicles should be needed when the level of sharing increases.

Parts (b) and (c) of Figure 7.10 reproduce the diagram for two scenarios with $\pi = 1{,}000$ and $\pi = 10,000$. The former is representative of medium density towns typified by Berkeley (California) and the latter by larger, denser cities such as San Francisco (California). Note how the two diagrams look similar to the diagram of part (a). This occurs only because the horizontal axes have been rescaled. In addition to the scale, two other differences are: (i) the *Conventional Transit* curves are comparatively lower; and (ii) the auto dots are now much closer to the Pareto frontier for transit.

Difference (i) is so pronounced that the Pareto frontier is well below the *DAR* curves in both scenarios, i.e., *DAR* is dominated by *Conventional Transit*. It is perhaps for this reason that *DAR* is mostly confined in the real world to serve the non-ambulatory poor — i.e., people who cannot take *Conventional Transit* and cannot afford *Taxi*. Other population groups can do better with either of these other modes.

Difference (ii) is actually more pronounced than shown in the figures because for the considered city types the time and cost associated with parking and looking for parking is no longer insignificant. Thus, the real position of the auto dot should be slightly displaced upward and to the right. In part (c) of the figure, it might well lie above the Pareto frontier. This placement seems in agreement with media reports that the proportion of auto trips in San Francisco now taken by *Taxi* and *Shared Taxi* rivals that of the private automobile. Given the economics, these auto trips probably include a significant fraction of long trips with one end outside the city (for which driving makes sense) and/or trips made by people unfamiliar or uncomfortable with the new forms of taxi service.

Also interesting is the increased size of the niche for *Conventional Transit* as the city size and density increases; and how (in part c) its level-of-service can nearly equal (or surpass if we consider parking) that of the automobile. Intriguing too is the reduction in the size of the *Shared Taxi* niche as the city size/density increases — this can be seen by comparing the niches' sizes for the three parts of the figure. This reduction is apparently driven by the increased efficiency of *Conventional Transit* which, for large/dense cities, can produce a level-of-service better than *Shared Taxi* for the same cost. The reason for the increased efficiency of *Conventional Transit* is that, as mentioned earlier in the chapter, the *DRT* modes have insignificant economies of scale compared with *Conventional Transit.* As a result, the latter shines when the demand is high.[22]

The main reason for the economies of *Conventional Transit* is that it is organized with established routes with fixed stations, which (for large, dense cities) allow people to congregate and be served in large groups. Although this rigidity requires people to walk, the resulting inconvenience is more than compensated by the ability to lay out transit routes in a direct way with few detours. Interestingly, it turns out that even for systems with rigid routes, a little bit of temporal flexibility can sometimes be introduced by eliminating the schedule and adapting the bus dispatches to the demand. This is particularly effective when much of the demand is concentrated at a route's terminus. This kind of adaptation is explored in the next section.

7.5. Demand Responsive Transit with Established Stations and Hubs: Jitneys

The discussion of collective flexible transit has assumed so far that demand is dispersed. In the real world, however, there often are hubs such as train stations, airports, and small CBDs; and if the concentration is sufficiently high it makes sense to focus service on these points. Section 6.4 explored the issue for conventional transit. It analyzed "trunk-and-branch" and "hub-and-spoke" network structures and showed that these structures were very well suited to serve a single hub. The main advantage of these types of networks is that they can serve the demand pinned at the hub directly, without a need for transfers or detours, and also serve the rest using transfers at the hub.

Demand responsiveness can and has been introduced into the hub-and-spoke structure. The prime example is jitney service. By this, we mean a hub-and-spoke system of formal routes, on which vehicles are dispatched adaptively. Jitneys have emerged spontaneously in many places as private operators take advantage of the natural demand concentration at hubs. Once operators have carved out corridors to/from the hub, more demand is attracted to the corridors, not necessarily going to the hub. This, in turn,

[22]This qualitative insight is consistent with real world evidence indicating that the larger and denser a city the larger the fraction of transit trips that are served by conventional transit. For example, whereas in San Francisco (population ~800,000) all taxi companies combined carry about 170 thousand trips (Bliss, 2018), which is comparable to the number of trips by buses and streetcars, the daily taxi ridership in NYC is about 550 thousand, which pales in comparison with both the daily bus ridership (2M) and the subway's (5.7M).

attracts more operators. This interaction creates a positive feedback cycle so the demand and service can organically co-evolve into extensive jitney networks — one example is the system of "publicos" in Puerto Rico. These private networks often compete for transit riders with the more structured official transit services. Startup companies like Chariot in San Francisco and Via in New York City are essentially trying to do the same thing in modern ways, leveraging smartphones and the internet.

For the most part, jitneys operate on fixed routes. The dispatching strategy is simple. Empty jitney vehicles queue at the hub in first-in-first-out fashion for the right to carry passengers. The vehicle at the head of the line accepts passengers until a set occupancy is reached and then is dispatched. The system is adaptive because when the demand is low dispatching intervals are long, and vice versa. Jitney services are found throughout the world because they are very easy to organize, requiring almost no coordination. Vehicles on any route can be run by enterprising individuals or different small operators. The main coordinating role of a public agency is enforcing order at the hub, and in some cases, establishing the routes. Ideally, a public agency should also regulate the target occupancy and pricing. Jitney service can be instituted with multiple hubs. In this case, vehicles are dispatched adaptively from the hubs on set routes that run from one hub to another, crisscrossing a region. This enables more people to travel without transfers and more direct travel for the rest.

Jitney service does not come without problems, however. Operators have an incentive to travel slowly so as to increase their headways and the number of paying customers they find along their route. This leads to vehicle pairing and leapfrogging, and to large gaps in service along the line. The mechanism is so common in South America that it has come to be known as "la guerra del centavo" (the war of the penny). Chapter 10 of this book discusses these instabilities in more detail. In this section, we shall only examine the consequences of adaptive dispatching. We will see that it leads to the same customer waiting times as a dispatching strategy with even headways, and that as a result, jitney routes can be planned with the tools of Chapter 6.

A second example of demand responsive transit with hubs is provided by the shuttles that supply "last mile" coverage from airports to local hotels and private residences. The terminals are the hubs. In the case of private residences, the routing is also adaptive. In both cases, however, the dispatching rule is similar to that of jitneys: shuttles wait for a "soft" vehicle occupancy target, and then leave. The target is soft because in practice it is often lowered when the waiting time becomes unacceptably long to the passengers already in the bus. Unlike jitneys, however, shuttles do not have set routes. Instead, operators partition the service region into districts and offer door-to-door service by routing their vehicles on-the-fly. To minimize the customers in-vehicle riding times they typically deliver customers on runs outbound from the airport and collect them on inbound runs. Run lengths, operator costs and passenger riding times can be easily estimated with the travel distance formulas of Appendix 7.A. These formulas can then be used to figure out how best to carve the service region into delivery districts. Detailed analyses can be found in Banks *et al.* (1982) and Daganzo *et al.* (1977). More general methods are also described in Daganzo (2005). Waiting times can be estimated with a similar method as for jitneys. We now turn our attention to this issue. Both hard and soft targets are discussed. We shall find that the latter can perform surprisingly well if properly designed.

7.5.1. *Waiting Times under Adaptive Dispatching*

We examine here the average waiting time of customers outbound from a hub when their arrival rate varies with time, both systematically and randomly. The customer arrival process is defined by a cumulative arrival curve $A(t)$, which is defined from an initial time $t_0 \equiv 0$ to an end time $t_{\max}$. The curve is known in real time only up to the present. We assume without loss of generality that $A(0) = 0$ and denote the total number of customers served in the interval by $A_{\max} \equiv A(t_{\max})$. We consider hard targets first.

7.5.1.1. *Hard Targets*

As a first step in the analysis, we look for the dispatching instants $\{t_n\}$ for $n = 1, 2, \ldots$ that arise from setting some hard occupancy target o. Figure 7.11 displays both a cumulative arrival curve and the dispatching instants that correspond to the strategy. Note that the departure curve of customers leaving the hub at the determined instants is a step function with variable headways but uniform heights. This enables us to estimate the average waiting time easily.

This is done for a long analysis period consisting of many weeks or months and including many headways, $\{H_i \equiv t_i - t_{i-1}\}$. We then classify the headways according to a discrete set of duration values $\{h\}$, and denote the (large) number of headways in each class by n_h.

Clearly, if arrivals are stochastic the average wait in each class should be approximately $n_h h/2$. This is reasonable because passengers do not know ahead of time when vehicles are dispatched so there is no reason to expect

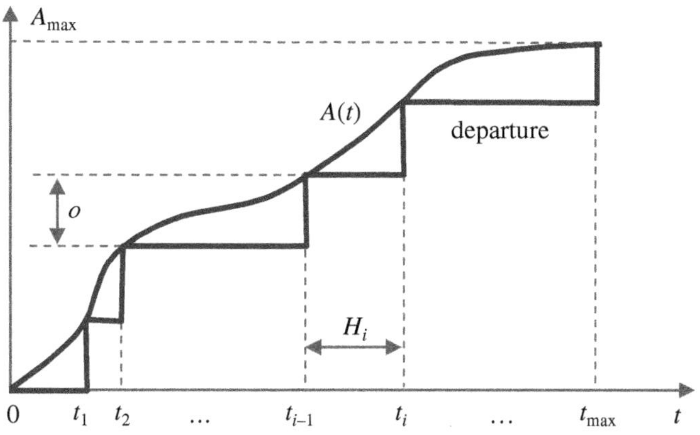

Figure 7.11. Cumulative arrival and departure curves with a fixed occupancy target.

different parts of a headway to receive more passengers than others. Since n_h is large this should be quite an accurate estimate. Thus, the expected wait over the complete period considering all headway durations should be $\frac{1}{2}(\sum_h hn_h/\sum_h n_h)$. Note that the numerator of the expression in parenthesis is the duration of the analysis period t_{max} and the denominator the total number of dispatches. Since the latter is A_{max}/o, we can write:

$$\text{Expected wait} = \frac{t_{max}}{2A_{max}/o}. \tag{7.43}$$

Next note that conditional on the total demand and the duration of the period, this expression is independent of the arrival curve. Therefore, it equals the average wait produced by a perfectly even demand curve with perfectly regular headways,

$$H = \frac{t_{max}}{A_{max}/o}. \tag{7.44}$$

This is fortunate from a design standpoint because it means that jitney systems can be designed with the methods of Chapter 6 *assuming that the demand rate is constant over time*. The optimal headway arising from that analysis can then be converted into a hard target occupancy by solving for o in (7.44).

A disadvantage of hard targets is that waits can unexpectedly be very long if it takes a long time for the last few passengers in a bus load to arrive. A second disadvantage is that the strategy does not take advantage of knowing in advance the variations in the average demand rate for different parts of the day.[23] Fortunately, as is shown below, both disadvantages can be alleviated with soft targets.

7.5.1.2. *Soft Targets*

The disadvantages of hard targets can be alleviated in two ways: (i) with hard targets that depend on the time of day; or (ii) with soft targets that depend on the passenger arrivals and time elapsed since the last dispatch. Method (i) requires that we know how the demand for different parts of the day and week, say during the peak, the off-peak, nights and weekends. To use it we would first estimate optimum headways for each of the periods separately as if the system was scheduled, e.g., as suggested in Section 6.5.1. We would then enter in (7.44) with the optimum headway, period duration and total demand specific to each period, and solve for the respective optimum occupancy targets. The appeal of this approach is that it is relatively simple to enforce. Disadvantages are that if the demand is wrongly predicted the results are suboptimal; and that the method still allows unexpectedly long waits.

Method (ii) prevents both problems. It is a bit more complicated to enforce but is conceptually simple. In each dispatching headway, the target is defined by a declining curve $o(\tau)$ that is the same for all headways, regardless of demand. Here τ is the elapsed time since the last dispatch. Vehicle $i+1$ is dispatched when the number of accumulated passengers since the departure of vehicle i, $A(t) - A(t_i)$, first reaches or exceeds the soft target of the moment, $o(t - t_{i-1})$. Figure 7.12 shows the idea graphically. It includes a dashed curve depicting the declining soft target, and an increasing solid curve displaying the accumulation observed in real time. The point of dispatch occurs when the two curves intersect.

Note, if the soft target is made to decline toward zero there is an upper limit to the wait even if only one customer arrives. Thus, the problem of unexpectedly long delays is eliminated. Furthermore, there is flexibility in choosing $o(\tau)$. This enables us to optimize it to achieve any desired goal. This is now done.

For single routes the optimum headway is known to vary inversely with the square root of the arrival rate; see (4.15) for the case of shuttles, and

[23] Recall from Section 4.2.2 that waiting times lower than (7.44) can be achieved if one adapts the headways to the average demand. Obviously, the same logic applies to adaptive dispatching. Therefore, overall efficiency should improve if average demand information is put to good use.

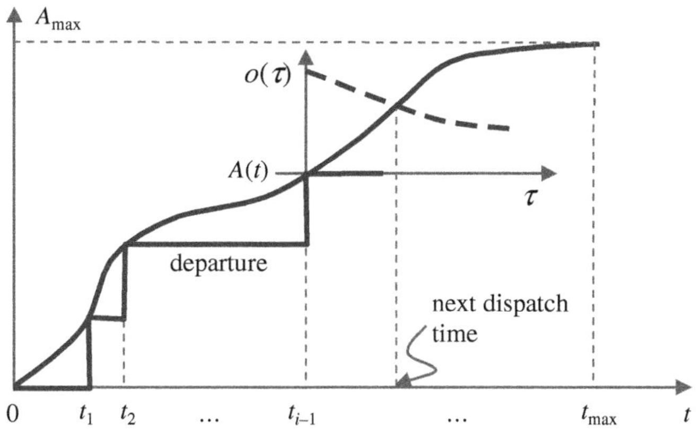

Figure 7.12. Cumulative arrival and departure curves with a soft occupancy target.

(5.24) for corridors. If most of the demand originates at the hub it makes sense to use the formula for shuttles, which is

$$H^* = \sqrt{\frac{2c_f}{\beta\lambda}}, \tag{7.45}$$

where λ stands for the average demand rate and c_f for the cost of a dispatch. We propose choosing an *o*-curve that will satisfy (7.45) for all the headways when one uses for λ the average demand rate experienced during the headway itself. This makes sense since the demand experienced in each headway is an excellent reflection of the current demand.

To do this, note from Figure 7.12 that at the moment of dispatch the headway is τ and the average demand in the headway is $o(\tau)/\tau$. Therefore, the stated condition can be rewritten in terms of the *o*-curve by entering with these values in (7.45). The result is: $\tau = \sqrt{2c_f/[\beta o(\tau)/\tau]}$. Now, solve for $o(\tau)$ to obtain:

$$o(\tau) = \frac{2c_f/\beta}{\tau}. \tag{7.46}$$

This is the proposed formula for the soft target.

Reasonably, this expression implies that higher targets should be used when covering the route is costly and when the passenger's value of time is low. Consideration also shows that this recipe automatically yields the optimum headways over long periods of time with very different demand rates, e.g., in the two-regime scenario that led to (4.15a) and (4.15b).

This concludes our analysis of DRT systems with hubs. We acknowledge that the discussion in both this section and the previous is not exhaustive because there are many ways in which DRT systems can be configured. For example, we have not considered systems with adaptive routing and check points, i.e., stops that are activated only if there is a user call. These kinds of systems are not used very often, at least for large systems, because for reasons that will become apparent in Chapter 10, they are very difficult to control. Conceptual analyses of some DRT extensions can be found in Daganzo (1984), Nourbakhsh and Ouyang (2012), and Chen and Nie (2017).

7.6. Closing Comments: Technology and the Future of Transit

Transportation is undergoing a soft revolution. The advent of smartphones, fast data transfer over the internet, cloud computing and vehicle tracking have facilitated the emergence of innovative companies such as Bird, Grab, Didi Chuxing, Jump, Lyft, Scoop, Scoot, Uber, Waze, and many others that offer improved transportation services. Examples have been mentioned in this chapter: the taxi industry is being disrupted by new ways of hailing crowd-sourced taxis; the car and bicycle rental business reshaped by vehicle sharing companies that leverage smartphones and the internet; and old carpooling methods are being improved by companies that organize matches for people using both taxis and owned automobiles.

Some of the taxi companies are now offering new cheaper ridesharing services (e.g., Uber Express) that require participants to walk short distances to locations designated on-the-fly so as to reduce or eliminate vehicle detours. In locales with high demand these locations can also serve to aggregate multiple people, so more can board a taxi at a single stop and conveniently share the ride. If the demand is sufficiently high then corridors may emerge spontaneously that would enable taxis to carry quite a few people making just a few stops. Something similar is also happening with other business models that require reservations (e.g., Chariot). If this comes to pass, prices can be dropped further encouraging more people to travel and solidifying corridors. This coevolution mechanism is akin to the one underlying the spontaneous emergence of informal private jitney services in the developing world. Thus, one may wonder whether the new cheaper ridesharing services are just a precursor of improved jitneys.

We have called what is going on a "soft" revolution because, as the above suggests, the changes that are taking place, at least so far, mostly involve improvements to old services: better taxis, better vehicle sharing, better ridesharing, and better jitneys. A harder revolution is on the horizon, however, if and when cheap robotic vehicles are added to the mix of innovations. Robots will allow taxi companies to offer mobility services more convenient than an owned automobile, and there is a reason to believe that in a not too distant future, individual urban transportation could be the exclusive domain of robo-taxis. This could enhance both city livability and travel because space consuming parking lots and curb parking would no longer be needed, and people would be relieved from the onerous task of driving. Could this rosy scenario mean that collective transit would become a thing of the past?

The answer is a resounding "no." Recall from Figure 7.10(c) that conventional transit has a large niche for large-dense cities. Conventional transit dominates the shared taxi mode and competes well with taxi even when the fleet sizes are allowed to be very large. Furthermore, as we now argue, large-dense cities are unlikely to have the street space needed to accommodate the very large fleet sizes that would be necessary to serve all the demand with robo-taxis.

In an ideal robo-taxi scenario much of the street parking would be eliminated, and lanes could be added to the streets.[24] Robo-taxis would carry an average of almost two people thanks to super-efficient matching. Jitney services would have evolved to be as efficient as current bus service, packing many people in each vehicle. However, all these services would be delivered on the city's existing streets. The main efficiency gained in the use of this space is that the traffic channels in each street could be widened slightly due to the elimination of parking; and that the traffic stream would include automobiles with slightly higher occupancy than today, mixed with buses with similar occupancies as today.

The technological improvements, however, do not change the physics laws of transportation movement and urban space described in Section 6.3.1. They only influence two parameters of the equations: a marginal increase in the capacity of a street, and another in car occupancy. Obviously, the qualitative analysis of Section 6.3.1 continues to be relevant. Large dense cities will still need regular, high-occupancy bus service, and even larger cities will continue to need subways. Of course, these high-occupancy transit modes will also benefit from the undergoing revolution as the technology already exists to better control and manage their vehicle fleets; and even to coordinate it with the new transportation modes. Robotic vehicles should eventually improve their controllability, overall efficiency, and safety.

---

[24]Some curbside space would have to be retained for robo-taxis to pick up and deliver passengers.

What are cities to do in view of all these changes? The new technologies impose demands on the cities' infrastructure and on non-users; such demands are not well understood, and are not internalized into the transactions with the users. For example, taxis relieve pressure on parking, but empty taxis add considerably to a city's traffic load. Taxis carrying multiple passengers may further alleviate traffic congestion — but not if these passengers would have otherwise taken mass transit. City transportation agencies need to understand these issues well in order to come up with reasonable regulations and pricing schemes that will encourage the new transportation companies to operate for the good of society.

Although most cities already regulate the number of taxicabs and parking spaces,[25] and allocate special lanes to high-occupancy vehicles, these tools are too blunt and do not capture well the new crowd-sourced transportation services. So far, a number of cities have taken baby steps to regulate the use of street space in more responsive, all-encompassing ways, mostly in the form of congestion pricing; see Lehe (2018) for a history. Most of these pricing schemes, however, are cordon-based, meaning that users pay once for the right to enter and travel in the region encircled by the cordon. This, unfortunately, is still too blunt an instrument. Clearly, a taxi driver benefits much more from this pricing scheme than a commuter, since the taxi pays once and can then spend the rest of the day traveling within the cordon and earning revenue.

Perhaps in recognition of this, tolls based on distance traveled are now being considered in Singapore, although they are not yet used. It should be intuitive that they should be a fairer and more efficient way of capturing the externalities of individual travel; see Daganzo and Lehe (2015) and Lehe (2017) for some discussion. However, even if optimally designed, these tolls still ignore the benefits of multi-occupant travel — since they would price equally a kilometer traveled by a taxi carrying zero, one or two passengers. Better tolls would account both for vehicle-kilometers traveled and passenger-kilometers served, or else include some proxy for the latter. These finer grained tolls have not yet been studied, but they deserve attention as they could encourage ridesharing. Also deserving of attention is the synergistic benefits that some forms of individual transit (one-way e-bike sharing for example) have with subways and regional transit by alleviating

[25]Cities are also trying to find ways of taxing TNC's, although this is a more difficult task; see Farmer (2018).

the "last mile" problem. These benefits, and creative ways of promoting them, can be evaluated with the analyses of this and prior chapters. Perhaps these analyses can spark further thinking and help cities do some of the catching up that seems necessary to keep up with technology.

References

Banks, J., Driscoll, W., and Stanford, R. (1982). "Design methodology for an airport limousine service." *Transportation Science*, 16, 127–148.

Barrios, J. A. (2011). "Flexible car sharing: Planning and operation," M.S. Thesis, Dept. of Civil and Environmental Engineering, U.C. Berkeley.

Bliss, L. (2018). "To measure the Uber effect cities get creative." *CityLab*, posted on January 12, 2018.

Chen, P. W. and Nie, Y. M. (2017). "Analysis of an idealized system of demand adaptive paired-line hybrid transit." *Transportation Research Part B: Methodological*, 102, 38–54.

Daganzo, C. F. (1977). "An approximate analytic model of many-to-many demand responsive transportation systems." *Transportation Research,* 12(5), 325–333.

Daganzo, C. F. (1984). "Check-point dial-a-ride systems." *Transportation Research*, 18B, 315–327.

Daganzo, C. F. (2005). *Logistics Systems Analysis* (4th edition), Springer-Verlag, Heidelberg, Germany.

Daganzo, C. F., Hendrickson, C. T., and Wilson, N. H. M. (1977). "An approximate analytic model of many-to-one demand responsive transportation systems," in *Proc. 7th Int. Symp. on the Theory of Traffic Flow and Transportation*, Kyoto, Japan, pp. 743–772.

Daganzo, C. F. and Lehe, L. J. (2015). "Distance-dependent congestion pricing for downtown zones." *Transportation Research Part B*, 75, 89–89.

Daganzo, C. F. and Ouyang, Y. (2018b). "A general model of ridesharing services." ArXiv: 1807.01140.

Daganzo, C. F. and Smilowitz, K. S. (2004). "Bounds and approximations for the transportation problem of linear programming and other scalable network problems." *Transportation Science,* 38(3), 343–356.

Farmer, L. (2018) "Governments increasingly tax Uber and Lyft for transit revenue." *Governing.* http://www.governing.com/topics/finance/gov-states-cities-taxing-ride-sharing-services-transit.html. Posted on April 4, 2018. Accessed on June 28, 2018.

Lehe, L. J. (2017). "Downtown tolls and the distribution of trip lengths." *Economics of Transportation*, 11–12, 23–32.

Nourbakhsh, S. M. and Ouyang, Y. (2012). "A structured flexible transit system for low demand areas." *Transportation Research Part B*, 46(1), 204–216.

Shoup, D. (2011). *The High Cost of Free Parking* (Updated Edition). Taylor & Francis Group, Routledge, Abington & New York.

Zha, L., Yin, Y., and Du, Y. (2017). "Surge pricing and labor supply in the ride-sourcing market." *Transportation Research Part B*, 117(B), 708–722.

Zha, L., Yin, Y., and Xu, Z. (2018). "Geometric matching and spatial pricing in ride-sourcing markets." *Transportation Research Part C*, 92, 58–75.

Homework

7.1 A transit company is currently using a route deviation system to serve a zone of fixed width that has a fine grid of orthogonal streets. Under the current strategy, buses travel along a fixed route through the middle of the zone and serve each passenger by deviating perpendicularly to the route as shown by the diagram below, on the left. As a consultant, you recommend doing away with the route by routing the buses directly between passengers, as shown by the diagram on the right. How much lateral travel distance in percentage terms can you claim to save?

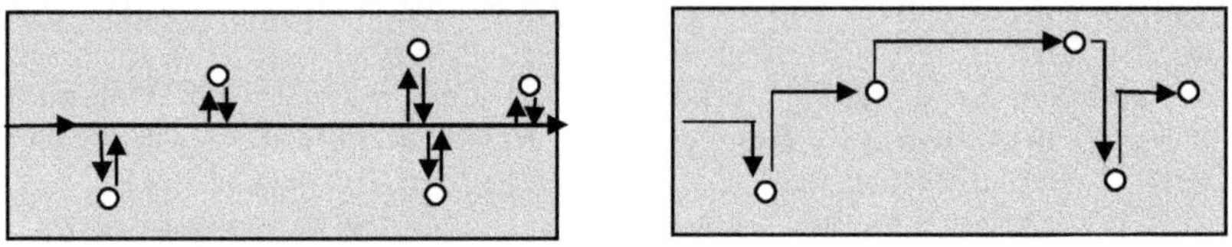

7.2 *Consider a flexible-route bus system operating on a grid street network. The bus lines form a larger square grid in the N-S or E-W directions; see dashed lines in the diagram below. In this imaginary system, the only fixed stops occur where the lines intersect. Elsewhere, passengers request service by declaring their locations with their mobile phones. To serve them, buses detour in a direction perpendicular to their travel direction, as illustrated in the diagram. The diagram also shows in gray the sub-region served by the N–S buses traveling between the two fixed stops shown. (Note, if a similar pattern is used between all adjacent stop pairs, the resulting sub-regions would tessellate the complete region.) You are asked to:

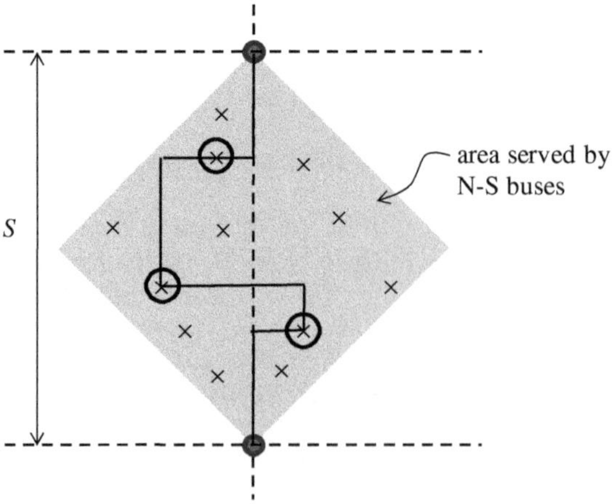

(a) Discuss the pros and cons of such a system as compared to its regular fixed-route counterpart from both the agency and passenger's perspectives; *extra credit*: discuss qualitatively the potential headway control problem.
(b) Derive the average access distance saved by the customer assuming all his/her locations are equally likely to be anywhere. Use S for the route spacing.
(c) Derive the distance added by the detour required to serve one passenger's trip end in the following two extreme cases: (i) the demand is so low that buses rarely make more than one stop per interval S; and (ii) the buses make a very large number of stops.
(d) Discuss the demand level under which our hypothetical flexible system could outperform its fixed-route counterpart if headway control was not an issue. Use ℓ for the representative trip length, v for the bus commercial speed, and v_w for the passenger walking speed.

7.3. Consider a corridor lying along the positive side of the x-axis; see diagram below. During the morning commute, a van travels back and forth between a residential area in the interval [0, 1] and the CBD at $x = 2$. In each trip, the van collects n passengers from a set of $N \gg n$ candidates. Candidates arrive uniformly in space and steadily in time, at the same rate as they are removed; so we assume that the residential area always contains N randomly and uniformly distributed candidates. Considering all this, please answer the following questions. You may also assume that the van's dwell time for each pickup or drop-off, t_s, is negligible.

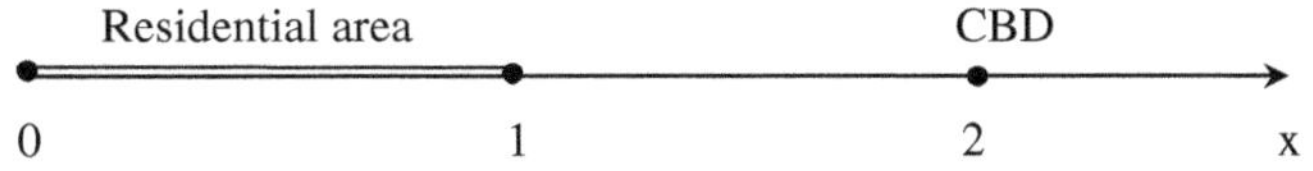

(a) If the van picks up passengers strictly on a first-call-first-served (FCFS) basis, what is the van's expected travel distance per round trip?
(b) Suppose now that upon returning from the CBD and reaching some intermediate point in the residential area far from its extremes, the van collects its passengers on the basis of proximity. What is the van's expected travel distance to collect an intermediate passenger (not the first or the last). [*Hint:* consider the logic of Appendix 7.A]
(c) Suppose the expected round trip time is T. What demand density [pax/dist-time] is needed to keep N constant over time?

7.4. You enjoy a monopoly to provide many-to-many paratransit taxi service in a region that has a dense square grid of streets. You plan to use m taxis — enough to serve the demand but not so many that taxis have to wait for passengers — and are interested in the steady state equilibrium of your system. For maximum profitability taxis always choose the closest passenger from the pool of waiting passengers. The region size is R [km^2]. Passengers arrive steadily with demand density λ [pax/h-km^2], and their origins and destinations are uniformly and independently distributed in the region. Taxis travel at a constant speed v [km/h] and the stop time to pick up and drop-off passengers is negligible. Then, ignoring the random fluctuations in taxi service times (i.e., assuming that taxis serve passengers at a steady rate consistent with the average time it takes to serve one passenger) do the following:

(a) Derive the equilibrium number of waiting passengers at home and their average total trip time as a function of m.

(b) Consider a city with the following parameters: $R = 400\,\text{km}^2$, $v = 20\,\text{km/h}$, $\lambda = 1\,\text{pax/h-km}^2$. Operating a taxi (including the driver wage) costs $50/h. If the taxi company is operated by a public agency that values customer time at $30/h, find the optimal fleet size it should provide.

(c) *Discuss what would happen if the agency was only interested in maximizing profit from charging a price of p [$/pax]. Consider a scenario where the demand density decreases with the passenger waiting time at home.

(d) *Assume now that passengers arrive as a Poisson process so the number of passengers waiting at home fluctuates over time around an equilibrium value. Do you know any queuing theory models that could roughly describe the system? Could you estimate the mean and the standard deviation of the number of passengers waiting at home? Could you use dimensional analysis, combined with simulation to derive accurate general formulae for this problem? Explain.

7.5. Equations (7.8), (7.11), and (7.16) describe the socially optimal performance of taxi service and one-way vehicle sharing, with and without pods. Consider now a city with the following parameters: $R = 400$ [km^2], $\ell = 15$ [km], $v = 20$ [km/h], $v_w = 3$ [km/h], and $\lambda = 2$ [pax/h-km^2]. Design the three mentioned systems (i.e., determine the number of vehicles and pods) so the social optimum is achieved, and compare their performance under the following scenarios:

(a) Benchmark case with a chauffeur: $\gamma = \$50/\text{h}$, and $\beta = \$30/\text{h}$.
(b) Self-driving vehicle case: $\gamma = \$20/\text{h}$, and $\beta = \$30/\text{h}$.

7.6. *We are charged with designing a demand responsive feeder system to serve a subway station during the evening peak. Minibuses synchronized with the train arrivals are to distribute the arriving passengers. Each train brings exactly N passengers of this type. Enough minibuses are provided so that there is always an available minibus ready to load passengers. Each minibus carries $n \ll N$ customers, so $\lceil N/n \rceil$ minibuses are needs to serve a train (the ceiling function is used because the number of buses must be an integer). The subway headway is H [h]. The service region is "round" with area R [km^2] and its streets form a square grid. The minibuses travel at speed v [km/h] and take t_s [h] to stop and deliver a passenger. Passenger destinations are uniformly and independently distributed in the service region. The queue discipline at the station is FIFO: when a train arrives, its passengers board the bus at the head of the line on a first come first served basis until its occupancy reaches n. Ensuing passengers then board the next bus, and the next, and so on until all the passengers are taken. Buses deliver passengers in the order that minimizes the distance traveled between visits to the station, and hence their cycle times. For this kind of system:

(a) Derive an expression for the bus cycle time C. With this information, derive an expression for the minimum fleet size, M, required to ensure there always are enough buses waiting at the station to carry all passengers. Remember that M must be an integer number of buses and that trains depart at regular headways. (*Hint:* relevant perhaps is the shortest time interval between successive trains that a bus can serve, i.e., the smallest multiple of H larger than a bus cycle time) You can use in your computation of cycle time the following well known approximation for the average distance of a "traveling salesman tour" that starts from a depot and includes n random stops: $L = 0.92\sqrt{R(n+1)}$. Discuss what happens to the buses and the passengers if you provide more buses than M.
(b) Derive an expression for the average riding time of the last passenger to be delivered.
(c) Assume there are no subsidies so the cost of running the buses must be recovered from the fare box. If the hourly cost of running a bus is $c_t = 30$ [\$/bus-h], use the answer from parts (a) and (b) to develop a table linking the necessary fare, F, to various design standards,

T_0 that limit the average riding time for the last passenger to be delivered. Use the following data: $R = 3\,\text{km}^2$; $N = 75\,\text{pax/train}$; $H = 8\,\text{min}$; $v = 15\,\text{km/h}$; and $t_s = 0.01\,\text{h}$.

(d) To gain some insight, repeat part (c) generically, but assume $n \gg 1$ ($n \approx n+1$), $t_s \to 0$ and $H \to 0$. Write an expression relating F to the choice of T_0 in terms of the problem's parameters.

(e) Repeat part (d) if we now assume that the service region has been partitioned into w wedges that are served by different sets of clearly labeled buses. The parameter w is a decision variable. The buses serving each wedge operate with the same FIFO rules as before. Is F still a function of the choice of T_0? (*Comment*: In solving this part you may assume that the traveling salesman formula can be used with the wedges. This, however, turns out to be somewhat inaccurate if $n < w^2$.)

Appendix 7.A. Expected Distance to the Closest of n Random Points

We look here for the expected distance from an arbitrary location in a region of area R to the closest of n independently and uniformly distributed random points in the region. The points are taxis in Section 7.1 and passengers in Section 7.4.

We start by evaluating the probability that our distance exceeds some arbitrary value x. Since this is the same as the probability that there are zero points in a disk of radius x around our location; and since the probability of a point not being in the disk equals the fraction of the region's area that is not covered by the disk, we can write

$$\Pr\{d_n \geq x\} \equiv \Pr\{\text{zero points in the disk}\} \approx \left(1 - \frac{\pi x^2}{R}\right)^n.$$

Now, since the integral of the complementary cumulative distribution function or a non-negative random variable is the expectation of the random variable, we can write

$$E[d_n] = \int_0^{\sqrt{\frac{R}{\pi}}} \left(1 - \frac{\pi x^2}{R}\right)^n dx.$$

To solve this integral, use the change of variable $y = \sqrt{\frac{\pi}{R}}x$, and rewrite it as

$$E[d_n] = \int_0^1 (1 - y^2)^n \sqrt{\frac{R}{\pi}} dy.$$

Now, use the second change of variable ($y = \cos\theta$) and the identity $(1-\cos^2\theta) = \sin^2\theta$ to reduce the integral to the known form:

$$E[d_n] = \sqrt{\frac{R}{\pi}}\int_0^{\frac{\pi}{2}} (\sin\theta)^{2n+1}d\theta = \sqrt{\frac{R}{\pi}}\frac{\sqrt{\pi}}{2}\frac{\Gamma(n)}{\Gamma(n+\frac{1}{2})}.$$

For $n > 3$,

$$\frac{\Gamma(n)}{\Gamma(n+\frac{1}{2})} \approx \frac{1}{\sqrt{n}},$$

which yields the result used in the text:

$$E[d_n] \approx \frac{1}{2}\sqrt{\frac{R}{n}}, \quad \text{for } n > 3.$$

For a region with where vehicles move along a rectilinear grid, the same method can be followed. The only difference is that instead of a circular disk, one must use a square of side $\sqrt{2}x$ oriented at 45° with the grid. One can then write

$$E[d_n] = \int_0^{\sqrt{\frac{R}{2}}} \left(1 - \frac{2x^2}{R}\right)^n dx,$$

and using the change of variable $y = \sqrt{\frac{2}{R}}x$, finally obtain

$$E[d_n] = \frac{\sqrt{\pi}}{2\sqrt{2}}\sqrt{\frac{R}{n}} = 0.63\sqrt{\frac{R}{n}}.$$

Caveat: The formulas just derived assume that the disks do not touch the boundary. Otherwise, they would have to be truncated. The assumption is reasonable if the distance from the location of interest to the boundary is larger than the radius of the disks that come into play. For random locations in round areas resembling squares or circles, this will occur if $E[d_n]$ is small compared with the region's diameter, i.e., if $n \gg 1$. Otherwise, our integrals are lower bounds. Thus, our formulas slightly under predict the actual distance. To give an idea of the error magnitude, we compare the known result for the distance between two random points in a unit square with a rectilinear grid, which is known to be $2/3 \approx 0.67$, and the prediction of the current formula for $n = 1$, which is 0.63. Thus, the error is quite acceptable for practical purposes.

Appendix 7.B. The Expected Fraction of Feasible Matches

This appendix derives the formulas for the expected fraction of feasible matches r provided in the text as equations (7.25), (7.26), (7.29), and (7.30).

The case of SRM: The derivation assumes that our seeker travels from his/her origin "O" to his/her destination "D" in the most direct way that requires only one turn. It is also assumed without loss of generality that (s)he travels horizontally first and vertically second; see Figure 7.B1. The horizontal distance between O and D is denoted by X and the vertical distance Y. The distances from point D to the right side and top side of the square are denoted by X' and Y', respectively. The side of the region will be denoted $\phi = R^{1/2}$.

We first look for the rate $r(z)$ at which the seeker encounters feasible takers at some point C after having traveled z distance units, conditional on (X, Y, X', Y'). For a taker's request to be feasible, it must be both origin-feasible and destination-feasible. By the former, we mean that the taker's origin A is in the locus of points that can be reached in time τ from C without adding to the trip's distance. By the latter, we mean that the taker's destination B must be aligned with D so no circuity is added to the path of either traveler. The rate $r(z)$ is the product of the rate at which origin-feasible takers are found when the seeker has traveled z distance units, $r_O(z)$, and the probability $p(z)$ that an origin-feasible taker is also destination-feasible: $r(z) = r_O(z)p(z)$.

Consider now $r_O(z)$. If C is in the horizontal segment of the route ($z \leq X$) the seeker can detour vertically in the direction toward D without increasing the distance traveled. Thus, the locus of feasible origins is a right isosceles triangle with perpendicular sides intersecting at point C, as shown by the dark triangle of Figure 7.B1.[26] These sides are of length $v\tau$ and the triangle's area is $1/2(v\tau)^2$. And since the call arrival density is λ, it follows that $r_O(z) = 1/2\lambda(v\tau)^2$ when $z \leq X$. Later in the trip, when $z > X$, the seeker is traveling vertically and has no routing flexibility. Thus, the origin-feasible catchment region vanishes, and therefore $r_O(z) = 0$ when $z > X$.

Consider now $p(z)$. This is only relevant when the seeker is traveling horizontally because otherwise we already know that $r(z) = 0$ (since

[26] If C is very close to the region's boundary (or if either of its coordinates is close to the corresponding coordinate of point D) this triangle is truncated. In the derivations that follow we optimistically ignore this truncation. This should not matter much when the triangle is small compared with the service region, which is what we are assuming.

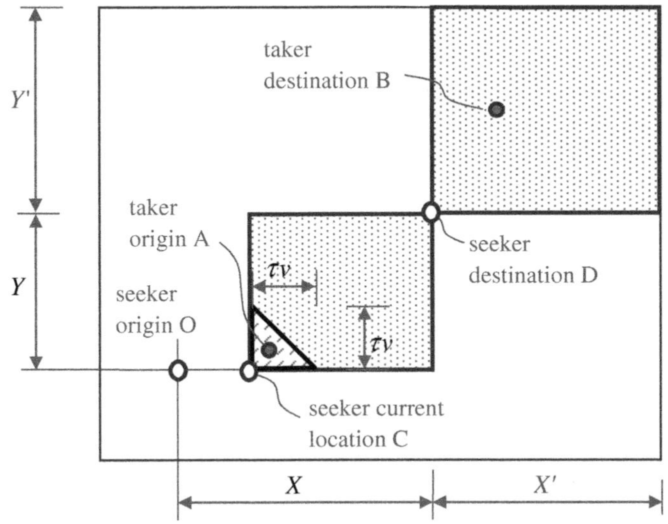

Figure 7.B1. A seeker's origin- and destination-catchment areas while in motion: the case of SRM.

$r_O(z) = 0$). To evaluate $p(z)$ for the horizontal portion of the trip, refer to Figure 7.B1 and note that a taker's request is destination-feasible if and only if the taker's destination B is either in the rectangle with vertical and horizontal sides that has one corner at A and another corner at D, or else in the rectangle formed by the extension of the two sides intersecting at D and the perimeter of the square. Figure 7.B1 illustrates the latter case. To simplify matters we optimistically assume that the corner of the first rectangle is at C and not at A. (This enlarges the rectangle, but not much when τv is small.) Thus, the area of the first rectangle is approximated by $(X - z)Y$. Since the area of the second rectangle is $X'Y'$, the area of the complete destination-feasible region is $[Y(X-z) + X'Y']$, and we have $p(z) \approx [Y(X - z) + X'Y']/\phi^2$ when $z \leq X$.

Now combine the expressions for $r_O(z)$ and $p(z)$ to obtain the final formula for the expected rate of feasible encounters conditional on (X, Y, X', Y'). It is

$$r(z) = r_O(z)p(z) \approx {}^1\!/\!_2\lambda(v\tau)^2[Y(X - z) + X'Y']/\phi^2 \quad \text{if} \quad z \leq X,$$
$$= 0, \text{otherwise}.$$

It follows that the expected number of feasible encounters conditional on (X, Y, X', Y') is

$$\int_0^X r(z)\frac{dz}{v} = \frac{\lambda v \tau^2}{2\phi^2}\left[\frac{YX^2}{2} + XX'Y'\right]. \tag{7.B1}$$

We know that X and Y are mutually independent, triangularly distributed random variables with the same p.d.f. and range: $f(w) = 2(\phi - w)/\phi^2$ for $w \in [0, \phi]$. Moreover, conditional on X and Y, the variables X' and Y' are independently and uniformly distributed on $[0, \phi - X]$ and $[0, \phi - Y]$, respectively. Hence, the expected number of feasible taker requests encountered by a random seeker is

$$\begin{aligned} n &= \frac{\lambda v \tau^2}{2\phi^2} \underset{X,Y,X',Y'}{E} \left[\frac{YX^2}{2} + XX'Y' \right] \\ &= \frac{\lambda v \tau^2}{2\phi^2} \underset{X,Y}{E} \left\{ \underset{X',Y'}{E} \left[\frac{YX^2}{2} + XX'Y' \middle| X, Y \right] \right\} \\ &= \frac{\lambda v \tau^2}{\phi^6} \int_0^\phi \int_0^\phi \left[yx^2 + x(\phi - x)(\phi - y)/2 \right] (\phi - x)(\phi - y) dx dy \\ &= \frac{\lambda \phi v \tau^2}{36} = \pi_1 \pi_2 / 36. \end{aligned} \tag{7.B2}$$

This is the result reported in the text.

If the demand is concentrated at the center of the square the derivations are identical, except now $p(z) = 1$ since everyone shares the same destination, and we only have to condition on (X, Y). With these changes, the expected number of feasible encounters conditional on (X, Y) becomes: $\int_0^X r_O(z) \frac{dz}{v} = \frac{\lambda v \tau^2}{2} X$. Since the random variable X is now uniformly distributed with mean $\phi/4$ we find that

$$n = \underset{X}{E} \left[\frac{\lambda v \tau^2}{2} X \right] = \frac{\lambda v \tau^2 \phi}{8} = \frac{\pi_1 \pi_2}{8}. \tag{7.B3}$$

The case of RRM: In the case of RRM, a seeker has the benefit of knowing all future calls so (s)he can scan for reachable callers in the whole region. Note that at every point in the region there is a time window of duration τ in which callers can be picked up in a temporarily feasible fashion — the windows of distant calls must, of course, occur later than those of nearby calls. And since the trip generation rate is time-independent and homogeneous, it follows that the spatial density of temporarily valid calls at every location is $\lambda \tau$. Then, integrating this density across the whole region we find the total expected number of valid calls: $\lambda R \tau = \pi_1$.

Next, let us derive the probability that a taker is both origin- and destination-feasible. To do this, consider a random seeker with origin and destination at points O and D. We assume without loss of generality that D is up and to the right from O (see Figure 7.B2). (No generality is lost

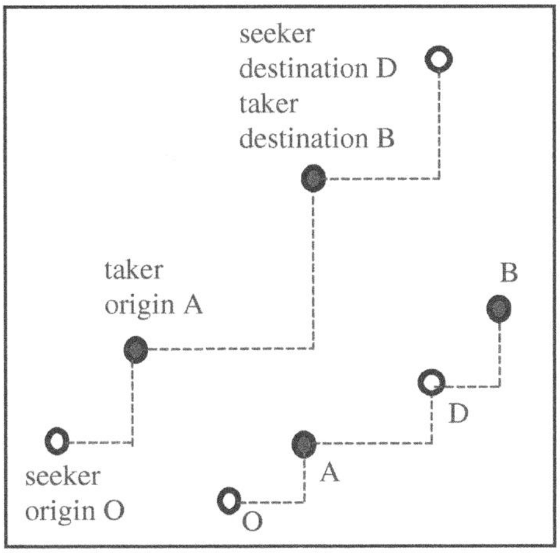

Figure 7.B2. Origin and destination feasibility for RRM.

because we can always rotate the square after the points are sampled.) Since the reachability domain is the whole region we then sample a random caller with origin A and destination B, and finally ask whether the two vectors align in the proper way.

The reader is invited to verify using figures such as Figure 7.B2 that a good alignment where the seeker is the provider exists if and only if: (i) both the x-coordinates y-coordinates of the four points in the figure are arranged in the order (OABD), or else (ii) both sets of coordinates are arranged in the order (OADB). Both of these patterns are shown in Figure 7.B2. Further consideration shows that all other orderings result in some detouring. Therefore, there are two good orderings. Now note that there are $4!/2$ possible orderings of each set of coordinates where O comes before D (one half of the possible permutations of four coordinates). Therefore the total number of orderings considering both sets of coordinates is $(4!/2)^2 = 144$. Finally, since all orderings are equally likely and only two are good, we see that the probability of a good alignment is $2/144 = 1/72$. Hence, the expected number of valid calls is $\pi_1/72$. This is the result reported in the text.

If the demand is concentrated, a seeker can pick up any caller with origin in the rectangle between the seeker's origin and the central location. The average area of such rectangle is $R/16$. Thus, the probability of a good alignment is $1/16$, and the expected number of valid calls is $\pi_1/16$.

The Case of SCP: The logic is the same as that for SRM. The only changes are that the call demand density is λf instead of λ and that the catchment

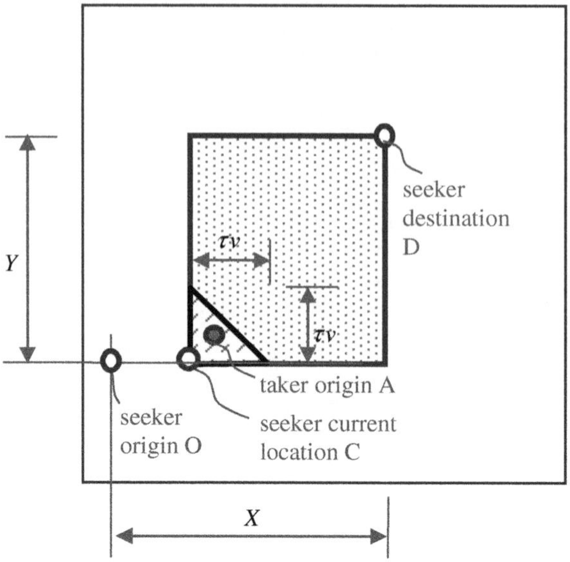

Figure 7.B3. A seeker's origin and destination-catchment areas while in motion: the case of SCP.

area for destinations does not include one of the rectangles; see Figure 7.B3. This happens because providers must finish their trips after their takers. With a smaller catchment area, $p(z)$ is now: $p(z) \approx [Y(X - z)]/\phi^2$ (when $z \leq X$). As a result, conditional on triangularly distributed random variables X and Y, the expected number of encounters becomes: $\int_0^X r(z)\frac{dz}{v} = \frac{\lambda f v \tau^2}{2}\left[\frac{YX^2}{2}\right]/\phi^2$ instead of (7.B1). And the same algebraic steps now yield the result of the text:

$$n \approx f\pi_1\pi_2/72. \tag{7.B4}$$

If the destinations are concentrated, both the seeker and the taker have the same destination. Then, $p(z) = 1$ as occurred with SRM, and the result is

$$n = \underset{X}{E}\left[\frac{\lambda f v \tau^2}{2}X\right] = \frac{\lambda f v \tau^2 \phi}{8} = \frac{f\pi_1\pi_2}{8}. \tag{7.B5}$$

The case of RCP: For dispersed destinations, the overall logic is the same as for RRM. The only changes are that the call demand density is now λf, and that pattern (ii) on the bottom right of Figure 7.B2 is impossible since the provider must end the trip after the taker. As a result, only 1 of the 144 possible orderings is feasible and therefore $n = f\pi_1/144$, as reported in the text.

Finally, if the demand is concentrated, the reachability area is the same as for RRM, and therefore $n = f\pi_1/16$.

Appendix 7.C. Proof that $T^* \approx c^2/\lambda Rm$ when $m \gg 1$

This appendix derives equation (7.40). To start, note that (7.34) allows us to write:

$$\frac{\kappa\sqrt{R}}{v}\left[\sqrt{\frac{1}{n_r}}+\sqrt{\frac{1}{n_w}}\right]=\frac{m}{\lambda R}. \tag{7.C1}$$

Now define c as in (7.40) and rewrite (7.C1) as

$$\left[\sqrt{\frac{1}{n_r}}+\sqrt{\frac{1}{n_w}}\right]=\frac{m}{c}. \tag{7.C2}$$

Note, this is an implicit way of writing the function $n_w(n_r, m)$ appearing in (7.39). Therefore, problem (7.39) can be rewritten as

$$\min_{n_r, n_w}\{T \equiv (mn_r + n_w)/\lambda R\}\,; \quad \text{s.t.}\left[\sqrt{\frac{1}{n_r}}+\sqrt{\frac{1}{n_w}}\right]=\frac{m}{c}. \tag{7.C3}$$

Now use ρ as a Lagrange multiplier for the constraint of (7.C3) and form the Lagrangian of the problem. Recall from Chapter 2 that a necessary condition for optimality is that the gradient of the Lagrangian with respect to both, the decision variables and the multipliers vanishes. In our case, we see that setting the partial derivatives of the Lagrangian with respect to n_r, m, and ρ equal to zero yields the following relations:

$$n_r = (2m/\lambda R\rho)^{-2/3}, \quad n_w = (2/\lambda R\rho)^{-2/3},$$
$$(2/\lambda R\rho)^{1/3}[1+(m)^{1/3}] = m/c.$$

The last equality can be rewritten as $(2/\lambda R\rho)^{1/3} = (m/c)\,[1+(m)^{1/3}]^{-1}$, or more conveniently for our purposes as $(2/\lambda R\rho)^{-2/3} = (c/m)^2\,[1+(m)^{1/3}]^2$. Now, insert this result into the right sides of the first two expressions to obtain formulas for n_r and n_w in terms of m; and then insert these two formulae into the objective function of (7.C3). The result is

$$T^* = \frac{c^2}{\lambda Rm}(1+(m)^{-1/3})^3 \equiv \frac{\theta(m)}{m}\frac{c^2}{\lambda R},$$

in agreement with (7.40).

Chapter 8

Management — Vehicle Fleets

A transit agency is, in essence, a mechanism for transforming money inputs from its users and the government into transportation services. Figure 8.1 illustrates the mechanism. The middle block is the transit agency, including its principal resources, and the arrows the agency's inputs and outputs. This book focuses on optimizing the outputs of this mechanism, with due consideration of its inputs, by managing the resources. The chapters just concluded did this by showing how efficient long term plans could be found.

We are now about to start a set of chapters that will explore how the agency should manage its resources in the medium and short terms to execute these long term plans. Since infrastructure resources cannot be changed easily, the focus of an agency's short and medium term efforts are on the agency's mobile resources: vehicles and personnel. Unlike planning decisions, successful managerial actions for the medium term can and should be invisible to the public. Thus, these actions can be taken whenever it is appropriate based only on the agency's needs. For successful operations an agency must also take short term (control) actions, to respond to unpredictable events such as traffic and vehicle malfunctions. These actions can and should be aggressively pursued on an ongoing basis because failures are highly visible to the public.

Because vehicles and personnel must be managed quite differently, medium term management strategies for these two resources will be discussed sequentially — vehicles in this chapter and personnel in Chapter 9. Chapter 10 will then address the short term control issues.

Although our attention in these three chapters will be focused on the organization of the middle box of Figure 8.1 we should not lose sight that this box is only part of the whole picture. A transit agency is also concerned with the arrows. Obviously, issues of goal-setting, funding and

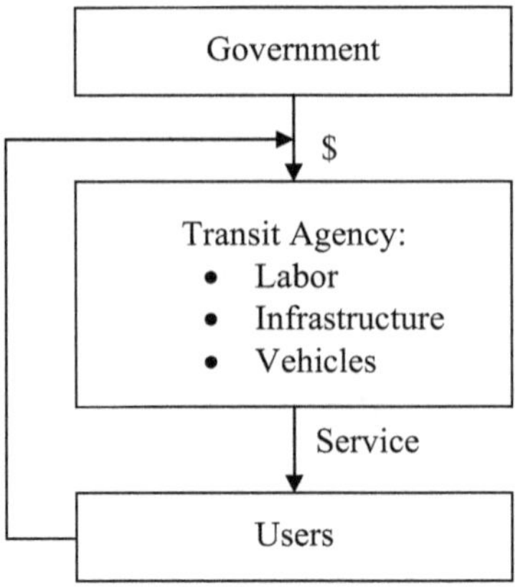

Figure 8.1. Schematic diagram of a transit agency.

governance (inbound arrows), and pricing, public relations and information dissemination (outbound arrows) play a very important role in the success of a transit operation. With the exception of funding and pricing, which are covered in Chapter 11, the remaining issues are not transit-specific and therefore will not be reviewed in this book.

In this and the ensuing chapters, it will be assumed that the schedules of all the system's routes are given, as are the routes themselves. A preliminary section in this chapter (Section 8.1) introduces basic concepts and terminology and explains how the management of vehicles and personnel interact. The rest of the chapter is specific to vehicles. It examines the vehicle resources and deployment strategies that can most efficiently cover the given schedules, disregarding for the moment their effect on drivers. As in the planning chapters, we shall start with simpler problems and use the acquired knowledge to address more complex ones. In this spirit, Section 8.2 discusses the case of a single route, and Section 8.3 discusses multiple routes.

8.1. Introduction

This section describes the overall management process for both vehicles and people and some of the terminology that will be used. We start with the latter. The terms defined will be used in both management chapters.[1]

[1]The terms for the concepts in this list have been chosen for their transparency in revealing what the concepts actually do. The terminology in the industry is somewhat more opaque and may be slightly different. For example, the term "bus run" in our list is sometimes called a "block" in transit parlance, and our "driver job" is sometimes called a "run."

Definitions

Schedule: set of timetables and/or frequency listings for all the routes, and associated maps, as advertised by the transit agency.

Depot: location where buses are stored without drivers.[2]

Terminus: part of a transit line (i.e., route) where buses are empty of passengers and buses can be changed. Routes often have two termini.

Interchanges: points on a transit line where drivers can be interchanged. Termini are usually interchanges, as is the depot. Because drivers can be changed with buses full, routes can also have interchanges that are not termini.

Bus task: contiguous part of the schedule that must be covered by a single bus; i.e., a part that is sandwiched between consecutively scheduled visits to the terminus or termini of a particular route.

Loop: part of the schedule sandwiched between consecutive visits to *the same* terminus of a particular route. If a route has only one terminus, every loop is a bus task. Otherwise, loops can be divided into tasks.

Bus run: time-space path of one specific transit vehicle from and returning to a depot during the course of a day. The vehicle needs a driver during the entire run. The run may include coverage of more than one transit line. The driver may change.

Deadhead (verb: refers to a bus): to travel out-of-service, usually between two termini, or between a terminus and a depot.

The next three definitions pertain to Chapter 9. They are provided here to give a sense of what the overall management problem needs to accomplish.

Driver task: indivisible part of a bus run that must be covered by the same driver. It usually consists of a bus trip between consecutive interchanges.

Job: sequence of driver tasks covered by one specific worker in a single day.

Worker type: classification characterized by pay rate and other conditions of the job such as its maximum duration.

Figure 8.2 illustrates some of these concepts using a time–space graph showing by means of nodes and arcs a depot and two single-terminus routes (A and B). The dashed curve in the figure is a run of a single bus. The

[2]For convenience and without loss of generality the term "bus" will be used generically, as in other parts of the book, to refer to any transit vehicle.

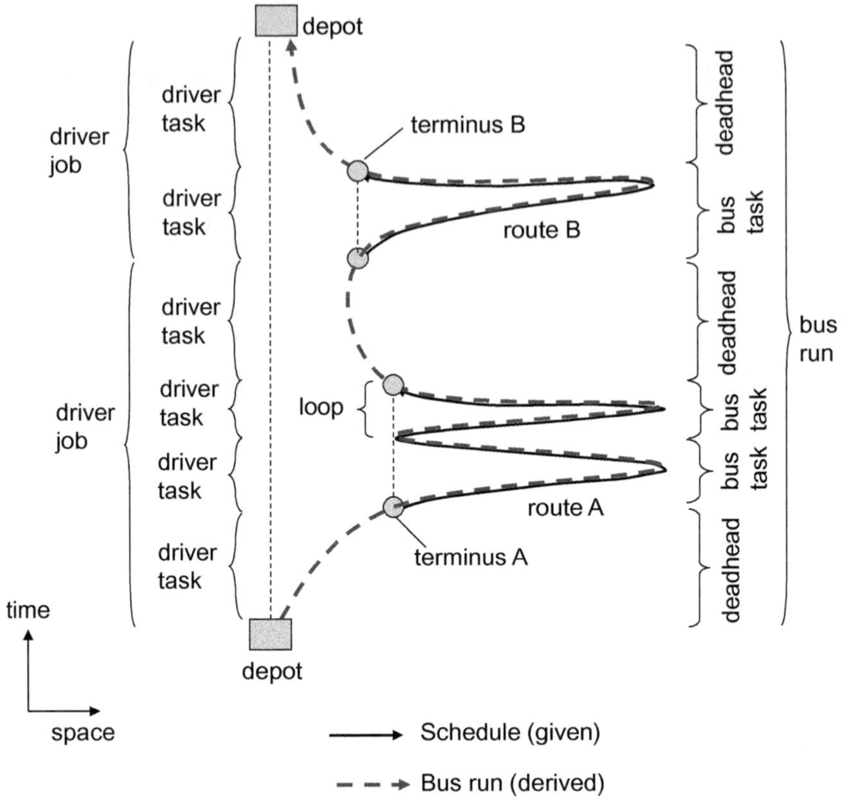

Figure 8.2. Spatiotemporal illustration of vehicle and personnel management concepts.

two solid curves overlapping with the dashed curve represent the part of the schedule covered by the run. Note the run covers two routes. The vertical dotted lines are the time–space trajectories of the depot and the termini.

Management plans for vehicles and personnel are usually found sequentially — with vehicles examined first. One first covers the schedule with bus runs and then populates these runs with drivers. These two steps are now briefly described:

(a) *Vehicle management*: The goal here is identifying a set of bus runs that cover a given schedule, and a specific bus for each run. Note that the number of runs in the set is the required fleet size. If the agency uses buses of different types and some portions of the schedule have to be covered by buses of a specific type (e.g., articulated) then a set of runs should be identified for each bus type. Once all the runs have been identified the process is completed by allocating a specific bus of the correct type (identified by its bus ID number) to each specific run. Since buses of the same type have similar capabilities there is much flexibility in this allocation, and the decisions are usually based on mechanical and maintenance considerations. For this reason, the allocation process will not be discussed in this chapter. Only methods for identifying the runs will be studied. These methods first divide the schedule into elemental tasks and then look with optimization algorithms for the fewest and shortest possible runs that cover these tasks. In doing this some vehicles may have to deadhead unproductively between termini, or even sit unused part of the time. Figure 8.3(a) illustrates these ideas with a time–space graph that shows part of a schedule for two routes, and the three bus runs that cover it. The given schedule is shown by thin black lines, as in the previous figure. Each run is a directed path from the depot node at the bottom to the depot node at the top. To avoid clutter, extra lines are not used to depict the parts of the runs that overlap with the schedule. But all other arcs are shown. So that the runs can be distinguished, deadheading arcs are displayed in a run-specific format. The remaining eight grey curved-vertical arcs do not change termini. They signify bus waiting.

(b) *Personnel management*: In this case, the process is very similar but a little more complicated because people are needier than inanimate objects. To start one looks for a specific set of jobs (perhaps classified by type) that cover the bus runs of step 1. This involves cutting the runs into elemental driver tasks and then identifying jobs to cover the tasks. In the end, some jobs may include idle portions. Figure 8.3(b) illustrates the ideas with the runs of part (a), which are now given and displayed by means of thin back lines. The arcs of these paths are the driver tasks. For clarity, only one job is shown, by means of a dashed grey line. This job starts at the depot with a run's first deadheading leg, which is a driver task. This is followed by the completion of three additional tasks on route B — a loop, some waiting, and another loop. Then, the driver moves to route A *without a bus* and without completing a task — the driver would travel on his/her own at some additional cost to the agency. The job continues with the completion of two more tasks: a loop on route A and a deadheading leg to route B — this time at no extra cost since the driver is accompanied by a bus and completing a task. The job ends with the completion of a task/loop on route B, and a final (unproductive) move without a bus to the depot. Once all the jobs have been identified in this way the problem is half-way done. The identified jobs then have to be populated by allocating specific workers

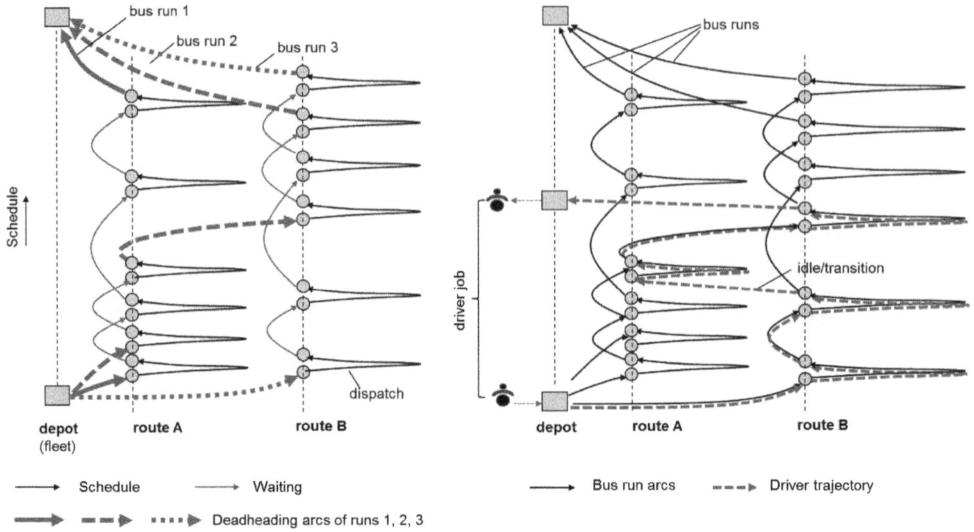

(a) Bus runs to cover a schedule

(b) Driver jobs to cover driving tasks

Figure 8.3. Two-step process to allocate vehicles and drivers.

(of the correct type) to specific jobs. This staffing process is not as simple as it is for buses because jobs start and end at different times, and may have other characteristics that are not equally appealing to all people. Therefore, the staffing process will be examined in considerable detail in Chapter 9.

In conclusion, note how the vehicle and personnel steps are parallel in structure. Both answer the overarching question: how many items are required to cover a set of requirements? We now turn our attention to vehicles, and what we learn will be applied to personnel in Chapter 9. Before we start, however, recall that an agency's yearly operating costs for vehicles are proportional to the vehicle-kilometers traveled, and those for personnel proportional to the number of vehicle-hours of service provided. In reality, however, buses must sometimes travel some extra distance deadheading, or sit idle without passengers. Thus, it is important to manage both vehicles and personnel in a way that will reduce this type of waste. Recall too that our planning recipes minimized these metrics, assuming there was no waste. Fortunately, we will see in the following sections that with good management this waste only needs to be a small fraction of the total. Thus, the assumptions made in the planning chapters turn out to be reasonable.

8.2. Schedule Covering: One Bus Route with a Single Terminus

As a preparatory step, this section examines the case of a single route with a single terminus. This simplification enables us to solve most of the problems with simple procedures that can be carried out by hand — and later to exploit this knowledge when tackling more complex problems. The section covers several topics. First, Section 8.2.1 describes a graphical procedure to estimate the fleet size required to serve the route. Then, Section 8.2.2 does the same numerically, and Section 8.2.3 extends the procedures to multiple bus types. Finally, Section 8.2.4 examines what happens if the terminus is moved to a different part of the route and Section 8.2.5 describes methods to identify a minimal set of runs to cover the schedule.

8.2.1. *Fleet Size: Graphical Analysis*

The data for our single-route, single-terminus problem (the schedule) can be represented in a time-space diagram showing each of the buses traveling along a route from a terminus (at $x = 0$) and looping back to it with headway H (see Figure 8.4). Each bus requires a cycle time T to make a full loop of length L and return to the terminal.

To keep track of the buses, we define a function $N(t, x)$ that gives the number of buses seen by time t at location x. The cumulative number of dispatches at the origin is $N(t, 0)$, and we also denoted it by $D(t)$. Similarly, the cumulative number of returns is $N(t, L)$. Note that $N(t, L) = D(t-T)$ if the cycle time is fixed.

We analyze this system as a queuing system from the perspective of the terminus — imagining for the time being that the depot is on top of the terminus and that the depot supplies buses when needed. We then ask: What is the minimum number of buses needed to sustain the schedule?

Each bus can be in one of two states: waiting in reserve at the terminus or circulating in service. They transition between states when they are dispatched and when they return to the terminus (see Figure 8.5).

A cumulative plot of buses made available $A(t)$, buses dispatched $D(t)$, and returned $R(t)$ shows graphically how the number of buses in reserve and in service evolves over time (see Figure 8.6). In this figure, curve D is given and the other two are derived. Curve R is obtained by shifting D to the right by T time units because buses return to the terminus T time units after they are dispatched. Analytically this is $R(t) = D(t-T)$. Curve A is obtained by shifting R up by M bus units, as shown in the figure, because the cumulative

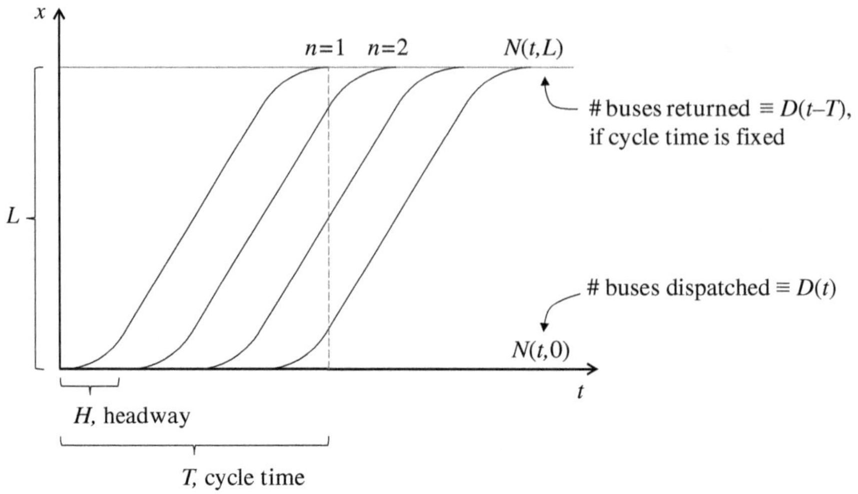

Figure 8.4. Vehicle dispatches and returns.

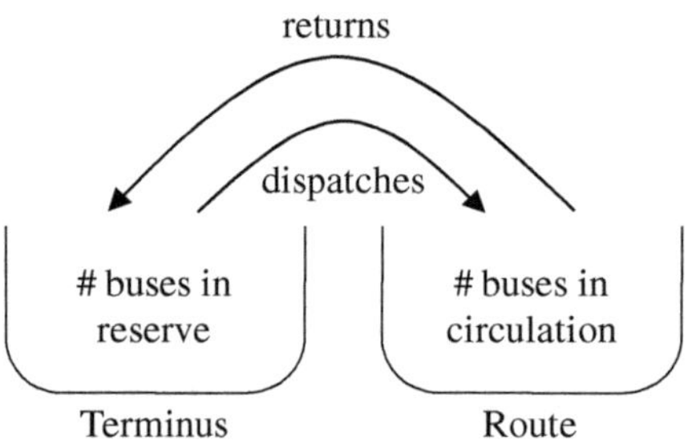

Figure 8.5. State transitions of a bus fleet.

number of buses made available is equal to those initially available, M, and those returned. This means that curve A can be obtained directly from D by shifting it to the right and up as per the vector shown in the figure. The analytic expression of these shifts is: $A(t) = M + R(t) = M + D(t - T)$, which encapsulates the relationship amongst the three curves.

The vertical separation between these curves at a given time gives the number of buses in a particular state. More specifically, the difference between curves A and D (buses made available and buses dispatched) is the number in reserve; the difference between D and R (buses dispatched and buses returned) is the number in circulation, which we shall denote $U(t)$; and the (fixed) difference between A and R (made available and returned) is the total number of buses in the system at the time, which is M by construction.

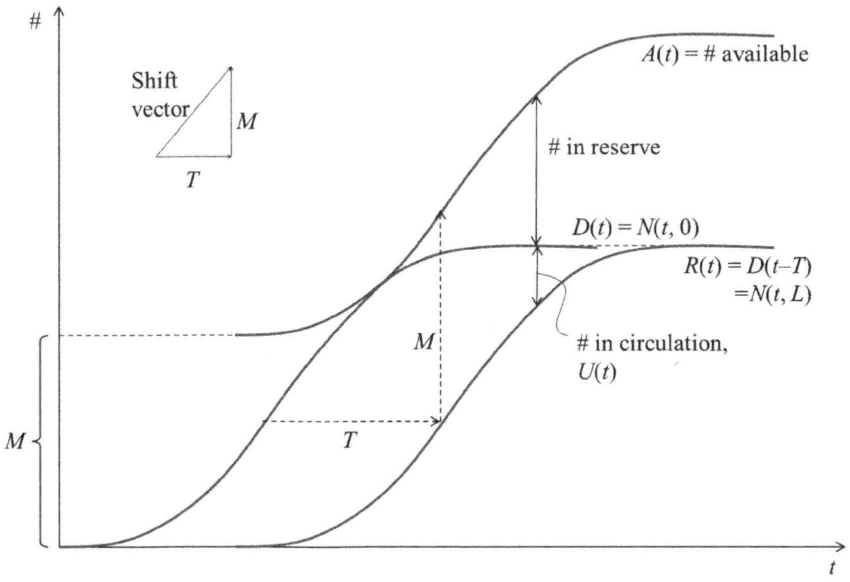

Figure 8.6. Cumulative plot of buses available $A(t)$, dispatched $D(t)$, and returned $R(t)$.

Now, since the number of buses in reserve cannot be negative, M should be chosen to be large enough to ensure that $A(t) \geq D(t)$. Thus, the minimum fleet size, $M = M^*$, can be obtained by shifting curve A vertically until $A(t)$ and $D(t)$ are tangent, as shown in Figure 8.6, and then measuring the vertical separation between A and R.

The minimum fleet size can also be obtained numerically. To see this note that at the time of tangency, t^*, there are no buses in reserve, so the number of buses in circulation equals the minimum fleet size; i.e., $U(t^*) = M^*$. Furthermore, since curves A and D are closest at that moment it follows that curves D and R are farthest. Therefore, $U(t) \leq U(t^*) \equiv M^*$, so that the fleet size can be obtained by solving:

$$M^* = \max_t\{U(t)\}. \tag{8.1}$$

Recall now that $U(t) = D(t) - R(t) = D(t) - D(t-T)$, and that both curve D and the cycle time T are given. Therefore, M^* can be expressed in terms of the problem's data as

$$M^* = \max_t\{D(t) - D(t-T)\}.$$

This expression can be evaluated numerically. It also yields simple bounds for M^* that do not require knowing the detailed shape of $D(t)$ — only the largest and shortest headways. To find these bounds note that the number of buses in circulation $U(t)$ must be maximized immediately after a bus dispatch because until the next dispatch the number of buses in circulation can only decline. The number of dispatches in the interval $(t-T,\, t]$ where t occurs immediately after a dispatch, i.e., the value of $U(t)$ at said instant, is at least 1. Therefore, since the number of dispatches in $(t-T,\, t]$ is inversely related to the average headway in $(t-T,\, t]$, we see that $U(t)$ must satisfy

$$\left\lceil \frac{T}{H_{\max}} \right\rceil \leq U(t) \leq \left\lceil \frac{T}{H_{\min}} \right\rceil, \quad \forall\ t,$$

and therefore, $M^* = \max\{U(t)\}$ also satisfies

$$\left\lceil \frac{T}{H_{\max}} \right\rceil \leq M^* \leq \left\lceil \frac{T}{H_{\min}} \right\rceil. \tag{8.2}$$

These are our bounds. Note that, since M^* is a maximum, it should be closer to the upper bound than the lower bound. You can also convince yourselves that if the headways vary slowly with times comparable with the bus cycle, then the upper bound is quite tight. In fact, in the time-independent case, where the headways do not vary at all, the lower and upper bounds coincide and we obtain the well-known result:

$$M = \left\lceil \frac{T}{H} \right\rceil.$$

The tightness of the bound means that the fraction of wasted vehicle-hours is small. Furthermore, since all the wasted time is spent at the terminus with the vehicle sitting idle, there should be no wasted vehicle-kilometers at all.

8.2.2. *Fleet Size: Numerical Analysis*

Here we show how to calculate M^* exactly when the cumulative dispatch curve $D(t)$ is known; i.e., we know the times $\{T_n\}$ of all dispatches ($n = 1, 2, \ldots$). This can be useful when the gap between the bounds of (8.2) is too wide for the purposes at hand.

To do this we shall first consider a tentative fleet size j and then determine if it is feasible. Note from Figure 8.7 that if $T_{n+j} - T_n \geq T$ for all n, then $A(t)$ is always to the left of $D(t)$ for all t, as we want so that j is feasible. This inequality merely ensures that every bus is available *before* it is dispatched again. The inequality holds if and only if there are no more than j dispatches in time T, which is the feasibility condition we used above to develop the headway-based bounds.

The minimum fleet size is the smallest feasible j. It can be determined with a spreadsheet structured as in Table 8.1 to check the feasibility

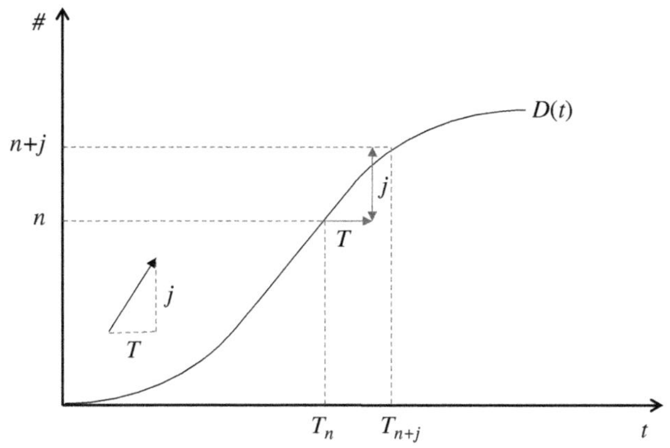

Figure 8.7. Checking fleet size feasibility.

Table 8.1. Numerical method to check fleet size feasibility.

		$j = 1$	$j = 2$	
n	T_n	$T_{n+1} - T_n - T \geq 0$	$T_{n+2} - T_n - T \geq 0$	...
0	time data			
1	time data			
2	time data			
⋮	⋮			

condition, $T_{n+j} - T_n \geq T$ (or $T_{n+j} - T_n - T \geq 0$), for different tentative fleet sizes, j. The lowest value of j corresponding to a column with all values greater than or equal to 0 is the sought fleet size. With the found fleet size the reserve of buses is never empty.

8.2.3. *Multiple Bus Types*

We have assumed in the above analysis that there is only one type of bus. In reality, even for one route, different bus types may be used. For example, a bus (subway) route could be allowed or required to use smaller buses (shorter trains) during the off-peak hours.

It should be clear that in the special case where every dispatch is required to use a bus of a specific type, the problem decomposes by type. To solve it one would construct separate D curves for the dispatches of each type, and then use the methods we just have described with each curve to determine the fleet size for the corresponding type.

In cases where some dispatches can use multiple bus types, the problem is more complicated since there is additional flexibility. A possible approach consists in first forcing a unique choice for every dispatch with some reasonable rule (e.g., as explained at the end of Section 8.3.2) so as to construct separate D curves using the choices made; and then solving the problem as we have just described. If desired, different choice sets can be tried and tested systematically with optimization tools.

8.2.4. *Terminus Location*

So far, we assumed that the terminus was at $x = 0$. Could it be possible to reduce the fleet size by locating the terminus in a different place along the route? It turns out that the minimum fleet size is insensitive to the location of the terminus if bus trajectories are the same through the day. Since this is usually the case, at least approximately, it follows that passenger considerations should drive termini location decisions.

Here is why the above is true. Let Δ_n be the travel time of bus n from an old to a new terminus (see Figure 8.8), and note that the time when bus n is dispatched from the new terminus if the schedule is not changed is given by: $T'_n = T_n + \Delta_n$. Now, if the bus trajectories are similar then $\Delta_n = \Delta$, and $T'_{n+j} - T'_n = T_{n+j} - T_n$ because the Δ's cancel out. This means that the feasibility condition for a fleet of size j with the new terminus ($T'_{n+j} - T'_n \geq T$) is equivalent to that of the old terminus ($T_{n+j} - T_n \geq T$).

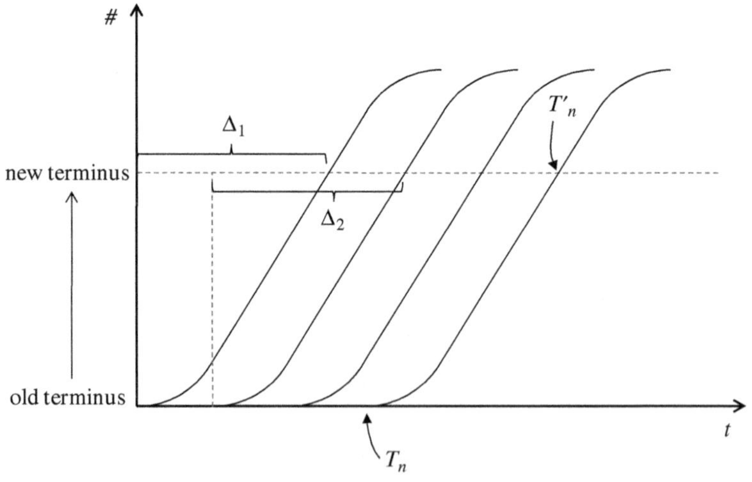

Figure 8.8. Effects of terminus location.

Therefore, j is either feasible in both cases or invalid in both. Since this is true for every j, the minimum feasible fleet size should be the same in both cases. This result is good because it allows agencies to put termini in locations where buses are nearly empty without an impact on the vehicle-kilometers traveled.

8.2.5. *Run Determination*

The bus scheduling problem consists of determining which bus is associated with each bus task (or loop since for the moment we have only one terminus), i.e., figuring out what each bus will do: the bus runs. We discuss two heuristic methods for determining the specific run for each vehicle, but there are many more — if there are multiple vehicle types and the problem has been decomposed as suggested in Section 8.2.3, then any of these heuristic methods can be applied to each one of the classes separately to solve the problem. Method 1 assigns buses to tasks and Method 2 does the reverse.

In Method 1 we initially place all buses in an imaginary pool of available buses from which they are to be selected to cover the tasks. We then step through time, and as tasks arise use a rule to choose a bus from those currently in the pool. Of course, when a task is completed its assigned bus is returned to the pool. The proposed method uses a last-in-first-out (LIFO) selection strategy such that new buses are only introduced when absolutely necessary. With LIFO, as we step through time and find that a task is scheduled to start, we assign it to the bus that completed its last task most recently. If there is no such bus then a bus is selected from those that have not yet been used. If no buses remain at any moment of need, the method has failed to find a feasible solution. This strategy is good because it keeps some individual buses running while others experience long periods of idling. The latter can be returned to the depot for driver relief.

Method 2 uses a greedy strategy that considers buses one at a time and a pool of unassigned tasks. The pool initially contains all tasks. When the first bus is considered, we assign tasks to it in order of increasing time (until the end of the day) always greedily choosing the unassigned task that starts the earliest after the current time — the net effect is that each bus is redeployed as early as possible after returning to the terminus. All assigned tasks are then removed from the pool. The next bus is then processed in the same way, and then the next, etc. until no tasks are left in the pool, or we run out of buses. In the latter case, the method has failed to find a feasible solution. The procedure is so simple that it can be implemented graphically by hand

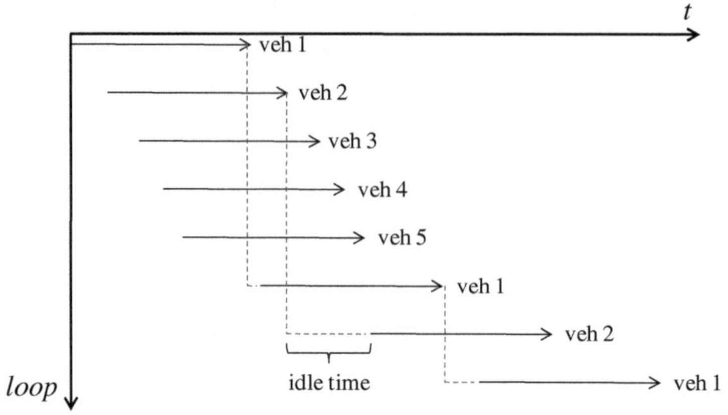

Figure 8.9. Assignment of bus loops to vehicles.

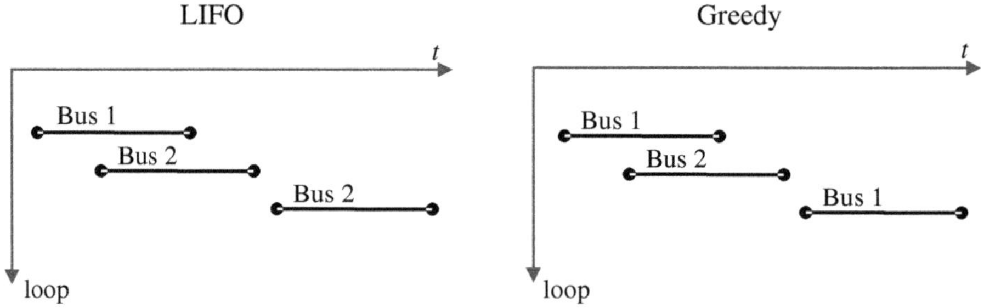

Figure 8.10. LIFO vs. greedy strategies for bus assignment.

by plotting each scheduled loop from the route's terminus against time as shown in Figure 8.9. Could you organize this in a spreadsheet?

It is perhaps intuitive, and can be proven (see Appendix 8.A), that both the LIFO and greedy methods find feasible solutions if the fleet size equals or exceeds the minimum fleet size given by (8.1).

Example 8.1 (*Qualitative Difference between LIFO and Greedy Methods*). The simple three-run, two-bus system in Figure 8.10 shows how/why the two methods differ. The dark bars are the loops, which are the same on both parts of the figure, and the labels denote the bus assignments, which are different.

Consideration of the selection procedures reveals that the LIFO method would assign the third loop to the bus that returned the most recently, which is Bus 2, as shown in the figure. By contrast, Method 2 would assign it to the lowest indexed bus, which is Bus 1.

The methods only differ in the bus that is selected from the pool of idle buses for the next run. Since we are assuming that all the buses of any given type are identical, the choice has no future repercussions on the availability of buses. In fact, one could have selected the bus at random, or with any other rule, and the strategy would perform similarly. Thus, the specific bus choice is usually made with other (non-bus) criteria in mind.

8.3. Schedule Covering for N Bus Routes

This section examines a more general problem involving N routes with multiple dispersed termini. Section 8.3.1 discusses a basic case in which the N lines have a common terminus that is close to the bus depot, and therefore there is no need for deadheading. Section 8.3.2 presents the general case, and Section 8.3.3 discusses qualitatively the effects of deadheading.

8.3.1. *Single Terminus Close to the Depot*

Imagine a map of many routes, all passing through a centrally located terminus (see Figure 8.11). We could imagine this is a bus station in the center of a city or at a busy rail transfer station. The terminus may be close or far from the depot. We imagine for now that it is close. This subsection explains how bus runs can be determined for such a system. The explanation assumes that there is only one bus type, but the ideas can also be applied to the separate sub-problems arising from any multi-type problem that has been decomposed by type as discussed in Section 8.2.2.1.

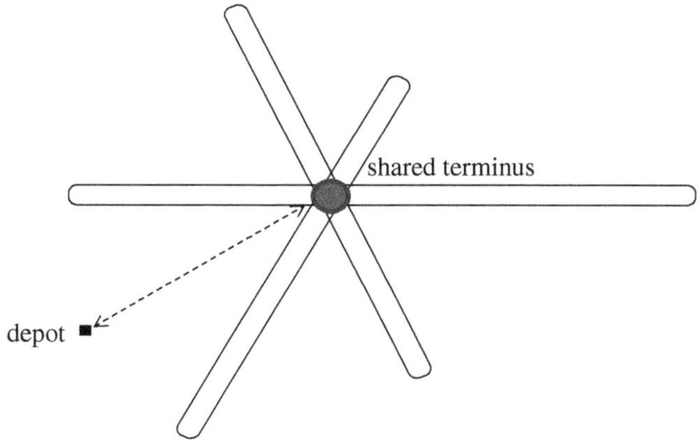

Figure 8.11. Multiple bus routes sharing one terminus.

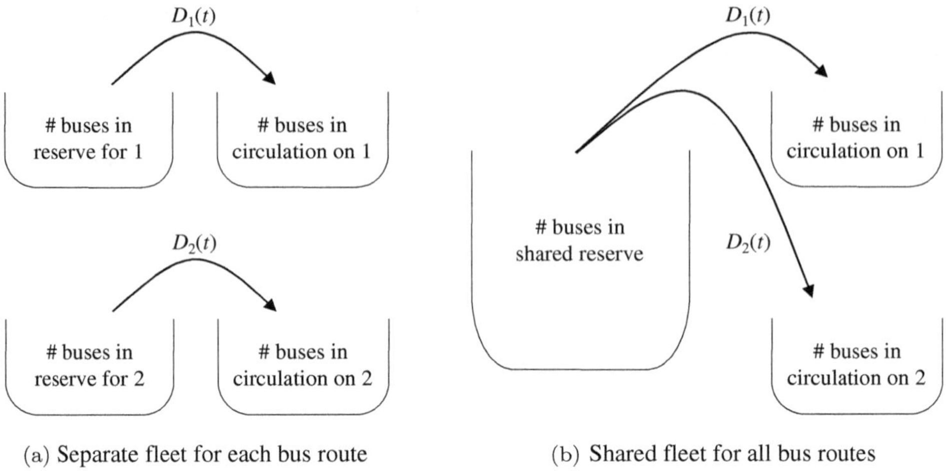

Figure 8.12. Fleet management strategies.

Although each route could be assigned a separate set of buses, and be treated independently as in Section 8.2, it stands to reason that efficiencies may be gained by sharing buses across routes. Sharing may allow us to reduce the overall fleet size. Figure 8.12(a) illustrates the separate case, with route-specific bus reserves, and Figure 8.12(b) the shared case. Note how in the latter case there is a unique reserve of buses available for all routes. Clearly, since a unique reserve allows for more operational flexibility it should produce better results.

The analysis of the shared strategy is only slightly different. The main change is that, because each route, i, is characterized by loops of different cycle times, the time until a dispatched bus returns, $T(i)$, is no longer a constant. It depends on the route. Recognizing this change, we can express the aggregated cumulative counts of dispatched and returned vehicles as

$$D(t) = \sum_i D_i(t), \tag{8.3}$$

$$R(t) = \sum_i D_i(t - T_i). \tag{8.4}$$

And since the fleet (M) is shared, the cumulative number of buses made available for collective use can still be expressed as

$$A(t) = M + R(t). \tag{8.5}$$

Curves $D(t)$, $R(t)$, and $A(t)$ can be plotted as before to determine the minimum fleet size, and the formula $M = \max\{D(t) - R(t)\}$ continues

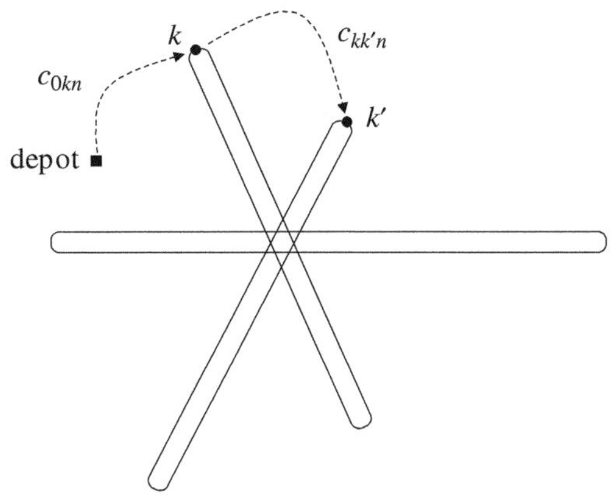

Figure 8.13. Dispersed termini and depot for multiple routes.

to hold. The only difference is that $R(t)$ is no longer related to $D(t)$ by a shift.

The main advantage of sharing is that the necessary fleet size is always less than or equal to the one without sharing. The reason is that the assignment obtained without sharing can also be achieved through sharing, but sharing allows us to do more things. This also means that the larger a system is the smaller its fraction of wasted vehicle-hours. And since this fraction was already small for a single route, it should be quite small for large systems. And, of course, if there is a single terminus, the wasted vehicle-kilometers should continue to be negligible.

8.3.2. *Dispersed Termini and Deadheading Heuristics*

Consider now the most general case where the termini and depot are dispersed and lines are allowed to have more than one terminus. Buses can be of different types. They may now have to deadhead across termini, adding vehicle-kilometers of waste. This is illustrated in Figure 8.13, where the termini are at the ends of the lines. So, let us now see how this waste can be minimized.

Recall that a bus task is the smallest part of the schedule, which cannot be subdivided anymore and must be covered by the same bus, i.e., a trip between consecutive termini and not necessarily a full loop. In what follows, k and k' index bus tasks. We shall use $k = 0$ for the task of being at the depot and $k = 1, 2, \ldots$ for the remaining bus tasks. If buses can be housed at multiple depots, stays at these depots would be indexed using consecutive negative integers.

The figure also illustrates by means of dashed lines some deadheading moves. Clearly, there may be some cost associated with transitioning a bus from task k to task k'. If the distance of the move is d, then the cost can be estimated as in the planning chapters by $c_d\, d$, where c_d is the cost of the vehicle and its driver per kilometer driven. Note, this product depends both on k, k' and on the bus type, n. For this reason, the costs are denoted, $c_{kk'n}$.

A simple heuristic method can be used to solve this problem approximately. To do this, the problem is reduced to the case of Section 8.3.1 by including as a part of each task a fixed setup time that is used to prepare the bus for the service about to be provided, including any necessary deadheading. This setup time should be long enough to allow for deadheading from any location where the bus may be. This framework adds time to the tasks but removes the need for considering deadheading. The results should be good if the set up times are small compared with the duration of the tasks. Otherwise, it produces inefficient but valid solutions. Even if that happens, these solutions can be used as starting points for further refinement via numerical search algorithms.

Intuitively, the shorter the setup times the better the result. Therefore, we propose choosing the shortest setup times possible. If we use $T_{(k)}$ for the duration of task k, $t_{k'k}$ for the transition times into k, and $\Delta_{(k)}$ for the setup time, the recommended choice is[3]

$$\Delta_{(k)} = \max_{k'}\{t_{k'k}\}.$$

Tasks are then redefined to start $\Delta_{(k)}$ time units earlier than originally without changing the ending time. Thus, the new durations are

$$T'_{(k)} = T_{(k)} + \Delta_{(k)}.$$

Since deadheading is included as part of the redefined tasks, we can treat this new problem (with the new starting times and durations) as we did in Section 8.3.1, with zero deadheading times. This solves the problem.

If the setup times are comparable to the durations of bus tasks, then the fleet sizes and sets of bus runs identified with the proposed method may not be very efficient. In this case, optimization algorithms should be used to improve them. Although the task is complex, the optimization problem at hand is similar to a famous problem that has been extensively studied — the

[3]If buses are not shared across routes, the maximum operation in the definition of $\Delta_{(k)}$ should only be taken over the set of termini from which buses can be pulled to begin task k; i.e., termini of the same route. Thus, the values of $\Delta_{(k)}$ will be generally smaller if buses are not shared than if they are shared.

vehicle routing problem (VRP). Thus, we do not have to develop a solution search method from scratch. The rest of this subsection describes the VRP and how it can help.

In the VRP, one looks for the best way of serving a set of customers with a fleet of vehicles. Given are the number of customers, a load size for each customer, and all the customer locations — or alternatively the distance between each customer pair. Also given is the location of a depot from which vehicles are dispatched and the types of vehicles that can be used — included as part of the data are basic properties such as the vehicle's capacity, cost per kilometer and cost per dispatch (if any). These data are then used to determine the cost of a leg between any two customers and the cost of any vehicle tour. The problem's objective is finding a set of vehicle tours starting and ending at the depot that will serve all the customers with the least possible cost.

Some variants of the VRP specify travel times and service times as part of the customer data and then allow for time-based constraints such as limitations to the duration of a vehicle tour and/or the provision of service in time windows. Appendix 8.B discusses some technical aspects of the VRP and presents a simple computer method to obtain good solutions. The method is generic and can be used with any variant of the problem.

Table 8.2 lays out the analogy between the VRP and our schedule covering problem (SCP). The analogy arises because both problems look for the least costly way to cover a set of spatially distributed customers/tasks. Note that the time-based features of the VRP are not used for the SCP — but they are included in column 1 to help with a problem that will arise in Chapter 9.

The solution algorithm described in Appendix 8.B works iteratively on an existing valid solution, and usually produces good results if the initial solution is reasonable. Therefore, we propose obtaining said solution with the simple heuristic algorithm with setup times that was described at the beginning of this subsection.

In the case of multiple bus types, also for purposes of initialization, we propose using the decomposition approach described in Section 8.2.2, i.e., associating with every task that can accept multiple bus types a *unique* bus type. This is useful because the initialization algorithm with setup times that we have just mentioned can then be applied to the class of tasks associated with each bus type separately. It produces a set of bus runs for each bus type, which together can be used as an initial valid solution.

The selection of a unique bus type for each task requires some care, however. If a consecutive portion of a route's schedule can be served by a single set of multiple bus types (e.g., route "5" in the midday can

Table 8.2. Analogy between the bus SCP and the VRP.

VRP	SCP
Customers i, j	Bus tasks k, $k\prime$
Customer load size, d_i	Bus task duration, d_k
Depot	Depot
Vehicle type m	Bus type n
Vehicle capacity	∞ (unless bus range is limited)
Cost for a leg, c_{ijm}	Cost penalty, $c_{kk'n}$ ($c_{kk'n} = \infty$ if impossible)
Fixed cost of a vehicle dispatch, c_m	Fixed cost of a bus dispatch, c_n (usually =0)
Vehicle types allowed to serve a customer	Bus types suitable for a task
Time for a stop, t_i	n/a
Time between stops t_{ij}	n/a
Maximum tour time, T_m	n/a
Time window for a dispatch $[T'_m, T''_m]$	n/a
Cost rate for exceeding T_m, c	n/a
Vehicle tour (customers served, in order)	Bus run (tasks covered by a bus, in order)

be served by either articulated or regular buses) then we should avoid chopping up any such part of a schedule into tasks with randomly varying bus types. This would be inefficient. Fortunately, this kind of inefficiency can be circumvented with a simple rule. First, number the vehicle types consecutively, $n = 1, 2, \ldots$ and then *allocate to each task the feasible type with the lowest number*. This rule ensures that all tasks with the same set of feasible types, e.g. tasks in the same route, are ultimately associated with the same bus type. If desired, one can also sort the bus type numbers according to certain preference (e.g., cost) so as to maximize the utilization of the preferred types.

8.3.3. *Discussion: Effects of Deadheading and Pooling*

Deadheading is useful because it allows buses to serve multiple lines when the terminals are dispersed, thereby allowing an agency to draw bus resources from a common pool. If done properly, deadheading should not add a significant fraction of vehicle-kilometers to the amount implicit in a schedule. Otherwise, the agency could always revert to treating routes individually as in Section 8.1, which was already quite efficient.

Deadheading also allows agencies to send buses back to the depot whenever the part of the schedule they serve exhibits an extended gap. At the depot they can stay without a driver, accruing no significant cost. This is important to ensure that the actual number of vehicles in service (with a driver and accruing cost) closely tracks the peaks and valleys in the number

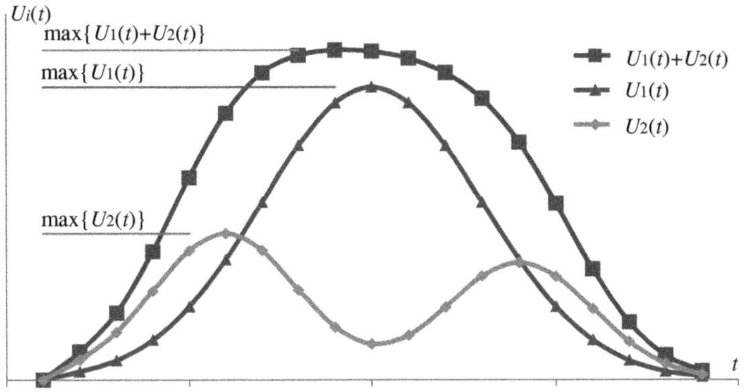

Figure 8.14. Number of circulating buses with and without pooling.

of vehicles scheduled for service during the course of a day. Furthermore, the pooling of resources that deadheading allows, can also help smooth out these peaks and valleys. This is desirable because the highest peak dictates fleet size.

To illustrate the benefit due to smoothing, suppose we have two bus lines with different peaking patterns, such as a commuter route running heavily in the morning and evening, paired with a route that is run most heavily during the middle of the day for something like an athletic event. Figure 8.14 shows the bus requirements implicit in the schedule for these two lines over the course of a day. Labeled $U_i(t)$, these requirements are shown both separately and pooled.

Now, let us compare the bus requirements if the routes are considered separately or pooled. In the former case, the required fleet size would be the sum of the maximum route requirements for the individual routes, i.e., the sum of the two peaks in Figure 8.14, or $\sum_i \max_t\{U_i(t)\}$ generally. If the routes are pooled, they would draw from the common pool and therefore the required fleet size would be the maximum of the dotted curve, or $\max_t \sum_i \{U_i(t)\}$ generally. As the figure clearly shows

$$\sum_i \max_t\{U_i(t)\} \geq \max_t \sum_i \{U_i(t)\}. \tag{8.6}$$

This is a famous inequality that holds in general. It shows that deadheading can reduce fleet size. The figure is exaggerated, but it shows that benefits are most significant if the system includes routes that peak at different times.

References

Ahuja, R. K., Magnanti, T. L., and Orlin, J.B. (1993). *Network Flows: Theory, Algorithms, and Applications*, Prentice Hall, Englewood Cliffs.

Daganzo, C. F. (2005). *Logistics Systems Analysis*, 4th ed., Springer-Verlag, Berlin, Heidelberg.

Dijkstra, E. W. (1959). "A note on two problems in connection with graphs." *Numerische Mathematik*, 1, 269–271.

Press, W. H., Flannery, B. P., Teukolsky, S. A., and Vetterling, W. T. (1987). *Numerical Recipes: The Art of Scientific Computing*. Cambridge University Press, New York.

Robuste, F., Souleyrette, R. R., and Daganzo, C. F. (1990). "Implementing vehicle routing models." *Transportation Research Part B*, 24(4), 263–286.

Homework

8.1. Consider one route with one vehicle type. How would you determine the required fleet size if you know that some buses will not be available during a given time interval? (*Hint*: M is no longer a constant but a varying function of time).

8.2. A commuter bus route extending from the suburbs to a CDB is operated from 5 a.m. to midnight. The route has 20 stops and is 10 km long in each direction. The bus cruising speed is 25 km/h, and each stop imposes a delay of 30 seconds. The headway is 10 min during the rush intervals (7 a.m.–9 a.m., and 3 p.m.–7 p.m.), and 30 min for the remainder of the service period.

(a) Please draw the cumulative bus dispatch curve $D(t)$, and the return curve $R(t)$.

(b) Determine the minimum fleet size M needed to serve this route.

The agency is also running a recreational shuttle route that shares the same terminus at the CBD. It is operated from 6 a.m. to 8 p.m., with 20-min headways from 8 a.m. to 4 p.m., and 45-min headways at other times. The route has no intermediate stops. It is 15 km long each way. The bus cruising speed is 25 km/h and the dwell time for passengers at each end of the route is 10 min.

(c) Determine the smallest fleet size that can provide this service.

(d) If the same pool of vehicles is used for the two routes, what is the minimal fleet size?

8.3. Consider two bus routes that operate from 6 a.m. to midnight from a common terminus. The bus round trip time is 120 min on both. Route A faces peak demand during the rush intervals, 7 a.m.–9 a.m. and 3 p.m.–7 p.m. During these intervals, it uses a headway of 20 min, and requires articulated buses. For the rest of the service period, the headway is 60 min and regular buses can also be used. Route B faces recreational demand that peaks from 10 a.m. to 4 p.m. During this peak the headway is 30 minutes and articulated buses are needed. For the rest of the service period, the headway is 90 min and buses of either type can be used. For the regular bus, the prorated capital cost per unit time (for acquiring the vehicle) is about the same as the operating cost per unit of working time (i.e., for driver wage, fuel consumption, and maintenance). Suppose now that an articulated bus costs about twice the prorated capital cost, and about 1.3 times the operating cost, as compared to a regular bus, but provides twice the capacity. Since running an articulated bus to serve the low demand of an off-peak period somewhat damages the agency's public image (i.e., making it appear that public resources are being wasted) an extra 10% penalty is added to the actual operating cost whenever articulated buses are used during these periods. Given all this information, determine the optimal bus fleet composition and the corresponding bus assignments, if we use (a) separate vehicle pools for each route; or (b) a common pool for both routes. Assume there is no deadheading cost. If there is a fixed cost per deadheading move (expressed as a percentage of a regular bus service trip) find the maximum cost that still makes (b) the lower cost option.

Appendix 8.A: Proof–LIFO and Greedy Methods use Fewest Buses

Any method that steps through time, such as the LIFO method, is feasible with only M^* buses because, as the text proved with queuing diagrams, the pool of idle buses always includes at least one bus. Thus, the focus here is on the greedy algorithm.

Consider the Gantt spreadsheet of the text with all the tasks plotted vs. time, and the set of times where a critical number of tasks, $M^* = \max\{U(t)\}$ (assumed to be an integer), have to be simultaneously performed. This set will be the union of one or more (critical) time intervals. Note that at least a new task must begin at the beginning of each one of these intervals for otherwise the interval could be lengthened. Therefore, because the algorithm is greedy, the first bus must enter the first interval with a task assigned to it (i.e., in a busy state) for if the bus was idle immediately before the start, it would be assigned the task that starts at the interval's beginning. Now, also note that if a bus enters an interval in a busy state it must remain busy until it exits because any task that ends within an interval must be immediately followed by another — otherwise the number of tasks in the interval would decline, which is impossible. Thus, on finishing a task the bus would continue with another because the algorithm is greedy. Therefore, the bus remains busy as it crosses the first interval, hence decreasing the number of simultaneous tasks in said interval by 1.

For the same reasons, the first bus must cross the second, and all other critical intervals, in a busy state. Thus, the bus also reduces the number of simultaneous tasks left to be processed in these intervals by 1. As a result, the number of critical tasks remaining after removing the tasks processed by bus 1 is $M^* - 1$.

When bus 2 is considered, the problem is analogous to the original albeit with $M^* - 1$ critical tasks. Thus, bus 2 also reduces the number of critical tasks by 1. The same happens for buses 3, 4, etc. so that when M^* buses have been processed, the number of critical tasks is 0, i.e., all the tasks have been completed. Therefore, the greedy algorithm is feasible with M^* buses.

Appendix 8.B: The "Vehicle Routing Problem" and Meta-Heuristics

This appendix describes a family of classic combinatorial optimization problems that include as special cases some of the problems encountered in this chapter and the next. The appendix also presents some solution methods.

8.B.1. *The Traveling Salesman Problem (TSP)*

Given are N nodes, $i = 1, 2 \cdots N$, and a matrix of distances $\{c_{ij}\}$ between every pair of nodes. In the TSP, we look for a closed-loop tour that visits all the points with the least total distance (or "cost").

Mathematically, we define the following decision variables: (i) a set of binary variables $\{x_{ij}, \forall i, j\}$ where $x_{ij} = 1$ if node j follows i in the tour, or 0 otherwise; and (ii) continuous variables $u_i, \forall i = 1, 2, \ldots, N$, which denotes the position of node i in the tour, and $1 \leq u_i \leq N$. The objective function can obviously be written as

$$\min \sum_{i,j} x_{ij} c_{ij}. \tag{8.B1}$$

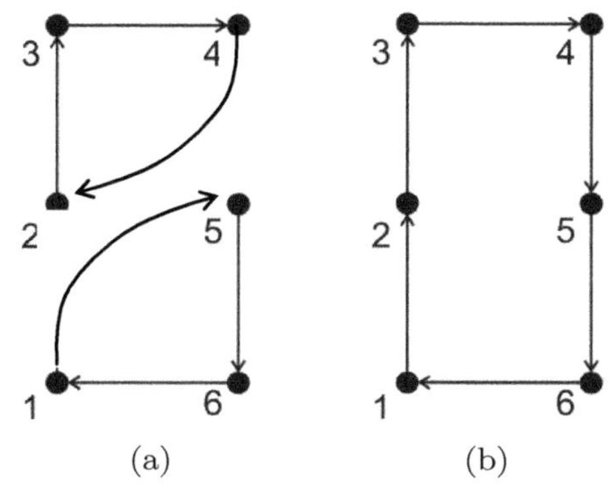

Figure 8.B1. Sub-tours vs. single tour.

To ensure that the tour is feasible, we first require that every node is visited once and only once during the tour, i.e., each node is arrived at from exactly one other node, and from each node there is a departure to exactly one other node. This can be satisfied by the following constraints:

$$\sum_j x_{ji} = \sum_j x_{ij} = 1, \ \forall i \,. \tag{8.B2}$$

The above constraints are however not enough to ensure feasibility yet. These constraints can also be satisfied by a set of multiple sub-tours that collectively cover all the nodes, e.g., as shown in Figure 8.B1(a). Note how every node is visited once and only once, but there is no way that a traveler can feasibly visit all nodes in one trip.

To prevent such sub-tours from happening, we add the following "subtour elimination constraints":

$$u_j \geq u_i + N x_{ij} - (N-1), \forall i, \forall j \neq 1; \tag{8.B3}$$

$$u_1 = 1. \tag{8.B4}$$

These constraints ensure that there is only a single tour, rather than two or more disjointed tours, that cover the nodes. For any $x_{ij} = 1$ and $j \neq 1$, we have $u_j \geq u_i + 1$; otherwise if $x_{ij} = 0$, we have $u_j \geq u_i - (N-1)$ which is always true. That is, these constraints require that the sequence of $\{u_i : i = 1, \ldots, N\}$ values must monotonically increase between every adjacent pair of nodes along the (sub)tour, except when node 1 is the successor. If any sub-tour exists but does not contain node 1, a contradiction arises because the value of $\{u_i\}$ cannot keep increasing in a closed loop. For example, these constraints will force the sub-tour (2, 3, 4, 2) in Figure 8.B1(a) to satisfy $u_2 < u_3 < u_4 < u_2$ strictly, which is impossible. Hence, any feasible solution must contain only one closed sequence of nodes, as shown in Figure 8.B1(b). It is feasible because we can simply assign the values of $u_1, \ldots, u_6$ to be $1, 2, \ldots, 6$.

8.B.2. *Local Searches and Meta-Heuristics*

The basic idea behind local search methods is to guess solutions that get increasingly better as the method proceeds. Solutions are characterized by a "state" which is a string of numbers. This can be illustrated with the TSP. Since the positions in which the cities appear in a tour are uniquely defined by an ordering of the first N integers (a permutation), any such ordering is a state of the TSP problem. The following example shows two possible states for a six-point TSP problem: (1, 2, 3, 4, 5, 6) and (1, 6, 3, 4, 5, 2). It is assumed in this example that costs are given by the Euclidean distances of the links. Thus, the cost of each state is the length of the tour one would measure with a ruler.

Any "local search" is based on perturbations that transform a state into a similar state, hopefully with lesser cost. For the TSP, a perturbation could be choosing two consecutive cities and swapping their order. For example, from (1, 2, 3, 4, 5, 6) we could go to (1, 3, 2, 4, 5, 6,) and from this to (1, 3, 4, 2, 5, 6). The set of states that can be reached in one step (one perturbation) is the state's local neighborhood. Perturbations should be simple (so they are easy to make and evaluate), but also comprehensive, in the sense that they should allow the system to reach any state from any other state. Consecutive city swaps have these two properties and are therefore acceptable perturbations for the TSP.

Given a current state, a "greedy" local search would evaluate the cost of all the states in its neighborhood and move to the one with the least cost if such a state exists; otherwise the search ends. This procedure is then repeated using this new state as the current state, and then repeated iteratively until the search ends because no improvement can be found. The termination point is called a "local" optimum. Local optima are generally not unique for the TSP. For example, you can verify that the two tours in Figure 8.B2 are locally optimal, even though tour (1, 6, 3, 4, 5, 2) on the right is quite bad.

In view of this, people have created "meta-heuristic" methods that in theory can avoid being trapped in local optima and converge to the global optimum. The simplest meta-heuristic method is called simulated annealing (SA). It differs from the greedy method in that it randomly chooses a *single* perturbation from the current state to identify a single new state. A coin is then flipped to see whether the new state is accepted and becomes the

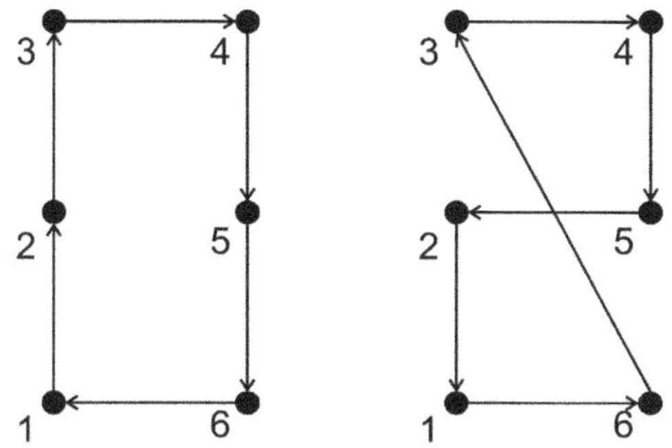

Figure 8.B2. Two TSP solutions before and after a swap of points.

new current state, or one stays put. The probability of success "p" is chosen to be the following function of the change in cost, Δe, and the iteration number, n: $p = 1$, if $\Delta e \leq 0$; but if $\Delta e > 0$ then $p = \exp\{-(n + a)\Delta e\}$, where "$a$" is a positive constant. Note that at the start of the search ($n = 1$) there can be a significant probability of accepting a more costly state (with $\Delta e > 0$) but this probability declines as the simulation progresses. This probabilistic feature of SA allows the algorithm to jump out of local optima and, given enough time, to converge to the global optimum. Unfortunately, convergence is slow for problems with more than (say) 100 points. Even in these cases, though, the method can be used to fine tune solutions obtained with other methods. A large value of "a" is normally chosen for this type of application.

8.B.3. *The Vehicle Routing Problem (VRP)*

The VRP arises in practice more often than the TSP, and many variants of it exist (e.g., with route length restrictions, time-windows, etc.). In its most basic forms, it seeks vehicle routes to serve a set of N customers distributed in space. Customers have items to be carried, which take up vehicle space. Available to serve these customers are M vehicles. Each vehicle, m, is assumed to have individual properties such as a vehicle capacity, cost per dispatch, etc. In essence, each vehicle is in a class of its own so we have M classes too. Note this 1:1 vehicle-class framework is general enough to allow us to model real problems with few vehicle types. To do this, we would assign the properties of each vehicle type to a large number of vehicles, and would repeat this for all types. For example, if there was only one vehicle type, one would assign the type's properties to all vehicles.

Given are:

N customer points, $i = 1, 2 \cdots N$
M vehicles, $m = 1, 2 \cdots M$
A depot at $i = 0$
A matrix of travel costs for all possible trip legs and vehicles, $\{c_{ijm}\}$
A vector of dispatching costs for all vehicles $\{c_m\}$
The subset of vehicles that are permitted to visit point i, $P_i \subseteq \{1, 2, \ldots, M\}$
A demand d_i for every point (node) i (in units of "quantity")
A vehicle capacity, V_m, for every vehicle m (also in units of quantity)

We look for: An allocation of customer points to vehicles and a set of vehicle routes ending and beginning at the depot that minimize the total cost.

Two sets of decision variables are used. The first set consists of binary decision variables that define the vehicle routes. The variables are $\{x_{ijm} : i, j = 0, \ldots, N, m = 1, \ldots, M\}$, where $x_{ijm} = 1$ if vehicle m travels directly from node i to node j, and 0 otherwise. The second set is auxiliary; it consists of continuous decision variables $\{u_i\text{: } i = 1, 2, \ldots, N\}$ that will be used to eliminate sub-tours as in the case of the TSP.

The objective is to minimize the total cost, i.e., the sum of the travel and dispatching costs:

$$\min \sum_{i,j,m} c_{ijm} x_{ijm} + \sum_{j,m} c_m x_{0jm}. \tag{8.B5}$$

These constraints play different roles. The following two sets of constraints ensure: (i) that every customer is visited by one and only one vehicle of the correct type (8.B6); and (ii) that if a vehicle arrives at a customer it also leaves that same customer (8.B7). In combination, the two sets of constraints ensure that the solution consists of non-overlapping closed loops, each executed by a specific vehicle.

$$\sum_{m \in P_i} \sum_{j=1}^{N} x_{ijm} = 1, \ \forall i = 1, \ldots, N \tag{8.B6}$$

$$\sum_{j=1}^{N} x_{jim} = \sum_{j=1}^{N} x_{ijm}, \ \forall i = 1, \ldots, N, \forall m = 1, \ldots, M \tag{8.B7}$$

These constraints, however, allow vehicles to execute any number of loops. More constraints are needed to ensure that vehicles either perform only one tour that starts and ends at the depot or are not used. This will be done with three additional sets of constraints. The first set (8.B8) ensures that each vehicle leaves the depot either 0 or 1 times:

$$\sum_{j=1}^{N} x_{0jm} \leq 1, \ \forall m = 1, \ldots, M. \tag{8.B8}$$

These constraints achieve the stated goal because if the inequality is strict the left side must be zero signifying that the vehicle leaves the depot 0 times (i.e., it is not used). And if the inequality is not strict then one and only one of the variables on the left side is 1, i.e., the vehicle leaves the depot exactly once.

The formulation is not complete because (8.B8) does not preclude vehicles that leave the depot 0 times from performing a tour. To correct this we should ensure that any sub-tours that do not include the depot are not allowed. This would conclude the formulation because then every vehicle has only two options: execute either a single depot-based tour or no tour at all.

Sub-tours are eliminated as with the TSP by forcing the auxiliary variables $\{u_i : i = 1, \ldots, N\}$ to increase along any tour links that do not touch the depot. Because a link belongs to a tour if and only if $\sum_m x_{ijm} = 1$ (and $\sum_m x_{ijm} = 0$ otherwise) the desired effect is achieved by rewriting constraints (8.B3) and (8.B4) as follows:

$$u_j \geq u_i + N \sum_m x_{ijm} - (N-1), \forall i, \forall j \neq 0; \tag{8.B9}$$

$$u_0 = 1. \tag{8.B10}$$

The above constraints (8.B6)-(8.B10) ensure that the VRP tours are geometrically feasible. To conclude the formulation we must also ensure that the capacity constraint is met for every vehicle, i.e., that all vehicle tours satisfy:

$$\sum_{i,j} x_{ijm} d_j \leq V_m, \forall m. \tag{8.B11}$$

The above formulation is complete for a "static" VRP with only the spatial dimension. Sometimes, the VRP problem also involves data and constraints related to the time dimension, e.g., specifying a maximum tour time, T_m, or requiring a vehicle dispatch window $[T'_m, T''_m]$. Other data for the problem include the time when vehicle dispatches can start (we assume $t = 0$ without loss of generality) and the relevant vehicle task duration measures: a service time s_{im} for each customer i and vehicle m, and a travel time t_{ijm} for each ordered customer pair $(i,\ j)$ and vehicle m. The time constraints can be "hard" or "soft." In the latter case violations are penalized at some rate, e.g., rate c of Table 8.2 for violating the tour time limit T_m.

Time-based constraints can also be introduced using a set of non-negative, continuous decision variables $\{v_{im} : i = 0, 1, \ldots, N; \forall m\}$ that represent the times at which vehicle m begins service at customer i. In the case of the depot $(i = 0)$ the variable shall represent the time at which the vehicle leaves the depot. The time-based constraints turn out to be simpler (linear in fact) if we formulate the problem more generally — allowing vehicles to hold for arbitrary amounts of time before beginning customer service, or leaving the depot.[4]

Feasible sets of start times are now defined by means of linear constraints. Note first that if $x_{ijm} = 1$ then the start time at j must satisfy $v_{jm} \geq v_{im} + s_{im} + t_{ijm}$. The slack in this inequality is the holding time. Also, note that if $x_{ijm} = 0$ then v_{jm} and v_{im} can take any non-negative values as they do not appear in any constraint. These conditions for $x_{ijm} = 0$ and $x_{ijm} = 1$ can be unified by defining a large number B (large compared with the largest possible v_{im}) and requiring instead:

$$v_{jm} \geq v_{im} + B(x_{ijm} - 1) + s_{im} + t_{ijm}, \quad \forall i, \forall j \in \{1, \ldots, N\}, i \neq j, \forall m \tag{8.B12}$$

$$v_{im} \geq 0, \quad \forall i, \forall m. \tag{8.B13}$$

These equations work because if $x_{ijm} = 0$ then (8.B12) is redundant with (8.B13) (as we wish) and otherwise $v_{jm} \geq v_{im} + s_{im} + t_{ijm}$.

We now use these definitional constraints to express the time when vehicle m returns and enters the depot. Again, we allow for holding time before entering the depot, so the entering time, $\tilde{T}_m$, is defined by:

$$\tilde{T}_m \geq v_{im} + (s_{im} + t_{i0m})x_{i0m}, \quad \forall i, m. \tag{8.B14}$$

These inequalities make sense for the last customer in the tour of vehicle m (i.e., where $x_{i0m} = 1$). The inequalities would make sense collectively if the remaining instances of (8.B14) could be ignored. Fortunately this is the case because when a customer is not the last then $x_{i0m} = 0$ so the right hand side of (8.B14) is just v_{im}, which can be (i) as low as zero if the customer is not in the tour (i.e., $\sum_j x_{ijm} = 0$), since as we have just seen this variable

[4]The formulation is more general because as we shall see, if one eliminates the holding times from any solution found with this method the solution remains feasible.

is not constrained by (8.B12), or (ii) surely smaller than $\tilde{T}_m$ if customer i is indeed in the tour (i.e., $\sum_j x_{ijm} = 1$), but not the last one, because (8.B12) ensures that v_{im} increases monotonically along the vehicle tour.

In view of the above, a hard constraint on the tour time can now be written simply as:

$$\tilde{T}_m \leq T_m, \quad \forall i = 1, \ldots, N, \forall m. \tag{8.B15}$$

If the constraint is soft, we should define the excess time $e_m = [\tilde{T}_m - T_m]^+$ and add to the objective function the term $c\Sigma_m e_m$. This is done with a common trick that uses the following linear inequalities as the definitional constraints:

$$e_m \geq \tilde{T}_m - T_m, \quad \forall m \tag{8.B16}$$

$$e_m \geq 0, \quad \forall m. \tag{8.B17}$$

The trick works because at optimality one of the two constraints associated with any m must be satisfied as a pure equality — otherwise, the term $c\Sigma_m e_m$ of the objective function could be reduced by lowering the corresponding e_m. And clearly, if one of the two constraints associated with an m is satisfied as a pure equality then $e_m = [\tilde{T}_m - T_m]^+$.

Finally, note that dispatch time windows $[T'_m, T''_m]$ can be enforced very easily. Simply, use

$$T'_m \leq v_{0m} \leq T''_m, \quad \forall m. \tag{8.B18}$$

The above mixed-integer linear program formulations of the VRP can be solved by construction heuristics and/or integer program based approaches such as column generation. Many commercial solvers may also be used directly based on such formulations. These techniques need to be adapted to the specific constraints used in a formulation. They may yield optimal results but only for medium/small problems with about 100 points.

The problems can also be effectively attacked by meta-heuristics such as simulation annealing (SA). These techniques are appealing because the formulation and solution method do not change much if the constraints or the objective function are changed. The same general principles apply. And the techniques give reasonable results for medium/large problems if reasonable initial solutions can be found with other methods. In the case of SA the TSP procedure needs little change. Instead of a single permutation, a "state" now consists of an ordered allocation of customers to vehicles and a tour starting time. One would then assign a cost to the state by evaluating the objective function, and verify the state's feasibility by checking whether

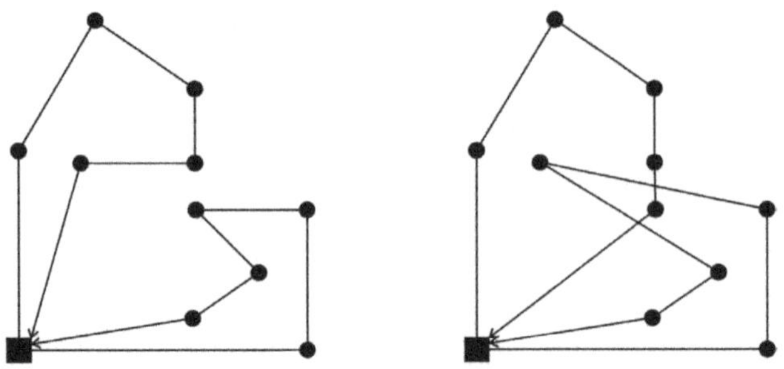

Figure 8.B3. Tours before and after a swap of points between tours.

the vehicle loads, and the tour start and ending times are in the allowed ranges. When the state is infeasible its cost is increased by a very large amount. These tasks are only a little more computationally expensive than those for the TSP.

The SA algorithm would work as before. One defines perturbations, which can be swaps of points (also called "customers") within a tour, or swaps of groups of customers among tours, or changes in the tour start times. Figure 8.B3 shows the result of swapping the last customer of the tour on the left with the middle customer of the tour on the right. It should be clear that any state whatsoever can be reached from any other state if one uses a proper sequence of swaps. Therefore, the SA approach with random swaps should (theoretically) work. In practice, experience with SA has been good with VRP problems as large as ~1,000 points. For larger problems, SA can be used as a fine-tuning tool with a large value of its parameter "*a*." A demonstration of this approach can be found in Robuste *et al.* (1990), which applied the SA annealing algorithm to a problem with about 200 points).

As is explained in the chapter's text and we'll further see in Chapter 9, many transit problems can be cast as a VRP-like problem that can be solved or fine-tuned with SA. This technique can be quickly mastered and applied. The case study in Robuste *et al.* (1990) took less than 1 week from conception to completion.

More Information

The following elementary readings could be of use. Section 10.9 of Press *et al.* (1987, pp. 326–334) describes simulated annealing in the context of the TSP, and shows some computer code. A short description can also be found in Appendix B of Daganzo (2005). Section 4.5.2 of this reference (*Fine-tuning Possibilities*) summarizes the case study in Robuste *et al.* (1990).

Chapter 9

Management — Staffing

We turn now our attention to driver management plans. These plans are usually developed in two phases: (i) job definition and (ii) staffing. The objective of the job definition phase is identifying a set of feasible driver jobs that cover the runs with as little wasted time as possible. Feasibility is dictated by legal and human factors requirements. The objective of the staffing phase is populating these jobs on an ongoing basis with actual drivers, i.e., designating a specific individual for each job, each calendar day. As occurred with buses we will find that with good management the amount of waste is relatively small. Therefore, the assumption made in the planning stage (that driver costs are directly proportional to the number of driver-hours in the schedule) is reasonable.

The job definition phase is technically very similar to the run identification process described in Chapter 8. Recall that the run identification process included two steps: a cutting of the schedule into bus tasks and a covering of these tasks with bus runs; see Figure 9.1(a). The job definition problem (JDP) is also solved in two parallel steps involving cutting and covering; see Figure 9.1(b) — one first cuts the bus runs into the smallest possible driver tasks and then covers these tasks with larger jobs.

Not everything is identical, however. A simplifying feature of the current problem is that workers, unlike buses, are assumed to be versatile enough to be able to cover any task. On the other hand, people's jobs (i.e., the set of tasks they do), unlike bus runs, cannot be of arbitrary duration. This complicates matters. The feature is modeled by introducing a menu of work *shifts*, i.e., time windows of specific duration that can be worked by single individuals for a set pay, and worker classes that can work different shift types. After some preliminaries in Section 9.1, Sections 9.2, and 9.3 explore how shift durations affect cost. The latter section will also show how the

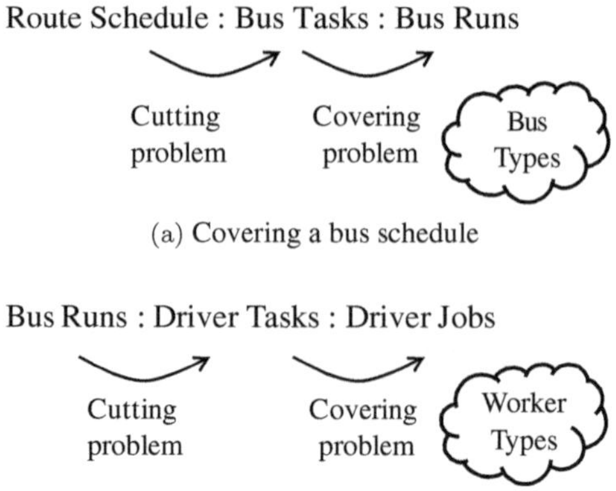

(a) Covering a bus schedule

(b) Covering the bus runs

Figure 9.1. Bus schedule covering and bus run covering.

numerical algorithms described in Chapter 8 can be slightly modified to define sets of driver shifts, and jobs within each shift, that cover all the tasks with as little waste as possible.

The staffing phase populates the identified shifts (and corresponding jobs) with specific individuals. This is considerably more complicated than populating bus runs with buses because people cannot work 7 days a week and individuals may like some weekly job combinations more than others. For this reason, staffing a set of shifts is not a trivial matter. Some strategies to solve the problem efficiently and fairly will be explored in Section 9.4.

9.1. Introduction

The result of the schedule-covering methods from the last chapter is a series of bus runs. These are assumed given in this chapter. Each run, $r = 1, 2 \ldots,$ is a continuous time–space trajectory beginning and ending at a bus depot. Given for each run is the time at which it begins, B_r, the time at which it ends, E_r, and the run's duration, $S_r = E_r - B_r$. These features can be displayed for all the runs, without reference to the runs' spatial aspects, with a Gantt-like chart (see Figure 9.2).

Also given is what drivers can do. It will be assumed that drivers can work continuously during their shifts. This is perfectly reasonable if the bus schedule is designed with enough slack at the termini (for driver breaks) to ensure that any combination of bus tasks would include the required rest. It shall also be assumed for the moment that individual drivers can only work a shift of a given type (indexed as type i), and that shift types are

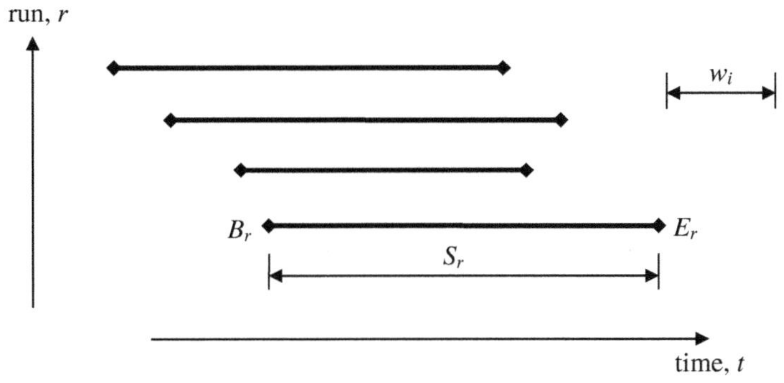

Figure 9.2. Multiple bus runs.

characterized by their durations, w_i. We will recognize that bus runs can be much longer than shifts, as Figure 9.2 schematically indicates. Often, the duration of a run is on the order of 16–20 h but drivers cannot work safely for more than 5–8 consecutive hours.

Given as well is how drivers are paid. It is assumed that drivers are paid by shift worked and that overtime is allowed. Each shift type i has as a corresponding wage rate (\$/time), which is paid for the full shift's even if the job takes less time. This wage rate is not affected by the time of day; only by the shift duration. If a job takes longer than the duration of the shift a premium is added to the standard rate for overtime.[1] When the wage rate and overtime premiums are the same for all shift types, we shall use monetary units so that the wage rate is 1. The overtime premium will be denoted $\pi > 0$, so that the overtime wage rate becomes $(1 + \pi) > 1$.

The next two sections examine how this compensation structure affects the agency's cost, compared to a lower bound in which drivers can be hired at the base rate "1" for any length of time. This pay scheme produces a lower bound because it allows the agency to cover every run at the lowest rate without any wasted idle time — in our units the lower bound is the run's duration. The comparison is of interest because the lower bound compensation structure is consistent with what we had in the planning chapters in this book, where it was assumed that the agency's cost is the product of the run time and the wage rate, as if there was no waste.

[1]The overtime part of the job can either be done by the worker assigned to the job or by another worker willing to work for a short while. This is common. Many transit agencies, such as the Champaign-Urbana District of Mass Transit, keep a pool of part-time on-call drivers who have flexible schedules and are paid at a higher rate.

9.2. Independent Runs

This section explores how the worker compensation structure affects the agency's cost in the suboptimal but simple case in which the agency determines driver jobs and shifts for each of its runs separately and independently. The scenario is of interest because it yields insights that apply generally. It will be found that if the agency has access to just a few shift types the agency's cost can be close to the lower bound.

9.2.1. *A Worst-Case Compensation Structure*

To start consider a generic run of duration S_r, assuming a compensation structure with only one shift type (e.g., $w = 8\,\text{h}$) and no overtime allowed ($\pi = \infty$). This is a worst case because the compensation structure affords no flexibility. Note that in this extreme case the cost of providing service increases with S_r as a step function with square steps of size w, i.e., cost is described by the curve: $w\lceil S_r/w \rceil$, where the half brackets denote the ceiling operation. The lower bound is S_r, which is the straight line tangent to the cost curve from the bottom. Therefore, the vertical separation between the two curves, $\$_w$, is the cost difference $w\lceil S_r/w \rceil - S_r$, which plotted as a function of S_r has the saw tooth shape of the curve in Figure 9.3. Note, it vanishes whenever S_r is an integer multiple of w. Since the wage rate is 1, this curve also represents the time drivers are not working and the cost wasted by the agency.

Now consider how this expression may affect an agency that has to cover multiple runs. As the figure shows, the extra cost of any of these runs will be low (high) if S_r is slightly below (above) a multiple of w. The figure also shows that the average extra cost across the whole range of run lengths is $w/2$. Therefore it is reasonable to expect the average extra cost across all of an agency's runs to be comparable with $w/2$ — since run durations are usually not designed with w in mind and therefore can lie anywhere along the time line. This estimate indicates that with the worst case compensation structure an agency would likely waste as idle time about one half of a shift per bus run. Since practical bus runs cannot be much longer than twice the duration of a shift, it becomes apparent that the percentage of time that drivers would be idle in this worst case scenario should be on the order of 20%.

Fortunately, this waste can be greatly reduced with two compensation plan features that are often used: overtime and multiple shift types. The effect of these improvements is examined in the next two subsections.

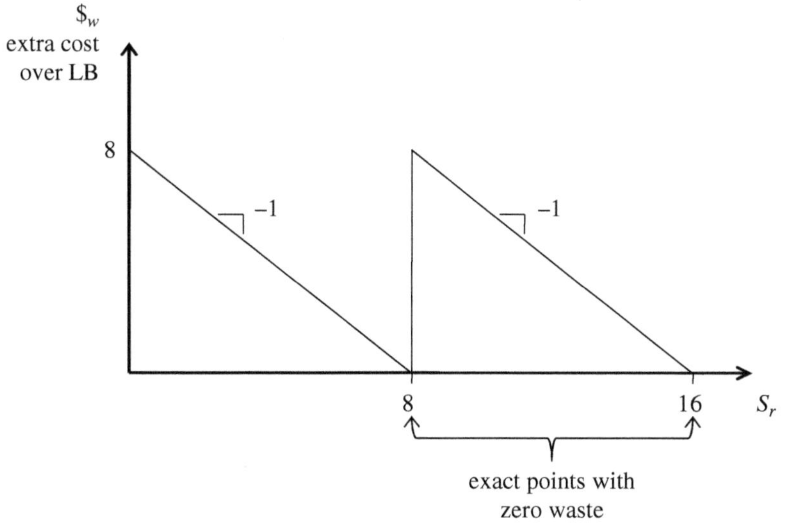

Figure 9.3. Extra cost due to driver shifts for the case; example with $w = 8$ h.

9.2.2. *Overtime*

It is now assumed that the compensation plan allows for jobs longer than a shift, where overtime is remunerated at the higher rate $(1 + \pi)$ [\$/time]. It is also assumed that the agency can fill very short jobs at this rate, perhaps with a separate pool of on-call drivers.

It will be convenient in what follows to use $s(x|y)$ for a generic saw tooth function such as the one of Figure 9.3 with amplitude and intervals of size y. In the figure $y = 8$ h. An expression for this function is

$$s(x|y) \equiv y\lceil x/y \rceil - x. \tag{9.1}$$

With this notation, we can write the extra cost of the previous section as $\$_w = s(S_r|w)$.

Let us now explore how the agency should exploit its added paying and hiring flexibility to cover a run of duration S_r. Again, the solution is trivial if S_r is an integer multiple of w because then S_r/w shifts cover the interval with no waste, and $\$_w = 0$. Therefore, we focus below on the non-trivial case.

To think about this it will be useful to define each solution of the problem by a partition of the run into a set of non-overlapping time intervals that are covered by ordinary workers at the basic rate, and a complement that is covered at the premium rate by either flextime workers or ordinary workers on overtime. For any such solution, the variable t $(t \leq S_r)$ shall denote the total amount of work time that is covered at the basic rate and $S_r - t$, the amount covered at the higher rate.

To start, let us find the optimum agency cost conditional on t. Note that for any given t, the cost of the ordinary work is minimized if this work is continuous; i.e., if it is timed to occur during a single interval without any gaps. The cost of the flextime drivers, on the other hand, is not affected by how or when their work occurs, only by the amount. Therefore, for any given t, the agency will achieve the least cost if the ordinary work is done continuously, e.g., at the beginning of the run during the time interval $[0, t]$. Then the cost of the ordinary workers is $t + s(t|w)$ and the cost of the flextime drivers, $(S_r - t)(1 + \pi)$. Thus, the combined cost for the given t would then be: $t + s(t|w) + (S_r - t)(1 + \pi) = S_r + s(t|w) + (S_r - t)\pi$. Now, since the lower bound to total cost is S_r, it follows that the optimum extra cost conditional on t is: $s(t|w) + (S_r - t)\pi$. In this expression, the first term corresponds to wasted idle time and the second to premiums due to flextime or overtime.

To conclude the optimization, let us now look for a value of t ($t \leq S_r$) that minimizes the conditional extra cost. This value is found by solving:

$$\min_t\{s(t|w) + (S_r - t)\pi : t \leq S_r\}. \tag{9.2}$$

A little thought (see below) shows that the optimum of this problem is either the largest multiple of w that does not exceed S_r, which we denote $t_1 \equiv w\lfloor S_r/w \rfloor$, or else at $t_2 \equiv S_r$. It can then be seen that the optimum extra cost is

$$\$_w(S_r) = \min\{s(S_r|w), [w - s(S_r|w)]\pi\}. \tag{9.3}$$

Proof: The result obviously holds if S_r/w is an integer as in that case the optimum solution is simply $\{t_1 = S_r; \$_w = 0\}$ which agrees with (9.3). Therefore, we now consider the case where S_r/w is not an integer. Note that the objective function in (9.2) declines with t except where t is an integer multiple of w. Therefore the minimum is either an integer multiple of w or $t = S_r$. Consider now the first possibility. Since t is now an integer multiple of w, $s(t|w) = 0$ and the first term of the objective function can be eliminated. Thus, the objective value is $(S_r - t)\pi$, which declines with t. The best t must then be the largest possible, i.e., $t_1 = w\lfloor S_r/w \rfloor$. We now rewrite this in terms of the ceiling function so (9.1) can be used. Since S_r/w is not an integer, the ceiling and floor functions satisfy $\lfloor S_r/w \rfloor = \lceil S_r/w \rceil - 1$. Thus, $t_1 = w\lceil S_r/w \rceil - w$. Now note from (9.1) that this is the same as $t_1 = S_r + s(S_r|w) - w$. Then, entering with this value in the objective function and remembering that $s(t_1|w) = 0$ we obtain the objective

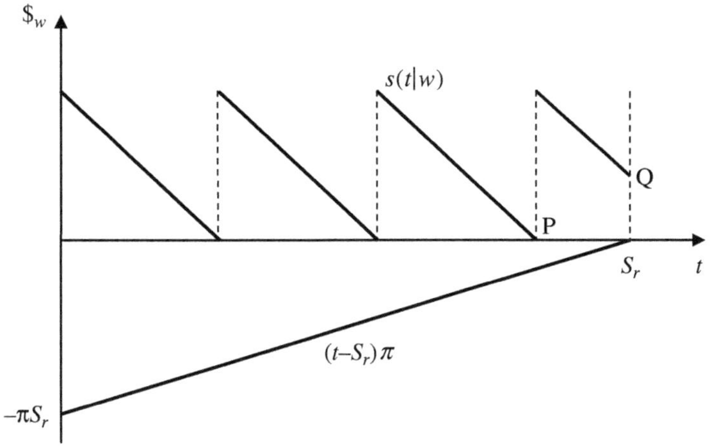

Figure 9.4. Wasted shift cost and overtime cost as functions of $t \le S_r$.

value: $[w - s(S_r|w)]\pi$. Now consider the second possibility, $t_2 = S_r$. In this case the objective value is $s(S_r|w)$. Thus, the overall optimum is $\$_w(S_r) = \min\{s(S_r|w), [w - s(S_r|w)]\pi\}$. □

The solution we have developed can also be understood and found graphically (see Figure 9.4). The top curve in the figure depicts the first term in the objective function of (9.2), $s(t|w)$. The bottom curve depicts the negative of the second term, $(t - S_r)\pi$. Accordingly, the objective function is the vertical separation between the two curves. Since only values of t in the feasibility interval should be considered the figure only depicts the curves for $t \in [0, S_r]$. The optimum occurs where the vertical separation between the two shown curves is minimal, i.e., at the first point of contact if the bottom curve is slid up until it touches the top curve. The figure clearly reveals that only two outcomes are possible, and that the particular point is determined by the slope of the bottom curve, π. More specifically, one can see that contact occurs first at the point P whose time coordinate is the largest multiple of w that does not exceed S_r (i.e., t_1) if π is low, and at the point Q whose time coordinate is S_r (i.e., t_2) if π is high. This matches the analytical result.

Note that the first term on the right side of (9.3) arises by covering the whole run with the least possible number of shifts, paying only ordinary rates. This is the same strategy as for the worst case scenario in Figure 9.3. As in that case, the excess cost term represents wasted idle time. The second term arises if one covers as much of the run as possible with complete consecutive shifts, and uses flex- or overtime work, paid at the premium rate, for the

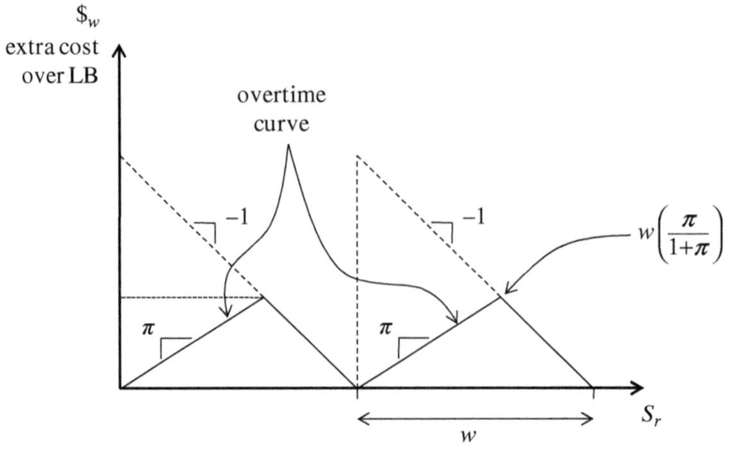

Figure 9.5. Extra cost under driver shifts and overtime.

rest. In this case, the cost term is the extra cost paid (due to premium rates).

Figure 9.5 overlays this second term, labeled "overtime curve," on the cost curve of Figure 9.3, which represents the first term. The actual cost is the lower envelope of the two curves. Dotted lines are used to denote the parts of the old curve that are improved by the additional flexibility.

Note from the vertical separation between the dashed and solid lines that flextime work reduces cost the most when the bus run slightly exceeds the duration of an integer number of shifts. Unless this integer is zero, i.e., the run is very short which is an unlikely occurrence, the required flex-work can be fulfilled by extending the last shift to include overtime. So there is no real need to have a pool of extra drivers in this case unless overtime is extremely unpopular with them.

The figure also reveals how the maximum and average cost of a run of random length is affected by the extra flexibility. The maximum extra cost is experienced at the cusps of the lower envelope. These occur at points where the two terms of (9.3) are equal. Solving for $s(S_r|w)$ we find that $s(S_r|w) = w[\pi/(1+\pi)]$ at the cusps. Since the two terms of (9.3) are equal, the found value is also the maximum extra cost — whereas before it was w. The average excess cost across many runs should be about 50% of the maximum for the same reasons as before. Thus, we see that the availability of overtime reduces the maximum and average agency costs per run by a factor of $\pi/(1+\pi)$.

This can be quite a significant reduction if π is smaller than 1. For example, if $\pi = 1$ and $w = 8\,\text{h}$, the average amount of waste per run would have been 4 h in the worst case scenario and 2 h if overtime is allowed. If bus runs were to average 12 h, waste would have been reduced from 1/3 to 1/6 of the time required. The next subsection examines how this kind of waste can be further reduced if the agency can use shifts of two (or more) different durations while still paying the basic wage rate. Section 9.4 will discuss how, perhaps counterintuitively, drivers can benefit from working shifts of multiple durations without being compensated at a higher rate. Since both the agency and the drivers benefit, strategies of this type are commonly used in practice.

9.2.3. *Multiple Shift Types*

Let us assume now that the agency can use multiple shifts of different durations (w_i, $i = 1, 2, \ldots, I$), and that the cost of a shift is the product of its duration and a common wage rate. As before, we choose the monetary unit so the numerical value of this basic rate is 1.

Before starting let us call the set of intervals in a run that are served by shifts of the same type a "part" (of the run). Obviously, there are at most I parts in a run and they cannot overlap. The subset of points in a run that are not included in any of its parts is called the "remainder." The remainder must be covered by overtime. The total duration of the part that is to be covered by shifts of type i (i.e., the sum of the interval lengths) is denoted t_i. By definition, $\Sigma_i t_i \leq S_r$ so the duration of the remainder is $S_r - \Sigma_i t_i$.

Fortunately for us, the problem simplifies because, as should perhaps be intuitive, there always is a minimum cost solution where every part is a single interval, i.e., where all shifts of the same type are served consecutively without gaps in between. This is true because (i) any solution with multi-interval parts has a corresponding solution with the same $\{t_i\}$ that is made up of single-interval parts; and (ii) the cost of covering this second solution cannot be any greater. Statement (i) is true because the single-interval parts can always be placed consecutively at the beginning of the run without exceeding the run's duration. Statement (ii) is also true because any combination of shifts and overtime that can cover a multi-interval part can also cover a single-interval part with the same t_i, since the latter is more compact.

The reduction of parts to single intervals is useful because the cost associated with any solution $\{t_i\}$ is given by formulas that have already been developed. Recall that the minimum cost of covering one of these intervals is $s(t_i|w_i)$ monetary units above the lower bound, t_i. Thus, the combined extra cost for all the intervals is $\Sigma_i\, s(t_i|w_i)$. And since the extra cost of covering the remainder with overtime is $(S_r - \Sigma_i t_i)\pi$, the total extra cost associated with our intervals is $\Sigma_i s(t_i|w_i) + (S_r - \Sigma_i t_i)\pi$. Thus, both the optimal set of intervals and the excess cost $\$_w(S_r)$ that arises, as a result, are found by solving:

$$\$_w(S_r) = \min_{\{t_i, \forall i\}} \{\Sigma_i s(t_i|w_i) + (S_r - \Sigma_i t_i)\pi : \Sigma_i t_i \leq S_r\}. \tag{9.4}$$

Note, in the special case where $I = 1$, problem (9.4) reduces to (9.2). Note as well that once the optimal $\{t_i\}$ has been obtained, the shifts allocated to each interval can be found with the method of Figure 9.4 by treating each interval as if it was a single run of duration t_i.

Given a run's duration and a set of shift durations problem (9.4) can be solved numerically, e.g., by first writing out its Lagrangian function as discussed in Section 2.3. It can also be solved graphically with a simple manual procedure that is based on the concept of subadditive functions (see Appendix 9.A). Analysis with this procedure shows that waste can be reduced quite dramatically by introducing just one extra shift type of relatively short duration, especially if it is not an integer submultiple of the larger shift.

This is now illustrated with an example solved with an intuitive heuristic method. The example has two shift durations, $w_1 = 8\,\text{h}$ and $w_2 = 3\,\text{h}$. The solution found in this manner turns out to be optimal since it is obtained as well with the exact graphical method (see appendix).

To start, enumerate all the possible run durations that can be covered exactly at zero extra cost with an integer number of shifts. Since these durations must be integer, we consider each integer value of S_r and see whether it can be formed by some shift combination. This approach quickly reveals the following run duration list (parenthetical numbers after each equal sign denote the shift combination): $3 = (3)$; $6 = (3 + 3)$; $8 = (8)$; $9 = (3 + 3 + 3)$; $11 = (8 + 3)$; $12 = (4 \times 3)$; $14 = (8 + 3 + 3)$; $15 = (5 \times 3)$; and every succeeding integer.

Run durations not included in this list require either overtime or idle time and will exhibit some extra cost. If the duration barely exceeds one of the times on the list then, as occurred in Figure 9.5, overtime is the better option; otherwise, idle time is better. Figure 9.6 depicts by means of a solid line the extra cost curve, $\$_w(S_r)$, obtained by comparing these two options. This solution happens to be the optimum. The dotted curve is the extra cost curve of Figure 9.5, without the short shifts. Note the considerable reduction in cost.

The average extra cost across all runs, $E[\$_w]$, is the average height of the extra cost curve weighted by the run durations. If we use $F(x)$ for the

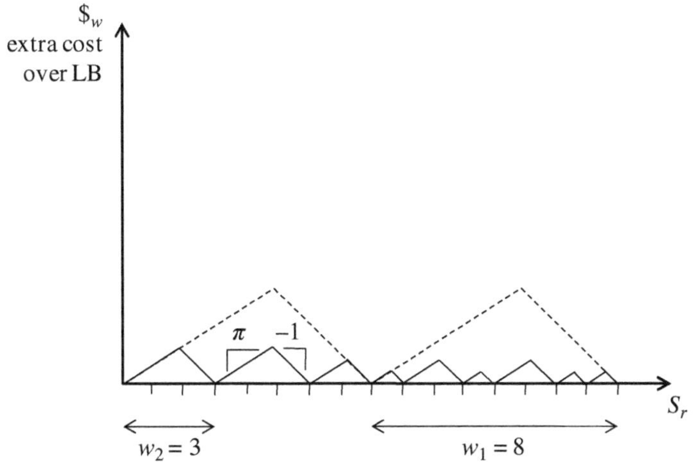

Figure 9.6. Effect of multiple shift types.

cumulative distribution function of these durations, and assume a 24 h day, the formula is

$$E(\$_w) = \int_0^{24} \$_w(x)dF(x), \tag{9.5}$$

where $\$_w(x)$ is the extra cost curve given by the solution of (9.4) and depicted as an example in Figure 9.6.

It can be seen from the figure that if the runs are contained in the range $S_r \in (8, 16)$, the reduction is very large regardless of $F(x)$. For example, if $F(x)$ is uniform in (8, 16) the average extra cost is simply the area under the curve in (8, 16) divided by 8, which is $0.75\pi/(1+\pi)$ hours. Numerically, this yields 0.375 h if $\pi = 1$, and 0.75 h if overtime is not allowed. Now, if bus runs were to average 12 h, waste would only amount to $0.375/12 = 3\%$ of the total worked time. And even if overtime was not allowed, waste would only amount to $0.75/12 = 6\%$ of the worked time.

This simple example illustrates that by introducing short shifts it is possible to almost completely eliminate waste. Thus, it is reasonable, as we did in the planning part of this book, to express total driver cost as a fixed multiple of the total bus-hours of service needed.

This favorable result arises even though the agency has not been allowed to define driver jobs that span more than one run. But this is not necessary. The next section relaxes this assumption by looking at all the runs together. This allows waste to be reduced even further.

9.3. Multiple Runs

In practice agencies should allow drivers to cover more than one run, duly recognizing the cost of switching, as this extra flexibility can reduce waste even further. The benefit arises from the pooling effect, much as it did in Chapter 8 when bus runs were allowed to cover multiple routes. Unfortunately, as occurred in Chapter 8, the problem now becomes more complex due to the costs associated with switching. For this reason, the section will present both, an algorithm to obtain a practical solution recognizing the full detailed complexity of the problem (Section 9.3.1), and a simplification that yields a lower bound to cost and some qualitative insights (Section 9.3.2). This lower bound can be considered in combination with upper bound (9.5) to get an idea of possible costs when the details are unknown, e.g., when exploring future possible compensation options.

9.3.1. *Solution Algorithm*

The process includes the two steps of Figure 9.1. We first cut the runs into indivisible driver tasks, indexed by p. These tasks are to be combined into jobs, which in turn must fit into shifts. For maximum flexibility these tasks should be as short as possible, recognizing the practical constraints that apply to the agency. Some agencies may allow drivers to be swapped anywhere along a route, whereas others may require switches to be done at designated interchange points. In either case, the process is rather straightforward. The resulting tasks can be displayed graphically in a Gantt chart, as we did for bus tasks in Chapter 8. Figure 9.7 shows an example.

The goal is to group the tasks of a problem into jobs (i.e., sets of tasks) that can be performed efficiently; i.e., using as few shifts and as little overtime as possible. Figure 9.7 displays as an example a job that could be covered without overtime (tasks 1, 7, and 10) and the corresponding shift. Data for the problem would include the information on the Gantt chart, including the task durations, t_p and the time gap, $t_{pp'}$ between the end of task p and the beginning of p'. Given too would be the basic parameters of the compensation plan, including the duration of each shift type, w_i, the total cost of each shift type due to wages, c_i (not necessarily at the same wage rate anymore), and the overtime rate, r_i (formerly, $1+\pi$). Note that flextime workers can be modeled in this framework by introducing a shift "0" of an infinitesimal duration. Also given is the non-wage cost $c_{pp'}$ of moving a driver from task p to task p'. This cost is set to infinity if the move is infeasible either because it cannot be accomplished in time or the driver is incapable of completing one of the

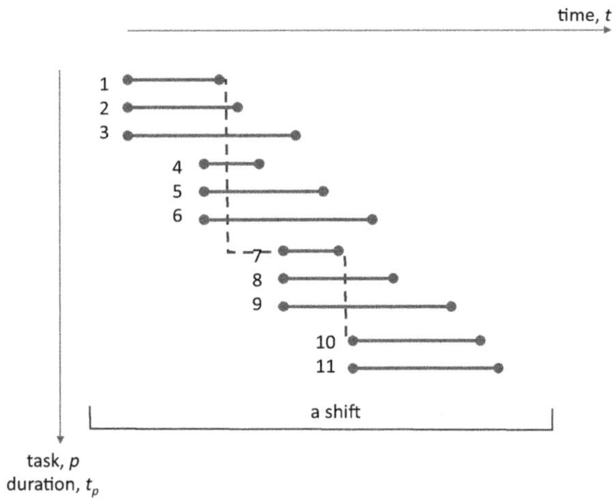

Figure 9.7. Gantt chart displaying multiple driver tasks, a driver job, and shifts.

tasks. Otherwise, the cost should capture any transportation costs required to move the driver when p and p' belong to different bus runs. The various shift types could also be restricted to start in specific time windows $[w_i', w_i'']$.

The problem's decision variables are the jobs, as well as the shift type and the shift time window associated with each job. A job starts when its first task begins and terminates when the last task ends. The job's duration may require overtime and therefore incur some extra wage cost. Extra switching costs may also arise if the job covers different runs. The goal is to find a set of jobs (and shifts) that cover all the tasks at the least possible cost. It should be intuitive from this set up that the JDP is very similar to the schedule-covering problem of the last chapter. And indeed, like the latter, it is a special case of the vehicle routing problem (VRP). Table 9.1 shows how the three problems are related.

Since the JDP is a special case of the VRP it can be approximately optimized with the simulated annealing (SA) method described in Appendix 8.B. The only change is in the procedure for finding a reasonably good initial solution for the SA method. We propose using the procedure described in Section 9.2.3 for single runs. To this end, we would define a common wage and common overtime rate for all shifts (e.g., by averaging the specific rates) and then solve one instance of problem (9.4) to find a set of jobs for each run. This can be done either numerically or graphically. Although this simplified problem uses some cost approximations, it does not relax any of the feasibility requirements of the original problem. Thus, the overall set of jobs obtained in this manner should be a reasonably good starting point for the SA procedure.

Table 9.1. Analogy among the schedule covering, job definition and VRPs.

Vehicle routing problem (VRP)	Schedule covering problem (SCP)	Job definition problem (JDP)
Customers i,j	Bus tasks k, k'	Driver tasks p, p'
Customer load size, d_i	Bus task duration, d_k	n/a
Depot	Depot	Depot
Vehicle type m	Bus type n	Shift type i
Vehicle capacity	∞ (unless bus range is limited)	∞
Cost for a leg, c_{ijm}	Cost penalty, $c_{kk'n}$ ($c_{kk'n} = \infty$ if impossible)	Non-wage cost of a move, $c_{pp'}$ ($c_{pp'} = \infty$ if impossible)
Fixed cost of a vehicle dispatch, c_m	Fixed cost of a bus dispatch, c_n (usually =0)	Cost (wages paid) for a shift of type i, c_i
Vehicle types allowed to serve a customer	Bus types suitable for a task	n/a (workers are versatile)
Time for a stop, t_i	n/a	t_p
Time between stops t_{ij}	n/a	$t_{pp'}$
Maximum tour time, T_m	n/a	Shift duration, w_i
Dispatch window $[T'_m, T''_m]$	n/a	Shift start window, $[w'_i, w''_i]$
Cost rate for exceeding T_m, c	n/a	Overtime rate, r or $(1 + \pi)$
Vehicle tour (customers served, in order)	Bus run (tasks covered by a bus, in order)	Driver job (tasks covered by a driver, in order)

9.3.2. *A Lower Bound to Cost*

This subsection shows how to obtain a lower bound to the optimum cost with a graphical procedure. The method shows at a glance where and why waste occurs, and the gains that can arise from using shifts of various durations. It also suggests how the set of bus runs can be changed to improve the final outcome. The agency can use this insight to tweak the service schedule.

The method uses two simplifications. It is assumed (i) that all tasks are infinitesimal, and (ii) that workers can switch tasks without any delay or transportation penalty. Although each of these assumptions can be realistically satisfied for some systems, the two are generally incompatible — the first assumption is satisfied only if the driver interchange points are very closely spaced along every route and the second only if there is a unique interchange point, common to all routes. Thus, the method necessarily produces optimistic results.

The benefit of these assumptions, however, is that workers can be reallocated instantaneously across runs. It is, therefore, not necessary to

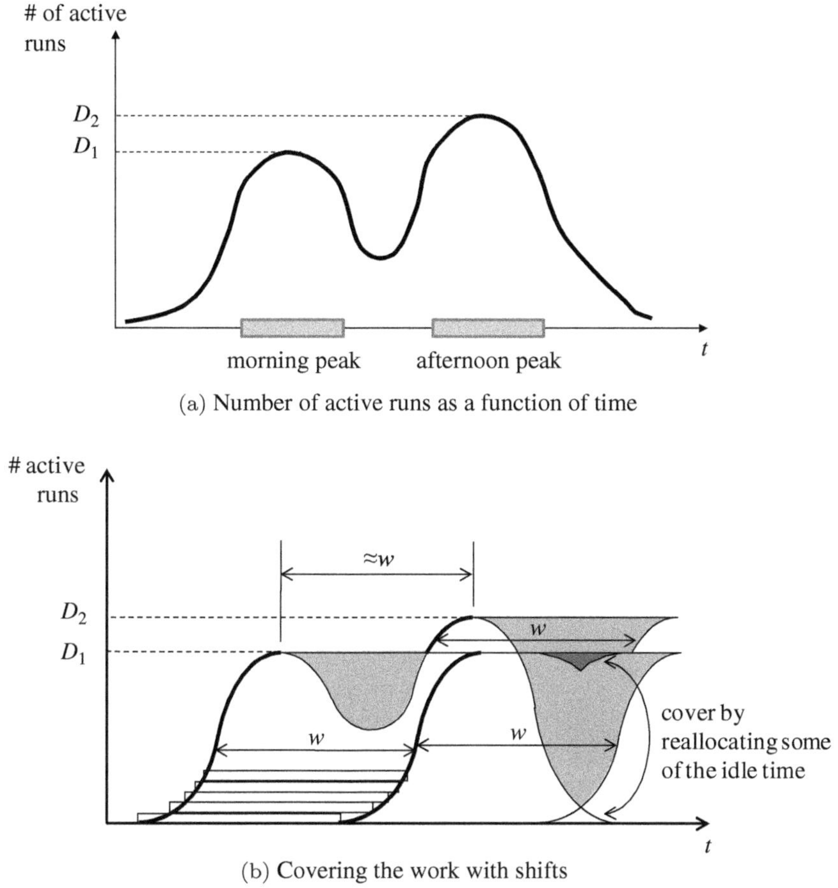

Figure 9.8. Graphical method for lower bound.

keep track of the specific jobs and the tasks they include. We can ignore the spatial distribution of tasks, and just look for a set of shifts that cover the bus runs temporally. This can be done by extracting from the Gantt data a curve with the number of active runs as a function of time (see Figure 9.8(a)), and then covering the area under the curve with rectangular strips with a height of one bus and a length of w time units to represent shifts.

The two-humped curve in Figure 9.8(a) arises from a typical service pattern with two daily peaks. In the figure, D_1 is the maximum number of active runs during the morning peak and D_2 the maximum number during the afternoon peak. Rising parts of the curve denote times when more runs start than end, and dropping parts the opposite. The area under the curve is the total amount of work (in driver-hours) that needs to be done.

Consider now how to cover this work with a single shift type and no overtime. A possible pattern is depicted in Figure 9.8(b). The pattern is characterized by the three dark curves, which mark the times when the shifts begin. It can be seen that the proposed solution uses a total of D_1+D_2 shifts, at a cost of $(D_1 + D_2)w$.

Because strips have a fixed length, w, and they must cover all the work, some have to extend beyond the dropping part of the work curve. The shaded portion of the graph represents these extensions, which correspond to wasted idle time. Note that there is a small curved triangle under the extreme right part of the work curve that could not be covered with two shifts and would, therefore, seem to require a dark rising curve of new shifts to be served. The figure introduces no such shifts because at the same time there are idle drivers that can be relocated to cover these tasks, e.g., those engaged in the jobs corresponding to the darker curved triangle directly on top.[2]

The conundrum for transit companies is that in most cases the separation between the curve's two daily peaks is typically comparable with the duration of a work shift. This necessarily leads to wasted time, since drivers who start their shifts at the beginning of the peaks must be idle for most of their shifts.

As occurred in the analysis of single runs, shifts of shorter duration can help ameliorate this problem. Figure 9.9 shows the benefit of introducing a second set of shifts with 50% duration ($w/2$). These half shifts allow us to cover the top parts of the humps where runs last less than $w/2$ time units with $w/2$ less waste per shift. Note that no driver is idle for more than $w/2$ time units even though most of the shifts are of regular length. Since much can be saved by adding just a few short shifts, some agencies use them. This is commonly done by introducing "split shifts," i.e., ordinary shifts that are split into two parts separated by an idle period. As the example illustrates, not many split shifts are needed to reap considerable savings. An agency can use these savings to raise driver wages and make the occasional split shift palatable. This idea is further explored in Section 9.4.1.2.

Finally, note that the savings should be even greater if one allows for overtime flexible work. To do this analysis, one should determine for each horizontal slice of Figure 9.9 whether the cost can be reduced by eliminating

[2]Some of the idle time at the end of the morning peak could in principle also be reallocated to cover some runs. For our particular data pattern, however, this action would merely postpone the start of some runs without eliminating any.

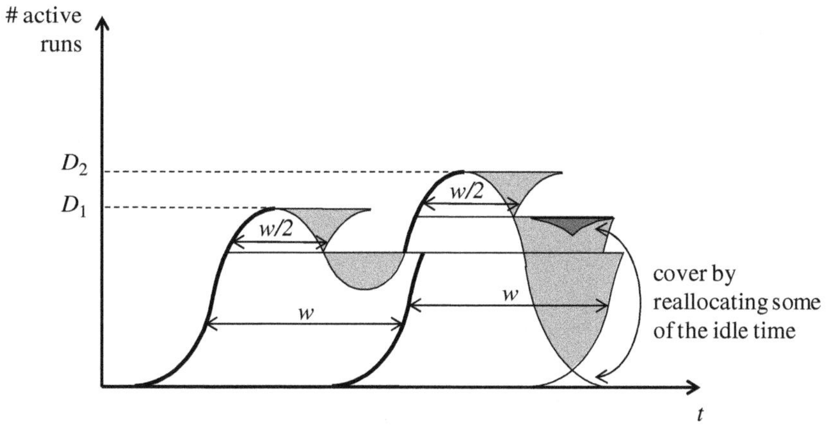

Figure 9.9. Lower bound cost with two shift types.

its last shift and replacing it with overtime. The tradeoff is the same as in Figure 9.5. Homework exercise 9.1 is used to explore it.

9.4. Staffing

Once a transit agency has determined the set of shifts that need to be worked every day of the week, it needs to populate these shifts with actual drivers recognizing that days of the week may differ. The job would be easy if drivers worked as contractors, paid by the shift, as then the agency would simply have to select from its contractor pool the drivers needed for each day.

In reality, though, it is cheaper for the agency and better for the drivers if a majority is hired as salaried employees. This creates a complication, however, because the agency must then find a way of covering a variable set of shifts with a fixed workforce. An extra complication is caused by absenteeism, e.g., due to sickness. The agency will usually plan for extra employees every day in anticipation of this problem, and then cover any shortfall using a pool of contractors that can be recalled on short notice — or by its ordinary employees under more lucrative conditions.

The purpose of this section is to show ways in which an agency can decide how to cover the shifts that need to be worked with salaried employees (and how many employees to hire), and how to deal with absenteeism. An understanding of the principles behind the methods can lead, not just to better staffing plans, but to the unveiling of compensation structures that can benefit both the agency and its drivers. Section 9.4.1 shows how to organize the salaried employees; and Section 9.4.2 how to plan for absenteeism.

9.4.1. *Populating Shifts with Salaried Employees*

We shall assume in this section that all the drivers are salaried employees as this situation poses the greatest challenge and illustrates the issues most cleanly. We shall assume that these drivers work a fixed number of hours each week, with two guaranteed off-days, and that they are paid a fixed weekly salary, so their pay is the same regardless of what they do. For simplicity, we shall also ignore vacations, although the ideas about to be presented extend trivially to that case.[3]

The set of shifts a driver is expected to work in a week is called an "assignment." The procedure defines a limited menu of assignment types, which are characterized by the type of shift worked each day of the week, and then selects enough drivers for each assignment type so the collection of shifts assigned to all drivers covers all the shifts that need to be worked in the week. We will assume that each menu entry stipulates five workdays, not necessarily the same for all entries. The pattern of days worked changes across entries. The only requirement is that all the conditions stipulated in the drivers' contract must be met by every entry. In this section, we shall assume that there are only two shift types; i.e., that each workday must include either a single continuous shift of the duration stipulated in the contract or two separate short shifts ("split shifts") that add up to said duration. This is a fairly common practice. The section will examine first a case with continuous shifts only, and then the case with split shifts.

9.4.1.1. *A Unique Shift Type*

We assume here that the JDP has been solved for every day of the week d ($d = 1, 2, \ldots, 7$) and that the solution applies to all the weeks in our problem's horizon. Thus, given are the number of shifts that need to be worked each weekday, N_d, and the times at which each shift starts.

To schedule our salaried drivers we first define the menu of assignment types, m, characterized only by the specific five days of the week that each assignment type covers. Assignment types are defined by indicator variables δ_{md} that are 1 if assignment type m includes day d and zero otherwise. Clearly, $\Sigma_d \delta_{md} = 5$ for all m. The goal is choosing the number of assignments that are to be used, by type, so the shifts made available each weekday

[3]Vacations can be handled by hiring a larger pool of people than the number obtained with the methods about to be presented and then encouraging all employees to stagger their vacations so the number of available employees stays close to the target number every day.

cover the desired target $\{N_d\}$ as closely as possible. We call this set of numbers the "staffing plan." It stands to reason that the broader the menu of assignment types the better the outcome of the staffing plan. Defining the menu, however, is more an art than a science — menus should be small, yet flexible and appealing to the work force.

The staffing plan consists of the number of assignments by type $\{a_m\}$ that are to be rostered with actual people. The sum $\Sigma_m a_m \equiv A$ is the total number of assignments and therefore the total number of people that need to be employed, i.e., it is proportional to the agency's wage cost. To find $\{a_m\}$ note that $\Sigma_m a_m \delta_{md}$ is the number of shifts supplied for day d. Now, since the required number of shifts for that day is N_d, the $\{a_m\}$ must satisfy the inequality $\Sigma_m a_m \delta_{md} \geq N_d$ for all d. And since an inequality constraint gives us the flexibility, we shall choose the $\{a_m\}$ that minimizes the agency's wage cost, i.e., minimize $A \equiv \Sigma_m a_m$. In other words, an optimal staffing plan is the solution of the following linear program:

$$\min\{\Sigma_m a_m : \Sigma_m a_m \delta_{md} \geq N_d,\ \forall\, d;\ a_m \geq 0,\ \forall m\}. \tag{9.6}$$

Note from (9.6) that we are not requiring the $\{a_m\}$ to be strictly integer. This is because agencies typically have to populate hundreds of shifts every day and therefore the optimal a_m should be large compared with 1.[4] The solution to (9.6) can always be rounded up without much impact on the weekly cost. Note as well that (9.6) always has a feasible solution.

Solutions that satisfy all the constraints without any slack are especially good. To see this note that for every menu $\{\delta_{md}\}$, the sum of the left hand sides of the constraints equals five times the objective function — since Σ_d $(\Sigma_m a_m \delta_{md}) = \Sigma_m (a_m \Sigma_d \delta_{md}) = \Sigma_m (a_m 5) = 5(\Sigma_m a_m)$. This means that any solution/menu combination that satisfies the constraints at equality has the lowest possible objective value across all possible menus. In other words, such a solution cannot be improved with other menus, as it has no slack. Sometimes it is more expedient to find a solution that satisfies this powerful condition than solving (9.6).

One important case where this occurs arises when one has only two types of days: workdays requiring N shifts, and weekend days requiring $n \leq N$ shifts. For this problem, a simple menu of assignment types with *consecutive* off-days turns out to achieve this feat — consecutiveness is stipulated since it should appeal to drivers.

The menu for this case includes seven assignment types characterized by the first day off. We use $m = 1$ for Saturday, $m = 2$ for Sunday, $m = 3$

[4]Recall the discussion in Chapter 3, Section 3.1.4.

for Monday, ... and $m = 7$ for Friday. Assignment 1 is "ordinary" in the sense that it gives people the usual weekend days off; the rest are called "special." Now note that the seven assignments from this menu, combined, jointly generate 5 shifts each day of the week including weekends. Thus, providing $n/5$ assignments of each type generates n shifts each day of the week. This shift supply covers the weekend without any slack; and it leaves $N - n \geq 0$ shifts to be covered on each workday. Fortunately, these extra workday shifts can be covered without any slack with a single set of $N - n$ ordinary assignments. Thus, the resulting staffing plan has no slack, which makes it not just optimal but perfect.

In terms of our variables, the staffing plan consists of $N-n+n/5$ ordinary assignments, and $n/5$ assignments for each of the special types, i.e., $\{a_1 = N - n + n/5$ and $a_2 = a_3 = \ldots = a_7 = n/5\}$. For example if $N = 800$ and $n = 350$, one would define 520 ordinary assignments and 70 assignments for each of the remaining types.

This concludes the staffing plan discussion. The next step, which we call "rostering," is refining the plan $\{a_m\}$ so it can be rostered with actual persons. The roster can be developed in two sub-steps. First, one takes every one of the A assignments in the staffing plan and, considering its type, matches its shifts to specific shifts from the JDP, so the shifts in the staffing plan are then endowed with start times. When done, there will be a 1:1 match between the shifts of the JDP and those of the A assignments. Since there is considerable flexibility in doing this, the agency should strive for something that drivers may like, e.g., make the shift start times within each assignment as similar to one another as possible. In addition, because some assignments may be very appealing and others not, the agency can consider partitioning the set of assignments into disjoint classes, c, of sizes A_c, with the similar combined appeal. Drivers could then be matched to classes rather than individual assignments and then rotated through the assignments in the class every A_c weeks.

Second, now that every assignment (or class) is fully defined, one must match each assignment (class) to a specific driver (set of drivers) who is (are) charged with fulfilling their fixed (rotating) assignments week after week. We call the resulting match the roster. This final match must be made in conjunction with the workforce, and this can be accomplished in a number of ways. Possibilities include lotteries; auctions; priority selection by merit and/or seniority; and combinations of these approaches. Because these processes are neither technical nor specific to transit, they are not discussed here.

9.4.1.2. *Two Shift Types*

We consider here ways of populating JDP solutions that include regular shifts (say 8 h) and halves (say 4 h) using a workforce that can drive the occasional split shift. Assignments will be assumed to include five workdays and 2 off-days, and each workday to consist of either a regular shift or a split shift. Furthermore, to ensure that all the half shifts in a day can be covered with split shits, the JDP solution is also assumed to have been constrained with its time window feature to produce an equal number of half shifts in the early and late parts of the day, with the condition that every early half shift ends before the beginning of every late half shift. Figures 9.8 and 9.9 illustrate that this should work well in practice. As before, we will first find a staffing plan and then a roster.

To find the staffing plan, one first extracts from the JDP solutions for each day of the week the numbers of regular and half shifts required each day, N_d and n_d. One then defines a menu of assignment types that differ by the positions and numbers of the ordinary and split shifts along the week. These assignment types are now characterized by *two* sets of indicator variables, δ_{md} and δ'_{md}. The former is 1 if assignment type m includes a regular shift on day d, and 0 otherwise. The latter is 1 if assignment type m includes a split shift on day d, and 0 otherwise. The problem to be solved is now an extended version of the linear program (9.6) that also includes the set of constraints: $\{\Sigma_m a_m \delta'_{md} \geq 2n_d, \forall d\}$. The factor "2" is needed in the right hand side because every split shift counted on the left side covers two half shifts. The solution to the problem is the staffing plan.

For the same reasons as in the previous subsection, if a solution satisfies all the constraints without any slack, then the solution is "perfect" in the sense that it is the best across all possible assignment menus. And fortunately again, the important special case with five identical weekdays and two identical weekend days turns out to have a simple solution of this type. This is demonstrated below for the special case where half shifts are used only on weekdays (and only sparingly) since this is sufficient to illustrate the idea. Homework exercise 9.2 covers other cases.

For our special case, it will be convenient to denote the numbers of regular and half shifts on workdays by N_R and N_H, respectively, and the number of (regular) shifts on weekends by n. It is assumed that both N_H and n are sufficiently small compared with N_R that $2N_H + n \leq N_R$. To do the analysis, it will be convenient to define the assignment types, m, by categories, q, with specific features. For our application, we define three categories.

The first category includes all assignment types that have five consecutive days with regular shifts and two consecutive off-days. There are seven possibilities, so this category includes seven assignment types. The pattern is summarized with the notation: $\{(R,\ R,\ R,\ R,\ R,\ 0,\ 0)_7\}$. The symbol R denotes a regular shift and the symbol 0 an off day. The first position in the list corresponds to a Monday and the last to a Sunday. The parenthesis denotes the rotation operation where all the symbols are shifted one position to the right and the last symbol is placed first. The subscript "7" indicates that the operation is done six times so the category includes seven assignment types. Note that when all seven assignments types in the category are combined into a group they produce five regular shifts every day of the week, including weekends.

The second category is $\{(R,\ R,\ R,\ R,\ S)_5\ 0,\ 0\}$, where the symbol S denotes a split shift. Note that this category includes five assignment types with the weekends off, and that all the assignment types in the category combined together produce four regular shifts and two half shifts every workday, and nothing on weekends. The third and final category is $\{(R,\ R,\ R,\ R,\ R)_1\ 0,\ 0\}$. It includes one assignment type that produces one regular shift every day of the week and none on weekends.

A set of assignment types including every member of a category q exactly once is called an assignment *group* of type q. The number of assignments in a group (the group's size) will be denoted $\{s_q\}$. For our example $\{s_1 = 7, s_2 = 5, s_3 = 1\}$. The staffing plan shall be defined in terms of assignment groups rather than individual assignment types, i.e., by the number of groups $\{b_q\}$ of each category rather than the $\{a_m\}$. Note that the number of assignments made in a plan, and therefore the number of workers, are both then given. The formula is:

$$A = \Sigma_q b_q s_q. \tag{9.7}$$

The rationale for the categories we have chosen is as follows. Category 1 is intended to cover the weekend shifts in their entirety while including some of the regular shifts every workday. Since this still leaves a very large surplus of uncovered regular shifts every workday, plus the original number of half shifts, categories 2 and 3 are then designed to be able to cover all of these *without affecting the weekends.* Category 2 is intended to take care of the half shifts and category 3 to mop up the remaining regular shifts. This logic illustrates the thinking process that can be used to create categories that will work for a specific case. For example if half shifts were more numerous it might have been useful to define a category such as $\{(R,\ R,\ S,\ R,\ S)_5$

0, 0}; or if half shifts also occurred on weekend days a category such as $\{(R, R, R, R, S, 0, 0)_7\}$.

So, let us now find the solution for our particular case. Since an assignment group from category 1 produces five regular shifts every day, it follows that $b_1 = n/5$ assignment groups of type 1 produce n shifts every day. Thus, these groups take care of the weekend, and leave $N_R - n$ and N_H shifts uncovered during workdays. Now, since a group of type 2 produces four regular shifts and 2 half shifts every workday, it follows that $b_2 = N_H/2$ groups of this type will cover the N_H workday shifts while producing $2N_H$ regular shifts. This leaves $N_R - n - 2N_H \geq 0$ shifts of this type to be covered during workdays with groups of type 3. Since each of these produces one regular run per workday, all the remaining shifts are covered by choosing $b_3 = N_R - n - 2N_H$. For example, if $N_R = 700$, $N_H = 200$ and $n = 250$ the solution would be: $\{b_1 = 50, b_2 = 100, b_3 = 50\}$. Note from (9.7) that $A = 900$.

Note as well that the generic solution to the problem has no slack and therefore cannot be improved by any other set of categories. It is perfect. Our example confirms this, since the total number of full shift equivalents that must be worked in a week $(800 \times 5 + (200/2) \times 5 + 250 \times 2 = 4,500)$ matches the full shift equivalents provided by the plan $(900 \times 5 = 4,500)$. This concludes our discussion of the staffing plan.

The rostering process, again, includes two steps. The first step matches the shifts in each group of assignments to specific shifts from the JDP so the former can be endowed with starting times. Regular shifts can be matched 1:1 with regular shifts arbitrarily, as happened for the one-shift problem. Split shifts, however, must be matched to pairs of half shifts that do not overlap. This is accomplished by only considering pairs of half shifts that include one *early* half shift and one *late* half shift from the JDP solution — this is always possible because the JDP solution includes an equal number of both, and the pairs can be formed beforehand. Furthermore, to ensure that the time gap between the two half shifts in every pair is as consistent as possible — and avoid huge gaps that could create inconveniently long work days — we recommend forming each pair by picking the earliest of the (not yet picked) half shifts of each type. In any case, once the pairs have been formed, they can be matched 1:1 with the split shifts in any way, as was done with the regular shifts.

The rostering process is concluded by matching people to assignments. And again this can be done in many ways. If equity is a concern, one can match people to sets of assignments that are balanced by difficulty. In our example, we could define a single set that would include the assignments in 1 instance of group 1, 2 instances of group 2 and 1 instance of group 3 for a total of $7 + 10 + 1 = 18$ assignments that could be rotated among 18 drivers over 18 weeks. Fifty such sets (requiring 900 drivers in total) would roster the complete set of $A = 900$ assignments of the problem. Of course, every situation is different, so the details need to be considered carefully. However, these issues are largely non-technical and therefore not covered in this book.[5]

9.4.1.3. *Dealing with Absenteeism*[6]

The above analysis assumes that the A drivers hired for the job show up for work reliably. This is not realistic, however, because human drivers get sick and experience emergencies. Furthermore, because these events are unanticipated and random, the actual number of drivers that show up for work varies from day to day. The number depends on the probability f that a driver does show up for work. The section will show that the agency cost is then approximately proportional to A/f, i.e., cost can model as a fixed multiple of the number of drivers, as was assumed in the planning chapters of this book.

To compensate for the number of absentees the agency could hire U drivers ($U \geq A$) and call them all every day. Of the U drivers, A would have shifts assigned, and the extra $(U - A)$ drivers would be designated as backups, entrusted with covering for no-shows.[7]

This strategy would work well in an ideal world with no day-to-day fluctuations (i.e., where the number of absent drivers equals the expectation on all days) as then the buffer could be precisely matched to the recurrent driver shortfall and the agency would not have to pay much for idleness. In reality, however, fluctuations exist. To deal with these the agency can increase the buffer size. Because excesses may still arise, we also assume that the agency can call additional individual drivers on very short notice at an extra cost. These drivers could be off-duty

---

[5]An analysis from a business perspective can be found in Coleman (1995).

[6]Much of this section is based on Muñoz (2002).

[7]These backups should be included in the staffing plan and rostering process. This can be easily done by introducing an extra shift of type "B," which means "serve as a back-up," and recognizing it in the developed list of assignment types (or assignment groups). Formulas for the staffing plan can be easily developed using the logic presented for the examples of Sections 9.4.1.1 and 9.4.1.2; see homework exercise 9.4.

salaried drivers willing to work overtime or come from a separate pool. They are assumed to be paid at a wage rate $r \equiv (1 + \pi)$ times greater than that of the salaried workers, but the agency only pays them if they work.[8]

The only degree of freedom in setting up this framework is the size of the buffer. Choose it too small and the premium wages paid to short term workers could skyrocket. Choose it too large and the cost of regular salaries would be large. Below we show that there is a happy medium and evaluate the ensuing extra cost — assuming that workers show up for work independently with probability f. This problem can be analyzed deterministically or stochastically. Below we will show that these two approaches yield very similar cost estimates.

Consider first the simpler approach where we ignore stochastic fluctuations; i.e., assume optimistically that the number of absentees is equal to its expectation every day. In this case we would hire $U = \lceil A/f \rceil$ salaried drivers. The buffer would then be $U - A = \lceil A/f \rceil - A$, e.g., if $A = 60$ and $f = 0.9$, we would need $U = \lceil 60/0.9 \rceil = 67$ drivers and the buffer would be 7. If we now use as the unit of currency the daily wage, then these expressions also represent the daily cost to the agency. This is our simple, deterministic cost prediction.

Now, to be more accurate and examine the impact of daily stochastic fluctuations, let P be the random number of drivers that show up for work on any given day. Given our assumptions, P is modeled as a binomial random variable with U trials and probability of success f. For a given U the expected agency cost across days would be

$$E[cost] = U + rE_P[(A - P)^+]. \tag{9.8}$$

For large U the binomial random variable P is approximately normal with mean Uf and variance $Uf(1 - f)$. Thus, the variable $(A - P)$ is itself approximately normal with mean $(A - Uf)$ and variance $Uf(1 - f)$.

Because $(A-P)$ is normal (9.8) can be expressed in terms of the standard normal p.d.f., $\phi(z) \equiv {}^1\!/_2(2\pi)^{1/2}\exp(-z^2/2)$. To do this, express $(A - P)$ in terms of a standard normal random variable Z whose pdf is $\phi(z)$, i.e., write $(A - P) = (A - Uf) + [Uf(1 - f)]^{1/2}Z$, and then insert this expression in the

[8]The notation "r" is introduced to avoid confusion with the number $\pi \approx 3.14$ that also appears in the section.

right hand side of (9.8). Then write the expectation as an integral to obtain[9]

$$E[\text{cost}] = U + r\left\{\int_{(Uf-A)/\sqrt{Uf(1-f)}}^{\infty}\left[A - Uf + z\sqrt{Uf(1-f)}\right]\phi(z)dz\right\}. \tag{9.9}$$

The value of U that minimizes (9.9), U^*, can be found numerically. The optimum value of (9.9) can then be compared with the simple deterministic prediction, $\lceil A/f \rceil$.

This is now illustrated with the previous numerical example. Recall that $A = 60$ and $f = 0.9$. If we now assume that the cost of on-call drivers is $r = 1.5$, a search for U^* using (9.9) yields $U^* = 65$. This is lower than the $\lceil A/f \rceil = 67$ drivers that would be hired in the deterministic scenario without fluctuations but higher than A. The resulting cost is 67.2 wage

Table 9.2. Cost comparison between the stochastic and deterministic approaches.

Input			Stochastic approach		Deterministic approach		
A	f	r	U^*	$cost^*$	U_{det}	$cost_{det}$	% cost diff
30	0.9	1.5	32	33.71	33	33	1.1
30	0.9	2.5	33	34.40	33	33	3.2
30	0.8	1.5	35	37.82	38	38	0.8
30	0.8	2.5	37	38.95	38	38	3.9
100	0.9	1.5	109	111.80	111	111	0.6
100	0.9	2.5	111	113.06	111	111	1.8
100	0.8	1.5	121	125.59	125	125	0.5
100	0.8	2.5	124	127.66	125	125	2.1
300	0.9	1.5	330	334.54	333	333	0.4
300	0.9	2.5	333	336.71	333	333	1.0
300	0.8	1.5	368	376.02	375	375	0.3
300	0.8	2.5	373	379.61	375	375	1.2
1,000	0.9	1.5	1,104	1,113.31	1,111	1,111	0.2
1,000	0.9	2.5	1,110	1,117.27	1,111	1,111	0.6
1,000	0.8	1.5	1,238	1,251.87	1,250	1,250	0.1
1,000	0.8	2.5	1,246	1,258.43	1,250	1,250	0.7

[9]The integral in (9.9) can solved and written in closed in terms of ϕ and its c.d.f., Φ, exploiting the fact that $\int z\phi(z)dz = -\phi(z)$. However, the result is a bit longer than the right side of (9.9) and therefore it is not given here. The resulting function turns out to have a unique minimum.

units, compared to the $\lceil A/f \rceil = 67$ units as estimated in the deterministic scenario. Thus, the cost increment caused by the random fluctuations, which is 0.2, amounts to under 0.3% of the deterministic cost. Of course, worse results would be obtained with smaller pools of drivers, but most real systems are considerably larger than what we have assumed in the example. Thus the results should be better. Table 9.2 displays this percentage for some representative values of A, f, and r.[10] The smallness of these percentages indicates that for planning purposes the salary cost, including the effect of random absenteeism, can be estimated deterministically with the formula:

$$E[\text{cost}] \approx \lceil A/f \rceil \approx A/f. \tag{9.10}$$

References

Coleman, R. M. (1995). *The 24-Hour Business*, AMACOM, N.Y.

Muñoz, J. C. (2002). "Driver shift design for single-hub transit systems under uncertainty," Ph.D. Thesis, Department of Civil and Environmental Engineering, U.C. Berkeley, CA.

Homework

9.1. Suppose a transit agency provides bus service from 5 am to 1 am daily. The number of active runs during the course of a day is shown in the following figure. Please estimate and compare the lower bounds on the driver costs in the three scenarios given below. Also, discuss the impact on the cost of variable shift lengths and overtime.

---

[10]Note from the table that if the size of the agency as measured by A is increased while holding the other two parameters constant the relative cost difference diminishes. This might be expected since the laws of large numbers tell us that the stochastic fluctuations should become negligible compared with A as $A \to \infty$, and therefore their effect should become negligible too. Note as well that if the overtime rate r is reduced ceteris paribus the relative cost also declines. This again should be expected since an optimum strategy for $r = 1$ (the lowest meaningful value) is to use flexible workers for all jobs in both the deterministic and stochastic scenarios, so there would be no cost difference between scenarios.

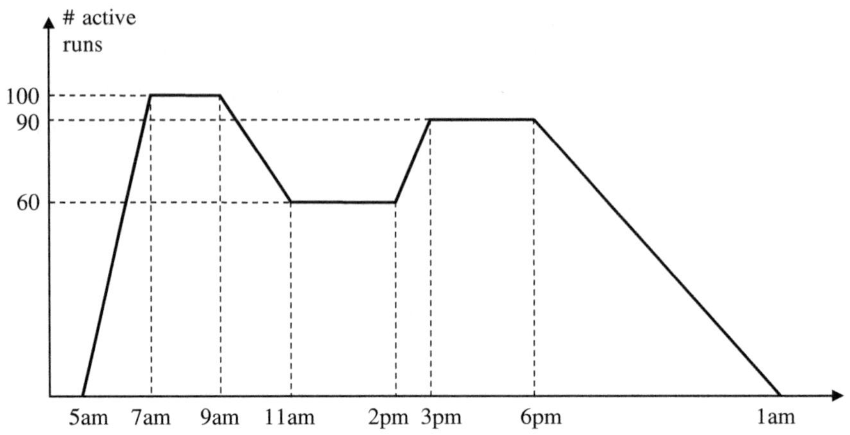

(a) All drivers work 8 h shifts and no overtime is allowed.
(b) Drivers can work either 8 h shifts or 4 h shifts, with no overtime allowed.
(c) Drivers work either 8 h shifts or 4 h shifts, and can also work overtime at pay rate $(1 + \pi) = 2.5$.

9.2. Consider a college town with such a rich football culture that it induces a large surge in transit demand every Saturday afternoon. As a result, the required numbers of 8-h and 4-h shifts are 36 and 36 on weekdays, 24 and 48 on Saturday, and 24 and 24 on Sunday. How should you staff your 40-h-per-week salaried drivers into rotations to cover all the shifts? You may need to first define a few reasonable shift types as input data and then solve a mathematical program.

9.3. Discuss how equation (9.8) would change in the following alternative scenarios, and solve for the optimal U^* numerically using the following parameters: $f = 0.8$ and 0.9, and $r = 1.5$ and 2.5.

(a) The $(P - A)^+$ idle drivers can work on other tasks or be sent home and generate a return equal to half of their salaries.
(b) In addition to the $(U - A)$ "backup" drivers that are paid regularly, additional individual drivers can be called instantaneously from a ready-pool at a higher cost rate of $r = (1 + \pi)$.

9.4. *Use the linear programming approach of Section 9.4.1.2 to define an optimal staffing plan considering absenteeism. To do this, introduce extra shift types for regular back-ups and on-call drivers. Please define all parameters and decision variables properly. Derive closed-form formulas for the staffing plan and the final cost, assuming no random fluctuations in the number of absentees.

Mini Project 3. Fleet and Crew Management[11]

The goals of this mini project are to cover a service schedule with buses, design bus runs, cover bus runs with drivers, and design driver tasks/shifts. AC Transit serves Berkeley, Oakland, and many of the surrounding areas with almost 200 routes and a fleet of over 600 buses. The area next to Berkeley's West Gate is a major transit hub served by a BART station and 15 different bus routes.

In this project, you will determine the optimal way to assign buses and drivers to three of the bus routes that serve this transit hub.

Route Characteristics

The three routes examined in this mini project are the 7, which provides service between Rockridge and El Cerrito Del Norte BART stations; the 18, which serves the Shattuck corridor and downtown Oakland; and the 52L, which provides service between El Cerrito and the UC Berkeley campus. Times to travel between their termini and to a depot are shown in Table MP3.1. The weekday schedules for these routes are shown in Tables MP3.2–MP3.4.

(a) For each route and direction determine the total *bus-hours in service* per day. This will be a lower bound used for comparison for the rest of the project. What are the peak service times for each route and direction? What is maximum number of bus loops over the day (the minimum number of buses needed) for each route?

(b) Assuming that the routes are treated independently, determine the minimal number of buses needed to cover the route if deadheading between termini is allowed. Determine the actual number of buses needed to cover the loops for each route using the greedy algorithm; identify the runs for each bus and the total bus-hours spent, including any dwell times not at a depot. If necessary, a bus can deadhead to the other end of its route to cover a loop. Compare the bus-hours and number of buses with the lower bounds from Part (a).

(c) In order to decrease the wasted bus-hours dwelling and the number of buses used, the three routes named above will use the same pool of buses to serve their routes. Determine the maximum number of buses

[11]Created by Josh Pilachowsky with input from Michael Cassidy. Relevant datasets can be downloaded from https://www.worldscientific.com/workscibooks/10.1142/10553#t=suppl.

Table MP3.1. Deadheading travel time between locations.

Deadheading Time	Rockridge BART	Del Norte BART	San Pablo/ Marin	Moraga/ Medau	Bancroft/ Telegraph	El Cerrito BART	
Rockridge BART	0:00:00	0:24:00	0:14:00	0:06:00	0:07:00	0:18:00	Route 7
Del Norte BART	0:24:00	0:00:00	0:07:00	0:25:00	0:18:00	0:08:00	
San Pablo/Marin	0:14:00	0:07:00	0:00:00	0:19:00	0:10:00	0:04:00	Route 18
Moraga/Medau	0:06:00	0:25:00	0:19:00	0:00:00	0:12:00	0:19:00	
Bancroft/Telegraph	0:07:00	0:18:00	0:10:00	0:12:00	0:00:00	0:14:00	Route 52L
El Cerrito BART	0:18:00	0:08:00	0:04:00	0:19:00	0:14:00	0:00:00	
Emeryville Depot	0:08:00	0:16:00	0:09:00	0:16:00	0:11:00	0:13:00	Depot

needed to cover the system if deadheading between all termini is allowed. Program and run an iterative covering algorithm to find the optimal arrangement of runs that minimizes the bus-hours required to serve the routes. Compare the bus-hours and number of buses with the lower bounds from Part (a).

(d) Cover the runs from your solution to Part (c). An asymmetric work rotation if offered such that each driver gets 4 days of 9-h shifts and 1 day with a 4-h shift (such that each day has a ratio of 9-h to 4-h shifts of 4:1). In addition, overtime is offered at double the wage rate. In order to avoid hiring more drivers than needed, you can also use 8-h shifts. Drivers can only be assigned to a bus at the Depot or at the Berkeley BART station. Determine the number of drivers needed, the job for each driver, and the driver-hours wasted (when a driver is not on a loop). Using a wage rate of \$25/h, what is the cost of covering the runs?

(e) Absenteeism can be a problem for transit agencies. If we assume that each driver has a 5% chance of skipping their assigned shift, determine how many backup drivers (who can cover any shift) the city must employ to ensure that there is 99% chance that all shifts will be covered for the work schedule you solved for in Part (d).

Table MP3.2. Service schedule of Route 7.

	Northbound				Southbound			
Route	Rockridge BART	Berkeley BART	El Cerrito Del Norte BART	Bus #	El Cerrito Del Norte BART	Berkeley BART	Rockridge BART	Bus #
7	6:22 AM	6:36 AM	7:08 AM		6:18 AM	6:52 AM	7:10 AM	
7	6:42 AM	6:56 AM	7:28 AM		6:36 AM	7:10 AM	7:28 AM	
7	7:05 AM	7:19 AM	7:51 AM		6:53 AM	7:28 AM	7:48 AM	
7	7:25 AM	7:41 AM	8:17 AM		7:14 AM	7:49 AM	8:09 AM	
7	7:45 AM	8:01 AM	8:37 AM		7:34 AM	8:09 AM	8:29 AM	
7	8:05 AM	8:21 AM	8:57 AM		7:57 AM	8:32 AM	8:52 AM	
7	8:24 AM	8:42 AM	9:19 AM		8:25 AM	9:00 AM	9:22 AM	
7	8:45 AM	9:03 AM	9:40 AM		8:55 AM	9:30 AM	9:52 AM	
7	9:05 AM	9:23 AM	10:00 AM		9:25 AM	10:00 AM	10:22 AM	
7	9:30 AM	9:48 AM	10:25 AM		9:54 AM	10:29 AM	10:51 AM	
7	10:00 AM	10:18 AM	10:55 AM		10:24 AM	10:59 AM	11:21 AM	
7	10:30 AM	10:48 AM	11:25 AM		10:54 AM	11:29 AM	11:51 AM	
7	11:00 AM	11:18 AM	11:55 AM		11:24 AM	11:59 AM	12:21 PM	
7	11:30 AM	11:48 AM	12:25 PM		11:54 AM	12:29 PM	12:51 PM	
7	12:00 PM	12:18 PM	12:55 PM		12:24 PM	12:59 PM	1:21 PM	
7	12:30 PM	12:48 PM	1:25 PM		12:54 PM	1:29 PM	1:51 PM	
7	1:00 PM	1:18 PM	1:55 PM		1:24 PM	1:59 PM	2:21 PM	
7	1:30 PM	1:48 PM	2:25 PM		1:53 PM	2:28 PM	2:50 PM	
7	1:57 PM	2:15 PM	2:52 PM		2:24 PM	2:59 PM	3:21 PM	
7	2:27 PM	2:45 PM	3:22 PM		2:50 PM	3:25 PM	3:47 PM	
7	2:56 PM	3:14 PM	3:51 PM		3:15 PM	3:50 PM	4:12 PM	
7	3:27 PM	3:45 PM	4:22 PM		3:37 PM	4:12 PM	4:34 PM	
7	3:58 PM	4:16 PM	4:53 PM		3:57 PM	4:32 PM	4:54 PM	
7	4:21 PM	4:39 PM	5:16 PM		4:19 PM	4:54 PM	5:14 PM	
7	4:46 PM	5:02 PM	5:38 PM		4:39 PM	5:14 PM	5:34 PM	
7	5:06 PM	5:22 PM	5:58 PM		4:59 PM	5:34 PM	5:54 PM	
7	5:26 PM	5:42 PM	6:18 PM		5:22 PM	5:57 PM	6:17 PM	
7	5:46 PM	6:02 PM	6:38 PM		5:44 PM	6:19 PM	6:39 PM	
7	6:06 PM	6:22 PM	6:58 PM		6:05 PM	6:40 PM	7:00 PM	
7	6:28 PM	6:44 PM	7:20 PM		6:35 PM	7:10 PM	7:30 PM	
7	6:55 PM	7:16 PM	7:52 PM		7:05 PM	7:43 PM	8:03 PM	
7	7:25 PM	7:46 PM	8:22 PM		7:35 PM	8:13 PM	8:33 PM	
7	7:55 PM	8:16 PM	8:52 PM		8:05 PM	8:43 PM	9:03 PM	

You are to submit a formal report describing your tasks and your findings. You may work in small groups of up to three people and submit a group report.

Table MP3.3. Service schedule of Route 18.

	Eastbound				Westbound			
Route	San Pablo Ave. & Marin Ave.	Berkeley BART	Moraga Ave. & Medau Pl.	Bus #	Moraga Ave. & Medau Pl.	Berkeley BART	San Pablo Ave. & Marin Ave.	Bus #
18	5:07 AM	5:28 AM	6:17 AM		5:29 AM	6:14 AM	6:29 AM	
18	5:38 AM	5:59 AM	6:48 AM		5:44 AM	6:29 AM	6:44 AM	
18	6:03 AM	6:26 AM	7:17 AM		5:59 AM	6:44 AM	6:59 AM	
18	6:19 AM	6:42 AM	7:33 AM		6:14 AM	6:59 AM	7:14 AM	
18	6:35 AM	6:58 AM	7:49 AM		6:29 AM	7:14 AM	7:29 AM	
18	6:50 AM	7:13 AM	8:04 AM		6:40 AM	7:30 AM	7:45 AM	
18	7:05 AM	7:28 AM	8:19 AM		6:55 AM	7:45 AM	8:00 AM	
18	7:20 AM	7:43 AM	8:37 AM		7:10 AM	8:00 AM	8:15 AM	
18	7:35 AM	7:58 AM	8:52 AM		7:21 AM	8:11 AM	8:26 AM	
18	7:50 AM	8:13 AM	9:07 AM		7:32 AM	8:28 AM	8:44 AM	
18	8:05 AM	8:28 AM	9:22 AM		7:48 AM	8:44 AM	9:00 AM	
18	8:20 AM	8:43 AM	9:37 AM		8:03 AM	8:59 AM	9:15 AM	
18	8:35 AM	8:58 AM	9:52 AM		8:18 AM	9:14 AM	9:30 AM	
18	8:46 AM	9:09 AM	10:03 AM		8:34 AM	9:30 AM	9:46 AM	
18	8:59 AM	9:22 AM	10:16 AM		8:51 AM	9:47 AM	10:03 AM	
18	9:17 AM	9:40 AM	10:34 AM		9:06 AM	10:02 AM	10:18 AM	
18	9:32 AM	9:55 AM	10:49 AM		9:21 AM	10:17 AM	10:33 AM	
18	9:47 AM	10:10 AM	11:04 AM		9:36 AM	10:30 AM	10:46 AM	
18	10:02 AM	10:25 AM	11:19 AM		9:51 AM	10:45 AM	11:01 AM	
18	10:17 AM	10:40 AM	11:34 AM		10:06 AM	11:00 AM	11:16 AM	
18	10:35 AM	10:58 AM	11:52 AM		10:15 AM	11:09 AM	11:25 AM	
18	10:50 AM	11:13 AM	12:07 PM		10:33 AM	11:27 AM	11:43 AM	
18	11:05 AM	11:28 AM	12:22 PM		10:48 AM	11:42 AM	11:58 AM	
18	11:20 AM	11:43 AM	12:37 PM		11:03 AM	11:57 AM	12:13 PM	
18	11:32 AM	11:56 AM	12:55 PM		11:18 AM	12:12 PM	12:28 PM	
18	11:47 AM	12:11 PM	1:10 PM		11:36 AM	12:30 PM	12:46 PM	
18	12:02 PM	12:26 PM	1:25 PM		11:51 AM	12:45 PM	1:01 PM	
18	12:17 PM	12:41 PM	1:40 PM		12:06 PM	1:00 PM	1:16 PM	
18	12:32 PM	12:56 PM	1:55 PM		12:21 PM	1:15 PM	1:31 PM	
18	12:47 PM	1:11 PM	2:10 PM		12:36 PM	1:30 PM	1:46 PM	
18	1:02 PM	1:26 PM	2:25 PM		12:53 PM	1:47 PM	2:03 PM	
18	1:17 PM	1:41 PM	2:40 PM		1:08 PM	2:02 PM	2:18 PM	
18	1:32 PM	1:56 PM	2:55 PM		1:23 PM	2:17 PM	2:33 PM	
18	1:47 PM	2:11 PM	3:10 PM		1:38 PM	2:32 PM	2:48 PM	
18	2:02 PM	2:26 PM	3:25 PM		1:53 PM	2:47 PM	3:03 PM	
18	2:17 PM	2:41 PM	3:42 PM		2:08 PM	3:02 PM	3:18 PM	
18	2:32 PM	2:56 PM	3:57 PM		2:23 PM	3:17 PM	3:33 PM	
18	2:47 PM	3:11 PM	4:12 PM		2:38 PM	3:32 PM	3:48 PM	
18	3:02 PM	3:26 PM	4:27 PM		2:53 PM	3:47 PM	4:03 PM	
18	3:17 PM	3:41 PM	4:42 PM		3:08 PM	4:02 PM	4:18 PM	
18	3:31 PM	3:55 PM	4:56 PM		3:23 PM	4:17 PM	4:33 PM	

(*Continued*)

Table MP3.3. (*Continued*)

	Eastbound				Westbound			
Route	San Pablo Ave. & Marin Ave.	Berkeley BART	Moraga Ave. & Medau Pl.	Bus #	Moraga Ave. & Medau Pl.	Berkeley BART	San Pablo Ave. & Marin Ave.	Bus #
18	3:51 PM	4:13 PM	5:10 PM		3:39 PM	4:34 PM	4:52 PM	
18	4:07 PM	4:29 PM	5:26 PM		3:54 PM	4:49 PM	5:07 PM	
18	4:23 PM	4:45 PM	5:42 PM		4:09 PM	5:04 PM	5:22 PM	
18	4:38 PM	5:00 PM	5:57 PM		4:24 PM	5:19 PM	5:37 PM	
18	4:53 PM	5:15 PM	6:12 PM		4:39 PM	5:34 PM	5:52 PM	
18	5:09 PM	5:31 PM	6:28 PM		4:55 PM	5:49 PM	6:07 PM	
18	5:25 PM	5:47 PM	6:44 PM		5:09 PM	6:03 PM	6:21 PM	
18	5:46 PM	6:08 PM	6:59 PM		5:25 PM	6:19 PM	6:37 PM	
18	6:04 PM	6:26 PM	7:17 PM		5:41 PM	6:33 PM	6:51 PM	
18	6:23 PM	6:45 PM	7:36 PM		5:56 PM	6:48 PM	7:06 PM	
18	6:41 PM	7:03 PM	7:54 PM		6:11 PM	7:01 PM	7:19 PM	
18	7:01 PM	7:23 PM	8:13 PM		6:27 PM	7:19 PM	7:37 PM	
18	7:20 PM	7:42 PM	8:32 PM		6:43 PM	7:29 PM	7:45 PM	
18	7:40 PM	8:02 PM	8:52 PM		6:58 PM	7:43 PM	7:58 PM	
18	8:00 PM	8:22 PM	9:12 PM		7:14 PM	7:59 PM	8:14 PM	
18	8:22 PM	8:44 PM	9:32 PM		7:33 PM	8:18 PM	8:33 PM	
18	8:42 PM	9:04 PM	9:52 PM		7:53 PM	8:37 PM	8:52 PM	
18	9:04 PM	9:26 PM	10:14 PM		8:12 PM	8:56 PM	9:11 PM	
18	9:24 PM	9:46 PM	10:34 PM		8:31 PM	9:15 PM	9:30 PM	
18	9:44 PM	10:06 PM	10:54 PM		8:51 PM	9:35 PM	9:50 PM	
18	10:04 PM	10:26 PM	11:14 PM		9:11 PM	9:55 PM	10:10 PM	
18	10:24 PM	10:46 PM	11:34 PM		9:31 PM	10:15 PM	10:30 PM	
					9:51 PM	10:35 PM	10:50 PM	
					10:13 PM	10:54 PM	11:09 PM	
					10:33 PM	11:14 PM	11:29 PM	
					10:53 PM	11:34 PM	11:49 PM	

Appendix 9.A. Combining Driver Types to Cover a Run

This appendix uses the concept of sub-additive functions to develop a systematic procedure for finding an optimal way of covering a run with more than one driver type. It is assumed that drivers can be interchanged anywhere along the run. In our notation, drivers of type $i = 0$ work flexible hours with wage $(1+\pi)$, and those of type $i = 1, 2 \ldots$ work a shift of duration w_i with wage 1. Given is the length of the run, x. We look for the minimum cost $F(x)$ of serving the run and for the tasks allocated to each driver.

Definition: Subadditive Function. A function $f(x)$ defined for $x \geq 0$ is said to be *subadditive* if $f(x_1 + x_2) \leq f(x_1) + f(x_2)$, $\forall x_1,\, x_2 \geq 0$.

Table MP3.4. Service schedule of Route 52.

Route	Northbound				Southbound			
	Bancroft Way & Telegraph Ave.	University Ave & Shattuck Ave	El Cerrito Plaza BART	Bus #	El Cerrito Plaza BART	University Ave & Shattuck Ave	Bancroft Way & Telegraph Ave.	Bus #
52L	6:00 AM	6:12 AM	6:41 AM		5:42 AM	6:21 AM	6:32 AM	
52L	6:27 AM	6:39 AM	7:16 AM		6:37 AM	7:09 AM	7:20 AM	
52L	7:00 AM	7:12 AM	7:41 AM		6:55 AM	7:27 AM	7:38 AM	
52L	7:30 AM	7:42 AM	8:19 AM		7:03 AM	7:35 AM	7:46 AM	
52L	8:00 AM	8:12 AM	8:41 AM		7:24 AM	7:56 AM	8:07 AM	
52L	8:30 AM	8:42 AM	9:13 AM		7:38 AM	8:10 AM	8:21 AM	
52L	9:00 AM	9:12 AM	9:43 AM		7:48 AM	8:20 AM	8:31 AM	
52L	9:30 AM	9:42 AM	10:13 AM		7:58 AM	8:30 AM	8:41 AM	
52L	9:58 AM	10:10 AM	10:41 AM		8:08 AM	8:41 AM	8:52 AM	
52L	10:30 AM	10:42 AM	11:13 AM		8:18 AM	8:51 AM	9:02 AM	
52L	11:00 AM	11:12 AM	11:43 AM		8:29 AM	9:02 AM	9:13 AM	
52L	11:30 AM	11:42 AM	12:12 PM		8:40 AM	9:13 AM	9:24 AM	
52L	12:00 PM	12:12 PM	12:42 PM		8:51 AM	9:24 AM	9:35 AM	
52L	12:30 PM	12:42 PM	1:12 PM		9:02 AM	9:35 AM	9:46 AM	
52L	1:00 PM	1:12 PM	1:42 PM		9:13 AM	9:46 AM	9:57 AM	
52L	1:30 PM	1:42 PM	2:12 PM		9:25 AM	9:58 AM	10:09 AM	
52L	2:00 PM	2:12 PM	2:51 PM		9:37 AM	10:10 AM	10:21 AM	
52L	2:32 PM	2:44 PM	3:15 PM		10:08 AM	10:41 AM	10:52 AM	
52L	3:02 PM	3:14 PM	3:47 PM		10:38 AM	11:11 AM	11:22 AM	
52L	3:29 PM	3:41 PM	4:14 PM		11:08 AM	11:41 AM	11:52 AM	
52L	3:44 PM	3:56 PM	4:29 PM		11:38 AM	12:11 PM	12:22 PM	
52L	3:54 PM	4:06 PM	4:39 PM		12:08 PM	12:41 PM	12:52 PM	
52L	4:04 PM	4:17 PM	4:49 PM		12:37 PM	1:10 PM	1:21 PM	
52L	4:15 PM	4:28 PM	5:00 PM		1:07 PM	1:40 PM	1:51 PM	
52L	4:25 PM	4:38 PM	5:10 PM		1:35 PM	2:08 PM	2:19 PM	
52L	4:35 PM	4:48 PM	5:19 PM		2:05 PM	2:45 PM	2:56 PM	
52L	4:45 PM	4:58 PM	5:29 PM		2:32 PM	3:12 PM	3:23 PM	
52L	4:55 PM	5:08 PM	5:39 PM		3:04 PM	3:37 PM	3:48 PM	
52L	5:05 PM	5:18 PM	5:49 PM		3:35 PM	4:08 PM	4:19 PM	
52L	5:15 PM	5:28 PM	5:59 PM		4:05 PM	4:38 PM	4:49 PM	
52L	5:25 PM	5:38 PM	6:09 PM		4:35 PM	5:08 PM	5:19 PM	
52L	5:36 PM	5:49 PM	6:20 PM		5:06 PM	5:39 PM	5:50 PM	
52L	5:46 PM	5:59 PM	6:30 PM		5:35 PM	6:08 PM	6:19 PM	
52L	5:57 PM	6:10 PM	6:41 PM		6:05 PM	6:38 PM	6:49 PM	
52L	6:10 PM	6:23 PM	6:53 PM		6:35 PM	7:08 PM	7:19 PM	
52L	6:20 PM	6:33 PM	7:03 PM		7:01 PM	7:41 PM	7:52 PM	
52L	6:29 PM	6:42 PM	7:12 PM		7:34 PM	8:14 PM	8:25 PM	
52L	6:40 PM	6:53 PM	7:23 PM		8:01 PM	8:34 PM	8:45 PM	
52L	6:55 PM	7:08 PM	7:38 PM		8:27 PM	9:00 PM	9:11 PM	
52L	7:10 PM	7:23 PM	8:01 PM		8:56 PM	9:29 PM	9:40 PM	

(*Continued*)

Table MP3.4. (*Continued*)

	Northbound				Southbound			
Route	Bancroft Way & Telegraph Ave.	University Ave & Shattuck Ave	El Cerrito Plaza BART	Bus #	El Cerrito Plaza BART	University Ave & Shattuck Ave	Bancroft Way & Telegraph Ave.	Bus #
52L	7:25 PM	7:38 PM	8:08 PM		9:26 PM	9:59 PM	10:10 PM	
52L	7:39 PM	7:52 PM	8:22 PM		9:56 PM	10:29 PM	10:40 PM	
52L	7:58 PM	8:11 PM	8:40 PM		10:26 PM	10:59 PM	11:10 PM	
52L	8:20 PM	8:32 PM	9:01 PM		10:58 PM	11:31 PM	11:42 PM	
52L	8:51 PM	9:03 PM	9:32 PM					
52L	9:25 PM	9:37 PM	10:06 PM					
52L	9:55 PM	10:07 PM	10:36 PM					
52L	10:27 PM	10:39 PM	11:08 PM					
52L	11:01 PM	11:13 PM	11:42 PM					

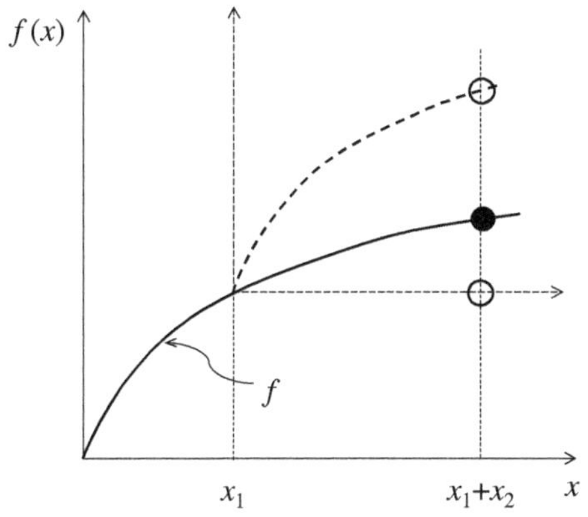

Figure 9.A1. Subadditive function.

If $f(x)$ represents the cost of producing a quantity x of something, subadditivity means that the production cost cannot be reduced by dividing the lot to be produced into parts. Another way of putting it is:

$$f(x) \leq f(z) + f(x - z), \quad \forall z \in [0, x].$$

In this appendix, all the arguments of our functions, such as x and z, are understood to be non-negative. Thus, non-negativity constraints such as x, $z \geq 0$ in the above definitions are omitted for brevity.

Subadditivity can be checked with the simple geometrical construction of Figure 9.A1. Imagine that the Cartesian axes and the graph of the function are copied on another sheet of paper and that this copy (dashed lines of

Figure 9.A1) is slid over the original figure, maintaining the origin of the copy over the original curve and without rotating the copy. Then, the function is subadditive if the copied curve never dips below the original at any point. You can convince yourselves with this method that concave-increasing functions are sub-additive and that convex functions are not.

What about the cost functions f_i shown in Figure 9.A2 for covering a task of duration x with a single driver type $i = 0, 1$? One finds by sliding each curve over itself that the slid curves never dip below the original. Therefore, both curves are subadditive. This means that if two (or more) tasks are covered by the same driver type, one can only benefit by combining them into a single task.

How to cover a task with two driver types: Assume now that two driver types i, j are available to cover a task of duration x. The solution to this problem is a cost-minimizing partition of the task into subintervals with a driver type assigned to each interval. However, because the single driver costs are subadditive, we know that if we combine all the intervals corresponding to each driver type into a single one then the total cost can only decline. This is useful because we only have to look for the total amount of time, x_i, covered by each driver type.

Thus, the optimum cost $F_{ij}(x)$ with drivers i and j is the solution of the following single-variable minimization problem:

$$F_{ij}(x) = \min_{x_i \leq x}\{f_i(x_i) + f_j(x - x_i)\} \tag{9.A1}$$

It turns out that this problem can also be solved graphically by the sliding method. This can be seen with the help of Figure 9.A3, which shows the case with $i = 1$ and $j = 0$. (Ignore the double lines for the moment). Curve

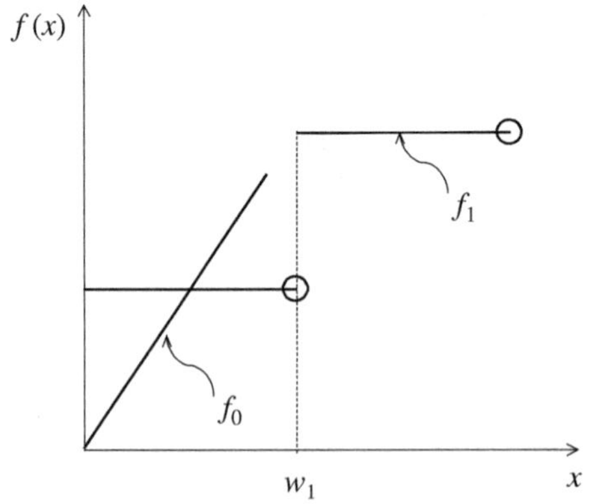

Figure 9.A2. Cost curves using a single driver type.

j is a straight line emanating from the origin, and curve i the step function shown. In the figure, the value of the argument of the minimum function in (9.A1) for the x_1 and x shown is the ordinate of the hollow dot. The position of this dot is obtained by sliding curve j over curve i until its origin is at point $(x_1, f_1(x_1))$ and then identifying where the slid curve intersects the vertical line at abscissa x.

To execute the minimization, slide curve i continuously over curve j allowing x_1 to sweep the interval $[0, x]$ and track the ordinate of the hollow dot at abscissa x. The lowest ordinate observed is the solution of (9.A1). In other words, $F_{ij}(x)$ is the ordinate of the lower envelope of all the slid curves evaluated at x. The optimum interval is given by the position of the origin of coordinates of the slid curve that formed the envelope at x.

Note that since the point x is arbitrary, the lower envelope itself identifies the values of $F_{ij}(x)$ for all x. Thus the recipe for $F_{ij}(x)$ is rather simple: slide curve j over curve i and take the lower envelope of all the curves that were slid. The result is a graph of $F_{ij}(x)$. This is shown by a double line in the figure. Note that this solution is lower in some places than either f_0 or f_1. This illustrates that combining driver types can reduce cost.

Now note from Figure 9.A3 that if we plot a graph of the function expressing the extra cost over the lower bound obtained after combining drivers, which is $[F_{01}(x) - x]$, we obtain the exact same curve as in Figure 9.5 of the text — as it should be.

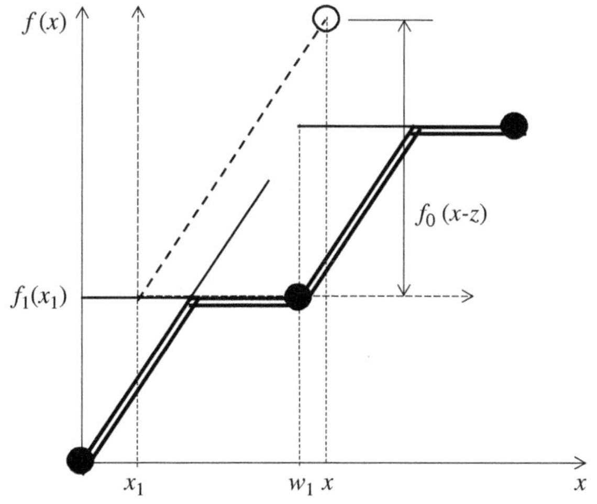

Figure 9.A3. Combining two driver types.

More driver types: We now show how to modify the procedure when there are more than two driver types.

Key to this extension is the realization that the function $F_{ij}(x)$ obtained with (9.A1) is itself subadditive. The reader can verify this graphically for the curve in Figure 9.A3. A general proof is given below.

Proof: Consider $F_{ij}(y) + F_{ij}(z)$, and an optimal partition of the two intervals: $y' \leq y$ and $z' \leq z$. Since these partitions are optimal, we can write: $F_{ij}(y) + F_{ij}(z) = f_i(y') + f_j(y - y') + f_i(z') + f_j(z - z')$. And since the f-functions are subadditive we can combine them and write: $f_i(y') + f_j(y - y') + f_i(z') + f_j(z - z') \geq f_i(y' + z') + f_j(z + y - z' - y')$. Note that $z' + y' \leq z + y$. Therefore, the value $(z' + y')$ is a feasible point in the minimization problem:

$$F_{ij}(y + z) = \min_{w \leq y+z} \{f_i(w) + f_j(y + z - w)\}.$$

Therefore, this minimum can be no greater than the feasible value: $f_i(y' + z') + f_j(z + y - z' - y')$. It follows that: $F_{ij}(y + z) \leq f_i(y' + z') + f_j(z + y - z' - y') \leq f_i(y') + f_j(y - y') + f_i(z') + f_j(z - z') = F_{ij}(y) + F_{ij}(z)$. □

To combine three driver types $(i,\ j,\ k)$ it suffices to solve

$$F_{ijk}(x) = \min_{x_i + x_j + x_k \leq x} \{f_i(x_i) + f_j(x_j) + f_k(x_k)\},$$

since the component functions are subadditive. This minimization has two degrees of freedom, so we want to simplify it. Note that the minimization can be performed in two steps as follows:

$$\begin{aligned} F_{ijk}(x) &= \min_{x_i + x_j + x_k \leq x} \{f_i(x_i) + f_j(x_j) + f_k(x_k)\} \\ &= \min_{x_k \leq x} \{f_k(x_k) + \min_{x_i + x_j \leq x - x_k} \{f_i(x_i) + f_j(x_j)\}\}; \end{aligned}$$

and that the inner minimization is the definition of $F_{ij}(x - x_k)$. Thus, we have:

$$F_{ijk}(x) = \min_{x_k \leq x} \{f_k(x_k) + F_{ij}(x - x_k)\}. \tag{9.A2}$$

Minimization (9.A2) has one degree of freedom. The equation expresses the composition of cost functions F_{ij} and f_k with rule (9.A1).

This shows that to combine three driver types, we can first combine any two and then combine the result with the third driver type. The procedure can be iterated with as many driver classes as desired. Graphically, at every step we solve a problem as in Figure 9.A3, using the lower envelope of the

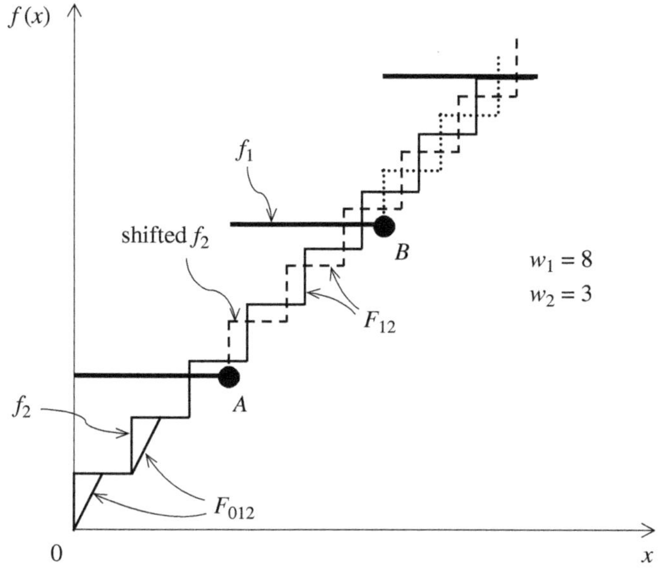

Figure 9.A4. Solving the problem with three driver types.

prior step as the curve over which one slides the curve of the new driver type.

Example 9.A1. This method is now applied to solve the problem leading to Figure 9.6 of the text. Recall that three driver types were combined: a driver type, $i = 0$, with flexible hours and two driver types with shift durations: $w_1 = 8$ and $w_2 = 3$. We start by combining the latter two. To do this slide curve f_2 (the thin-solid step function of Figure 9.A4) over f_1 (the thick-solid step function). Figure 9.A4 only shows three instances of curve f_2 anchored at points O, A and B because these anchors dominate any anchors to their lefts; so the latter is redundant. The anchored curves are depicted by means of solid-thin, dashed and dotted lines. The lower envelope of these three curves is the intermediate solution with driver types 1 and 2: curve F_{12}.

To find the final solution $F_{012}(x)$, we now slide the straight ray with slope $(1 + \pi)$ corresponding to the flexible drivers over curve F_{12} to determine a new lower envelope. This is the final solution. Sliding the ray has the effect of chopping the steps as occurred in Figure 9.A3. To keep the figure "clean," the result $F_{012}(x)$ is only shown for the first two steps in Figure 9.A4.

The wasted time over the lower bound achieved with this solution is $[F_{012}(x) - x]$. Note that this function is zero whenever curve F_{012} touches the line with slope 1 passing through the origin, and is bilinear in between these touches, just as the function in Figure 9.6. Note from Figure 9.A4 that the zero-points of F_{012} are: (3,6,8) in the first 8 h period; (9,11,12,14,15,16) in the second 8 h period; and (17,18,19,20,21,22,23,24) in the third. This matches exactly the result in Figure 9.6, therefore proving that it is optimal as had been conjectured.

Chapter 10

Operations — Reliable Transit Service

The previous two chapters described how to plan the deployment of a transit agency's mobile assets (vehicles and drivers) to cover a schedule. This chapter will examine how to manage these assets in the very short run to provide reliable service despite random disruptions such as traffic congestion, vehicle breakdowns, passengers with special needs, or simply bursts of passenger demand. Because these disruptions are not known beforehand the management methods, unlike the plans of Chapters 8 and 9, have to be adaptive in real time.[1]

Also recall from Chapter 1 that reliability refers to the predictability of vehicle arrival times and headways and that throughout this book we have assumed that the planned vehicle headways and trip times can be faithfully reproduced in the field. This chapter will show that it is indeed possible to approximate this assumed ideal of reliable transit operations.

Reliable service is highly desirable because when transit users are asked to name their concerns, reliability is usually at the top of the list. This should not be surprising — when vehicles deviate from a schedule or provide irregular headways passengers experience longer waits; and if door-to-door travel times cannot be predicted within a reasonable margin, passengers cannot conveniently use transit for travel to scheduled appointments.

[1]As a result of this adaptation the chapter must use some control theory concepts that, being somewhat specialized, were not reviewed in Chapter 2. They are covered instead in Section 10.2. Furthermore, because technology is inducing rapid changes in this field, the chapter also attempts to present the latest advances. For all these reasons, the chapter is more complex than previous chapters in this book and may only be appropriate for advanced students.

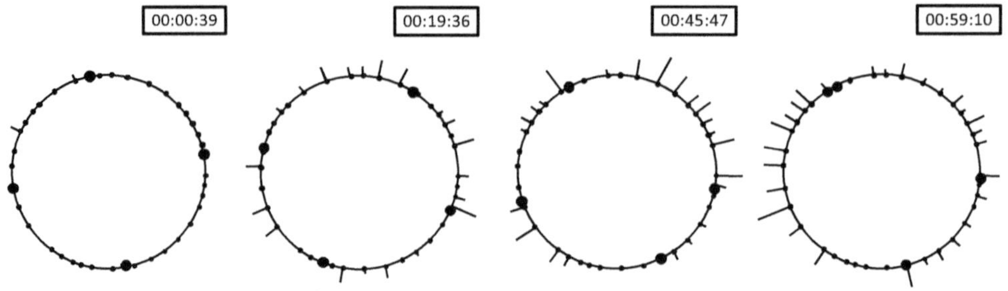

Figure 10.1. Simulated process of bus bunching.

Unfortunately, real-world disruptions can severely hinder reliability. Disruptions force transit vehicles to deviate from their desired time–space trajectories. We will see early in this chapter that if nothing is done these deviations naturally amplify over time until vehicles ultimately bunch or pair, creating undesirable large gaps in service and a severe loss of reliability. Figure 10.1 shows snapshots at the times shown in the boxes from a simulation that illustrates the phenomenon for a closed-loop bus line.[2] The circle is the loop. The dots are buses that move clockwise. The little rays emanating outward from the circle are passengers waiting at the stops. Although the buses are initially evenly spaced at the beginning, these equal spacings gradually get disrupted as time passes until the buses pair up.

Compensating for disruptions is challenging not just because they are pervasive but because the aforementioned deviations turn out to grow at an accelerating rate and propagate interactively to other vehicles. Accelerating growth means that deviations should be detected early and resolved quickly with some form of adaptive control. Their interactive propagation implies that control actions should treat the operating vehicles as a system of interacting parts and not individually. Clearly, both detection and control require considerable diligence.

Although control algorithms have been developed, detection is currently an issue. Most transit agencies have automatic vehicle location (AVL) and communication systems that enable dispatchers at their control centers to see where their buses are located and whether they are early or late related to their schedule. The systems enable two-way communication between dispatchers and drivers. But the typical AVL system refreshes every minute

[2]The full animation can be found at: https://www.worldscientific.com/worldscibooks/10.1142/10553#t=suppl.

or longer so it is jumpy and does not deliver an accurate real-time picture to dispatchers. Moreover, even if the AVL resolution were to be perfect, dispatchers — being human and limited in number — can only notice large disruptions. Thus, early detection is out of the question. In the authors' opinion, dispatchers should not be a critical link in the detection process.

In this spirit, some AVL systems have delegated the responsibility for taking action directly to the drivers by giving them some location information, e.g., showing their deviation from the schedule (if there is a schedule) and/or their position relative to other nearby buses. The presumption is that drivers are experienced enough to know what to do with the information given. Unfortunately, this only works for systems with a schedule, and where the vehicle interactions due to the bunching tendency are weak, e.g., for rail systems run on a segregated right-of-way. Otherwise, the problem is too complex and the information passed too incomplete for any human to react properly. In the authors' opinion, these AVL systems unfairly burden drivers with tasks they cannot do.

Other existing systems delegate the control task to ground inspectors who are in contact with the control center and stationed at control points distributed along the routes. The ground inspector at a control point is responsible for telling drivers whether to leave or how long to hold based on the recommendations of control algorithms. This control architecture works better but requires more personnel — some transit companies will deploy comparable numbers of inspectors and drivers. The approach is also appealing because it separates the tasks of driving and thinking about the control, and because inspectors can also monitor wayward drivers. On the other hand, the ground personnel approach still has a strong human element and therefore suffers from the early detection syndrome. Furthermore, it can only control vehicles at the limited locations where personnel is stationed, often after deviations have grown, so the control can be a bit rough.

Things should improve if detection is completely automated with smart onboard devices, and specific control actions are continuously recommended to the drivers via these devices — as well as to dispatchers and inspectors. Systems of this type, which we call "AVC" for automated vehicle control, have been successfully piloted but are not yet prevalent. AVC systems act as a "dispatcher in the bus" that continuously and calmly tells drivers how to gently respond to ongoing disruptions with simple feasible suggestions. As a result, drivers manage ordinary disturbances without stress or dispatcher input. Dispatchers and inspectors are only charged with monitoring errant drivers and managing the extraordinary. If robotic buses are put into transit service one day they will have to be equipped with AVC systems of some type.

In view of the above, the chapter is organized assuming that control is applied either with ground inspectors or an AVC system. And because the control problem at hand is both difficult and varied the chapter will only introduce fundamental ideas, starting with the most basic. In this spirit, Section 10.1 quantifies the effect of unreliability on passengers, building on the qualitative discussion in Chapter 1; Section 10.2 includes background on analysis tools for systems dynamics; and Section 10.3 examines the dynamics of uncontrolled transit systems, including the phenomenon of bus bunching. The remainder of the chapter discusses control. Section 10.4 explains a mildly effective form of control for systems with a schedule that treats vehicles individually; Section 10.5 then describes and evaluates more effective strategies that consider vehicle interactions, and Section 10.6 discusses more drastic remedial approaches that can be used to deal with bus bunching after it has happened.

10.1. How Unreliability Affects Transit Users

This section will quantify how irregularities in the trajectories of a route's buses affect transit users, considering people with and without scheduled appointments at their destinations. The effects include increased average waits at origin stops, as well as increased variability in both the waiting and riding time. The evaluation metric will be the travel time that a passenger would save by a system without irregularities, averaged over many trips. The results turn out to depend on whether passengers have appointments or not, and whether the transit system advertises a schedule or not. Since the results turn out to be fairly different for the four possible combinations they are discussed in four separate subsections. We start with the least structured case.

10.1.1. *Passengers without Appointments, Unscheduled Service*

When the transit schedule is not advertised, or service is so frequent that passengers do not pay attention to it, passengers arrive randomly and independently of the timetable. As a result, and as the snapshots of Figure 10.1 illustrate, long headways capture proportionately more passengers than short headways. This means that the distribution of headways experienced by randomly arriving passengers at a bus stop is "length-biased" compared with the distribution seen by an observer stationed at the stop. To avoid confusing these two random variables, we will use h for the latter and h' for the former. Consideration shows that their probability density functions must be related by $f_{h'}(x) = xf_h(x)/E[h]$.[3] As a result, the moments of h and h' are related by

$$E[h'^k] = E[h^{k+1}]/E[h]. \tag{10.1}$$

A passenger's expected waiting time conditional on the length of the headway in which (s)he lands is one half of the headway's duration, i.e., $E[\text{waiting time} \,|h'] = h'/2$. The passenger's unconditional expected wait is then the average of this expression, which can be evaluated using (10.1) with $k=1$, i.e.,

$$E[\text{waiting time}] = \tfrac{1}{2}\, E[h'] = \frac{1}{2}E[h^2]/E[h] = \frac{1}{2}(E[h] + \mathrm{Var}[h]/E[h]).$$

This is the same expression obtained in Section 2.2.3 with a different method. Note that the final expression reaches a minimum equal to its first term, $\frac{1}{2}E[h]$, when $\mathrm{Var}[h] = 0$, i.e., when service is perfectly regular. Hence, the second term is the added wait due to unreliability:

$$E[\text{added waiting time}] = \frac{1}{2}(\mathrm{Var}[h]/E[h]). \tag{10.2}$$

Since the right side of (10.2) is proportional to the headway variance it tells us that the marginal benefit for reducing the standard deviation of the headways by a fixed amount is larger when the standard deviation is already high than when it is low, i.e., that efforts to improve headway regularity should be focused on the routes with the worst performance.

Unreliability also increases the average riding time if the bus system is not well controlled. The reason is that uncontrolled buses following a long headway have more work to do and therefore travel more slowly.[4] And since passengers are more likely to land in a long headway than a short headway,

---

[3]This is a basic theorem in the theory of statistical sampling. To see it, note that $f_h(x)\,dx$ is the probability of the observer seeing a headway of length $x{\sim}(x{+}dx)$, and the number of passengers arriving in that range of headway is proportional to $x f_h(x)\,dx$. So the probability for a random passenger to see that range of headway, $f_{h'}(x)\ dx$, equals $x f_h(x)\,dx / \int x f_h(x)\,dx = xf_h(x)\,dx/E[h]$.

[4]This comment (and the riding time formula that ensues) does not apply to well-controlled systems because the main lever of control an agency has, and the basis for all control algorithms, is slowing buses that are going too quick so as to equalize the commercial speed of all buses regardless of how much work they have to do. Riding time formulas for controlled systems that consider this systematic slowing are presented later in this chapter for different forms of control. These formulas, together with (10.2), are then used to optimize the control.

they are also more likely to travel in a slow bus than a fast bus. It turns out as we show below that the expected added riding time is given by an expression similar to (10.2). Therefore, efforts to improve the regularity of uncontrolled routes when considering riding time should still focus on the worst routes.

The formula for the riding time penalty when buses are uncontrolled is now derived. For simplicity of notation, this is done for a homogeneous system in which all the stops are alike. The difference between the expected time needed by a bus with headway h to cover one stop, and the time needed by a bus with a standard headway, $H \equiv E[h]$, can be approximately expressed as $\beta[h - H]$, where $\beta \ll 1$ is an experimentally determined unitless constant.[5] It can be interpreted as the extra travel time between successive stops that a bus must endure on average per unit of excess headway. (The retardation is due to the excess work the bus has to do, e.g., boarding the extra passengers accumulated in the excess headway.)

In view of the above, if a passenger travels on average S stops and the headway on which (s)he lands, h', does not change much during these stops, then the difference in his/her travel time relative to the standard must be approximately $\beta S[h' - H]$.[6] Now compare this with the increase in average waiting time for any such passenger, which would have been $\frac{1}{2}[h' - H]$, and note that the two expressions differ only in the coefficient of $[h' - H]$. Hence, since the average of $\frac{1}{2}[h' - H]$ across passengers was found to be $\frac{1}{2}(\text{Var}[h]/E[h])$ as per (10.1), it follows that the average of $\beta S[h' - H]$ across passengers must be $\beta S(\text{Var}[h]/E[h])$. Thus, we have

$$E[\text{added riding time}] = \beta S(\text{Var}[h]/E[h]). \qquad (10.3)$$

Finally, to express the total time added by unreliability, add (10.2) and (10.3). The result is

$$E[\text{added total time}] = (\tfrac{1}{2} + \beta S)(\text{Var}[h]/E[h]). \qquad (10.4)$$

This concludes the discussion of the first case.

---

[5]This constant is unrelated to the parameter β that was used in the planning chapters to denote the users' value of time.

[6]This is a reasonable, conservative estimate. Section 10.3 will show that the headway experienced by this passenger at his/her downstream stops should roughly grow according to a geometric series, with a factor of $(1 + \beta)$. A more accurate expression for the extra travel time experienced after S stops is hence $[(1 + \beta)^S - 1][h' - H]$. But this quantity is approximately equal to $\beta S[h' - H]$ when $\beta \ll 1$.

10.1.2. *Passengers without Appointments, Scheduled Service*

Consider now service with an advertised schedule and a fixed headway H, still for non-appointment passengers. We assume that buses never depart a stop ahead of schedule and can only be late. Accordingly, we call the schedule deviations "latenesses," and denote them by the symbol ε. We focus on the relevant case where these latenesses are small compared with H — otherwise, if latenesses were large, passengers would not use the schedule and (10.4) would apply. Also, note that in the ideal case with perfectly reliable service ($\varepsilon \equiv 0$) passengers can eliminate all wait; thus, for the systems we are now studying, the added wait due to unreliability coincides with the actual wait.

A good, albeit not necessarily optimal, strategy for our passengers is postponing their arrival as much as possible without missing the bus, i.e., arriving on schedule.[7] With this kind of behavior the wait matches the lateness, so the average (excess) wait is the average lateness, which we denote $\mu = E[\varepsilon|\varepsilon \geq 0]$. Clearly, a reduction in μ by one unit reduces average travel time by one unit regardless of the route's original reliability. It follows that transit agencies should pay similar attention to all their scheduled routes regardless of their reliability status and focus effort on the routes that can be improved most easily.

Curiously, unreliability in service does not increase the expected riding time, unlike it occurred with unscheduled systems. The reason is that if transit service is scheduled and most passengers use the schedule they will arrive early enough to meet their bus despite variability in bus arrivals. As a result, the number of boarding passengers should be unaffected by the bus arrival times so that a bus's commercial speed should itself be unaffected by past latenesses. Obviously then, a passenger's average riding time should not depend on the distribution of the lateness ε, i.e., it should not rise due to unreliability. Thus, the added total time due to unreliability assuming that most passengers use the schedule is simply

$$E[\text{added total time}] = \mu. \tag{10.5}$$

[7]The optimal arrival time should balance the costs of arriving too early and missing the bus (see Turnquist, 1978). It can be shown, however (see homework exercise 10.1), that if the pdf, $f(x)$, and the cdf, $F(x)$, of the lateness satisfy $(H - x)f(x) > 1 - F(x)$ for all $x > 0$, then arriving on time is optimal. Note that this inequality is easier to meet when H is large relative to the typical lateness, i.e., when service is moderately unreliable, as we are assuming.

Let us now turn our attention to passengers with time-sensitive appointments that cannot be missed. Reliability is even more of a factor for these people because they must consider not just their expected waiting and riding times but the actual times, which they do not know beforehand. This uncertainty inflates the duration of their journeys considerably, as we shall now see.

10.1.3. *Passengers with Appointments, Unscheduled Service*

Again we consider unscheduled transit service first. To do this easily we shall assume that passengers behave in a simple but rational way: that they estimate travel time assuming the worst, and then execute the journey so as to arrive on time if the worst comes to pass. Since passengers are then never late unless something extraordinary happens, the estimated travel time essentially matches the actual travel time, including waiting at the destination.

To derive a formula for the increase in travel time the following is assumed. Expecting bad luck, passengers anticipate that they will arrive at the origin station near the beginning of a long headway of a specific duration, H_c, which also due to bad luck is a high percentile of h. Thus, they factor in a wait of H_c time units at the origin. Passengers also recognize from experience that for this headway their in-vehicle stop time would be βSH_c. Thus, they anticipate that the total stopped time for the complete journey is $(1 + \beta S)H_c$. Now, if service were perfectly reliable passengers would still assume a worst-case scenario; but now they would anticipate a total stopped time of $(1 + \beta S)H$ time units. Comparing these two expressions we see that unreliability increases a passenger's worst-case estimate for the stop time by $(1 + \beta S)[H_c - H]$ units. Passengers also realize that the bus motion between stops is not affected by H_c, or by the reliability level Thus, the above formula is also the increase in the worst-case estimate for travel time. Since the worst-case estimate for travel time is also the actual travel time including wait at the destination, it follows that the actual increase in travel time due to unreliability is also $(1 + \beta S)[H_c - H]$.

To get an order of magnitude idea of this effect while keeping things simple, let us assume that passengers choose H_c to be equal to the mean plus two standard deviations of the headways they experience, h', and derive a formula in terms of the moments of h'. To this end, recall from (10.1) that $E[h'] = E[h^2]/H$ and $E[h'^2] = E[h^3]/H$. Therefore, $\text{Var}[h'] = \left[\frac{E[h^3]}{H} - \left(\frac{E[h^2]}{H}\right)^2\right]$, and $H_c = \frac{E[h^2]}{H} + 2\left[\frac{E[h^3]}{H} - \left(\frac{E[h^2]}{H}\right)^2\right]^{1/2}$. Now, insert this expression into the formula for added total time, $(1 + \beta S)[H_c - H]$, to find

$$E[\text{added total time}]$$
$$= (1+\beta S)\left\{\frac{E[h^2]}{H} - H + 2\left[\frac{E[h^3]}{H} - \left(\frac{E[h^2]}{H}\right)^2\right]^{1/2}\right\}. \tag{10.6}$$

Although this formula is rather ugly, application to different distributions of h reveals how (10.6) and (10.4) compare for passengers with and without appointments. This is further explored in homework exercise 10.2, which reveals that appointment passengers always suffer considerably more from unreliable operations. For example, in the special case where h is uniformly distributed with mean H and range $\Delta \ll H$, equation (10.4) reduces to $(1/2+\beta S)(\frac{\Delta^2}{12H})$, whereas (10.6) becomes $(1+\beta S)(\frac{\Delta^2}{12H}+\frac{\Delta}{\sqrt{3}})$, which is more than an order of magnitude larger.

10.1.4. *Passengers with Appointments, Scheduled Service*

Unreliability turns out to be a less serious problem for passengers with appointments when transit service is scheduled. This occurs because passengers with appointments only need to consider for their worst-case trip planning the largest likely deviations from the schedule *at the destination* since they can be assured of catching any chosen bus by arriving at the origin on schedule. To quantify the consequences, consider first perfectly punctual service. A generic passenger will experience no waiting at the origin, a fixed travel time, and some time waiting at the destination, $A \in [0, H]$, which is equal to the time difference between the ideal arrival moment for meeting the appointment and the actual (earlier) arrival moment. Now, if service is unreliable the same passenger will consider a bus's maximum likely lateness at the destination, denoted μ_m (where μ_m should be several times larger than μ but smaller than H), and factor it into his/her plans. If $A > \mu_m$ then the passenger will choose the same bus as if service had been reliable, and experience the same total time between arrival at the origin and the appointment. Unreliability adds no penalty in this case. If on the other hand $A < \mu_m$, then the passenger will choose the previous bus, departing from the origin one headway earlier. This adds H time units to his/her total travel time on average. Now, since the event $\{A < \mu_m\}$ happens with probability μ_m/H across all passengers, it follows that the expected time penalty due to unreliability across all passengers is $H(\mu_m/H) = \mu_m$. Thus, in this case

$$E[\text{added total time}] = \mu_m \approx 2\mu \sim 3\mu. \tag{10.7}$$

Note this is several times larger than (10.5), but not an order of magnitude larger as occurred for unscheduled service.

The discussion for the above four cases has shown that (mildly unreliable) transit serves passengers without tight deadlines the best. This suggests that such service is more convenient for the evening than the morning commute, e.g., that people may car pool in the morning and return home on transit or parents may drive their kids to school and have them return home by bus. The above formulas also show that scheduled operations are much better for passengers when headways are long. And under these conditions, scheduled operations are also better for operators as bus trajectories are then less affected by deviations from the schedule, so the pairing tendency should be weaker.

On the other hand, if service has to be very frequent it is very difficult for an agency to provide small latenesses compared with the headways, so the system will function as if it was unscheduled, regardless of whether it has a schedule or not. Therefore, an agency does not have to advertise a schedule when service is frequent — just the service frequency. This is good for the agency for several reasons: (i) buses can be put in and out of service much more flexibly; (ii) the agency does not have to plan schedules (which are difficult to adjust to future demand and unable to anticipate temporary disruptions such as work zones); and (iii) it is operationally easier to maintain even headways than to ensure schedule punctuality, since the former can be achieved without the latter. Thus, ideally one should one would operate by headway when service is frequent and by schedule when service is infrequent. The analysis in this chapter will touch on both approaches.

10.2. Systems of Systems

This section introduces a modeling framework that will be used throughout this chapter. In this framework, a transit system is viewed as a "system of systems" (SoS), i.e., as a system of interconnected systems (known as agents) that use decentralized, agent-specific rules to interact. Below we show how to quantify the macroscopic behavior of transportation-related SoS based on the individual rules governing their agents.

The rules governing the behavior of a particular agent (indexed by i) are allowed to depend on the current (and past) state, x_i, of the particular agent,

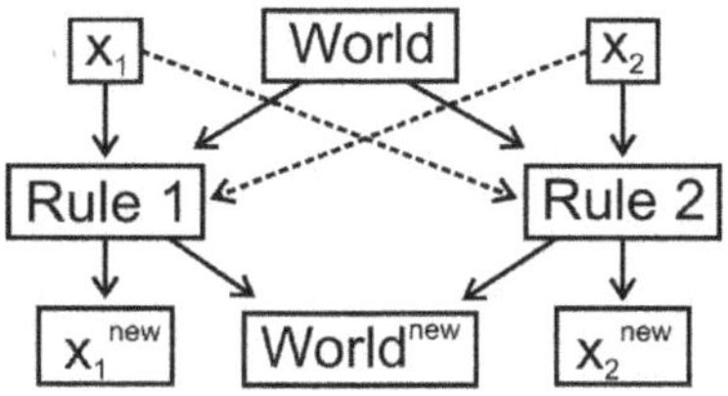

Figure 10.2. One step in the evolution of a two-agent system.

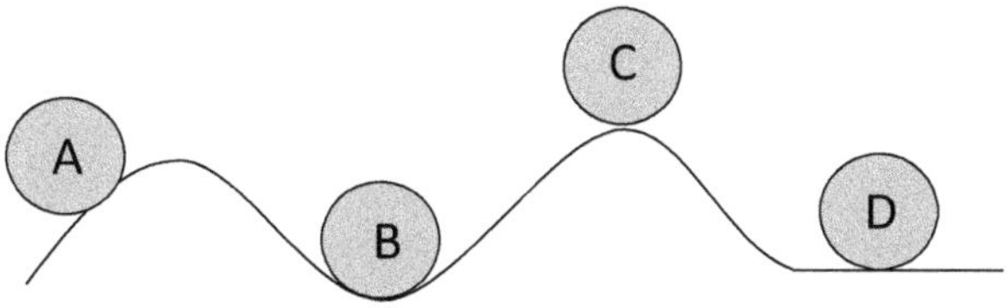

Figure 10.3. Illustration of stable and unstable states.

on outside factors representing the world and on the state of other agents with which the agent interacts. Figure 10.2 displays graphically the generic structure of a two-agent system. Agent 1 is on the left and agent 2 on the right. Arrows denote the inputs and outputs of each agent's rules. An SoS is characterized by the mathematical functions embodying these rules. These functions are called the "dynamic equations" of the system.

In transportation applications, we typically encounter lightly connected systems in which each agent only interacts with a limited number of counterparts with relatively simple rules. Despite this simplicity, the overall system behavior may be quite complex, and this is why it needs to be studied. Can you think of examples of SoS in the transportation field and what the agents and rules would be for them? Consider, for example, buses in a corridor, cars in traffic, airplanes nearing an airport, etc.

Since SoS can behave in complex ways we need to understand their behavior over time. Important questions about this behavior include: If the world is fixed in a steady state, does the SoS have an equilibrium state, i.e., a state that would be invariant in time? And if so, is this equilibrium unique? And is it stable?

An equilibrium state is said to be stable if the system it describes tends toward equilibrium when the initial state $\mathbf{x} = (x_1, x_2, \ldots)$ is close to it. This concept is graphically illustrated by the analogy in Figure 10.3.

The figure displays four possible locations of a ball on an uneven surface. The ball and the surface are the SoS, and the ball's location along the x-axis the state. The rules of interaction are the laws of physics: gravity and (slight) friction. Obviously then, ball A is not at an equilibrium state because from its current location it will slide down and to the left. Balls B, C, and D, on the other hand, are in equilibrium because their positions would remain invariant unless externally disturbed — ball D is on a horizontal part of the surface. However, only ball B is in a stable equilibrium because it is the only ball that would return to its original position if displaced a little in any direction: ball C would roll away and ball D would stay put.

These three questions (existence, uniqueness, and stability) can usually be answered in three steps: 1) identify the variables $\mathbf{x}$ that define the state and write the SoS dynamic equations embodying the rules that change $\mathbf{x}$ into $\mathbf{x}^{\text{new}}$; 2) solve the equations to find the equilibrium states if they exist; and 3) determine which of the found equilibria are stable.

The ease with which questions 2 and 3 are answered depends on the character of the dynamic equations. If the dynamic equations are linear they can be analyzed easily with the rules of linear algebra as shown by the examples below. For example, linear algebra tells us that if a problem has multiple equilibria they must form a convex set; and since convex sets are simply connected none of these equilibria can be stable. On the other hand, if the problem does not have multiple equilibria, then an equilibrium state may or may not exist. If it does it must be unique and isolated, and perhaps stable. The nonlinear case is more complicated. Nonlinear SoS can have multiple isolated equilibria; some stable and others not. It is also possible for SoS not to have any equilibrium points and for the state to wander over the state space in strange chaotic ways. In this chapter, we will mostly deal with linear systems.

So far we have talked only about deterministic systems, but in this chapter, we will also deal with systems where some of the inputs are random. In this case, the ideas of equilibrium and stability need to be generalized, as we will see in this section. Fortunately, the changes are conceptually minor. To get some feel for all of this and the type of analyses that will be done let us examine some examples, starting with simple systems that consist of a single agent.

10.2.1. *Single-Agent Systems*

Example 10.1 (A Stable Single-Agent SoS). A parking lot with a fixed input flow is periodically inspected. The input flow is $\lambda = 1,000$ vehicles per period. Parking durations are such that 10% of the vehicles in the lot at the beginning of a time period leave by the end of the period. We now show that this parking lot can be modeled as a single-agent SoS. The agent is the parking lot and the world is the entity supplying the demand. Our three steps are as follows.

Step 1 (identify state and write dynamic equations): We propose that the state of the system is the number of vehicles in the parking lot at the beginning of any time period t, $x(t)$. This will be a proper state if one can determine the next state from it using the given rules for the world and the agent. The resulting formula, expressing how $x(t)$ changes, is the dynamic equation. In our case $x(t)$ is a proper state: the number of vehicles parked in the lot at the beginning of time period $t+1$ is simply equal to the number of vehicles in the lot at the beginning of time period t, plus the number that enter during the time period, minus the number that leave. Therefore, the dynamic equation is

$$x(t+1) = x(t) + 1{,}000 - 0.1x(t). \tag{10.8}$$

Note the equation is inhomogeneous because it has a constant term.

Step 2 (identify all possible equilibria): At equilibrium, the number of vehicles in the lot will not change with time. Therefore, the equilibrium solution is found by setting $x(t) = x(t+1) = x_{eq}$.

$$x_{eq} = x_{eq} + 1{,}000 - 0.1\, x_{eq} \tag{10.9}$$

This is a linear equation with a unique solution: $x_{eq} = 10{,}000$. Since the solution is unique it can be stable.

Step 3 (stability analysis): to determine if the equilibrium is stable, we examine what happens when $x = x_{eq} + \varepsilon$ for any small ε. The equilibrium will be stable if $x \to x_{eq} = 10{,}000$. To see graphically what happens, plot the state of the system at time $t+1$ as a function of the state of the system at time t as per (10.8). This is the dark line in Figure 10.4.

The equilibrium point is the point on the dark line where $x(t+1) = x(t)$, i.e., the thick point where the dark line intersects the dotted line $x(t+1) = x(t)$. Imagine now that the system is perturbed by a value ε from the equilibrium state as shown in the figure. The arrows in the figure show how the state would evolve from the perturbed state according to the dynamic equation — the abscissa of the first point on the dark line is the initial state and its ordinate the final state after the first step; the two arrows emanating in sequence from this point identify a second point on the dark line whose initial state equals the final state of the first step; this concludes the first step, which is then iterated to obtain a sequence of points.

Clearly, our point sequence converges toward the equilibrium. Consideration shows that is also true if $\varepsilon < 0$, and that it continues to be true for any

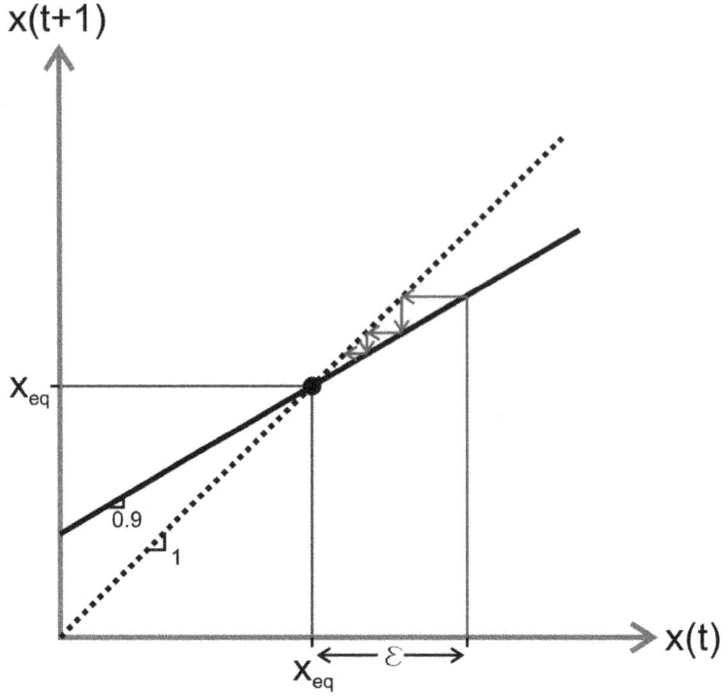

Figure 10.4. Graphical analysis of a stable system.

value of ε (check it for yourself.) Thus, since the system returns to x_{eq} after any minor initial perturbation, we conclude that our equilibrium is stable.

Stability can also be determined analytically by performing the same steps algebraically. To do this, define the residual perturbation after t steps: $\varepsilon(t) \equiv x(t) - x_{eq}$. The dynamic equation can then be rewritten in terms of $\varepsilon(t)$ by subtracting (10.9) from (10.8). The resulting equation is

$$\varepsilon(t+1) = 0.9\varepsilon(t). \tag{10.10}$$

The reason for re-expressing the dynamic equation in this way is to remove the independent constant from (10.8) as then the dynamic equation becomes homogeneous and easier to analyze. This is a standard simplification that is always made. In our particular case, it handily reveals that $\varepsilon(t) = 0.9^t\varepsilon(0)$, and therefore that $\varepsilon(t) \to 0$ for $t \to \infty$ regardless of $\varepsilon(0)$, i.e., that the equilibrium is stable.

Example 10.2 (An Unstable Single-Agent SoS). Consider now a deterministic queuing system in which the customer service time increases with queue length. This could happen for example if the queue interfered with the server, reducing its efficiency. In our example, customers arrive at a rate of 1,000 per time step. The server can process 2,000 customers at each time step, but this amount is reduced by 10% of the number of customers in the queue at the beginning of the time step. The server works as hard as possible; thus, if a step's initial queue is so small that the server can process more customers than will become available the server will slow just enough to leave an empty queue. This SoS will be modeled as in Example 1, with the world supplying the demand and the queue as the unique agent.

Step 1 (*identify state and dynamic equations*): We propose using the number of customers in the queue at the beginning of each time step, t, as the state, $x(t)$. This works because one can write $x(t+1)$ from $x(t)$ using the information of the problem. Note that the number of customers available for processing in step t is $x(t) + 1,000$, and the number that can be served $2,000 - 0.1x(t)$. Since the server works as hard as possible, the dynamic equation is

$$x(t+1) = [(x(t) + 1,000) - (2,000 - 0.1x(t))]^+ = [1.1x(t) - 1,000]^+. \tag{10.11}$$

Step 2 (*identify equilibria*): This is always done in the same way. Replace $x(t)$ and $x(t+1)$ with x_{eq} and solve the resulting equation. Consideration shows that it has two isolated solutions: $x_{eq} = 0$ and $x_{eq} = 10,000$. This outcome is possible because the resulting equation is nonlinear through the truncation operation $[\ ]^+$. Figure 10.5 shows the dynamic equation and the two equilibria using the same conventions as Figure 10.4.

Step 3 (*stability analysis*): All the equilibria are now checked for stability. Because our equation is nonlinear this is done by considering neighborhoods of its two equilibrium points where the equation is approximately linear. This is done in two sub-steps: (i) linearize the dynamic equation; (ii) perform the same linear analysis as in Example 1. We do this first for $x_{eq} = 10,000$. Start by rewriting (10.11) in a linear form that is approximately valid in the neighborhood of this equilibrium. We can see from Figure 10.5 that in our case no approximation is necessary; we can simply ignore the truncation and write:

$$x(t+1) = 1.1x(t) - 1,000, \quad \text{for } x \approx 10,000. \tag{10.12}$$

Next, eliminate the constant as we did to obtain (10.10) — by writing the difference between (10.12) and its equilibrium version ($x_{eq} = 1.1x_{eq} -$ 1,000, for $x_{eq} \approx 10,000$) in terms of the deviations from equilibrium,

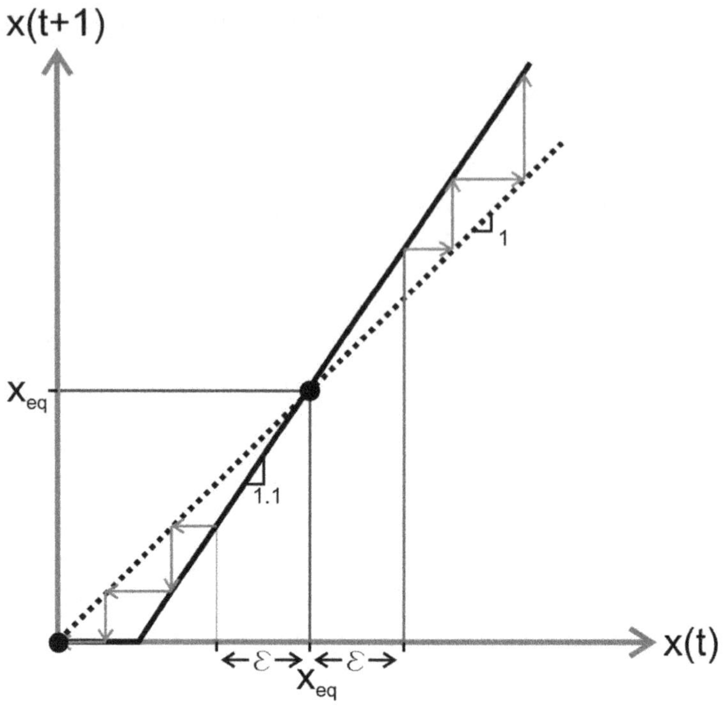

Figure 10.5. Graphical analysis of an unstable system.

$\varepsilon(t) = x(t) - x_{eq}$. The result is

$$\varepsilon(t+1) = 1.1\varepsilon(t), \quad \text{for } x \approx 10,000. \tag{10.13}$$

We can now readily see that in this case, $\varepsilon(t) = 1.1^t\varepsilon(0)$. Thus, it is clear that any perturbation in the neighborhood of the equilibrium point will grow in size, and the system will diverge from the equilibrium state. Therefore, this equilibrium is unstable. This is also confirmed with the graphical method of Figure 10.4 (see Figure 10.5). It can be readily seen that minor perturbations move the system away from the $x_{eq} = 10,000$ equilibrium.

The same analysis should be repeated for $x_{eq} = 0$, but we will not do it here because the figure clearly shows that the system moves towards the equilibrium $x_{eq} = 0$ if one starts anywhere close to it, so this particular equilibrium is stable.

The linearization method used in the example is pretty standard for nonlinear problems. It can help determine which equilibrium points are stable and which are unstable. But there is a caveat. Unlike in the linear case, a determination of stability does not guarantee a return to equilibrium if a perturbation is large. In our particular example, the figure clearly shows that the system does not return to its stable equilibrium ($x_{eq} = 0$) if a perturbation from this point exceeds x_{eq}.

Example 10.3 (A Single-Agent SoS with Stochastic Noise). Here we generalize the previous two examples by adding an uncertain input. This is done because the resulting model will come up later in this chapter. The following generalization of the examples' dynamic equations is considered:

$$x(t+1) = \nu + \phi x(t) + \gamma(t), \tag{10.14}$$

where ϕ is a fixed coefficient that we call the "gain," and $\gamma(t)$ is a stochastic noise term of independent and identically distributed (i.i.d.) disturbances with zero mean and standard deviation σ.

Even though problem (10.14) is stochastic, the solution steps are similar. First, we eliminate the independent constant ν by working with deviations from the deterministic equilibrium that would arise without noise. This equilibrium is found by ignoring the noise term in (10.14) replacing the state variables by x_{eq} and solving the resulting equation:

$$x_{eq} = \nu + \phi x_{eq}.$$

Obviously a deterministic equilibrium exists if and only if $\phi \neq 1$, or else $\phi = 1$ and $\nu = 0$. We can then subtract this equation from (10.14) and express the result in terms of the deviations from equilibrium. The result is

$$\varepsilon(t+1) = \phi\varepsilon(t) + \gamma(t). \tag{10.15}$$

This is similar to the deterministic homogeneous equations of the previous examples, except for the stochastic noise term.

We are interested in seeing whether (10.14) has a *statistically* stable stochastic equilibrium at x_{eq}, i.e., whether $\varepsilon(t)$ remains *statistically* close to 0 if (10.15) is iterated starting with $\varepsilon = 0$. The specific criterion we use is that $\varepsilon(t)$ should have zero mean and bounded variance as $t \to \infty$.

To evaluate these moments we take expectations and variances on both sides of (10.15). Then, using $\mu_\varepsilon(t)$ and $\sigma_\varepsilon^2(t)$ to denote the mean and variance of $\varepsilon(t)$, we find

$$\mu_\varepsilon(t+1) = \phi\mu_\varepsilon(t), \quad \text{and} \tag{10.16a}$$

$$\sigma_\varepsilon^2(t+1) = \phi^2\sigma_\varepsilon^2(t) + \sigma^2. \tag{10.16b}$$

We now solve these equations using the initial conditions corresponding to $\varepsilon = 0$. These are $\mu_\varepsilon(0) = 0$ and $\sigma_\varepsilon^2(0) = 0$. Clearly, the solution of (10.16a) is

$$\mu_\varepsilon(t) = 0. \tag{10.17a}$$

Equation (10.16b) is solved by applying it recursively, starting with $\sigma_\varepsilon^2(0) = 0$. Note how the recursion generates the following results: $\sigma_\varepsilon^2(1) = \sigma^2$, $\sigma_\varepsilon^2(2) = (1+\phi^2)\sigma^2$, $\sigma_\varepsilon^2(3) = (1+\phi^2+\phi^4)\sigma^2, \ldots, \sigma_\varepsilon^2(t) = (1+\phi^2+\phi^4 \cdots + \phi^{2t})\sigma^2$. Since the general expression for $\sigma_\varepsilon^2(t)$ is the partial sum of a geometric series with common ratio ϕ^2 and initial value σ^2, it can be reduced to

$$\sigma_\varepsilon^2(t) = \sigma^2 \left[\frac{1-\phi^{2t}}{1-\phi^2}\right], \text{ if } \phi \neq 1; \quad \text{and} \quad \sigma_\varepsilon^2(t) = \sigma^2 t, \text{ if } \phi = 1 \quad \text{and} \quad \nu = 0. \tag{10.17b}$$

Moreover, note from the linearity of (10.15) that if $\gamma(t)$ is Gaussian then $\varepsilon(t)$ is also Gaussian.

Now let us analyze stability. It turns out to depend on the gain, ϕ. If $\phi < 1$, as occurred in the stable deterministic system of Example 10.1, then (10.17b) converges from below to the finite quantity: $\sigma^2/(1-\phi^2)$. And since the mean of $\varepsilon(t)$ converges to zero, the equilibrium is statistically stable. On the other hand, if $\phi \geq 1$, as in Example 10.2, the variance given by (10.17b) increases toward $+\infty$. Thus, the system is statistically unstable.

Finally note that stability in our example does not depend on σ; only on the gain, and that this dependence is of the same form as in the deterministic case. In other words, the system is stable in the stochastic case if and only if it is stable in the deterministic case.

10.2.2. *Multi-Agent Systems*

The ideas in these three examples can be extended to multi-agent systems using concepts of linear matrix algebra and control theory. This is only explored briefly here because this chapter will only use single-agent models — the multi-agent systems that come up in our discussion can either be reduced to single agent systems or be treated with simpler ad hoc methods. This is not to say that multi-agent ideas are not useful. They can, for example, shed light on some cases that are not discussed in this chapter, such as transit systems consisting of a few buses serving a short loop. The interested reader can get a taste of what is required by solving homework exercises 10.3, 10.6, and 10.7. Otherwise, the rest of this subsection can be skipped without loss of continuity.

The key when dealing with a multi-agent system is to express the system state by a (column) vector of variables, $\mathbf{x}(t)$, and then write the system dynamics equations as a single matrix equality, which is then manipulated with the exact same steps of the earlier examples.

In the deterministic case, we follow the steps of Examples 10.1 and 10.2. The matrix equation is

$$\mathbf{x}(t+1) = \mathbf{L}\mathbf{x}(t) + \mathbf{L}_0,$$

where $\mathbf{L}$ (a square matrix) and $\mathbf{L}_0$ (a column vector) are given and constant. By definition, an equilibrium point must satisfy $\mathbf{x}(t+1) = \mathbf{x}(t) = \mathbf{x}_{eq}$. Therefore, if one exists, it is found by solving

$$\mathbf{x}_{eq} = \mathbf{L}\mathbf{x}_{eq} + \mathbf{L}_0.$$

As before, these two equations are then subtracted on both sides to recast the dynamic equation in terms of the (vector) perturbations from equilibrium, $\boldsymbol{\varepsilon}(t) = \mathbf{x}(t) - \mathbf{x}_{eq}$. The new expression is

$$\boldsymbol{\varepsilon}(t+1) = \mathbf{L}\,\boldsymbol{\varepsilon}(t).$$

Note the similarity of this result to (10.10) and (10.13). As in Examples 10.1 and 10.2, the new expression is linear and homogeneous, and therefore easy to solve by iterating it over time. The solution is now

$$\boldsymbol{\varepsilon}(t) = \mathbf{L}^t\boldsymbol{\varepsilon}(0).$$

Note again the similarity.

The system is defined to be stable if $\boldsymbol{\varepsilon}(t)$ in the above expression tends to the zero vector regardless of the initial conditions. As in the examples, this occurs if and only if the coefficient matrix of $\boldsymbol{\varepsilon}(0)$ in the above expression tends to the zero matrix, i.e., if $\mathbf{L}^t \to 0$ as $t \to \infty$. This is a bit more difficult to check than in the earlier examples, but not much. Linear algebra tells us that the condition holds if the absolute values of all the eigenvalues of $\mathbf{L}$ (which may be complex numbers) are less than 1.

In the stochastic case, we would follow the steps of Example 10.3. The dynamic equation is then $\mathbf{x}(t+1) = \mathbf{L}\mathbf{x}(t) + \mathbf{L}_0 + \boldsymbol{\gamma}(t)$, where $\boldsymbol{\gamma}(t)$ is a noise vector that is independent across time slices, and has a fixed covariance matrix and zero mean. As in the example, this dynamic equation is first reduced to the form $\boldsymbol{\varepsilon}(t+1) = \mathbf{L}\,\boldsymbol{\varepsilon}(t) + \boldsymbol{\gamma}(t)$ by considering the difference between the original equation and the deterministic equilibrium equation. One then takes means and (co)variances of the reduced equation using matrix notation to obtain the equivalent of (10.16). Finally, one steps through time to obtain the matrix version of the expression $\sigma_\varepsilon^2(t) = (1+\phi^2+\phi^4+\cdots+\phi^{2t})\sigma^2$. Linear algebra (Jordan canonical form) is then used to simplify the series expression and find a condition for statistical stability. As in the deterministic case, the system turns out to be statistically stable if every eigenvalue of $\mathbf{L}$ (either real or complex) has a norm inferior to 1.

10.3. Uncontrolled Bus Motion

This section examines the behavior of a transit line in which drivers are dispatched at regular intervals and then operate individually and as expeditiously as possible without the aid of a schedule or any form of external control. It will be shown using concepts from the previous section that the deterministic equilibrium of this system (with even headways for every bus at every stop) is unstable and therefore the system cannot stay close to it. Instead, there is a natural tendency for buses to form pairs and bunch. This is why external aids are necessary.

10.3.1. *The Ideal Deterministic Operation*

Consider the ideal situation shown in Figure 10.6 below, which displays three bus trajectories as they cross the locations of three (inactive) control points — for simplicity the trajectories of the buses have been smoothed so their stop-and-go motion at the intermediate stations is not shown. The situation is "ideal" because the bus trajectories are equidistant and parallel so that an even headway is maintained everywhere throughout the trip of every bus. There is no exogenous random noise. In the figure, and throughout the remainder of this chapter, the buses are indexed by n and the control points by s.

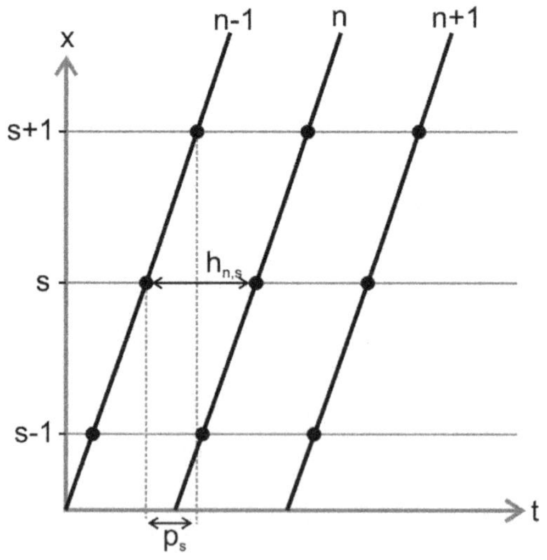

Figure 10.6. Ideal trajectories of consecutive buses in a time–space diagram.

The ideal deterministic operation is defined by the collection of bus arrival times at succeeding control points, $\{t_{n,s}\}$. These times are related to one another by the following equations:

$$t_{n,s+1} = t_{n,s} + p_s, \qquad n, s = 0, 1, 2\ldots \tag{10.18a}$$

$$t_{n,s+1} = t_{0,0} + nH + \sum_{i=0}^{s} p_i, \qquad n, s = 0, 1, 2\ldots \tag{10.18b}$$

where H is the target headway and p_i is the ideal travel time from control point i to $i+1$, including stops. The first term on the right-hand side of (10.18b) is the time at which the first bus arrives at the control point at the origin. The second term is the time separation between the first and the nth bus at the origin. The last term is the bus travel time from the origin to $s+1$.

Transit companies use equations like (10.18) to develop their schedules. In order to improve schedule adherence with the forms of control discussed later in this chapter, agencies usually include some "slack" in p_i, i.e., they require drivers to operate a bit more slowly than possible. To avoid confusion, we shall use c_s for the expeditious time (without slack) and $p_s = c_s + d_s$ for the scheduled time (with slack) where d_s is the amount of slack.

But of course, if no attempt at control is made there is no point in including slack, and this is the situation that we now examine. Can uncontrolled buses running expeditiously maintain a schedule with no slack?

10.3.2. *Uncontrolled Bus Motion*

In our analysis we treat the buses, n, as agents, the control points, s, as "time," and the actual arrival times, which we denote $a_{n,s}$, as the state of the agents. We want to see if $a_{n,s}$ can stay close to $t_{n,s}$ as the buses proceed forward, for $s \to \infty$.

Let us now start the analysis. As a preliminary step, note that the real uncontrolled travel time for bus n between stops s and $s+1$, $u_{n,s} \equiv a_{n,s+1} - a_{n,s}$, should obey

$$\begin{aligned} u_{n,s} &\equiv a_{n,s+1} - a_{n,s} = c_s + \beta_s(h_{n,s} - H) + \gamma_{n,s} \\ &= c_s + \beta_s(a_{n,s} - a_{n-1,s} - H) + \gamma_{n,s}, \end{aligned} \tag{10.19}$$

where $h_{n,s} \equiv a_{n,s} - a_{n-1,s}$ is the headway ahead of bus n at control point s, c_s is the expeditious mean travel time (without slack but including stops) when all buses are on-time, β_s is the station-specific version of the experimentally determined dimensionless constant β that was introduced in Section 10.1, and $\gamma_{n,s}$ is a random noise term representing variations in the mean travel time of bus n due to traffic and boarding passengers.

The noise terms $\gamma_{n,s}$ have units of time and are modeled as zero-mean i.i.d. random variables with standard deviation σ. Recall from Section 10.1 that the constants β_s express the average increase in bus travel time from arrival at s to arrival at $s+1$ due to a unit increase in headway time. Their values typically range from 0.01 to 0.1 if the distance between control points is on the order of 1 km.

Equation (10.19) is now used to write the dynamic equations for the system. Since the station number plays the role of time, these equations shall produce the arrival times of buses at station $s+1$ from their arrival times at station s. This is indeed possible, since the second and fourth members of (10.19) yield:

$$a_{n,s+1} = a_{n,s} + c_s + \beta_s(a_{n,s} - a_{n-1,s} - H) + \gamma_{n,s};$$
$$n = 1, 2 \ldots; s = 0, 1, 2 \ldots \tag{10.20a}$$

However, for these equations to be iterated we need to define initial conditions that include the trajectory of the first bus, $n = 0$, and the times at which all buses are dispatched from $s = 0$. To be as favorable as possible we shall assume that the first bus follows the deterministic ideal without any noise and that the remaining buses are dispatched regularly with headway H. Thus, if we place the origin of time at the moment of the first dispatch, the initial conditions become:

$$a_{0,s+1} = a_{0,s} + c_s; \qquad s = 0, 1, 2, \ldots \tag{10.20b}$$

$$a_{n,0} = nH; \qquad n = 0, 1, 2, \ldots \tag{10.20c}$$

Note that if there is no noise, the dynamic equations (10.20) reproduce the deterministic trajectories of (10.18) with c_s instead of p_s. Thus, (10.18) can be thought of as a deterministic equilibrium of (10.20), much as we did in Example 10.3 for a single-agent example. However, if there is noise and the noise terms are given, equations (10.20) produce a realization of bus arrival times for a given day that will differ from the ideal. Furthermore, the noise terms change every day so the bus trajectories also change. We would like to determine if these real trajectories can stay close to the ideal day after day.

To do this, as we did in Example 10.3, we eliminate the independent constants of (10.20) by subtracting from (10.20) the deterministic equilibrium (10.18) and expressing the result in terms of the deviations from the deterministic ideals $\{t_{n,s}\}$. The deviations from these values are denoted, $\varepsilon_{n,s} = a_{n,s} - t_{n,s}$. For the sake of intuition, Figure 10.7 displays two

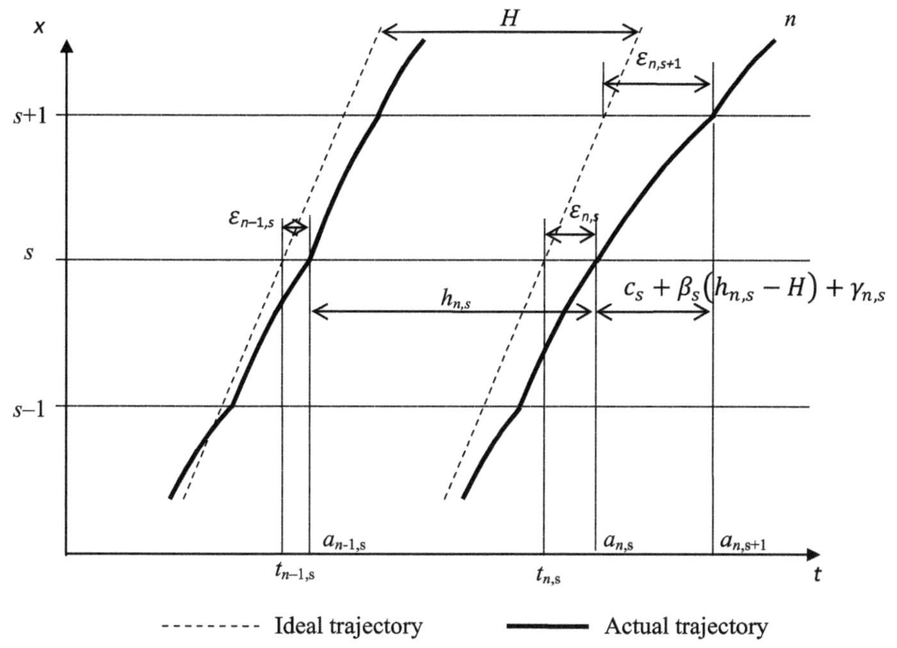

Figure 10.7. Possible trajectories of consecutive buses in a time–space diagram.

bus trajectories, the corresponding deterministic ideals and the involved variables. The analytic result is

$$\begin{aligned}\varepsilon_{n,s+1} &= \varepsilon_{n,s} + \beta_s(\varepsilon_{n,s} - \varepsilon_{n-1,s}) + \gamma_{n,s}\\ &= (1+\beta_s)\varepsilon_{n,s} - \beta_s\varepsilon_{n-1,s} + \gamma_{n,s}, \quad n, s = 0, 1, 2, \ldots \quad (10.21)\end{aligned}$$

As in Example 10.3, this new set of dynamic equations is very close to being homogeneous, except for the noise. The terms $\beta_s\varepsilon_{n-1,s}$ in these equations represent inter-agent interactions, which complicate the analysis. However, for bus $n = 1$ this term vanishes because the initial conditions imply that $\varepsilon_{0,s} = 0$. Thus, for $n = 1$, (10.21) is of the same form as (10.15) with a gain: $\phi = 1 + \beta_s > 1$. So we see that the bus trajectory is unstable, and the variance of its deviations from the schedule increases exponentially with s as per (10.17b).

The instability persists even if we ignore the noise. The first bus obeys then

$$\varepsilon_{n,s+1} = (1+\beta_s)\varepsilon_{n,s}, \quad \text{for } n = 1 \text{ and no noise.} \quad (10.22)$$

Still with a gain greater than 1, this is qualitatively the same recursion as in Example 10.2. Clearly, the $\varepsilon_{n,s}$ do not tend to zero for increasing s. Thus, the system is unstable even without noise.

What's the interpretation of this instability? The dynamics of Example 10.2 illustrate the physics. The bus is an agent that processes work (overcoming distance between control points while serving passengers) at a commercial pace that depends on its deviation from the schedule. Because large positive deviations imply more passengers to be served and therefore slower commercial paces they further imply further increases in the deviations. This is why (10.22), like Example 10.2, exhibits a "positive feedback." Note both physically and mathematically that this positive feedback also takes hold if the bus is initially ahead of schedule. In this case, the bus gets farther and farther ahead of schedule until it catches up with its predecessor, becoming a pair. This corresponds to the stable equilibrium at the origin for Example 10.2.

Since pairing occurs even under super favorable noiseless conditions and the ideal initial conditions we have chosen, it should not be surprising that it occurs so frequently in reality. Therefore, something must be done to counteract it. The next two sections explore possible remedies. None are perfect in practice, but they can alleviate the problem.

10.4. Control by Schedule

This section will examine the simplest control method used in practice. The approach has been in use for a long time not just because of its simplicity but also because it is cheap and does not require complex telematics. The method only requires a schedule, which is advertised to the public and used by the drivers. Drivers apply the control at specific control points. The control rule treats buses in isolation: drivers are simply told to leave each control point as soon as possible but not before the scheduled time, regardless of what other buses are doing.

For this to work, the schedule must be designed so buses can arrive in time at the control points most of the time. Otherwise, buses would be often late, and this would defeat the purpose since these buses would essentially run uncontrolled. Their headways could then grow unchecked as we just saw. For this reason, the amount of slack should be carefully considered. The following subsection describes the system dynamics under schedule control. Section 10.4.2 then shows how to choose the slack and the control points, and what might be expected as a result.

10.4.1. *System Dynamics with Schedule Control*

Recall that the scheduled travel time between arrivals at the control points is $p_s = c_s + d_s$, where d_s is the amount of slack. This decision variable should be selected to be greater than the typical disturbances arising during the bus trip between the control points. For example, if $\gamma_{n,s} \sim N(0, \sigma_s)$ then d_s can be set to $4\sigma_s$ so that the bus will be able to arrive in time over 99% of the time.

To start the analysis, let us rewrite Equation (10.18a) for the equilibrium bus travel times in a way that shows the slack:

$$t_{n,s+1} = t_{n,s} + c_s + d_s, \quad n, s = 0, 1, 2, \ldots \tag{10.23}$$

We assume that buses are allowed to run free between control points but are held immediately before the control points so they will not arrive ahead of schedule. Thus, buses arrive and open their doors at the later of two instants: their actual arrival time without holding, or the scheduled arrival with some holding. Figure 10.8(a) illustrates this control strategy in the ideal situation.[8]

With this control strategy the dynamic equations are

$$a_{n,s+1} = \max\{t_{n,s+1}, a_{n,s} + u_{n,s}\}, \quad n, s = 0, 1, 2, \ldots \tag{10.24}$$

We now express (10.24) in terms of deviations from the schedule as we did in the derivation of (10.21). To do this, subtract (10.23) from (10.24). The result is

$$\varepsilon_{n,s+1} = \max\{0, \varepsilon_{n,s} + u_{n,s} - (c_s + d_s)\}, \quad n, s = 0, 1, 2, \ldots \tag{10.25}$$

Now express $u_{n,s}$ in terms of the deviations by replacing the factor $(a_{n,s} - a_{n-1,s} - H)$ in the last member of (10.19) by its equivalent in terms of the deviations: $(\varepsilon_{n,s} - \varepsilon_{n-1,s})$ to obtain

$$u_{n,s} = c_s + \beta_s(\varepsilon_{n,s} - \varepsilon_{n-1,s}) + \gamma_{n,s}, \quad n, s = 0, 1, 2, \ldots \tag{10.26}$$

Next, insert this expression in (10.25). The result is

$$\varepsilon_{n,s+1} = [(1 + \beta_s)\varepsilon_{n,s} - \beta_s\varepsilon_{n-1,s} + \gamma_{n,s} - d_s]^+, \quad n, s = 0, 1, 2, \ldots \tag{10.27}$$

This expression is analogous to (10.11) if we treat not just $\gamma_{n,s}$ but also $\varepsilon_{n-1,s}$ as an input from the "world." You can convince yourselves that the resulting relation between $\varepsilon_{n,s+1}$ and $\varepsilon_{n,s}$ is nonlinear

[8]If the control point is a stop, the slack can be introduced at the stop itself while the bus doors are open. In this case, the published departure time would be the scheduled arrival time plus the expected time to load passengers. If every stop is a control point (and *vice-versa*) then the expected stop time at stop s is $\beta_s H$; thus, the advertised departure time would be $t_{n,s} + \beta_s H$.

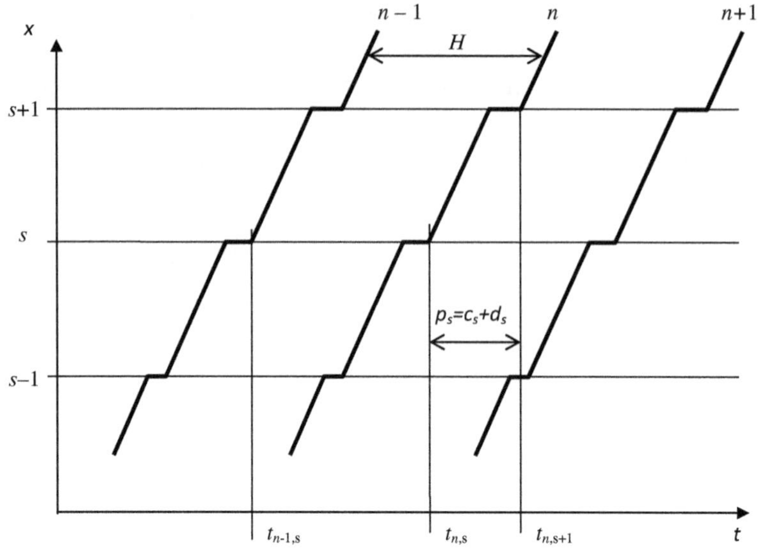

(a) Ideal bus trajectories

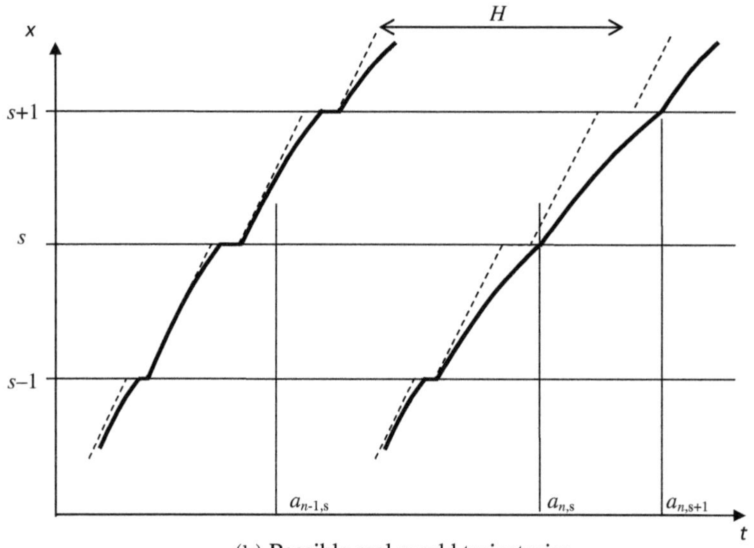

(b) Possible real-world trajectories.

Figure 10.8. Vehicle trajectories with built-in slack.

with a similar graph (see Figure 10.9) and similarly located equilibria. The coordinates of the equilibrium points are now: $\varepsilon_{n,eq} = 0$ (stable) and $\varepsilon_{n,eq} = d_s/\beta_s > 0$ (unstable). You can verify this algebraically from (10.27).

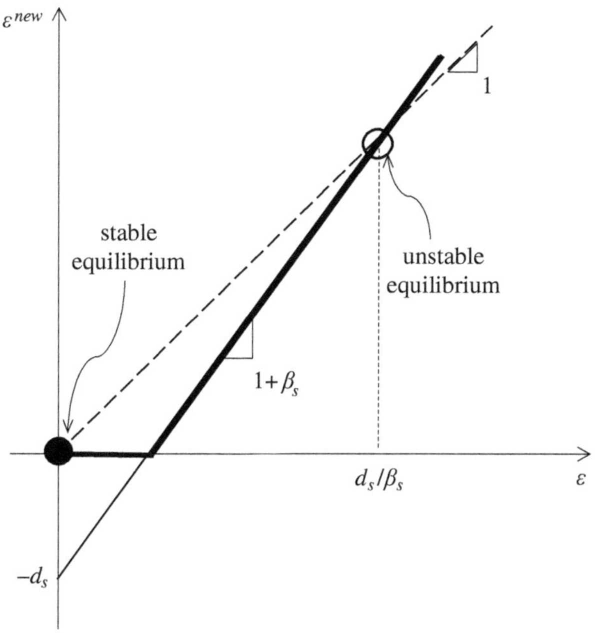

Figure 10.9. Two equilibrium points under schedule control.

Consider now the stable equilibrium at $\varepsilon_{n,eq} = 0$, and note from (10.27) that even if we allow for perturbations in $\gamma_{n,s}$ and $\varepsilon_{n-1,s}$ that are small compared with d_s the term $[\cdot]^+$ should still equal zero if $\varepsilon_{n,s}$ is small. This is illustrated by the bus trajectory on the left-hand side of Figure 10.9(b). It means that under small perturbations bus n returns to schedule immediately after passing the control point. This is good.

Unfortunately, as occurred in Example 10.2, perturbations larger than d_s/β_s spell trouble as they land beyond the unstable equilibrium on the increasing part of the system dynamic equation; see Figure 10.9 and note how it resembles Figure 10.5, which depicted a similar unstable equilibrium. That trouble is also illustrated by the bus trajectory on the right-hand side of Figure 10.8(b), which escapes from the region of stability close to the ideal trajectory and forever diverges. Like the bus shown, any bus that fails to arrive in time due to a large perturbation will not be able to recover and will fall further behind schedule. This escape behavior shows that conventional schedule control is not resilient to large perturbations, such as those caused by bus breakdowns. It explains why it is so difficult for transit agencies to keep buses on schedule, despite the agencies' best efforts, and why improved methods are necessary.

Before we look at these other methods, however, let us examine how d_s and the separations between successive control points should be chosen to improve overall performance under small disturbances.

10.4.2. *Setting the Slack and the Control Point Separations*

For maximum transparency, we consider in this section a homogeneous line with evenly spaced control points. Their separations are quantified by the number m of intervening stops. Our goal is determining values of m and the slack time d that are used at each control point in order to prevent bunching while maximizing the level of service provided.[9] This level of service will be quantified by an average passenger's trip time, including waiting. In what follows, we shall first choose d for a given m and then optimize m.

To prevent bunching, the slack $d(m)$ that we choose for any given value of m should ensure that drivers can arrive at the control points early most of the time. Accordingly, $d(m)$ is set equal to some multiple of the standard deviation of the travel time noise, which we denote $\sigma(m)$, i.e., $d(m) = k\sigma(m)$. Reasonable values for k are in the range of 2 to 4. Lower values allow for a higher commercial speed (which saves money for the agency and is good for the riders on board) but also increase the risk that more buses will run uncontrolled. Because models of transit systems with some uncontrolled buses are complex and dependent on the peculiarities of a particular system, we propose that the value of k should be selected by trial and error in the real world. A value close to 3 seems like a reasonable starting point.

To analyze the problem, let us first derive an expression for the average passenger's in-vehicle time. If the average passenger rides for r stops, his/her trip time will be rp, where p is the time that a bus needs to cover a single stop and travel to the next. An expression for p can be written on noting that the scheduled travel time between control points is mp, which can be expressed as $mp = mc + k\sigma(m)$, where c is the expeditious time per stop (without control) and the second term is the slack. If we now use σ^2 to denote the known variance of the travel time noise between successive stops, and assume this noise is independent across stops, we have, $\sigma(m) \approx \sigma\sqrt{m}$. (The independence assumption breaks down for large m as then buses begin to pair, but simulations in Daganzo (2009) show that it is fair as long as $m \leq 0.25/\beta$.) Therefore, our expression becomes[10] $p = c + k\sigma/\sqrt{m}$, if $m \leq 0.25/\beta$,

[9] The subscript s is not used in this subsection because the line is homogeneous.
[10] If $m > 0.25/\beta$, the value of p would be larger.

and we can write

$$\text{passenger riding time} = r(c + k\sigma/\sqrt{m}), \quad \text{if } m \leq 0.25/\beta. \tag{10.28}$$

Consider now the waiting time. The amount of wait depends on how many passengers use the schedule. We shall assume here that headways are sufficiently short for people to ignore the schedule; homework exercise 10.4 explores the alternative case.

Since people do not use the schedule, their wait at a particular stop depends on the headways. On average it equals $H/2$ plus the penalty due to unreliability given by (10.2). The variance of the headway appearing in (10.2) should be close to zero at the control points, but larger at the intermediate stops, $m' = 1, 2, \ldots m - 1$. To quantify this variance let $\sigma^2_{m'}$ be the variance of the bus arrival time at a stop m', and note that $\sigma^2_{m'} \approx m'\sigma^2$. Since a headway is the difference in the arrivals of two fairly independent buses, its variance should be $2\sigma^2_{m'} \approx 2m'\sigma^2$. Therefore, the average waiting time at bus stop m', $\bar{w}_{m'}$ should be

$$\bar{w}_{m'} = H/2[1 + 2m'\sigma^2/H^2]. \tag{10.29}$$

This expression should now be averaged across all $m' \in [0, m - 1]$ to obtain the overall average, $\bar{w}$. The result is

$$\bar{w} \approx H/2[1 + (m - 1)\sigma^2/H^2] = H/2 + (m - 1)\sigma^2/(2H). \tag{10.30}$$

Now, combine this with (10.28) to obtain the total trip time, T:

$$T = r(c + k\sigma/\sqrt{m}) + H/2 + (m - 1)\sigma^2/(2H), \quad \text{if } m \leq 0.25/\beta. \tag{10.31}$$

We are almost done, but before this expression can be optimized with respect to m a second constraint must be introduced. This is necessary in order to ensure that buses do not run into each other by chance at the intermediate stops. This is a definite possibility if m is large and H is small. The most likely locale for this accidental bunching would be right before the control point, where the headway variance is largest: $2\sigma^2(m) \approx 2m\sigma^2$. To ensure that accidental catchups are extremely rare, we use the constraint: $4\sigma(m) \leq H$, which in terms of m becomes $16m\sigma^2 \leq H^2$, i.e., $m \leq (H/\sigma)^2/16$.

We now solve the mathematical program resulting from the combination of this constraint and (10.31). The procedure is simplified by using H as the unit of time as this allows us to eliminate H from the formulation. In this system of units, the standard deviation of the noise will be labeled $A \equiv \sigma/H$. And instead of using T/H as the objective function, we use $z \equiv T/H - [rc + H/2]/H$. This is the extra time above the lower bound,

$[rc+H/2]/H$, which occurs under ideal/deterministic conditions. After a few simple manipulations, the resulting mathematical program reduces to

$$z = \min\left(\frac{A}{2}\right)\left\{-A + \frac{2kr}{\sqrt{m}} + Am\right\} \text{ s.t. } 1 \le m \le \min(1/(16A^2),\ 0.25/\beta)$$

The solution of this mathematical program gives the optimal length of the control segment. You can verify that the unconstrained solution is $m_u^* = (kr/A)^{2/3}$, and that the actual solution recognizing the constraints is $m^* = \text{middle}\,[1,\ \min\{1/(16A^2), 0.25/\beta\}, (kr/A)^{2/3}]$.

For small β, $k = 4$ and $A = 0.1$ (a moderately high value for the noise), we find that $m^* = 6.25 \approx 6$ for all r, and z^* is 0.5 for $r = 3$, 1.2 for $r = 7$ and 3.3 for $r = 20$. Recall that the unit of time is H so it looks like the control increases travel time above the ideal by 1 or 2 headways for reasonable values of r. This would be quite unacceptable if headways were long, but recall that the analysis assumes short headways. So the result is not completely discouraging. On the other hand, the performance hinges on applying the control regularly (every six stops); and this requires resources and discipline. Furthermore, as we saw earlier, the control method is not resilient to large disturbances so it will fail every once in a while — sometimes spectacularly. Therefore, improved control schemes should be sought. The following section describes some ideas in this direction.

10.5. Multi-Bus Control Strategies

Schedule control is dynamic and adaptive but rigid, and buses are controlled individually. It should be intuitive that performance might be improved if buses are allowed to respond more flexibly, not just to their deviations from a schedule but also to what other buses are doing, e.g., by holding a bus at a control point beyond its scheduled time when the bus is too close to the preceding bus or too far ahead of the following bus. To derive these benefits, an investment in telematics equipment and software is necessary. Fortunately, advances in information technologies should reduce the associated costs over time. As of this writing, multi-bus control strategies are not widely used, although they have been piloted. We believe this is the way of the future. Therefore, this section will discuss different types of multi-bus control algorithms, the roles that they can play and the results of these pilot tests.

The first part of this section (Section 10.5.1) discusses basic algorithms. These are simple and can be implemented in *ad-hoc* ways without much field personnel or significant telematics investment. Basic algorithms, however, are not resilient to large disturbances. Section 10.5.2 is then devoted to robust control. These more advanced algorithms ensure resiliency and perform better overall, but they require high fidelity telematics able to provide finely grained information with low latency.

10.5.1. *Basic Algorithms*

The idea behind all algorithms is the same: a bus n is artificially delayed on its leg from control point s to control point $s+1$ by a dynamically determined time $D_{n,s}$ that depends on the status of the buses when bus n arrives at s. The status of the buses is quantified by the time separations between bus n and all other buses, and if a schedule is used, on the deviations from the schedule, $\varepsilon_{n,s}$, too. For example, the dynamic delay for the schedule control algorithm of the previous section was $D_{n,s-1} = (-\varepsilon_{n,s})^+$, i.e., if $\varepsilon_{n,s}$ is negative (the bus arrives early at s) the bus is held at the end of its trip from $s-1$ to s by the amount of earliness; otherwise the bus is not held. This recipe is appealing from an operator's perspective because buses do not require information from other buses or precise location information to implement it. Thus, there is no need for high capacity communications channels, ground personnel or GPS tracking. Buses are self-sufficient.

In this subsection, we shall explore more general control laws that still allow the buses to be self-sufficient.[11] The recipes use as an input the forward headway of the bus in question $h_{n,s}$, and if a schedule is used $\varepsilon_{n,s}$ as well, i.e., the recipes are of the form $D_{n,s} = D_{n,s}(h_{n,s}, \varepsilon_{n,s})$. Buses can be self-sufficient because computers onboard can estimate $h_{n,s}$ from the observed duration of the bus stop time at s — since the average of this duration is $\beta_s h_{n,s}$. In practice, to improve the efficiency of this estimator one may use the average time spent at a few preceding stops, or even more simply the elapsed travel time between these stops.

The present subsection only considers (truncated) linear control laws, i.e., of the form $D_{n,s} \equiv [k_{0,s} + k_{1,s}h_{n,s} + k_{2,s}\varepsilon_{n,s}]^+$, because their stability can be studied with methods similar to those we have described. We will

[11]This subsection and the rest of the chapter focus on simple control methods that are easily scalable, require little information and can still be used when the unpredictable happens. There is also an extensive academic literature that uses rolling-horizon optimization methods to choose the holding times. While this is interesting, this approach requires deterministic forecasts of things that are tough to know such as future demand and driver behavior; and the approach increases rapidly in complexity as a system becomes large. Therefore, these methods are beyond the scope of this book.

find that the law is stable to small perturbations only if the coefficients $k_{1,s}$ are negative — this should not be surprising since intuition tells us that one should delay a bus when its forward headway is small. Unfortunately, this negativity also means that the law cannot work when extraordinary conditions such as bus malfunctions, bad weather or unusual traffic increase the headway in front of some buses so much that $D_{n,s} = 0$. Then, as occurred with schedule control, the affected buses run uncontrolled and suffer from the escape problem with increasing headways.

The sub-subsection below examines the situation where a schedule is not provided. This is appropriate for high frequency lines as then passengers do not have much use for a schedule. Eliminating a schedule is also advantageous for an operating agency because (i) a schedule does not have to be designed; (ii) when conditions change unexpectedly the buses can continue to operate without tweaking the schedule; (iii) punctuality does not have to be maintained (only headway regularity); and (iv) buses can be inserted and withdrawn from service much more flexibly.

10.5.1.1. *Frequency-Based Systems*

Here we shall do the following: propose a control law; derive its dynamic equations; analyze their stability; fine-tune and evaluate the algorithm's performance, and discuss some practical issues.

The control law: Because a schedule is not used, this subsection examines a control law based on headways. It is assumed to be of the form

$$D_{n,s} \equiv [k_{0,s} + k_{1,s}h_{n,s}]^{+}, \tag{10.32}$$

where $k_{0,s}$ and $k_{1,s}$ are parameters that can be freely chosen. This strategy was proposed in Daganzo (2009), where more details can be found. This section will suggest reasonable values for $k_{0,s}$ and $k_{1,s}$, and examine how the law performs.

The law is evaluated for an open ended line with many stops. Buses are introduced at the upstream end (stop 0) at regular intervals H. The goal is keeping succeeding headways at downstream stops as even as possible. Although a schedule is not used it will be convenient for analysis purposes to define a "virtual" schedule like (10.18) that is not advertised to the passengers or used by the control law. This virtual schedule is only a point of reference for the analysis.

Recall from Section 10.3 that the uncontrolled motion of a bus was unstable because the gain coefficient in (10.21) (the coefficient of $\varepsilon_{n,s}$) was greater than 1. This suggests that the control law should aim at canceling the β_s constant in this coefficient so as to reduce it below 1.

One can find by working backwards that the control law that changes the gain coefficient from $(1 + \beta_s)$ to any desired value in (0, 1) is

$$D_{n,s} = [d_s - (\beta_s + \alpha)(h_{n,s} - H)]^+, \quad \text{where} \quad d_s > 0 \quad \text{and} \quad \alpha \in (0, 1). \tag{10.33}$$

This corresponds to (10.32) with control parameters:

$$k_{0,s} = d_s + (\beta_s + \alpha)H, \quad \text{where } d_s > 0 \text{ and } \alpha \in (0, 1); \text{ and} \tag{10.34a}$$

$$k_{1,s} = -(\beta_s + \alpha), \qquad \text{where } d_s > 0 \text{ and } \alpha \in (0, 1). \tag{10.34b}$$

The constants d_s and α appearing in these expressions can be chosen at will as long as they satisfy the stated constraints. We call them "tuning variables" since they determine how the control law performs. Note from (10.34) that the only data needed *a priori* to implement this control are the β_s values, which can be measured experimentally. The ideal headway H is given. Figure 10.10 shows how the law affects the bus travel times. The thin rising line is the expected travel time without control, and the thick broken line the travel time with control. The vertical separation between the two lines is the dynamic holding delay. Note how it decreases linearly with the headway until it reaches zero, where there is a truncation point. For larger headways there is no holding.

***The dynamic equations*:** We shall now verify that the linear regime of this control law reduces the gain as we desired. We assume d_s is so large that (10.33) is almost always in this regime. Thus, we look for the system's dynamic equations in their linear form, assuming no truncations.

In what follows (including Section 10.6) it is assumed that the trip time of bus n in the segment from s to s+1 is the sum of the holding time $D_{n,s}$ and an unrestrained time equal to the uncontrolled time $u_{n,s}$ given by (10.19) using the headway at s; see Figure 10.10. In other words, it is assumed that the amount of holding in that segment does not affect the unrestrained travel time in it. Clearly, for this to happen, buses must be held at the downstream end of the segment. Otherwise, the headways in the segment's stations would change, and so would the travel time.[12]

[12]Section 10.6 discusses other holding possibilities and their consequences.

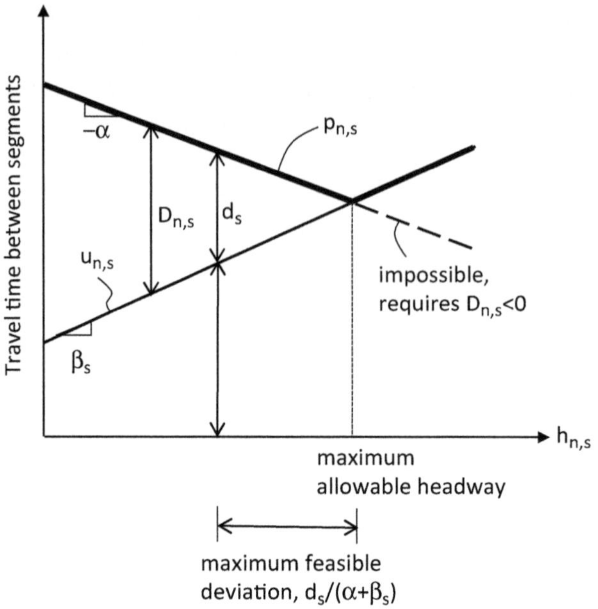

Figure 10.10. Control law under the forward-looking strategy.

In view of the above, the dynamic equations are simply written by adding $D_{n,s}$ to the uncontrolled version of the equations. Note that the latter can be written in terms of headways by rearranging terms in the second and third members of (10.19), i.e., we have

$$a_{n,s+1} = a_{n,s} + c_s + \beta_s(h_{n,s} - H) + \gamma_{n,s}. \quad \text{(uncontrolled)} \qquad (10.35a)$$

Thus, the controlled version of the dynamic equations is

$$\begin{aligned} a_{n,s+1} &= a_{n,s} + c_s + \beta_s(h_{n,s} - H) + \gamma_{n,s} + D_{n,s} \\ &= a_{n,s} + c_s + \beta_s(h_{n,s} - H) + [d_s - (\beta_s + \alpha)(h_{n,s} - H)]^+ + \gamma_{n,s}. \\ &\quad \text{(controlled)} \end{aligned} \qquad (10.35b)$$

On removing the truncation the β-terms cancel out and we obtain the linearized equation for the controlled system:

$$a_{n,s+1} = a_{n,s} + c_s + d_s - \alpha(h_{n,s} - H) + \gamma_{n,s}, \quad n, s = 0, 1, 2, \ldots \qquad (10.36)$$

Now, express this in terms of deviations from the schedule by substituting $(\varepsilon_{n,s} - \varepsilon_{n-1,s})$ for the factor: $(h_{n,s} - H)$. The result is

$$a_{n,s+1} = a_{n,s} + c_s + d_s - \alpha(\varepsilon_{n,s} - \varepsilon_{n-1,s}) + \gamma_{n,s}, \quad n, s = 0, 1, 2, \ldots \qquad (10.37)$$

And finally, since the virtual schedule satisfies $t_{n,s+1} = t_{n,s} + c_s + d_s$ as per (10.23) we can subtract this equality from the above to find our target equation:

$$\varepsilon_{n,s+1} = (1-\alpha)\varepsilon_{n,s} + \alpha\varepsilon_{n-1,s} + \gamma_{n,s},$$
$$\text{for } \alpha \in (0,1) \quad \text{and} \quad n,s = 0,1,2,\ldots \tag{10.38}$$

As claimed, the gain coefficient is changed from $(1+\beta_s)$ to an expression $(1-\alpha)$ that can take any value in (0, 1).

***Stability analysis*:** Let us now turn our attention to stability. If we could show that the deviations from the schedule arising from (10.38) had zero mean and a bounded variance then the headways would have zero mean and a bounded variance. Then we would be done. Unfortunately, it can be shown (see Daganzo, 2009) that the schedule deviations have infinite variances. Thus, we must analyze the headways themselves.

Fortunately, this can be done without complex mathematics, albeit only approximately, with simulations and dimensional analysis — a more formal analysis is provided in Daganzo (2009). Let us assume that the noise terms are Gaussian and i.i.d. As in the discussion before (10.28), we use σ^2 to denote the known variance of the independent travel time disturbances (i.e. the noise) between successive stops, and m be the number of stops between two neighboring control points. Hence, the variance of the noise term in (10.38) is $\sigma(m) \approx \sigma\sqrt{m}$. Then, the problem of finding $E[h_{n,s}]$ and $\text{Var}[h_{n,s}]$ from (10.38) (where $h_{n,s} \equiv \varepsilon_{n,s} - \varepsilon_{n-1,s}$) is defined by only two parameters ($\sigma(m)$ with units of time, and α which is dimensionless). Since our unknowns have units of [time] and [time^2], respectively, Buckingham π Theorem (discussed in Chapter 2) tells us that the mean and sample variance formulae must be of the form: $E[h_{n,s}] = \sigma(m)g_{n,s}(\alpha)$ and $\text{Var}[h_{n,s}] = \sigma^2(m)f_{n,s}(\alpha)$. Therefore, we can estimate $g_{n,s}(\alpha)$ and $f_{n,s}(\alpha)$ for all n and s, by repeatedly simulating (10.38) with $\sigma(m) = 1$ for a fixed value of α, and then using the sample mean and the sample variance of the headways observed across simulations for each $(n,\ s)$ as our estimators of $E[h_{n,s}]$ and $\text{Var}[h_{n,s}]$, for each (n, s). On doing this for different α we obtain estimates for $g_{n,s}(\alpha)$ and $f_{n,s}(\alpha)$. Numerical results closely fit the relations $g_{n,s}(\alpha) = 0$ and $f_{n,s}(\alpha) \approx 1/[\alpha(1-\alpha)]$ for large n and s, i.e., as found in Daganzo (2009), $E[h_{n,s}] = 0$ and:

$$\text{Var}[h_{n,s}] \approx \frac{\sigma^2(m)}{\alpha(1-\alpha)}; \quad \text{for } \alpha \in (0,1); \text{ as } n,s \to \infty. \tag{10.39}$$

***Fine-tuning and evaluation*:** Let us now see how to choose the two tuning variables, d_s and α. Note from Figure 10.10 that the maximum deviation one should allow to ensure that the control is in force (no truncation) and therefore guarantee good performance is $\frac{d_s}{\alpha+\beta_s}$. Since the headway standard deviation σ_H is given by (10.39) as $\sigma(m)/\sqrt{\alpha(1-\alpha)}$, we propose choosing d_s and α so the inequality $\frac{d_s}{\alpha+\beta_s} \geq 3\sigma_H = 3\sigma(m)/\sqrt{\alpha(1-\alpha)}$ is satisfied for all s.This ensures that the system will remain in the controllable regime most of the time. Because d_s is the expected in-vehicle delay riders experience for every control segment, we choose it to be as small as possible, i.e., $d_s = \frac{(\alpha+\beta_s)3\sigma(m)}{\sqrt{\alpha(1-\alpha)}}$. This makes sense because d_s does not influence the out-of-vehicle-delay, since it does not influence the headway variance.

The parameter α influences both types of delay, though. For a very long line, the variance of the headway at the control points will approach (10.39), so the average waiting time across all stops can be roughly approximated using (10.2) by: $\frac{H}{2}\left[1+\frac{\sigma^2(m)}{\alpha(1-\alpha)H^2}\right]$, and hence the extra waiting due to unreliability is $\frac{H}{2}\left[\frac{\sigma^2(m)}{\alpha(1-\alpha)H^2}\right] = m\sigma^2/(2\alpha(1-\alpha)H)$. Furthermore, if passengers travel for an average of r stops, or approximately r/m control points, the average riding delay will be approximately: $(\frac{r}{m})(\frac{(\alpha+\bar{\beta})3\sigma(m)}{\sqrt{\alpha(1-\alpha)}}) = (\frac{r}{\sqrt{m}})(\frac{(\alpha+\bar{\beta})3\sigma}{\sqrt{\alpha(1-\alpha)}})$, where $\bar{\beta}$ is the average of β_s across the route. The sum of these two expressions is the total extra delay passengers have to endure. One should choose the values of α and m that minimize it subject to $0 < \alpha < 1$ and the constraints on m introduced in Section 10.4.2: $1 \leq m \leq \min\{\frac{H^2}{16\sigma^2}, \frac{0.25}{\beta}\}$. Numerical examples show that the method has some promise; see homework exercise 10.5.

***Practical considerations*:** Before we turn our attention to other control methods, two practical issues should be discussed: reducing the variability of the holding times, and application to closed loops with a given number of buses, where H is endogenous.

Let us consider the variability in holding times across control points now. Expression (10.33) reveals that this variability could be considerable even if the headway of bus n stays fixed. This happens because d_s and β_s vary across control points. If holding times change considerably from one control point to the next drivers may be asked to hold at some points and not others, sometimes in alternating fashion. This could confuse drivers and be annoying to passengers; especially if the control points are closely spaced. For this reason, it makes sense to replace the constants d_s and β_s in the control law by route-wide averages, which we denote d and β, and use the following instead of (10.34):

$$k_0 = d + (\beta + \alpha)H, \qquad \text{where } d > 0 \text{ and } \alpha \in (0, 1); \text{ and} \tag{10.40a}$$

$$k_1 = -(\beta + \alpha)h_{1,s}, \qquad \text{where } d > 0 \text{ and } \alpha \in (0, 1). \tag{10.40b}$$

This simplification is also useful because it reduces the number of tuning variables and the data needed to apply the rule. This enables an operator to fine-tune the control based on actual field performance. Fine tuning of this type is highly desirable not just because our estimates of β can include some error but also because all drivers may not hold for the recommended amounts accurately, and it may be wise to compensate for any bias. To do this, we propose choosing initial values for the tuning variables as was just discussed (using a reasonable tentative value for H) but then, after using the resulting control law for a while, one should fine tune them experimentally in the real world over a period of weeks to see if the system performs any better.

Closed loops also need special mention because instead of an initial headway, H, the system is initialized with a fixed number of buses M and the headways are endogenous. Therefore, to determine an initial value for k_0 with (10.40a) we suggest using for H in this formula the average headway that can be expected for the given M and a chosen d, assuming an uncontrolled but stable system; see homework exercise 10.6. Fine tuning is more important in this case because the performance of the control law for closed loops has not been studied systematically.[13]

10.5.1.2. *Scheduled Systems*

Let us now examine systems with a published schedule like (10.18), presumably with long headways. The schedule is assumed to be given and to contain some slack. The primary goal of the control is adhering to the schedule.

We anticipate that the control law about to be presented shall generate deviations from the schedule that are small and unbiased. As such, they will be unrestricted in sign and often negative. Since the latter is undesirable — early bus departures defeat the purpose of the schedule — we propose using

[13]The difference in performance between open and closed routes should disappear as the number of stops and the number of buses tends to infinity. Thus, the fine tuning step is most important for systems with few buses.

for control a hidden schedule that is pushed back in time relative to the advertised schedule. A shift of about two standard deviations of the schedule deviations should eliminate the problem for all practical purposes. Since this subsection deals with control, it only uses the hidden schedule. Therefore, the word "hidden" will be dropped.

Given that there is a (hidden) schedule, the family of control laws we consider can be more general than (10.32). We shall use the form:

$$D_{n,s} \equiv [k_{0,s} + k_{1,s}h_{n,s} + k_{2,s}\varepsilon_{n,s}]^+, \tag{10.41}$$

which uses three control parameters and includes one dynamic term for the headway and another for the lateness. These two dynamic variables still allow buses to be self-reliant. The control law about to be described was proposed and analyzed in Xuan *et al.* (2011). To analyze it quickly using the concepts we have presented, we first introduce some simple conditions that guarantee (10.41) to be stable, and the performance formulas that arise under these conditions; then present the control law as a special case of (10.41) that satisfies these conditions; next show how to choose the tuning variables; and finally how the control law can be generalized to systems with multiple interacting routes.

***Sufficient conditions for stability and performance formulas*:** As before, our goal is choosing values of the control parameters that will reduce the linearized version of the dynamic equations to a favorable form. Since we have one more control parameter to play with, we shall shoot for an even more favorable form than (10.38) — a form that does not include the interaction term involving the previous bus, i.e.,

$$\varepsilon_{n,s+1} = \varepsilon_{n,s}\alpha_0 + \gamma_{n,s}, \quad \text{where } \alpha_0 \in [0,1) \quad \text{for } n, s = 0, 1, 2, \ldots \tag{10.42}$$

By removing the interaction term involving $\varepsilon_{n-1,s}$, equation (10.42) completely eliminates the influence of the previous bus. Thus, as long as deviations remain small, a controlled bus operates in isolation from the rest, i.e., without any bunching tendency.

Note as well that (10.42) is of the form (10.15) with gain $\phi = \alpha_0$. Therefore, the system is statistically stable, and the moments of the deviations from the schedule are given by (10.17b). In the limit of $s = \infty$, the moments are

$$E[\varepsilon_{n,\infty}] = 0, \quad \text{and} \tag{10.43a}$$

$$\text{Var}[\varepsilon_n, \infty] = \frac{\sigma^2}{(1-\alpha_0^2)}, \quad \text{where } \alpha_0 \in [0,1). \tag{10.43b}$$

Recall that this variance is an upper bound to the variances at every s.

Also important is the standard deviation of the headways because, as we have seen, it is used to define a lower bound on the scheduled headway H that ensures buses rarely catch up with each other. To evaluate this standard deviation recall that $\text{Var}[h_{n,s}] \equiv \text{Var}[\varepsilon_{n,s} - \varepsilon_{n-1,s} + H]$. And since buses operate in isolation under our form of control, the two latenesses appearing in this expression are independent. Thus, $\text{Var}[h_{n,s}] = \text{Var}[\varepsilon_{n,s}]+\text{Var}[\varepsilon_{n-1,s}] = 2\text{Var}[\varepsilon_{n,s}])$. Clearly, in the limit we have

$$\text{Var}[h_{n,\infty}] = \frac{2\sigma^2}{(1-\alpha_0^2)}. \tag{10.44}$$

Finally, note from (10.43b) and (10.44) that the smaller α_0 the better. However, we will find that if α_0 is too small the bus is more likely to run into the truncated part of the control regime with zero holding times, where (10.44) does not apply and escaping is a looming possibility.

The control law: To explore these ideas let us now examine the control law that achieves (10.42). Derivations similar to those of the previous subsection show that the law is

$$D_{n,s} \equiv [d_s - \beta_s(h_{n,s} - H) + (\alpha_0 - 1)\varepsilon_{n,s}]^+,$$
$$\text{where } \alpha_0 \in [0,1) \quad \text{for} \quad n,s = 0,1,2,\ldots \tag{10.45}$$

This corresponds to (10.41) with control parameters:

$$k_{0,s} = d_s + \beta_s\, H; \quad k_{1,s} = -\beta_s; \quad \text{and} \quad k_{2,s} = \alpha_0 - 1. \tag{10.46}$$

Note that the first two parameters are fixed and the third can be tuned with α_0.

For the control law to perform as designed and obey (10.43) and (10.44), the slack d_s in the planned schedule should be large enough for (10.45) to be rarely truncated, i.e., it should satisfy $[d_s \geq \beta_s(h_{n,s} - H) - (\alpha_0 - 1)\varepsilon_{n,s}]$ most of the time. Since the right hand side of this inequality has zero mean, this will happen if $d_s \geq 3\sigma_D$, where σ_D is the standard deviation of said right-hand side.

To put this condition to use let us express σ_D in terms of knowable constants. First rewrite the right-hand side in terms of the schedule deviations as $\beta_s(\varepsilon_{n,s} - \varepsilon_{n-1,s}) - (\alpha_0 - 1)\varepsilon_{n,s}$. Then, remembering that $\varepsilon_{n,s}$ and $\varepsilon_{n-1,s}$ are independent and using (10.43b), we find

$$\sigma_D \geq \sigma\sqrt{\frac{(1+\beta_s-\alpha_0)^2+\beta_s^2}{1-\alpha_0^2}}. \tag{10.47}$$

Thus, the slack in the schedule should satisfy

$$d_s \geq 3\sigma_D = 3\sigma\sqrt{\frac{(1+\beta_s-\alpha_0)^2+\beta_s^2}{1-\alpha_0^2}}. \tag{10.48}$$

Taken together, the conditions in (10.45) and (10.48) ensure that the control law (10.45) stays in the linear regime, and as a result (10.43) and (10.44) apply.

***Fine-tuning and evaluation*:** For $\alpha_0 = 0$ (the value that maximizes reliability) the right side of the inequality in (10.48) is $3\sigma_D \approx 3\sigma$, since β_s is usually much smaller than 1. Although typical schedules do not include so much slack, (10.48) can often be met without changing the slack by increasing α_0. This is particularly effective for small β_s. For example, if $\beta_s = 0.03$, the right-hand side of (10.48) drops almost linearly from a value of 3.1σ at $\alpha_0 = 0$ to a minimum of 0.82σ at $\alpha_0 = 0.95$; and if $\beta_s = 0.1$ the right-hand side also drops gradually from 3.3σ at $\alpha_0 = 0$ to a minimum of 1.5σ at $\alpha_0 = 0.86$. Therefore, if the available slack is above these minimum right-hand side values there should be a range of values of α_0 consistent with (10.48). And of course, one should choose the smallest of these, as this will minimize (10.43b) and (10.44), i.e., it will maximize reliability. Unfortunately, if the provided slack is beneath these minimum right hand side values then no feasible α_0 exists. Then, the control strategy cannot be made to work as predicted, and the escape problem will persist.

When choosing α_0 in practice one should recognize that drivers may not follow instructions exactly, adding to the noise, and that their compliance level may be negatively affected by highly volatile holding times. For this reason, the parameter σ appearing in (10.48) should incorporate driver error. Because this is hard to predict we recommend fine-tuning the choice of α_0 over a period of weeks with real world observations. Section 10.7 discusses how this was handled in a pilot study.

Control law (10.45), called the "simple control" in Xuan *et al.* (2011), turns out to be quite efficient. This reference examines not just the choice of α_0 but also the choice of slack at the planning level. It takes as given a desired reliability level quantified by a standard lateness variance (i.e., an upper bound), $\widetilde{\sigma_\varepsilon^2}$, and looks for least slack that can achieve the standard. Formally the reference looks for the solution of the following problem:

$$\min_{\alpha_0}\left\{d = 3\sigma\sqrt{\frac{(1+\beta-\alpha_0)^2+\beta^2}{1-\alpha_0^2}} : \frac{\alpha^2}{(1-\sigma_0^2)} \leq \widetilde{\sigma_\varepsilon^2}\right\}, \tag{10.49}$$

where the subscript s has been omitted since using an averaged demand rate is a practical thing to do. The solution to this problem for homogeneous problems with $\beta \leq 0.1$ and moderate standards, $\widetilde{\sigma_\varepsilon^2} \leq 2\sigma_\varepsilon^2$, reveals that in all cases the slack is reduced below $d = 1.7\sigma$ sometimes substantially. This can be compared with schedule control (the case with $\alpha_0 = 0$) which as we have seen yields $d \geq 3\sigma$.

The reference also uses a generalized cost framework incorporating waiting and riding time to evaluate the optimum level of service that the simple control can provide. It finds that the simple control cannot be improved significantly by more complex algorithms as long as the deviations stay small.

***Generalization to multiple routes*:** The simple control strategy we have described pertains to single lines but it can be generalized. Its control law (10.41) has two dynamic terms: a term, $k_{1,s}h_{n,s}$, that cancels the influence of the preceding bus, and a second term, $k_{2,s}\varepsilon_{n,s}$, that compensates for lateness. This works because the cancellation term removes the influence of other buses, so the deviations of the controlled bus can be treated in isolation from the rest. Argote *et al.* (2015) show how this cancellation-compensation idea can be extended to lines that share stations with other lines. This requires some care because in this case some passengers may be willing to board more than one line and as a result buses can be affected by buses from different lines — a bus's dwell time will then depend, not just on the headway within its own line, but also on the time gaps with buses from other lines that might syphon off passengers. To capture this effect, Argote *et al.* (2015) propose modifying the cancellation term of (10.41) to read: $\mathbf{k}_{1,s}\mathbf{h}_{n,s}$, where $\mathbf{h}_{n,s}$ is a column vector of time gaps with relevant lines and $\mathbf{k}_{1,s}$ is a row vector of constants. The reference discusses how the relevant lines can be selected and the constants estimated.

The cancellation feature of the simple control also has an important side benefit. By isolating buses, it confines errors caused by errant buses to the problematic buses, so the feature endows the system with a bit of robustness. On the other hand, this feature, and all other features we have discussed so far assume small disruptions. Consider this. If large disruptions create long headways and large latenesses the simple control returns $D_{n,s} = 0$ (no control), which allows the escape problem to persist. Thus, a more robust form of control should be sought.

The next section shows that stability under large perturbations can be achieved if the control law also includes the headway behind the bus "the backward headway." Because this information can no longer be estimated by a bus alone, frequent communications with the control center are needed. Therefore, robust control requires an investment in telematics; and depending on the telematics architecture also in ground personnel.

10.5.2. *Robust Control*[12]

This is perhaps the most pressing need in transit system operation. Today there doesn't seem to be any systematic way to restore order to a transit system that has been knocked so out of whack that the schedule is no longer useful. Severe weather and traffic often provide this impetus, but so can unanticipated surges in demand, bus malfunctions and even the course of time during normal operations when schedules have too little slack. So the quest for robust control laws should be a priority.

Since schedules do not always work, for a law to be robust it should not use a schedule and should instead focus on headways. It should regularize headways no matter how irregular, and do this even if the buses in operation are changed anywhere along the route at any time during the day.

The ability to operate on headways without exposure to the escape problem is appealing to transit agencies not just for the robustness it affords but also because it provides considerable flexibility — it allows the agency to insert and withdraw buses from a route anywhere instead of specific locations, and in so doing reduce unproductive deadheading kilometers when buses have to be moved across routes. The ability also relieves the agency from planning schedules that can rapidly become obsolete as the demand changes while providing a service that automatically adapts to the demand.

Unfortunately, it is very difficult to operate on headways in real life. Dispatchers have to monitor all buses in real time then determine what to do without the guidance of a schedule, and finally, communicate decisions to drivers in a timely and continued way. Obviously, the more frequent the communications and the smaller the requested adjustments the better the results, but this places a huge load on busy dispatchers and drivers. The authors are aware of only one instance where this type of control was tried as an experiment with existing technology, and it could not be made to work (AC Transit, 2015). We are not aware of any successful implementations on any transit line with many stops and buses.[13] What is needed is a control algorithm that produces simple feasible instructions for

[12]The narrative in this section builds on unpublished and non-refereed research notes (Daganzo, 2017). Therefore there is a chance for errors. The ideas, however, have been developed by Via Analytics into a CAD/AVL control tool, which has been tested in the field. We believe that the topic is so important that it needs to be presented, even if imperfectly.

drivers to follow, and the telematics architecture to deliver these instructions through onboard devices. This solution would eliminate ad hoc decisions and recurrent dispatcher interventions.

To our knowledge, no systems of this type are currently used by major transit agencies, although one was successfully piloted with the Oahu Transit Services (OTS) in Honolulu, Hawaii (Via Analytics, 2016). This test used the control algorithm about to be described. Onboard smart tablets told drivers how long to hold at the route's stops. During the tests, OTS was able to insert and withdraw buses at will from the routes tested without any dispatcher follow ups and despite these drastic disruptions achieved a similar level of headway regularity as when a schedule was being used.

Although the mathematical derivations below are general, our verbal description implies that control is applied at every stop. This is how it should be applied in practice for two reasons. First, the holding time formulas about to be derived do not include independent constants that systematically penalize buses every time they are held. And second, frequent holding allows the control law to compensate for disturbances when they are small, with tiny adjustments, avoiding the severe remedies that might otherwise be required.

Our overview starts with the control law, follows with stability analysis and concludes with some comments on fine-tuning and evaluation.

The control law: The holding times for the algorithm in question are of the form:

$$D_{n,s} \equiv [k_{0,s} + k_{1,s}(h_{n,s} - H) + k_{2,s}(\eta_{n+1,s} - H)]^+, \qquad (10.50)$$

where $\eta_{n+1,s}$ is the latest observed headway of bus n+1 (at some station upstream of s) when bus n arrives at s, i.e., the most recently observed "backward headway" of bus n. Although it would be simpler to use $h_{n+1,s}$ instead of $\eta_{n+1,s}$ in (10.50), this is not possible in practice because $h_{n+1,s}$, is not known at the time of calculation.[14]

The backward headway is included in (10.50) because it helps with the escape problem. Recall that this problem arises when headway grows so large

[13]Some systems show drivers their relative position to buses nearby, and then delegate the control decisions to drivers. While this approach can reduce the communications traffic, the systems are passive — they do not tell drivers what to do and do not let dispatchers know if what drivers are doing is counterproductive. Therefore performance cannot be guaranteed and the systems do not prevent the escape problem.

[14]On the other hand, since headways evolve slowly in transit systems, a simpler version of (10.50) with $h_{n+1,s}$ instead of $\eta_{n+1,s}$ should behave similarly as (10.50). Therefore analyses of the simpler equation can be useful.

that its trailing bus cannot keep up with the ensuing passenger loading work. Clearly, to address the problem the headway must be shortened. Since the trailing bus cannot go any faster, this can only be achieved if the lead bus slows down; and for this to happen, the lead bus must be aware of what is happening behind.

Since the purpose of our control law is providing robust control, i.e., dampen all perturbations no matter how large, the law must be evaluated without removing the truncation from (10.50), fully recognizing its nonlinear character. This is hard to do but has a silver lining: since we can operate freely in the nonlinear regime we no longer need to constrain $D_{n,s}$ to be positive most of the time. As a result, we are allowed to consider smaller $k_{0,s}$ and benefit from the shorter holding times that can ensue.

***Stability analysis*:** So let us now look at stability under large perturbations and the parameter values that achieve it. Intuition tells us that the longer a bus's forward headway the shorter the required holding (so we expect $k_{1,s} < 0$); and conversely, the longer its backward headway the longer the holding (so $k_{2,s} > 0$). There is a good reason to expect these choices to prevent the escape problem because there is no upper limit to the latter type of holding so we can, in theory, slow the lead bus of a long headway as much as necessary to prevent the headway from growing further. Indeed, as homework exercise 10.8 shows, if the control parameters satisfy the following constraints then the longest headway at any station is always expected to decrease in the next time instance:

$$k_{0,s} \geq 0, \quad \text{and} \quad -k_{1,s} = k_{2,s} \geq \beta_s > 0. \tag{10.51}$$

Note in choosing these constants that it is highly desirable to eliminate the target headway in (10.50) because said headway is hard to predict;[15] and that the equality $-k_{1,s} = k_{2,s}$, achieves this feat. Thus, the following two-parameter rule in which $k_{1,s}$ has been eliminated is proposed:

$$\begin{aligned} &D_{n,s} \equiv [k_{0,s} - k_{2,s}h_{n,s} + k_{2,s}\eta_{n+1,s}]^+ \\ &\text{with} \quad k_{0,s} \geq 0, \quad \text{and} \quad k_{2,s} \geq \beta_s > 0. \end{aligned} \tag{10.52}$$

[15]For closed loops the headway is not just endogenous but susceptible to uncertain traffic and demand conditions. Furthermore, as buses are inserted and withdrawn from a route of any type, the ideal headway would change. It would have to be recalculated, creating an extra hassle and a chance for errors.

Simulations of a broad set of scenarios with highly uneven starting headways confirm that (10.52) indeed restores order (Daganzo, 2017; Anderson and Daganzo, 2018).

Fine-tuning and evaluation: It should also be noted that setting the independent constant as low as possible is desirable because this reduces the duration of each hold. Therefore, we recommend choosing $k_{0,s} = 0$. The OTS pilot project used the same parameters for all control points: $k_{0,s} = 0$ and $k_{2,s} = 0.3 > \beta_s \approx 0.03$.

Note that law (10.52) does not use any information whatsoever about the system: not the number of buses; not the schedule; not a target headway; not the length of the route; nothing. Thus the law can continue to be used unchanged when the unpredictable happens and still achieve the desired result. The only requirement is that $k_{2,s}$ should exceed in absolute value the dimensionless demand parameter. But this is easy to ensure and adjust. We show below that setting $k_{0,s} = 0$ and $k_{2,s} = 0.2$ or 0.3 at all stations works well for any situation.

A *linearized* version of (10.52) with a large $k_{0,s}$ was tested in Xuan *et al.* (2011). It performed almost as well as the simple control, which used a schedule, and considerably better than the basic forward-headway method.[16] This evaluation, however, severely underestimates the real value of the robust control since it did not test it for low values of $k_{0,s}$.

The robust control turns out to heal the system fairly quickly from disruptions. Simulations of (10.52) with $k_{0,s} = 0$ and $k_{2,s} = 0.3$ were conducted for a route with similar demand and variability as OTS's Route 8: $\beta = 0.03$ and $\sigma = 0.25$ min (Daganzo, 2017). The simulations tracked a platoon of 100 buses that were dispatched from the route's terminus (stop 0) at random intervals with a mean of $H = 5$ min. The route was extended to include 1,000 stops to explore what happens in the extreme long run.

Results showed that order improved quickly and substantially after a few stops, and then more slowly, even if the initial headways were chaotic. For example, when the initial headways were endowed with a large standard

[16] A historical note is in order. Although the ideas in this section are based on Daganzo (2017), there are earlier related works. Daganzo and Pilachowsky (2011) proposed a space-based, two-way looking control rule that was very similar to (10.52) — it used spacings instead of headways in the control formula. The reference showed that this rule allowed buses to travel faster and with more regular headways than forward-looking control. Bartholdi and Eisenstein (2012) proposed a similar rule to (10.50) in which $k_{1,s} = 0$ and demonstrated that it solved the escape problem for low demand, but this rule slows buses considerably (see Xuan *et al.*, 2011).

Table 10.1. Simulated performance of robust vs. forward-headway control (Daganzo, 2017).

	Robust control (10.52)			Forward-headway control (10.34)		
k_2	σ_H/σ	Average holding/σ	Max holding/σ	σ_H/σ	Average holding/ σ (i.e., d/σ)	Max holding/σ
0.2	2.9	0.5	4.2	2.5	1.7	–
0.3	2.7	0.8	7.1			

deviation (4 min) and nearly 20% of the buses were dispatched as pairs, the control never succumbed to the escape problem. Instead, after many repetitions of the simulation, at station 1 the robust control invariably had eliminated nearly all pairs and reduced the standard deviation to 3 min (in a ratio of 1.3); at station 10 the control always eliminated all pairs, and further reduced the standard deviation to 2.0 min on average (in a ratio of 2.0). The standard deviation continued to decline, albeit more gradually — it averaged about 1.5 min by station 40 (ratio of 2.7), and about 0.7 min (ratio, 5.7) after hundreds of stations, close to the steady state. Similar ratios were found for other values of H and σ.

Just as importantly, the steady state statistics turned out to be rather favorable. This is illustrated in columns 2 to 4 of Table 10.1, which were obtained for the values of the k_2 shown in column 1.[17] Performance degraded for larger values close to 1 and smaller values close to β. As a point of comparison, the last three columns show corresponding statistics for the forward-headway control. They were obtained by applying the formulas of Section 10.5.1.1 to the same route.

Note the robust control achieves comparable headway regularity while delaying the buses two to three times less. Speeding up the buses is quite important. For example, if $\sigma = 15$ s per station the total amount of holding during 40 stops (a reasonable route length) would be 17 min for the basic strategy, but only 5 to 8 min for the robust control. And the robust control prevents the escape problem.

[17]These statistics are representative of steady state operations. As such they do not capture the effect of large disruptions such as the insertion and withdrawal of buses at inconvenient locations. As we have seen, when headways are disrupted performance degrades while the system returns to the steady state; and this can take a while.

To be fair, the results in the table are optimistic because they do not factor in real world difficulties like time-dependent demand, driver errors, measurement errors and communication malfunctions. But the magnitude of the improvement strongly suggests that the use of smart algorithms and advanced telematics has much potential. Consequently, if an agency has these tools at their disposal then any route that is operated on headways should probably be operated under some form of robust control all the time. Routes operated on a schedule should use the "simple control," but they should be monitored closely so that when/if the simple control fails (or is about to fail) one can flip a switch, ditch the schedule and fail-safe with a pre-determined robust control law for the rest of the day.

10.6. Practical Considerations

We describe here how to estimate the input data required by the control strategies described in this chapter, and the ways in which they can be modified to improve their effectiveness.

Input data are needed to properly set the $k-$parameters of our control laws. You can see from expressions (10.34), (10.46), and (10.50) that we need estimates of the dimensionless demand parameters β_s; and in the case of the simple control (10.46) also estimates for the slack times d_s included in the schedule.[18] Fortunately, these parameters can be estimated very accurately by direct observation of the buses.

Consider β_s first. For each station s, this parameter can be estimated by seeing how the observed arrival headways $\{h_{n,s}; n = 1, 2, 3, \ldots\}$ of the various buses influence the times $\{\tau_{n,s}; n = 1, 2, 3, \ldots\}$ the buses take to traverse a cell centered on the station — in order not to miss any part of the delay caused by a stop, this cell should be big enough (say 100 m) to accommodate the portions of the buses trips where they decelerate and accelerate on arrival and departure from the station, but also not so big that traffic conditions near the station would introduce significant unnecessary noise. Recall that β_s represents the rate at which the mean of $\tau_{n,s}$ conditional on $h_{n,s}$ increases with $h_{n,s}$. To the extent that this relationship is approximately linear, as was assumed in our models, it makes sense to estimate β_s as the slope coefficient,

[18] Note that the equilibrium headway H appearing in (10.34a) is endogenous if the route is a closed loop operated with a fixed number of buses. Side calculations should then be performed to estimate it. Great accuracy is not needed because H does not affect $k_{1,s}$ and, as can be seen from (10.34a), only constrains $k_{0,s}$ through the inequality $k_{0,s} > (\beta_s + \alpha)H$.

$\widehat{\beta}_s$, of a linear regression with the data set: $\{(\tau_{n,s,}, h_{n,s}); n = 1, 2, 3, \ldots\}$. Note that these data can be automatically harvested if the onboard equipment is sufficiently sensitive, as was the case in the aforementioned OTS pilot tests. Thus any initial estimation (say lasting a week or two to get accurate estimates for different times of the day and week) can and should be automatically updated on an ongoing basis.

Consider now the slack, which can be estimated using the definition: $d_s \equiv p_s - c_s$. The scheduled travel time p_s is given by the schedule (10.18), and the expeditious travel time c_s can be estimated from (10.19) as explained below, using the observed bus arrival times. We recommend gathering these observations while the system is in uncontrolled mode as one cannot be assured that drivers will drive expeditiously when in control mode (e.g., if they are repeatedly asked to hold). To derive our estimator, solve for c_s using the first and third members of (10.19) and write: $c_s = u_{n,s} - \beta_s(h_{n,s} - H) - \gamma_{n,s}$. Now note that every variable and constant on the right-hand side of this equality, except $\gamma_{n,s}$ is either observed or known. Fortunately, this noise term can be eliminated. The reason is that the equality is true for every n and therefore it must also be true for its average across n, which does not include $\gamma_{n,s}$, i.e., the following is true: $c_s \approx \bar{u}_s - \beta_s(\bar{h}_s - H) \approx \bar{u}_s - \hat{\beta}_s(\bar{h}_s - H)$. Thus, an estimator for the slack is $\widehat{d_s} \approx p_s - \bar{u}_s + \widehat{\beta}_s(\bar{h}_s - H)$.

We now examine how the control may be enhanced by selecting where to hold the buses. There are three possibilities. Buses can be held: (i) after loading all the passengers at the last station before reaching $s+1$; (ii) while traveling from that station to $s+1$; or (iii) immediately before loading the passengers at station $s+1$. There are advantages and disadvantages to these possibilities.

Approach (i) is convenient. One holds the buses after loading the passengers, and to put on-board passengers at ease, the driver can keep the doors open to avoid the false impression of an immediate departure. On the other hand, the bus occupies the station while holding, which increases the chances of interference with a bus from another line. The approach performs as predicted by our dynamic equations.

Approach (ii) reduces the chances of interference and completely hides the holding from on board passengers. In this case, too, our equations are accurate. The approach is more complicated, however. It requires sensitive onboard devices operating with low latency in order to continuously inform drivers whether they should drive more slowly. And it requires more extensive driver training. The approach can be used together with (i), for example by setting a limit for holding while cruising to (say) 10 or 15 s and holding in place for the excess, if any. A similar hybrid method (Argote-Cabanero *et al.*, 2015) was used in the abovementioned pilot study.

Approach (iii) also performs as predicted. Although the method may inconvenience passengers (unless the control points are not located at stops) it is intriguing because by postponing holding to the last minute the controller can also postpone the calculation of $D_{n,s}$ and have an opportunity to measure the noise. This opens the door for noise-canceling control and improved reliability.

Noise-canceling control[19]**:** To illustrate this idea as clearly as possible, consider an extreme case where the noise measurement is exact, i.e., we know $\gamma_{n,s}$ when the holding time is calculated and applied. Therefore, the noise can be introduced in the formula for $D_{n,s}$. To do this it is useful to write the original control law as $D_{n,s} = [L_{n,s}]^+$, where $L_{n,}$ is the untruncated part of the formula, and the noise-canceling law as $D_{n,s} = [L_{n,s} - \gamma_{n,s}]^+$. The reader can verify that this modification cancels out the noise term from the linearized dynamic equations, e.g., (10.38) and (10.42). This is useful because without noise ($\sigma = 0$) the system would then become perfectly reliable.

In practice, this is impossible of course because any noise measurement must include some error, e.g., due to human holding imperfections. However, any such error should have a smaller standard deviation than σ and therefore translate into improved reliability — recall that the standard deviation of all our reliability measures are directly proportional to σ.

Although the noise-canceling idea has not been piloted, it is worth considering. The noise appearing in the holding time formula $D_{n,s} = [L_{n,s} - \gamma_{n,s}]^+$ should be estimated immediately prior to the moment of arrival at $s+1$, when control is about to be applied. This can be done by solving for $\gamma_{n,s}$ in the system's dynamic equations. For example, in the case of (10.35b) the estimator for $\gamma_{n,s}$ would be

$$\widehat{\gamma_{n,s}} = \widehat{a_{n,s+1}} - a_{n,s} - \widehat{c_s} - \widehat{\beta}_s(h_{n,s} - H) - [\widehat{d_s} - (\widehat{\beta}_s + \alpha)(h_{n,s} - H)]^+. \tag{10.53}$$

The estimation is possible because at the moment of estimation $\widehat{a_{n,s+1}}$ is known accurately, and every other term in the right side of this equation — or the equivalent equation for other control methods — is also approximately known. Note that the design parameters H and α are given; the constants

[19]This subsection can be skipped without loss of continuity.

$\widehat{c_s}, \widehat{d_s}$, and $\widehat{\beta_s}$ have been estimated; and the dynamic variables $h_{n,s}$ and $a_{n,s}$ have been previously observed.

Note that the noise-canceling method can also be applied if the bus is held at the last station upstream of station $s+1$. The only thing that changes is that the estimates $\widehat{a_{n,s+1}}$ and $\widehat{\gamma_{n,s}}$ will then contain some error. However, if control is not applied at every station the error in the latter, $(\widehat{\gamma_{n,s}} - \gamma_{n,s})$, should have a standard deviation $\sigma' < \sigma$ since only part of the noise remains unrealized at the moment of estimation. This should be helpful since the standard deviation of our reliability measures is in theory proportional to σ'.

10.7. Field Studies and Human Factors

So far we have examined how transit systems respond to control assuming that drivers faithfully follow instructions, with small random errors. What about reality, however? Since drivers are not robots (yet) it is important to understand how control instructions should be relayed for maximum clarity, and how compliant drivers can be expected to be. One can then fine tune the control algorithm to compensate for the drivers' behavior, and also estimate the real improvement in performance that can be achieved *vis-à-vis* the theoretical improvement. This section will discuss what we currently know about these issues. It will present evidence from two multi-month pilot projects that tested the simple control and the robust control law.

10.7.1. *The Simple Control: Evidence from Dbus (San Sebastian, Spain)*[20]

The first pilot study was carried out in San Sebastian (Spain) by *Via Analytics* with the cooperation of the local bus agency, *Dbus*, which is internationally recognized for its excellent service and adoption of novel technologies. The experiment site was a long, busy corridor served by two overlapping lines: one with equal headways of 20 min and the other with alternating headways of 6 and 8 min.

Since much of the passenger traffic was confined to the common corridor, *Dbus* analysts had devised a clever schedule that interweaved both lines to provide short (4 and 6 min) headways in the corridor. They were keenly aware that to achieve these short headways with regularity, their buses would have

[20]This subsection is based on Argote-Cabanero *et al.* (2015), which contains additional details.

to be kept tightly on schedule. They teamed up with *Via Analytics* to try a solution based on the simple control.[21]

Called *Tempo* this solution operates in the cloud. Onboard equipment relays position information to cloud servers. This information is then fed into the control algorithm, which updates the current instructions for each driver, much as if a dispatcher was in the bus. The algorithm also refreshes web-based maps with layers of relevant information for dispatchers and field inspectors.

In the case of *Dbus*, Android tablets (Samsung Galaxy Tab 2) mounted on the buses' dashboards served as the onboard equipment, and the control algorithm was the simple control. The tablets displayed two types of information: cruising guidance icons while moving; and a countdown clock with any remaining holding time when stopped at a station. The display format was optimized in collaboration with drivers. Easily seen colored bars that expanded up and down like mini-blinds were used as cruising guidance icons (see Figure 10.11). Also provided were numerical scores ranging from +5 (when the display is intense red signifying to slow down) to −5 (for intense green signifying to speed up safely). On arrival at a station, this display automatically changed to a large digital countdown clock and, upon expiration, a large green arrow signifying "go."

To rigorously stress-test the system, its performance was evaluated at the last station of the common corridor since this is where the largest schedule deviations occurred. The root mean square deviation from the scheduled arrival time at this station was the benchmark. This statistic is a better

(a) behind schedule (b) ahead of schedule

Figure 10.11. On-board device interface for cruising guidance.

[21] As of this writing, *Dbus* has adopted this solution and is expanding it to its entire system.

measure than the standard deviation because it also captures the possibility of bias, which could exist for reasons not captured in our theories.

The following three control scenarios were compared:

(1) *Static control: the status quo (simulated and real).* This captures Dbus' original performance with static schedule control, and buses held only at the terminal stations. The dashed horizontal line of Figure 10.12 shows the benchmark in seconds. This type of control was also simulated assuming that drivers follow the control recommendations perfectly. This is shown by the solid horizontal line of the figure. Note performance improves slightly in the simulations, presumably because real drivers do not follow the control instructions as well as the ideal. The closeness of the lines suggests that the simulations were reasonable.

(2) *Simple control: control at every station (simulated).* This set of simulation experiments used the same parameters as before but buses were held at every station as per (10.41) and (10.46). The strategy was simulated for many different values of α_0, each for 50 days. The solid dark curve of Figure 10.12 displays the result. At its lowest, this curve improves the original benchmarks by more than 35%. The shape of the curve also agrees with (10.43), confirming that there is a (theoretical) benefit to reducing α_0, especially when α_0 is close to 1.

(3) *Simple control: control at every station with cruising guidance (simulated and real).* This is the same as the above but including cruising guidance. It was evaluated in the field and with simulations. In the field, the cruising guidance bars were tuned to reflect the current deviation from the schedule in minutes, $-\varepsilon_{n,s}$. The performance was then observed for two control settings: $\alpha_0 = 0.5$ and 0.9. The results are depicted by the square and the triangle of Figure 10.12. Performance was also simulated for many other values. To this end, drivers were assumed to respond to cruising guidance linearly, as estimated from many weeks of field data. The simulated results including this feature are shown by the dashed sloping curve in the figure. Not surprisingly the new curve is lower than the old. Curiously, however, the curves do not match well the real results for $\alpha_0 = 0.5$. The good news is that the observed performance for $\alpha_0 = 0.9$ (the square) is about 22% better than the original benchmark. This square is even below the solid curve without cruising guidance, suggesting that cruising guidance helps real drivers perform even better than holding-only ideal drivers. On the other hand, the results for $\alpha_0 = 0.5$ were not as close to the simulated curves. Argote *et al.* (2015) attribute this to the fact the holding times were then larger,

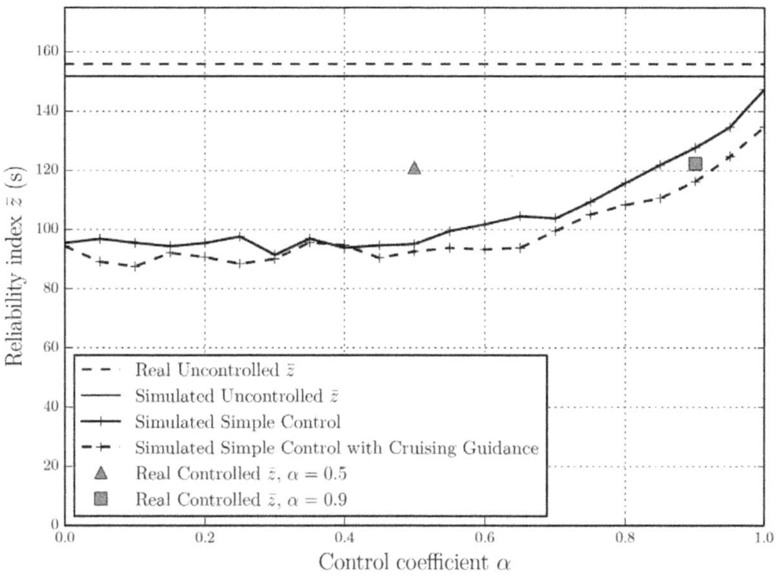

Figure 10.12. Simulated and observed root-mean-square errors of the simple control (*source*: Argote-Cabanero *et al.*, 2015).

highly variable and more difficult for the drivers to match — see the ensuing discussion on human factors.

Human factors: The experience with Dbus revealed insights about both, driver compliance with the holding clocks and adaptation to cruise guidance, which can be used for both training and design.

Driver compliance was directly observed at the control points. *Tempo* detected the instants of departure from every holding point and compared these instants to the corresponding expirations of the driver countdown clocks. Departures before expiration were labeled as violations, measured in seconds and associated with the driver at the wheel.

Compliance levels were first evaluated for $\alpha_0 = 0.9$. The dataset included 355 trained drivers who received at least 100 holding instructions over the course of 5 months. Figure 10.13(a) shows the results. Each data point represents one driver. The histogram on the top reveals that drivers were quite different but good overall: the median driver violated instructions only 13.8% of the time, and the worst driver by 26%. The scatter plot on the bottom part shows that the drivers' average violation values ranged from -6 s to -12 s, and were higher for the less compliant drivers.

The system was then evaluated for $\alpha_0 = 0.5$ for a period of three weeks without retraining or notifying the drivers. This data set included 224 drivers

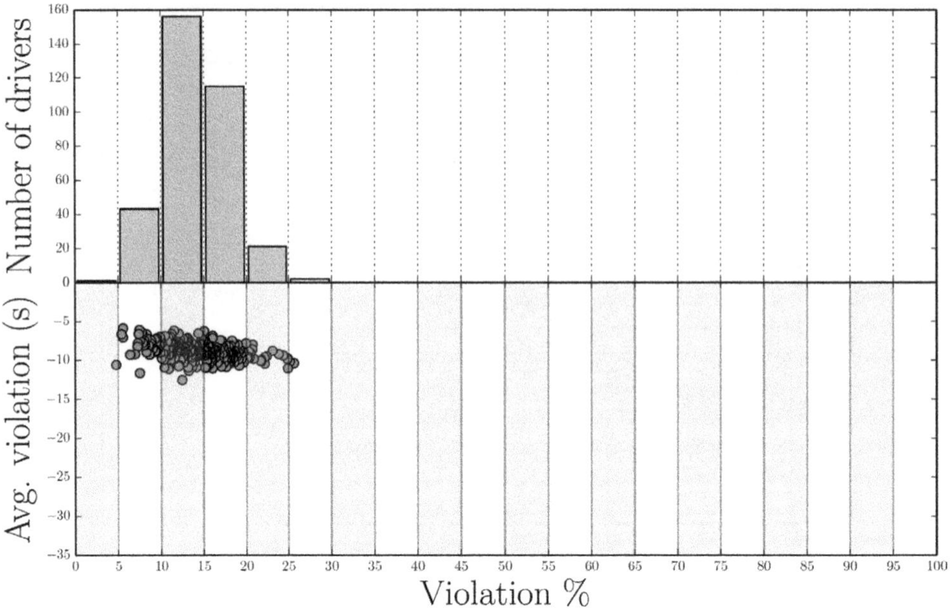

(a) $\alpha_0 = 0.9$

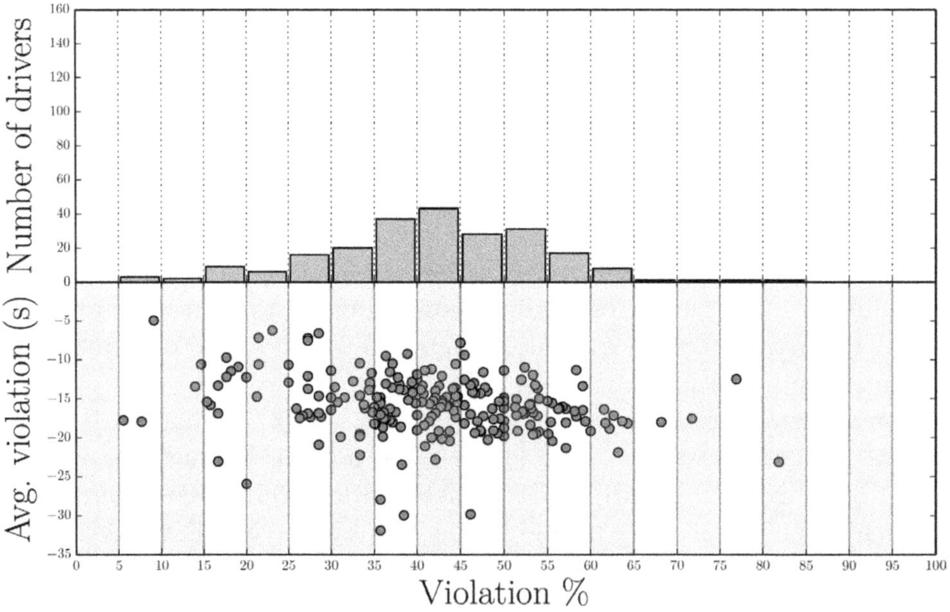

(b) $\alpha_0 = 0.5$

Figure 10.13. Driver compliance for different α_0 : each data point corresponds to one driver.

with a minimum of 10 holding instructions each. Compliance turned out to be significantly lower, as shown in Figure 10.13(b). The median driver violated holding instructions 42.1% of the time, the best 5%, and the worst 80%. The average magnitude of these violations ranged from −5 s to −32 s with an average of −15.8 s. The considerable drop in compliance explains the unexpectedly high positioning of the triangle in Figure 10.12 relative to the curves simulated with ideal drivers. The notable difference between parts (a) and (b) of Figure 10.13 also indicates that drivers tend to disobey holding times that are highly variable. This problem can probably be alleviated with a combination of more training, rigorous compliance monitoring and a system of rewards. And of course, the problem should not arise at all if/when robotic buses are used in the future.

Drivers, on the other hand, seem to respond well to cruising guidance. This was established by relating the observed cruising times from departure to arrival at consecutive stations, $c^o_{n,s}$, to the driver's stimulus, $-\varepsilon_{n,s}$. This relationship was found to be highly dependent on location, which should not be surprising since some street segments are more conducive to saving time than others. Figure 10.14 displays scatter plots from two of the segments with a strong dependence. Darker tones indicate higher measurement densities. The figure shows that drivers decreased their average cruising time on these segments by about 5% for every minute of lateness. As reported in Argote-Cabanero *et al.* (2015), the effect across the complete line was noticeable even though many street segments exhibited no correlation.

To close this discussion we now show that this stimulus-response effect can act as a stabilizing force that improves reliability, not just in our particular

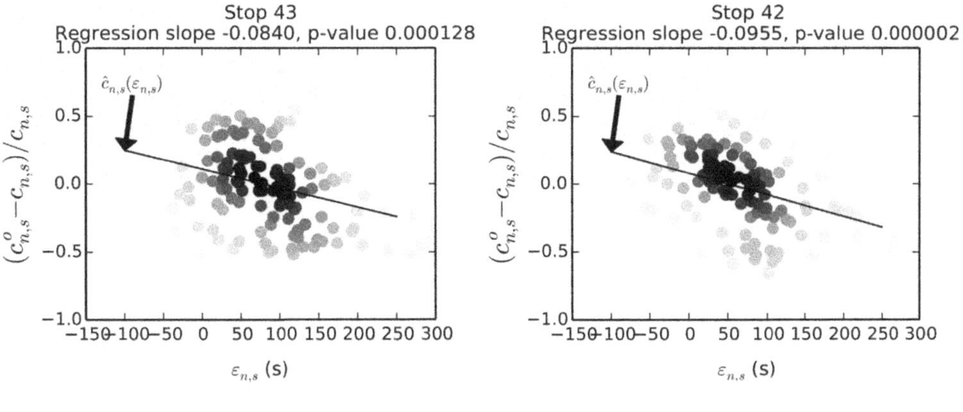

Figure 10.14. Schedule deviation vs. relative cruising time difference.
Source: Argote-Cabanero *et al.* (2015).

case study as exemplified by the simulated curves of Figure 10.12, but in general. To see this note that cruising guidance changes the travel times in the dynamics equations (10.19), (10.26), (10.37) from being constants c_s to becoming functions of the deviations at the start of the segments, i.e., to $c_s - \tau_s \varepsilon_{n,s}$, where we expect $0 \leq \tau_s \ll 1$. As a result of this change the extra term $-\tau_s \varepsilon_{n,s}$ must be added to the right side of the system dynamics equations. This addition reduces the original coefficient of $\varepsilon_{n,s}$, which is good for stability. For example, (10.42) becomes

$$\varepsilon_{n,s+1} = (\alpha_0 - \tau_s)\varepsilon_{n,s} + \gamma_{n,s+1}, \text{ where } \alpha - \tau_s \in (0,1). \tag{10.54}$$

Note that this expression has the same structure as the original, except the gain coefficient $(\alpha_0 - \tau)$ is smaller. Therefore, reliability improves. What's good is that this reduction in gain is achieved without reducing α_0, and therefore without adding to the volatility of holding times, which as we have seen hinders compliance. This shows that cruising guidance should be included in a control system when at all possible.

10.7.2. *The Robust Control: Evidence from OTS (Honolulu, Hawaii)*[22]

A second pilot study was conducted in Honolulu (Hawaii). The goal was eliminating the schedule on one of *Oahu Transit Services* routes (Route 8), and see if by operating it on headways regularity could be improved. As it turns out other benefits emerged.

Route 8 serves a busy corridor, which includes the Ala Moana shopping center and much of Honolulu's beachfront. The line is long and simultaneously served by many buses, so the task was challenging. OTS was interested in a headway-based operation because OTS likes to temporarily borrow buses from Route 8 to fill in gaps in other nearby routes. If successful, headway operations would allow OTS to do this kind of interlining adaptively and very efficiently — it could pull any bus from Route 8 service at any time, and return it anywhere, without any sequencing or scheduling considerations.

The pilot test used the same tablet-based *Tempo* system of Section 10.7.1 and the robust control law. Unfortunately, two technical difficulties stood in the way of a rigorous evaluation of regularity improvements: (i) Route 8

[22] This subsection is based on an unpublished final report and presentation to OTS (VIA Analytics, 2016) and on numerous private communications with OTS engineers and management.

often included interlined buses from other routes that were not equipped with *Tempo* tablets; and (ii) the amount of interlining (and the percentage of unequipped buses) increased substantially after the transition to headway-based operations.

To deal with (i), the control algorithm was modified to include a "phantom" bus where an unequipped (and therefore unobservable) bus was supposed to be. Phantom buses in a gap between two equipped buses were assumed to divide the gap evenly into equal "phantom" headways. These headways were then used in the control algorithm of the equipped buses. Phantom buses were not controlled. As you can imagine this form of operation is far from ideal since unequipped buses can bunch and misdirect the control of the remaining buses.

The system was benchmarked for two weeks starting August 16, 2015. During this time OTS operated Route 8 as usual, with schedule control and relying heavily on dispatchers to manage large disruptions. Headway-based tests with the robust control were then conducted for 12 weeks from September to November, 2015. During this period, the percentage of equipped vehicles fluctuated around 70%.

Headway data was gathered before-and-after. Obviously, only headways between equipped buses could be measured throughout the study. Records show that a statistic of interest to OTS ("headways in good range") increased from 17% to 25% after the system was transitioned to the robust control; and more generally, the variance of the headways decreased by nearly 30% (from 72 to 50 min^2). This is good but not as large as one might have expected in a more controlled environment without phantom buses.

Perhaps the most important takeaway of this experiment, however, is not that reliability improved but that operation by headways in a real world setting involving a long route with many buses and significant overlaps with other routes was *feasible*. The control algorithm also proved to be *resilient* to the severe interlining demands imposed on it. According to OTS, the ability to do this allowed them to increase significantly the utilization of the buses involved, which impacts their bottom line. OTS also reported that both, dispatchers and drivers felt that their workload was reduced despite the more complex operation. All of these benefits are qualitative but tangible. They should be considered alongside the reliability improvements.

In summary, the OTS pilot established that headway-based operations are feasible on a large scale and that they can be resilient, economical and worker-friendly. The results also suggest that the robust control can improve reliability, although the exact amount has yet to be pinpointed.

10.8. Remedial Measures

This section discusses *ad-hoc* measures that can and are often used in practice to intervene when the basic control mechanism has failed to prevent large gaps in service. All these measures focus on reducing the sizes of these gaps by either narrowing or filling them. Our review below groups them into three sets. Measures in the first set are applied at termini; those in the second increase the speed of lagging buses; and those in the third use more aggressive actions.

10.8.1. *Terminus Strategies*

These strategies exploit the special nature of termini, where buses can be held for a long time without affecting passengers. Long holding times are also useful because they allow drivers to get extra rest while staying with their buses.[23] Given the importance of rest, our strategies will maximize orderliness while guaranteeing minimum rest periods.

Notation that highlights the special role of termini will be used. Since the only control point is the terminus, the sub-index s referring to the control point will be omitted. The bus sub-index n will be retained, however. We shall also introduce g_n as the time at which bus n is scheduled to depart from the terminus (if there is a schedule) and d_n as the actual departure time. We will use T_o for the minimum guaranteed rest, and $T > T_o$ for the target dwell time at the terminus. The difference $T - T_o$ is slack that drivers can enjoy as extra rest time. If there is a schedule with arrival times t_n, the target dwell time is $T \equiv g_n - t_n$. If there is no schedule, T is a control variable. The actual arrival times will be denoted a_n and the actual dwell times $R_n = d_n - a_n$.

Operation with a schedule: Let us now consider the situation with a schedule. In this case, it makes sense to release each bus as early as possible but not before its driver has rested the minimum amount, and not before the scheduled time, i.e., we set the departure time as

$$d_n = \max\{a_n + T_o, g_n\}. \tag{10.55}$$

[23]Rest can also be given in other ways. For example, a relay method (contemplated by OTS) decouples rest from the dwell time using a pool of X resting drivers. It works as follows: when a bus arrives at the terminus its driver hands it off to the member of the pool who has rested the longest, then the arriving driver joins the pool, rests for X headways and takes the next handoff.

This rule makes sense because it restores perfect order ($d_n = g_n$) if the slack is sufficiently generous, i.e., if $(T - T_o) > \varepsilon_n$ for all n. To see this, replace a_n by $(t_n + \varepsilon_n)$ and g_n by $(t_n + T)$ in (10.55). This yields: $d_n = \max\{t_n + \varepsilon_n + T_o, t_n + T\} = (t_n + T_o) + \max\{\varepsilon_n, T - T_o\}$. Now, if $(T - T_o) > \varepsilon_n$ for all n then the last member of our string of equalities becomes $t_n + T \equiv g_n$. Thus, $d_n = g_n$.

This insight should be used to select T in the planning process. The value $T \approx T_o + 2\sigma_{\varepsilon}$. seems reasonable since it should restore nearly perfect order under normal operation. Larger values should be seriously considered, however, since they provide insurance against the unexpected.

Operation without a schedule: Let us now consider operations without a schedule. The goal now is to regularize the departure headways as much as possible by allowing the dwell times R_n to fluctuate dynamically around the target value T while satisfying $R_n \geq T_o$. The proposed control rule is

$$d_n = a_n + R_n = a_n + T + \alpha\,(\eta_{n+1} - h_n). \tag{10.56}$$

It is assumed that the calculation is done at the moment of arrival, although there might be a slight advantage in postponing the calculation to get a more updated value for η_{n+1}.[24] In any case, we expect $\eta_{n+1} \approx h_{n+1}$ so our analyses will use the approximation:

$$d_n \approx a_n + T + \alpha\,(h_{n+1} - h_n). \tag{10.57}$$

The last term, in these two expressions, is the dynamic adjustment to the target rest time. These adjustments must be sufficiently small to ensure that the dwell times $R_n = T + \alpha\,(\eta_{n+1} - h_n) \approx T + \alpha(h_{n+1} - h_n)$ match or exceed T_o with a high probability. Since this constraint automatically forces $d_n \geq a_n$ it obviates the need for the truncation operation in (10.57).

We now derive the mean and variance of the dwell times. These moments are needed to set a reasonable value for T. Consider the mean first. We assume that the system is in a steady state with bounded headways, as would occur with the robust control. Then, the sum of the dynamic terms $(h_{n+1} - h_n)$ across an arbitrarily large number of consecutive buses is the difference between the backward headway of the last bus and the forward headway of the first. This is a bounded quantity independent of the number of buses. Thus, the ratio of this quantity and the number of buses (i.e., mean of the headway differences) tends to zero as the number of buses is increased.

---

[24]For example, one could reevaluate d_n while the bus is waiting every time η_{n+1} changes, and dispatch the bus as soon as the latest computed d_n is in the past.

Thus, in the steady state $E[h_{n+1} - h_n] = 0$; and hence, the expectation of $R_n \approx T + \alpha\,[h_{n+1} - h_n]$ is $E(R_n) \approx T$.

Now consider the variance. Since $R_n \approx T + \alpha\,(h_{n+1} - h_n)$, we have: $\text{Var}[R_n] \approx \alpha^2\ \text{Var}[h_{n+1} - h_n] = \alpha^2\{\text{Var}[h_{n+1}] + \text{Var}[h_n] - 2\text{Cov}(h_n,\, h_{n+1})\}$, where $\text{Cov}(h_n,\, h_{n+1})$ is the covariance. Now, if we use σ_H for the standard deviation of the headways and ρ for the lag-1 correlation of the headway sequence, the above can be rewritten as $\text{Var}[R_n] \approx \alpha^2[(\sigma_H)^2 + (\sigma_H)^2 - 2\rho(\sigma_H)^2]$, which reduces to

$$\text{Var}[R_n] \approx 2\alpha^2(1 - \rho)(\sigma_H)^2. \qquad (10.58)$$

We expect ρ to be negative, most likely in the range $(-0.3,\, 0)$.

The driver rest constraint ($T_o \leq R_n$) is satisfied with high probability if $T_o \leq E[R_n] - 2(\text{Var}[R_n])^{1/2} \approx T - [8\alpha^2(1 - \rho)]^{1/2}\sigma_H$. If we use $\alpha \approx 1/3$ in this expression (a value that will turn out to minimize the variance of the outbound headways) then the term in brackets ranges from 0.94 to 1.33 for negative values of ρ; and is close to 1 for the most likely values between -0.3 and 0. Then the constraint reduces to $T_o \leq T - \sigma_H$, so it makes sense to choose: $T = T_o + \sigma_H$. This formula shows that the required slack to ensure minimum rests is comparable with a single standard deviation of the headway.

Let us now examine the outbound headways, $h'_n \equiv d_n - d_{n-1}$. We shall find that the control strategy preserves their mean and reduces their variance by a factor of 3. To see how the inbound and outbound headways are related we subtract the two consecutive instances of (10.57) corresponding to n and $n-1$ because then the left-hand side is what we are looking for: $d_n - d_{n-1} \equiv h'_n$. The result of the subtraction is $h'_n \approx a_n - a_{n-1} + \alpha(h_{n+1} - 2h_n + h_{n-1})$, where $a_n - a_{n-1} \equiv h_n$. Thus, we have

$$h'_n \approx (1 - 2\alpha)\,h_n + \alpha\,(h_{n+1} + h_{n-1}). \qquad (10.59)$$

As before, we now consider the system after it has reached its steady state and take means and variances on both sides of (10.59).

For the mean, we find

$$E[h'_n] \approx (1 - 2\alpha)E[h_n] + 2\alpha E[h_n] = E[h_n]. \qquad (10.60)$$

This shows that the control strategy approximately preserves the average headway.

Now let σ'_H denote the steady state standard deviation of the outbound headways, and take variances on both sides of (10.59). In doing this we assume (reasonably) that the steady state correlations for the inbound headways are negligible for lags greater than 1. The result of this exercise is

$$(\sigma'_H)^2 \approx [(1-2\alpha)^2 + 2\alpha^2 + 2\alpha(1-2\alpha)\rho](\sigma_H)^2. \tag{10.61}$$

The factor in brackets is minimized by $\alpha^* = (2-\rho)/(6-4\rho)$, which is very close to $1/3$ for all $\rho \in [-1, 0]$. Thus, a good approximation for the optimum of (10.61) is its value for $\alpha = 1/3$. This yields:

$$(\sigma'^*_H)^2 \approx [(3+2\rho)/9](\sigma_H)^2 \leq (\sigma_H)^2/3, \text{ for } \rho \in [-1, 0]. \tag{10.62}$$

This expression shows the steady state standard deviation of the headways can be reduced by a factor of $\sqrt{3}$ or greater at the terminus, whereas this is impossible at an ordinary station.[25] This happens because the target slack at a terminus can be set to be very large without disturbing either the passengers or the drivers, and when this is done the control law never transitions into the inefficient truncated regime that arises at ordinary stops.

General comments: Because all terminus control methods change the bus latenesses and headways so much, any holding time calculations at other stations that need some input from the terminus should use the latest information available. The issue arises with the robust control when the backward headway of a bus n is estimated at some station s downstream of the terminus when bus $n+1$ is still in the terminus. In this case $\eta_{n+1,s}$ would ordinarily be defined by the headway h_{n+1}, observed when bus $n+1$ arrived. However, at the current moment of estimation d_{n+1} has already been computed, so an estimate for the departure headway h'_{n+1} is already available. If one can assume (reasonably) that drivers are fairly punctual this estimate, which we denote $\hat{h}'_{n+1}$, will be very accurate. Therefore, it makes sense to use $\eta_{n+1,s} = \hat{h}'_{n+1}$ instead of the arrival headway. And of course, one should use $\eta_{n+1,s} = h'_{n+1}$, this time without error, if bus $n+1$ has left the terminus but not yet reached the next control point. Although the issue does not arise with the simple control, the idea could be relevant for other forms of control by schedule.

To define a complete control strategy for a transit route with termini, and predict what to expect, it is necessary to use the ideas in this section and those of Section 10.5 together. Mini-project 4 asks the advanced student to

[25] If the arrivals have reached the long term equilibrium then ordinary stations maintain the same standard deviation. And if the standard deviation happens to be larger (e.g., due to upstream disruptions) then ordinary stations can reduce it by up to a factor of about 1.3, as we have seen, but this is not as much.

do this for a particular route, using both the simple control and the robust control.

10.8.2. *Speeding-up Individual Buses*

If for whatever reason some of a route's buses are severely lagging behind both, the schedule and their predecessors, and are not responding well to conventional control, it may make sense to escalate the response. Three often used strategies are discussed below: "denying boardings," "skipping stops" and "short-turning."

The first two have a similar objective: reducing the duration or even eliminating dwell times in order to make up for lost time. The first strategy is the least drastic. It denies boardings but serves onboard passengers normally. Its only disadvantage is that it extends the wait of bypassed patrons by one headway. The second strategy denies both boardings and alightings. Therefore, it also penalizes all onboard passengers whose stops are bypassed, since they then must either extend their walks or wait for the next bus. When either of these strategies is used all affected customers should be informed of what to expect, i.e., which stops are about to be skipped for onboard passengers, and when the next bus is expected for people waiting outside. At the very least the bus should display an appropriate sign such as "Next Bus Please" or "Not in Service."

When a bus is in boarding-denial mode, it skips a stop if no onboard passenger wishes to alight and otherwise stops but refuses boardings.[26] This mode of operation eliminates some dwell times and reduces the duration of the others. When a bus is in skip-stop mode it skips all stops and therefore eliminates all associated dwell times. Thus, it goes even faster.

These two modes of operation are usually activated and deactivated by the control center based on the experience of its dispatchers.[27] Their decisions are driven by two factors: need and cost (to affected passengers). A simple, easily visualized proxy for need is the schedule lateness and/or the forward headway. A good proxy for cost is the size of the backward headway — since the lower the headway the lower the cost to bypassed patrons. The strategies should be applied if the need is urgent and the cost is low.

---

[26] Exceptions are sometimes made at selected stops such as transfer points.

[27] The boarding-denial mode is sometimes triggered naturally by overcrowding. In these systems buses with long headways get full. As a result they refuse boardings and travel faster. Somewhat perversely, in such cases, overcrowding turns out to be a stabilizing force for schedule control.

This philosophy can be automated by modifying slightly the control rules we have discussed in Sections 10.5 and 10.6. To do this we would quantify the need and the cost by two dynamic proxies, and set trigger levels for both. Since humans are not involved, the proxy for need can be as sophisticated as desired. For all control algorithms it makes the most sense using the calculated value of the holding time *before truncation*, $L_{n,s}$, since negative values of this variable indicate the need and large absolute values indicate that the need is urgent. The cost can be quantified by the backward headway, $\eta_{n+1,s}$. Critical values of these two proxies could then be defined to trigger and deactivate each mode of operation. These critical values should be optimized through simulation rather than analytically because the system dynamics in these modes of operation are nonlinear, and ultimately they should be fine-tuned in the field. To the authors' knowledge, automatic triggers have not yet been tested in the field.

The third and most drastic measure is "short-turning." This is the act of skipping the terminus and some of its adjoining stops by taking a shortcut in "out-of-service" mode after asking all passengers to alight (see Figure 10.15). When short-turning, a bus would stop short of the terminus at a stop "a," release its passengers, turn around and move through a shortcut to a nearby stop "b" on the portion of the route outbound from the terminus; it would then wait for some time to give the driver some rest and get itself back on schedule.

Note that short-turning is a special case of stop-skipping where the skipped route section includes many stops and the terminus — the only difference is that the skipped section is wholly determined at the moment of activation and covered very fast. Since the strategy is similar to stop-skipping, the need can be evaluated as we explained above. The cost can also be assessed with the backward headway as before, but the ratio of cost to headway is now higher because *all* onboard passengers are now affected and most will choose to wait for the next bus rather than take a long walk.

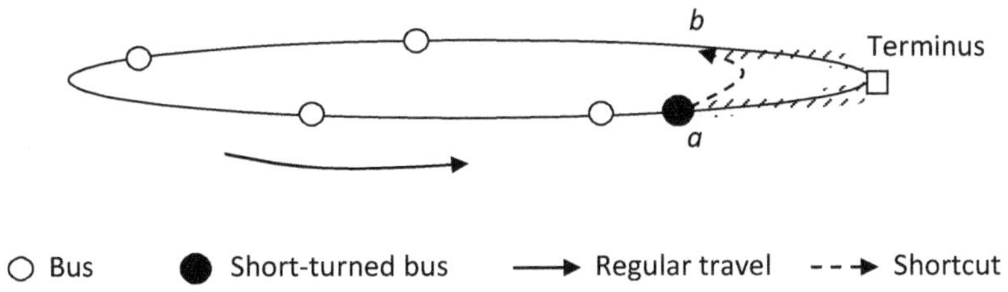

Figure 10.15. Short-turning operation.

It follows from all this that stop skipping can be triggered with the same method as before, although perhaps decreasing a bit the critical value for the backward headway proxy for the cost to reduce the likelihood of triggering it.

The only thing that remains to be determined is the route segment to be skipped. We recommend setting a menu of (a, b) pairs and then choosing an appropriate pair at the moment of activation. The choice depends on the amount of time that needs to be saved. Any given pair can save at most the difference between the time required to cover the skipped portion of the trip including the minimum driver rest T_o at the terminus, and the time for taking the shortcut also including the minimum rest at station b. (A minimum rest is assumed at the terminus because the bus needs to be sped up.) Since the two rest times are the same, we see that the maximum savings are just the difference in the two transportation times: along the skipped section while providing service; and along the shortcut in out-of-service mode. These savings can be easily calculated and associated with each (a, b) so dispatchers can easily pick items from the menu depending on need.

10.8.3. *More Drastic Strategies*

We consider here three adaptive measures that involve more than one bus: "interlining," "leapfrogging" and "station-splitting."

A bus is said to be interlining if it is deadheaded from one line to another. These moves are often preplanned and part of a bus's job, but they can also be done adaptively to fill gaps in service. For example, if a bus in an important route is behind schedule and exhibits a long headway, the operating agency can correct the problem by taking a bus from a neighboring donor line and inserting it in the gap at a proper time and place. This is adaptive interlining. The strategy is typically used when the interlined bus is not being productive (e.g., is part of a pair) and/or the donor route is secondary. (OTS' Route 8, described in Section 10.7.2, was used as a donor line.)

To be successful, the interlined bus should be inserted well ahead of the lagging bus and then directed to start service at the best moment possible, i.e., when indicated by the original schedule if one is used, or else after the desired headway has elapsed. Obviously, the duration of the deadhead move must be factored in when choosing the insertion station in order to make sure that the move is feasible, i.e., that the interlined bus can arrive before it has to start service. In general, one should start the intervention as soon as possible. Therefore, it makes sense to choose the insertion locale that provides the earliest feasible start. The station can be identified by comparing two tables, both indexed by station number: (i) the ideal times at

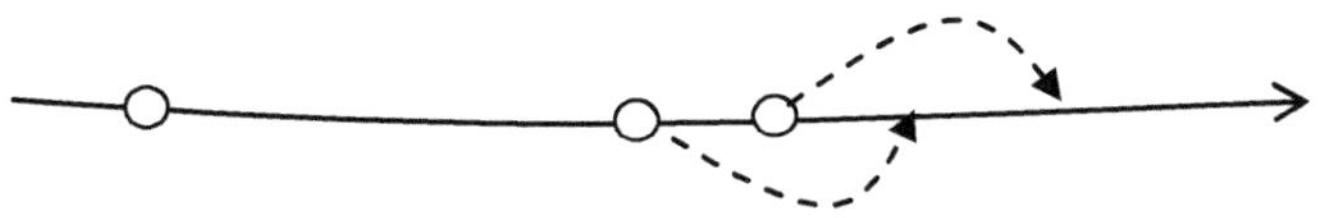

Figure 10.16. Leapfrogging operation.

which service should start if the bus is inserted at station s, $\{i_s\}$, and (ii) the earliest times at which the interlined bus can reach station s, $\{e_s\}$. We then identify the station by solving: $\min_s \{i_s | i_s \geq e_s\}$. Homework exercise 10.9 expands on this idea.

If the bus being replaced is redundant, e.g., because it is being closely followed by another bus, it may make sense to send it back to the donor line. This can be instantly accomplished by asking all its passengers to disembark and board the following bus, or gradually by putting the bus in boarding-denial mode until all its passengers have disembarked.

Many operating agencies keep a small pool of standby buses for adaptive interlining. Sometimes called "interlining with reserve buses" or "bus substitution," the strategy works as described above except buses are drawn from this pool instead of a neighboring donor line; see Petit *et al.* (2018) for an analysis. The strategy is appealing because it simplifies the decisions (one does not need to look for a suitable bus and a donor line) and because it can be carried out without collateral damage to donor lines. On the other hand, because the pool is fixed in space the deadheading distances may be longer. This is of particular concern if the operating agency serves a large geographical area. In these cases agencies should consider keeping multiple, spatially scattered pools, so every critical part of the system can be reached by a reserve bus in a reasonable time.

The second remedial strategy (leapfrogging) is used to speed up bus pairs without attempting to separate them. The strategy makes sense for example if a third bus is closely following the pair. The strategy is a cousin of stop-skipping and works as follows; see Figure 10.16. The two buses leapfrog each other alternately stopping and skipping the stations. All the stations are served by the pair but each bus serves only a half. As a result, the pair's trip time is cut down considerably.

Note that leapfrogging is predicated on the ability to pass safely. This may not always be possible, e.g., on single track rail systems or if a single separated lane is used to segregate buses from traffic and stations are not recessed from the lane. Fortunately, a similar effect can be achieved in these cases by a closely related strategy that we call "station-splitting." The idea is to split the stations on the line into two alternating sets that are separately covered by each bus, and to do this adaptively by instructing the first bus driver to skip every other station, and the second to follow as closely as possible while covering the skipped stations.[28] A minor complication with adaptive station-splitting is that gaps will open between the two buses, so when the driver of the second bus loses sight of the lead bus (s)he will need to be reminded of which stations to serve. This can be achieved either with the help of attentive dispatchers or smart telematics.

Leapfrogging and station-splitting are good ways of operating headway-based systems when control has failed. Let's see why. If control fails and these strategies are not used, the unique stable equilibrium is a single bus bunch with the lead bus doing all the work and the rest following in its wake. The headway is the cycle time, and the cycle time is long because the lead bus travels slowly. Thus, passengers experience long waits and slow service. If the strategies are implemented, however, the buses still travel as a bunch in the stable equilibrium but they split the workload evenly and travel faster as a result.[29] Thus, passenger waiting and riding times are both reduced. Of course, even better is the solution with control — in this case, buses share the load and therefore travel at the same speed as if they were leapfrogging, but they are now evenly spaced along the route and therefore provide shorter headways and further reduce passenger wait. Homework exercise 10.10 explores these ideas quantitatively.

References

Anderson, P. A. and Daganzo, C. F. (2018). "Effect of transit signal priority on bus service reliability." arXiv:1806.09254 [math.OC].

Argote-Cabanero, J., Daganzo, C. F., and Lynn, J. W. (2015). "Dynamic control of complex transit systems." *Transportation Research Part B*, 81(1), 146–160.

Barnett, A. (1974). "On controlling randomness in transit operations." *Transportation Science*, 8(2), 102–116.

Bartholdi III, J. J. and Eisenstein, D. D. (2012). "A self-coordinating bus route to resist bus bunching." *Transportation Research Part B*, 46(4), 481–491.

Daganzo, C. F. (1997). "Schedule instability and control," in *Fundamentals of Transportation and Transportation Operations* (C. F. Daganzo, editor), Elsevier, New York, N.Y., pp. 304–309.

[28]Non-adaptive versions of the strategy are used by New York City's subway; this and other static versions of station-splitting have been studied in Munoz *et al.* (2013) and Gu *et al.* (2016).

[29]That traveling in a bunch is a stable equilibrium can be experimentally seen from the panels of elevator systems of tall buildings such as hospitals, which invariably show strong bunching and leapfrogging behavior.

Daganzo, C. F. (2009). "A headway-based approach to eliminate bus bunching." *Transportation Research Part B*, 43(10), 913–921.
Daganzo, C. F. (2017). Unpublished research notes, VIA Analytics, Berkeley, California.
Daganzo, C. F. and Pilachowski, J. (2011). "Reducing bus bunching with bus-to-bus cooperation." *Transportation Research Part B*, 45, 267–277.
Eberlein, X. J., Wilson, N. H. M., and Bernstein, D. (2001). "The holding problem with real-time information available." *Transportation Science*, 35(1), 1–18.
Gu, W., Amini, Z., and Cassidy, M. J. (2016). "Exploring alternative service schemes for busy transit corridors." *Transportation Research Part B*, 93(1), 126–145.
Newell, G. F. and Potts, R. B. (1964). "Maintaining a bus schedule," in *Proc. 2nd Australian Road Research Board*, Vol. 2, pp. 388–393.
Newell, G. F. (1977). "Unstable Brownian motion of a bus trip," in *Statistical Mechanics and Statistical Methods in Theory and Applications* (U. Landman, editor), Plenum Press, pp. 645–667.
Newell, G. F. (1974). "Control of pairing of vehicles on a public transportation route, two vehicles, one control point." *Transportation Science*, 8(3), 248–264.
Muñoz, J. C., Cortés, C. E., Giesen, R., Sáez, D., Delgado, F., Valencia, F., and Cipriano, A. (2013). "Comparison of dynamic control strategies for transit operations." *Transportation Research Part C*, 28, 101–113.
Petit, A., Ouyang, Y., and Lei, C. (2018). "Dynamic bus substitution strategy for bunching intervention." *Transportation Research Part B*, 115, 1–16.
Turnquist, M. (1978) "A model for investigating the effects of service frequency and reliability on bus passenger waiting times." *Transportation Research Record*, 663, 70–73.
Via Analytics. (2016). "Performance evaluation of headway-based control," Final Report submitted to Oahu Transit Services (Honolulu, Hawaii), Berkeley, CA.
Xuan, Y., Argote, J., and Daganzo, C. F. (2011). "Dynamic bus holding strategies for schedule reliability: Optimal linear control and performance analysis." *Transportation Research Part B*, 45, 1832–1845.

Homework

10.1. When using a bus route that publishes a schedule but exhibits random latenesses, a "rational" passenger will choose an arrival time that optimally balances the costs of arriving too early and missing the bus. Assume two consecutive buses in one such route are scheduled to arrive at $t = 0$ and $t = H > 0$. Further assume that the actual bus latenesses are independent draws of a non-negative random variable with pdf $f(x)$, and cdf $F(x)$, i.e., the bus never departs from the stop early. Formulate an optimization problem to find the optimal arrival time, τ, for a passenger who wishes to catch the $t = 0$ bus. Prove that if $(H - x)f(x) > 1 - F(x)$ for all $x > 0$, then arriving on time at $\tau = 0$ is optimal. What if the bus is allowed to depart early from the stop with possibly negative lateness?

10.2. Show that passengers with important appointments always suffer considerably more from unreliable operations than passengers without appointments. (*Hint:* We have already seen this in the special case where h is uniformly distributed in the range $[H - \varepsilon/2,\ H + \varepsilon/2]$ with $\varepsilon \ll H$. In this case (10.4) reduces to $(1/2 + \beta S)(\frac{\varepsilon^2}{12H})$, whereas (10.6) becomes $(1+\beta S)(\frac{\varepsilon^2}{12H} + \frac{\varepsilon}{\sqrt{3}})$, which is more than an order of magnitude larger.)

10.3. In this exercise, you will use the concepts of Section 10.2.2 to determine the stability of a multi-agent SoS. The SoS consists of two queues in series, which are our agents, as shown in the figure below. The upstream demand λ is a constant, and the servers work as hard as possible leaving an empty queue if there is not enough work. Their service rates depend on the queue lengths as follows: $\mu_1 = x_1 + \gamma(x_1 - x_2)$ and $\mu_2 = x_2 - \gamma(x_1 - x_2)$, where $\gamma > 0$. The γ-terms indicate that work processing resources are constantly being moved from the small pile to the large pile, presumably to balance them. One example of such mechanism could be airport personnel being moved between the airline check-in counter and the boarding gate. For such an SoS, please do the following:

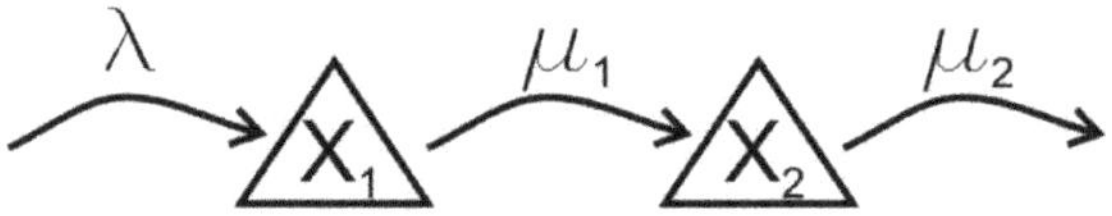

(a) Define state variables and formulate dynamic equations. Note that the system dynamics might be nonlinear due to non-negativity truncations.
(b) Solve all possible equilibrium points, and assess stability of each. Please remove independent constant terms in the system dynamics equations by expressing the state variables in terms of deviations from the equilibrium.
(c) Check if the system is unstable for $\gamma = \lambda = 1$. Can you determine a range of values of λ and γ for the system to be stable? Recall from Chapter 2 how the eigenvalues of a 2-by-2 square matrix are related to the matrix elements.

10.4. Re-derive equation (10.29), which gives the average passenger's waiting time at a generic station between control points, assuming now that the headways are sufficiently large for people to use the schedule.

10.5. Here you will compare the best performance of control by schedule, as given by the formulas at the end of Section 10.4.2, with the optimized performance of control by forward headway (10.32). In considering the latter, note that the parameter α and the number of stops per control point m influence both, the average extra waiting time, $m\sigma^2/(2\alpha(1-\alpha)H)$, and the average riding delay of a passenger traveling r stops, $(\frac{r}{\sqrt{m}})(\frac{(\alpha+\bar{\beta})3\sigma}{\sqrt{\alpha(1-\alpha)}})$. To optimize performance, choose values of α and m that minimize the summation of these two delay terms. Do this numerically, recognizing the constraints mentioned in the text. Compare the results with those of control by schedule using $r = 7, H = 20\,\text{min}, \bar{\beta} = 0.03$, $c_s = 5\,\text{min}$ and $\sigma = 1\,\text{min}$. Discuss your findings.

10.6. This exercise is intended to solidify some of the concepts in Section 10.2. We will study the dynamics of two buses ($i = 1, 2$) that serve a one-way, closed-loop route of length L =10 distance units. When uncontrolled, the average speed of bus i in time period $t = 1, 2, \ldots$ [i.e., the distance it advances which we denote $y_i(t)$] should decline with the spacing in front of the bus at the beginning of the time period, which we denote $x_i(t)$. Assume the relationship is $y_i(t) = 1 - x_i(t)/10 + \varepsilon_i(t)$, where $\varepsilon_i(t)$ is a random disturbance. Also assume that the system is in equilibrium for $t < 0$, with $\varepsilon_i(t) = 0$ and $x_i(t) = 5$ for all $t < 0$ and $i = 1, 2$. Finally assume that $\varepsilon_i(t) = 0$ for $t \geq 0$ and $i = 1, 2$, except that $\varepsilon_i(t) = 1$ for $t = 0$ and $i = 1$. Using this information, do the following:

(a) Write the dynamic equations for the system, i.e., express the vector $\mathbf{x}(t+1)$ as a function of its history $[\mathbf{x}(t), \mathbf{x}(t-1), \ldots]$ and the history of the disturbances $[\boldsymbol{\varepsilon}(t), \boldsymbol{\varepsilon}(t-1), \ldots]$. Express them in as simple a form as possible. You should eliminate the y's from the expression, and also reduce the equations to scalar form by eliminating one of the x's. (*Hint:* the sum of the two spacings always equals the length of the loop.).

(b) Verify from the dynamic equations that the system is in equilibrium for $t < 0$. What happens for $t > 0$? How long until the two vehicles become a pair?

(c) Assume vehicles are given information about the current position of the other one. Suggest a modified rule of motion that will alleviate the positive feedback problem. Note that there are many possible answers. However, you are only allowed to slow vehicles down, never speed them up. Justify your proposal qualitatively and demonstrate its effectiveness with the disturbance data given above.

(d) Would the control strategy proposed in part (c) be difficult to implement in the field? What would be the technological requirements?

Assume you want to model a general urban bus line with the relation $y_i(t) = a - bx_i(t) + \varepsilon_i(t)$, where $\varepsilon_i(t)$ has mean zero and standard deviation, c. What units of measurement (dimensions) would you use for all the variables of this equation? What real-life factors would determine the values of a, b, and c? Give rough estimates for the values that would pertain to a typical bus in your city.

10.7. *This exercise uses the matrix-based method in Section 10.2.2 to analyze a closed-loop line with three buses. In doing so it extends the ideas of uncontrolled buses in Sections 10.3. The route in question has $s = 1, 2, \ldots, S$ control points and is served by three buses. These buses are labeled consecutively, $i = 1, 2, 3$ (with the understanding that bus 1 follows bus 3). When bus i is uncontrolled, its lateness at control point $s+1$ only depends on its own lateness at s and the lateness of the preceding vehicle, i.e., the system dynamic equations resemble (10.21), except the equations now pertain to a closed-loop. With this in mind, please do the following:

(a) Write the dynamic equations for the entire system, properly defining all the state vectors and the coefficient matrices. Express them in as simple a form as possible. Find the equilibrium and show that it is unstable. (*Hint:* you should assume that $\eta_{n+1,s} = h_{n+1,s}$ and then write the dynamic equation in terms of deviations from the virtual schedule. Your equation should be of the form $\boldsymbol{\varepsilon}(t+1) = \mathbf{L}\boldsymbol{\varepsilon}(t) + \gamma(t)$ as explained at the end of Section 10.2.2.)

(b) Devise a robust control law in the form of (10.52) that holds the potential to eliminate instability. Write the control law and dynamic equations for the controlled system. Find the equilibrium state and prove that it is stable.

(c) Conduct a numerical simulation with properly chosen parameters to verify the system stability results with and without control.

10.8. *Please analyze the stability of control law (10.50) under large perturbations. Assume its parameters satisfy (10.51), i.e., $k_{0,s} \geq 0$, and $-k_{1,s} = k_{2,s} \geq \beta_s > 0$. Also, assume that we can replace $\eta_{n+1,s}$ by $h_{n+1,s}$ in (10.50) without a significant error. Using this simplification, write the system's dynamic equations and show that the longest headway across stations is always expected to decrease over time. Confirm your findings with a simulation.

10.9. City streets form a very dense orthogonal grid with negligible spacing. An east-west bus line runs between termini $(-a, 0)$ and $(a, 0)$, with S intermediate stops uniformly distributed in each direction (see part (a) of the figure below). The agency has decided at time $t = 0$ to use an interlining strategy to help a lagging eastbound bus that is at $(-a/2, 0)$ at that moment. Assume for simplicity that the late bus will not become further delayed. Bus insertion can occur at any stop along the east–west bus line that is at least 1 stop spacing ahead of the lagging bus. Any bus in service, including the lagging bus, has a commercial speed, v. Interlining buses run faster, with speed $2v$.

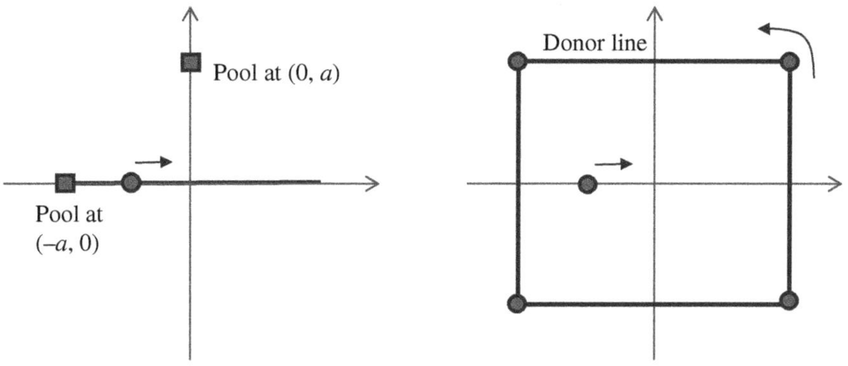

(a) Fixed pools (b) Donor line

Please answer the following questions using: $a = 10\,\text{km}$, $S = 9$, and $v = 20\,\text{km/h}$.

(a) If a pool of reserve buses is kept at $(0, a)$, what is the earliest possible time (and the associated location) for bus insertion? What if another pool of reserve buses are kept at $(-a, 0)$ as well?

(b) Now consider a slightly different situation as shown in part (b) of the figure. Interlining buses must come from a donor line along which four buses run counterclockwise along a square loop with corners at $(-a, -a)$, $(a, -a)$, (a, a), and $(-a, a)$. These buses are mostly empty so they can be interlined at any time. At t =0, the four buses happen to be at these four corners. What is the earliest possible time (and the associated location) for bus insertion?

(c) Now, suppose traffic congestion in the city causes random travel delays to any interlining bus. For simplicity, assume the travel time over any distance $l > 0$ follows a Gaussian distribution with mean $l/(2v)$ and variance-to-mean ratio $\gamma = 1/16\,[\text{h}]$. Stochastic delays may cause an intended interlining insertion to fail. Considering this, what is the earliest possible time (and the associated location) for bus insertion from the donor line that will ensure a 95% probability of success? Run a simulation to verify your result.

10.10. Consider a one-way, closed-loop bus route of length L, similar to that in Figure 10.1. The route has S evenly distributed stations. A fleet of N buses operate on this route without a schedule. The passenger arrivals at each stop follow an independent Poisson process with rate λ [pax/time], and also the random number of stop spacings through which a passenger rides is uniformly distributed between 1 and r. The average bus speed including the part of the dwell time at stations due to acceleration/deceleration and opening/closing doors is v. The remaining part of the dwell time is proportional to the number of boarding or alighting passengers by δ time units each. Think of plausible leapfrogging and station-splitting strategies for completely bunched vehicles, assuming that there is no headway control. Passengers walk at a low speed of v_w. Then, derive the average passenger waiting, riding, and walking times under the following scenarios: (i) all buses are bunched together and nothing is done (no control); (ii) leapfrogging; (iii) station-splitting; and (iv) as the benchmark, when headway-based control is perfectly effective. Discuss your findings and confirm the results with a series of simulations. You may use reasonable parameter values for the simulation (e.g., $L = 10\,\text{km}$, $S = 20$, $N = 6$, $\lambda = 5\,\text{pax/h}$, $r = 10$, $v = 20\,\text{km/h}$, $v_w = 3\,\text{km/h}$, $\delta = 1\,\text{s}$).

Mini Project 4. Designing Control Strategies to Mitigate Bus Bunching[30]

The Champaign-Urbana Mass Transit District (CUMTD) serves the towns of Champaign-Urbana and the campus of the University of Illinois with

[30]Created by Antoine Petit.

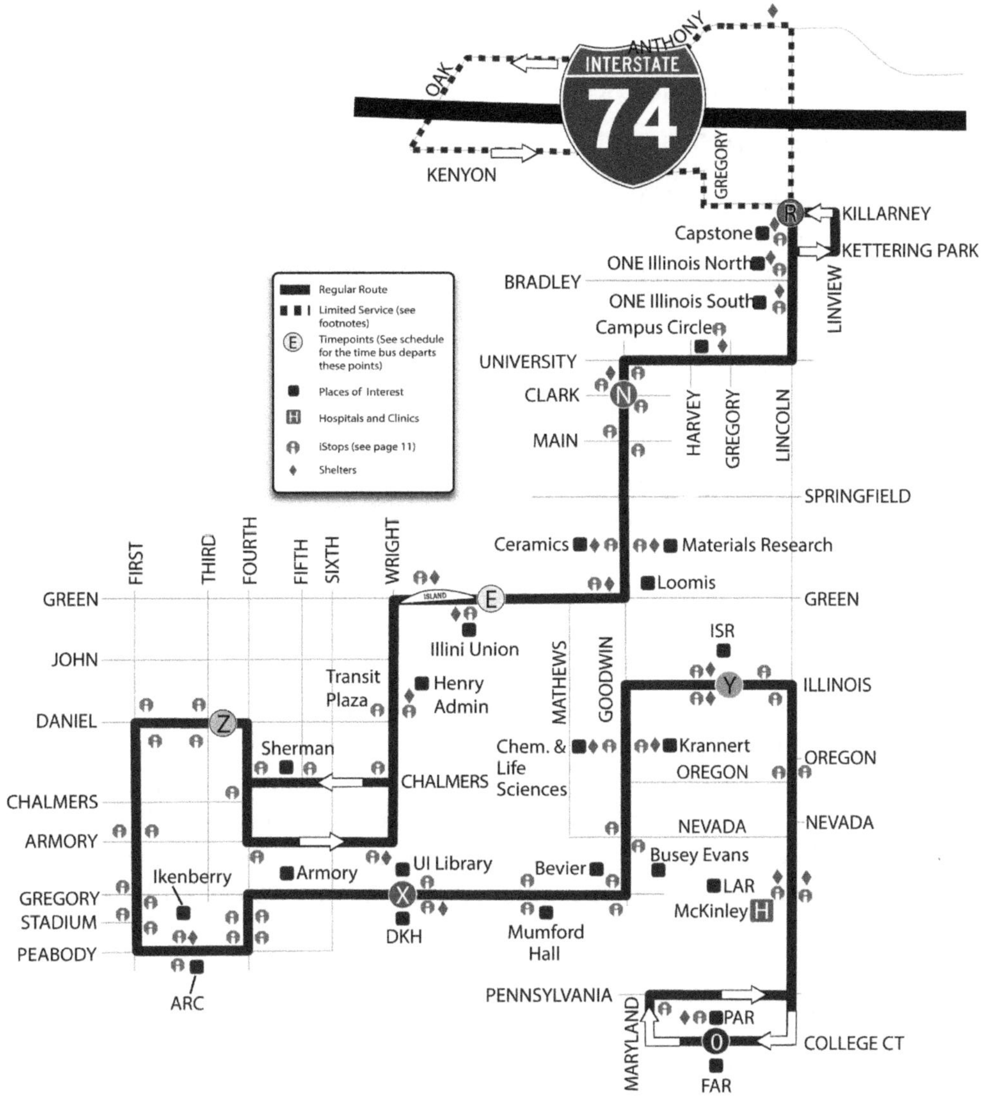

Figure MP4.1. Map of the Illini bus line.

hub-and-spoke networks of almost 20 day-time routes. The 22 Illini bus line is a popular route that provides service through most of the campus area between PAR/FAR (South Campus) and W Killarney St. (North Urbana); see Figure MP4.1.

This mini-project focuses on the south part of the route between Transit Plaza (a transit hub) and PAR/FAR, which is known to be quite vulnerable to bunching. This mini-project will examine and compare different strategies that improve service reliability of the existing bus line against random disturbances (e.g., traffic congestion, excessive signal delays, pedestrian interference and driver vagaries).

Route characteristics

The route of interest is 6 km in each direction and has a headway of 10 min. It has 23 stops (including PAR and Transit Plaza). The cruising speed of every bus averages 20 km/h, but it is subject to disturbances. The expected total travel time from PAR to Transit Plaza is 25 min.

(a) The boarding demand density in each direction is constant and shown in Table MP4.1. Assume that boarding and alighting are simultaneous, such that the influence of alighting passengers can be ignored. The boarding time per passenger is 2 s. Based on all this information, build a feasible bus schedule depicting the bus arrival times at all stops.
(b) The travel time from one stop to the next includes a random noise term, caused by the aforementioned exogenous disturbances. This term is approximated as a random variable with mean $\mu = 0$ min/km and standard deviation $\sigma = 0.45$ min/km. Write the dynamic equation for the actual arrival time of bus n at stop $s + 1$.
(c) Suppose that the first bus in service, indexed 0, is perfectly on schedule, and bus 1 is dispatched on time from the beginning of the route but experiences an unexpected delay of 30 s at the first stop. All following buses follow the dynamics equation from Part (b). How much time, in terms of expectation, does it take for the first two buses to bunch, i.e., to reach headway under 30 s?

CUMTD implemented the traditional schedule-based control method. The current control points are denoted by capital letters in Figure MP4.1. The agency aimed to add slack to a set of selected control points so as to achieve about 99.7% on-time arrivals in the system.

(d) Find the optimal number of control points in each direction, as well as the minimum amount of slack that should be added at each control point. Compare the optimum number of control points with the current number. Estimate the total dwell time added to the passenger travel times under the control strategy you have designed, as well as the total waiting time experienced by all passengers for one hour.

Assume that a small incident occurs to bus 2 between stop s =0 (Transit Plaza) and stop s =1 (Wright & Chalmers), which resulted in an additional

Table MP4.1. Boarding passenger rate (pax/h) along the route.

Stop	Southbound	Northbound
Transit Plaza	66	0
Wright & Chalmers/Armory & Wright	30	42
Fifth & Chalmers/Armory & Fourth	2	10
Fourth & Chalmers	2	6
Third & Daniel	22	10
First & Daniel	26	36
First & Armory	18	14
First & Gregory	1	10
First & Stadium	6	10
ARC	20	20
Fourth & Peabody	12	8
Fourth & Gregory	2	12
Gregory at Library	22	36
Gregory at Mumford Hall	1	2
Goodwin & Gregory	6	24
Goodwin & Nevada	0	20
Krannert Center	8	32
Illinois St. Residence Hall	2	28
Lincoln & Illinois	1	22
Lincoln & Oregon	1	8
LAR	2	54
Penn & Maryland	–	2
PAR/FAR	0	7

travel time of 2 min. Recall from the text that in case of large perturbations, the conventional schedule-based control method may fail, i.e., buses will bunch.

(e) Show, through a simulation, that the described incident will cause buses to bunch despite the schedule-based control scheme you obtained from Part (d).

You are now hired by CUMTD to propose a better strategy to deal with bus bunching.

(f) Discuss the benefits of designing a headway-based control scheme. Design a simple control law, following the forward-looking strategy, to ensure stable headways along the bus line. Would the proposed method mitigate the impact of the incident that occurred between stop $s = 0$ and stop $s = 1$? Derive the expected waiting time and additional dwell time, under your control law for a passenger traveling from Transit Plaza to PAR/FAR. Compare the result with that from Part (d).

A follow-up study after implementing the forward-looking strategy you proposed finds that most transit users are still frustrated because the terminus (PAR/FAR) is a popular destination, and passengers do not like to experience additional dwell time at each stop due to the slack. In addition, the study finds that large disturbances still cannot be absorbed with the forward-looking method.

(g) Design a more robust control law (e.g., a two-way-looking strategy) that would prevent bus bunching and outperform the control law proposed in Part (f). Show the benefits of using your new strategy over the forward-looking strategy.

In reality, CUMTD is experiencing bunching on several bus lines. Suppose now that CUMTD is willing to use a few additional buses as standby backups.

(h) Propose a possible strategy to make use of those additional standby buses to prevent bunching on multiple lines. This strategy must have the potential to mitigate the impact of events such as the one mentioned in Part (e).

You are to submit a formal report describing the tasks and your findings. You may work in small groups of up to three people and submit a group report.

Chapter 11

Epilogue: Economics and Pricing

Although this book has centered on planning, design, and operations, these are not the only issues that need to be addressed to create and maintain a successful transit system. The system also needs to be economically sustainable and politically viable. Earlier in this book, Section 3.3 argued that economic and political issues could be separated from the more concrete planning, design, and operations questions if some assumptions were met. We called the latter questions, the "design" problem — the focus of this book — and the former the "pricing" problem. It was shown that if some restrictive assumptions on the pricing structures were met then, conditional on the user demand, the design problem can be reduced to an optimization problem of the "standards" type. Chapters 4–10 were then devoted to the formulation and solution of various versions of this problem and its key components. But economics and pricing were not discussed.

Although another book could be written about economics and pricing, some of the ideas are so essential to making transit viable that this book would not be complete without at least a mention. Fortunately, this can be done fairly briefly using the lingo we have been using so far, and this is the purpose of this epilog. We try to do two things. First, we circle back to the decomposition idea of Section 3.3 and show with slightly more involved arguments that the decomposition simplification applies to all commonly used pricing systems and not just the special cases of Section 3.3. Then, under the same modeling framework, we turn our attention to economics, including budgeting and pricing. More specifically, we examine the different outcomes achieved when the transit agency is run by an ideal non-profit agency that has society's benefits at heart, as well as when it is run by real agencies that also consider what is dear to them and to their employees.

We then show how certain market mechanisms can be used to make for-profit franchises behave like the ideal.

11.1. Decomposition

Let us now extend the decomposition idea of Section 3.3 and prove under more general conditions that transit systems can still be optimally designed by solving an optimization problem of the "standards" type. We do this in four steps. First, we generalize the formulation of the joint design-pricing problem to capture the general conditions. Second, we re-express it in a more useful way with a change of variable that removes the demand variables. Third, we further simplify the problem by relaxing one of its constraints and proving that the resulting problem has the same set of optimal decision variables as the original. And finally, we show that this final version can be decoupled into a "standards' design problem and a pricing problem.

So, let us now generalize the formulation. In Section 3.3, the complete pricing and design problem had been formulated as follows. We assumed that the numbers of users $\boldsymbol{\alpha}$ of different classes were endogenous, i.e., that these numbers were ultimately determined from a balance of demand and supply. (As in Section 3.3 boldface designates row vectors.) In addition to $\boldsymbol{\alpha}$, we looked for the optimum system design, $\mathbf{x}$, and set of fares for the different user classes $\mathbf{p}$. We also introduced the travel disutilities, $\boldsymbol{\pi}$, associated with the various user classes resulting from combining the fares $\mathbf{p}$ and the travel times $\mathbf{T}$ they experience. The ($\boldsymbol{\alpha}$, $\mathbf{p}$, $\boldsymbol{\pi}$, $\mathbf{T}$) vectors include one element per population class and therefore have the same dimension. Vector $\mathbf{x}$ was unrelated. The complete design/pricing problem with endogenous demand was then initially expressed as the following design problem (DP):

$$
\begin{aligned}
\text{(DP)}\quad & \max\,\{S(\boldsymbol{\pi},\boldsymbol{\alpha}) - \$(\mathbf{x},\boldsymbol{\alpha})\} && \text{(society's benefits minus agency costs)}\\
& \text{s.t.}\ \ \mathbf{T}(\mathbf{x},\boldsymbol{\alpha}) \le \mathbf{T_0} && \text{(standards on system performance units)}\\
& \qquad \boldsymbol{\alpha} = \mathbf{D}(\boldsymbol{\pi}) && \text{(endogenous demand function)}\\
& \qquad \boldsymbol{\pi} = \mathbf{G}(\mathbf{p},\mathbf{T}) && \text{(disutility function).}
\end{aligned}
$$

It was then assumed in Section 3.3, somewhat restrictively, that the pricing scheme had enough flexibility to be able to produce any desired set of disutilities $\boldsymbol{\pi}$ by manipulating the prices $\mathbf{p}$. Although this assumption holds for systems such as BART in the San Francisco Bay Area that use a different fare for each origin destination pair, it rules out systems such as Muni or AC Transit with more origin destination pairs than distinct fares.

Therefore, we now relax the assumption of fully flexible pricing. This is done by adding the constraint $\mathbf{p}\boldsymbol{\Delta} = \mathbf{0}$ to problem DP, where $\boldsymbol{\Delta}$ is a square matrix of constants that define the pricing structure. For example, a matrix of zeroes allows for arbitrary fares; and if one inserts 1's in the main diagonal and -1's in the lower sub-diagonal one forces a unique fare for all classes. We shall also add the non-negativity constraint $\mathbf{p} \geq 0$, which was omitted in DP but used implicitly.

As a further generalization, we shall also add a budget constraint stipulating that the agency's cost cannot exceed the revenue. This revenue has two components: the revenue form fares, which we denote F, and the income from subsidies. If we use a dot for the vector inner product then the fare revenue is $F = \mathbf{p} \cdot \boldsymbol{\alpha}$. We also assume that the income from subsidies cannot exceed an upper limit, which is given and denoted B. Thus, our budget constraint is

$$\$(\mathbf{x}, \boldsymbol{\alpha}) \leq B + F = B + \mathbf{p} \cdot \boldsymbol{\alpha}.$$

This concludes the first step. Before taking the second step, we introduce some final assumptions and notational conventions. As is logical, we assume that both S and $\mathbf{D}$ are non-negative functions, that both functions are non-increasing in π, and that S is non-decreasing in $\boldsymbol{\alpha}$. For the sake of tractability and without much loss of generality we shall also assume that \$ and $\mathbf{T}$ are independent of $\boldsymbol{\alpha}$. This is reasonable. For a given design, $\mathbf{x}$, the demand $\boldsymbol{\alpha}$ affects \$ because it influences wear and tear on the vehicles and infrastructure, but this is clearly a second order effect. And the demand affects $\mathbf{T}$ only when it is so large that the system becomes crowded and inefficient, but the assumption is reasonable otherwise. These assumptions allow us to eliminate the argument $\boldsymbol{\alpha}$ from the functions giving the agency's cost and the system performance times, which now become $\$(\mathbf{x})$ and $\mathbf{T}(\mathbf{x})$. We shall also express the system performance times $\mathbf{T}$ in monetary units, and then assume that a user's disutility is the sum of the user's time and fare, i.e., that the disutility function is $\boldsymbol{\pi} = \mathbf{p} + \mathbf{T}(\mathbf{x})$.

We are now ready for step two — removing the demand variables. We eliminate both $\boldsymbol{\alpha}$ and the constraint $\boldsymbol{\alpha} = \mathbf{D}(\boldsymbol{\pi})$ from the problem by inserting $\mathbf{D}(\boldsymbol{\pi})$ into the two remaining places where $\boldsymbol{\alpha}$ appears: (i) in the benefit function of the objective function, which becomes $S(\boldsymbol{\pi}, \mathbf{D}(\boldsymbol{\pi}))$; and (ii) in the budget constraint, which becomes: $\$(\mathbf{x}) \leq B + \mathbf{p} \cdot \mathbf{D}(\boldsymbol{\pi})$. To further simplify the notation, we define the single-argument function: $S(\boldsymbol{\pi}) \equiv S(\boldsymbol{\pi}, \mathbf{D}(\boldsymbol{\pi}))$. Thus, the generalized design/pricing problem with endogenous demand (denoted GDP) becomes:

(GDP)	$\max\,\{S(\boldsymbol{\pi}) - \$(\mathbf{x})\}$	(society's benefits minus agency costs)
	s.t. $\mathbf{T}(\mathbf{x}) \le \mathbf{T_0}$	(standards on system performance units)
	$\boldsymbol{\pi} = \mathbf{p} + \mathbf{T}(\mathbf{x})$	(disutility function)
	$\$(\mathbf{x}) \le B + \mathbf{p} \cdot \mathbf{D}(\boldsymbol{\pi})$	(budget constraint)
	$\mathbf{p}\boldsymbol{\Delta} = \mathbf{0}$	(pricing structure)
	$\mathbf{p} \ge \mathbf{0}$	(pricing structure).

We seek optimum values of ($\mathbf{x}$, $\mathbf{p}$, $\boldsymbol{\pi}$). Note that a transit agency only has to choose $\mathbf{p}$ and $\mathbf{x}$; vector $\boldsymbol{\pi}$ is an auxiliary variable that emerges after the fact from the second constraint and is a derived quantity. Also note from our earlier assumptions about $\mathbf{D}(\boldsymbol{\pi})$ and $S(\boldsymbol{\pi}, \mathbf{D}(\boldsymbol{\pi}))$ that the new function, $S(\boldsymbol{\pi})$, in the objective function must be non-increasing with respect to all its arguments.

Let us now take our third step. We construct a relaxed version of GDP that we denote GDP' and then show that both problems have the same set of optimal solutions (which may not be unique) $\{(\mathbf{p^*}, \mathbf{x^*})\}$. The relaxation is achieved by replacing the equality constraint $\boldsymbol{\pi} = \mathbf{p} + \mathbf{T}(\mathbf{x})$ by the inequality $\boldsymbol{\pi} \ge \mathbf{p} + \mathbf{T}(\mathbf{x})$, so the new problem is

(GDP′)	$\max\,\{S(\boldsymbol{\pi}) - \$(\mathbf{x})\}$	(society's benefits minus agency costs)
	s.t. $\mathbf{T}(\mathbf{x}) \le \mathbf{T_0}$	(standards on system performance units)
	$\boldsymbol{\pi} \ge \mathbf{p} + \mathbf{T}(\mathbf{x})$	(disutility constraint)
	$\$(\mathbf{x}) \le B + \mathbf{p} \cdot \mathbf{D}(\boldsymbol{\pi})$	(budget constraint)
	$\mathbf{p}\boldsymbol{\Delta} = \mathbf{0}$	(pricing structure)
	$\mathbf{p} \ge \mathbf{0}$	(pricing structure).

The equivalence is established by the following proposition.

Proposition 11.1. *Problems GDP and GDP' have the same set of optimal solutions* $\{(\mathbf{p}^*, \mathbf{x}^*)\}$.

Proof: It suffices to show that for any optimal solution of GDP' where $\pi_i^* > p_i^* + T_i(x^*)$ for some i there is a corresponding optimal solution with the same ($\mathbf{p}^*$, $\mathbf{x}^*$) and $\pi_i^{*\prime} = p_i^* + T_i(x^*)$, which of course is a solution of GDP too. To see what we are trying to show, first note that the corresponding solution continues to be feasible — since the right hand side of the budget constraint cannot decline when $\boldsymbol{\pi}$ declines to $\boldsymbol{\pi}'$ because $\mathbf{D}$ is non-increasing. Moreover, since S is also non-increasing, the objective value associated with $\boldsymbol{\pi}'$ cannot be smaller than the one associated with $\boldsymbol{\pi}$. Thus the corresponding solution is feasible and matches the optimal objective. Thus, the solution is optimal as we set to show. □

As the fourth and final step, we now show how GDP' decomposes. To this end, consider a conditional version of GDP' where $\boldsymbol{\pi}$ and $\mathbf{p}$ are held fixed at some feasible values. For this version of the problem, which we call CDP, the benefit term in the objective function becomes a constant and therefore can be dropped. Moreover, the last two constraints are automatically satisfied and can also be dropped. Finally, the first and second constraints can be combined to read: $\mathbf{T}(\mathbf{x}) \leq \min\{\mathbf{T_0}, \boldsymbol{\pi} - \mathbf{p}\}$. Thus, the CDP becomes:

$$
\begin{array}{llll}
\text{(CDP)} & \min_{\mathbf{x}} & \{\$(\mathbf{x})\} & \text{(agency cost)} \\
 & \text{s.t.} & \mathbf{T}(\mathbf{x}) \leq \min\{\mathbf{T_0}, \boldsymbol{\pi} - \mathbf{p}\} & \text{(revised standards constraint)} \\
 & & \$(\mathbf{x}) \leq B + \mathbf{p} \cdot \mathbf{D}(\boldsymbol{\pi}) & \text{(budget constraint).}
\end{array}
$$

We now show that if this problem is feasible its budget constraint is redundant and can be dropped, i.e., we can solve instead:

$$
\begin{array}{llll}
\text{(CDP}'\text{)} & \min_{\mathbf{x}} & \{\$(\mathbf{x})\} & \text{(agency cost)} \\
 & \text{s.t.} & \mathbf{T}(\mathbf{x}) \leq \min\{\mathbf{T_0}, \boldsymbol{\pi} - \mathbf{p}\} & \text{(revised standards constraint).}
\end{array}
$$

Proposition 11.2. *If CDP is feasible, every optimum of CDP' solves CDP.*

Proof: Note that CDP' is feasible because it is a relaxed version of a feasible problem (CDP). Therefore both CDP and CDP' have optimal solutions. Now let $\mathbf{x}^*$ be an optimum of the former, and $\mathbf{x}^{*'}$ of the latter. It suffices to show that $\mathbf{x}^{*'}$ is an optimum for CDP, i.e., (i) that $\mathbf{x}^{*'}$ satisfies the budget constraint, and (ii) that $\$(\mathbf{x}^{*'}) \leq \$(\mathbf{x}^*)$. Inequality (ii) holds because CDP' is a relaxed version of CDP, so we just need to establish (i). This is true, however, because $\$(\mathbf{x}^*) \leq B + \mathbf{p} \cdot \mathbf{D}\ (\boldsymbol{\pi})$ (since $\mathbf{x}^*$ is feasible for CDP) and by virtue of (ii) we have $\$(\mathbf{x}^{*'}) \leq \$(\mathbf{x}^*) \leq B + \mathbf{p} \cdot \mathbf{D}(\boldsymbol{\pi})$. □

This proposition establishes the decoupling principle because CDP' is a conventional design problem of the standards type, with (revised) standards, $\mathbf{T}_0' \equiv \min\{\mathbf{T_0}, \boldsymbol{\pi} - \mathbf{p}\}$. In fact, if as is usual the demand function is invertible so that $\boldsymbol{\pi} = \mathbf{V}(\boldsymbol{\alpha})$ then these standards can be expressed in terms of the demand and price vectors as was done in Section 3.3, i.e., as: $\mathbf{T}_0' = \min\{\mathbf{T_0}, \mathbf{V}(\boldsymbol{\alpha}) - \mathbf{p}\}$. In either case, for any given set of feasible disutilities (or demands) and fares the design problem can be treated with the standards approach used in the book, or alternatively with its Lagrangian counterpart.

The final outcome is a family of optimal designs $\mathbf{x}^*(\boldsymbol{\pi}, \mathbf{p})$ that can then be used to solve the pricing problem. To figure out the optimal prices, simply solve the GDP after replacing the variable $\mathbf{x}$ by the functions $\mathbf{x}^*(\boldsymbol{\pi}, \mathbf{p})$. This can be done analytically or numerically. Homework exercise 11.1 shows how the process works in a simple case. Pricing is however a complex issue, entangled with the economics of budgeting, and politics, so a basic optimization formulation for the fares does not do it justice. In recognition of this complexity, this epilog now concludes with its second theme: economics.

11.2. Economics

We shall consider here the behavior of realistic agencies, using as a basis for comparison the performance of an ideal public agency that chooses not just the prices but also its external subsidy, and does so for the good of society. This is obviously the best for society, so it is a reasonable benchmark. We will first examine this benchmark and then the behavior of realistic agencies.

11.2.1. *An Ideal Agency*

We first show that to maximize society's net benefit, two things should happen: (i) the subsidies should be amply available, in the sense that B should be large enough to ensure that the budget constraint is not binding, and (ii) the agency should provide free transit, i.e., choose $\mathbf{p}^* = \mathbf{0}$. The first condition should be clear because any time one relaxes a binding constraint the problem's feasible region enlarges and its optimum can only improve. The second condition requires some thought, however. To prove that transit should be free, consider the previously formulated GDP assuming that $B = \infty$ and fix $\mathbf{x}$ at any feasible value. We call this the large-subsidy pricing problem (LSPP). We shall show that the LSPP is optimally solved by $\mathbf{p}^* = \mathbf{0}$ for every $\mathbf{x}$, i.e., that condition (ii) holds.

The LSPP is just a pared down version of the GDP without the budget constraint (since $B = \infty$) and where the only decision variable is $\mathbf{p}$ (since $\mathbf{x}$ is fixed). Given this structure, it can be further streamlined. First, since $\mathbf{x}$ is fixed at a feasible value, we can drop the agency cost from the objective function and eliminate the standards constraint. And second, we can also eliminate the decision variable π and the disutility function constraint by substituting the right hand side of said constraint for the argument of S in the objective function. With these simplifications, the LSPP becomes

$$\begin{array}{lll} \text{(LSPP)} & \max_{\mathbf{p}} \{S(\mathbf{p} + \mathbf{T}(\mathbf{x}))\} & \text{(society's benefit)} \\ & \text{s.t.} \quad \mathbf{p}\boldsymbol{\Delta} = \mathbf{0} & \text{(pricing structure)} \\ & \qquad \mathbf{p} \geq \mathbf{0} & \text{(pricing structure).} \end{array}$$

It is now easy to see that $\mathbf{p}^* = \mathbf{0}$. Note from the non-negativity constraint that every feasible $\mathbf{p}$ satisfies $\mathbf{p} + \mathbf{T}(\mathbf{x}) \geq \mathbf{T}(\mathbf{x})$. Now, since S is non-increasing over all its arguments, it follows that $S(\mathbf{p} + \mathbf{T}(\mathbf{x})) \leq S(\mathbf{T}(\mathbf{x}))$, i.e., that the objective value for every feasible $\mathbf{p}$ is bounded above by $S(\mathbf{T}(\mathbf{x}))$. Obviously then, the solution $\mathbf{p}^* = \mathbf{0}$ is optimal since it achieves this upper bound. Note as well that if S is strictly monotone as one would expect in most applications then only $\mathbf{p}^* = \mathbf{0}$ reaches the upper bound, i.e., $\mathbf{p}^* = \mathbf{0}$ is the unique optimum.

Qualitatively, this all means that there is no societal downside for increasing the subsidy limit and reducing fares. This should be intuitive: the fares and subsidies paid are zero-sum transactions within members of society that do not change society's benefit; and clearly, the higher the budget and the lower the fares the larger the number of people who benefit from transit. In economics parlance, the marginal benefit of increasing subsidies (and reducing fares) is positive.

11.2.2. *Realistic Agencies*

In the above, we have shown that in an ideal world transit should be amply subsidized and provided free of charge. In reality, however, this is not the case. This subsection explores the reasons for this discrepancy, and the next subsection how to overcome it.

The first reason is that large subsidies to provide free transit are likely to be met with resistance by the government because there usually are other demands on public funds such as education, health care, and public infrastructure. Besides, transit subsidies imply a considerable transfer of wealth from non-transit users to transit users. This transfer can be politically unpopular if non-users do not realize the external benefits they receive when other people use transit. Therefore, for most practical applications budget constraints should be included in the problem — and in most cases, they will be binding. Consideration shows that when this occurs the solution with $\mathbf{p} = \mathbf{0}$ is no longer necessarily optimal. Thus, positive fares are justified, as we see in real life. To gain additional insights into this idea see homework exercise 11.2. It examines a simple scenario with a single user class and a single design variable.

The second reason is that non-profit transit agencies may be far from ideal. After all, human nature suggests that the parties involved in decision-making (management and unions) should both be selfishly interested in maximizing the size of the operation, and not just society's welfare. This is reasonable because a large operation provides jobs and enhances the stature of managers. A proxy for the size of the operation is the agency's total revenue/expenditure, which we denote R. Thus, to improve the realism of our objective function one can add to it a term proportional to R.

The total revenue/expenditure is the sum of the fare-box revenue, $F = \mathbf{p} \cdot \mathbf{D}(\mathbf{p} + \mathbf{T}(\mathbf{x}))$, and the actual subsidy income which we denote by I, i.e., $R \equiv I + F \equiv I + \mathbf{p} \cdot \mathbf{D}(\mathbf{p} + \mathbf{T}(\mathbf{x}))$. It should be stressed that the income I is a decision variable, distinct from the subsidy *limit* B, which is given, and that the two are related by the inequality $I \leq B$.

It is informative to consider how a totally selfish agency with only R in the objective function would choose I and $\mathbf{p}$ for a given design $\mathbf{x}$. We call this the selfish-agency pricing problem (SPP). By definition, such an agency would choose I and $\mathbf{p}$ to maximize $R = I + F = I + \mathbf{p} \cdot \mathbf{D}(\mathbf{p} + \mathbf{T}(\mathbf{x}))$, subject to $I \leq B$, and to the two pricing structure constraints of the LSPP.

The resulting problem can be simplified. Note that the objective function increases with I and this variable only appears in the constraint $I \leq B$. Therefore, it follows that $I^* = B$, i.e., the agency chooses the maximum subsidy possible. This result allows us to eliminate I from the formulation, replacing it by B; and also to eliminate the constraint $I \leq B$. The objective function then becomes $R = B + F$, where B can be dropped for optimization purposes. Thus, the SPP reduces to

$$
\begin{array}{llll}
\text{(SPP)} & \max_{\mathbf{p}} & \{F = \mathbf{p} \cdot \mathbf{D}(\mathbf{p} + \mathbf{T}(\mathbf{x}))\} & \text{(fare-box revenue)} \\
 & \text{s.t.} & \mathbf{p}\boldsymbol{\Delta} = \mathbf{0} & \text{(pricing structure)} \\
 & & \mathbf{p} \geq \mathbf{0} & \text{(pricing structure).}
\end{array}
$$

Curiously, the solution $\mathbf{p} = \mathbf{0}$ is not just suboptimal for this problem, but a *minimum* of it — since its objective value (zero) is a lower bound to all feasible SPP objective values. This illustrates that selfish behavior perversely drives the solution in the opposite direction of the ideal. Of course, this assessment may be overly pessimistic because public pressure should force decision-makers to include level-of-service measures in the objective function, but on the other hand, it would be naïve to assume away agencies' yen for increasing fare revenue. Homework exercise 11.3 explores how different agency objectives affect outcomes for the public, for the agency and for society at large. The next subsection examines ways to encourage better behavior.

11.2.3. *How to Encourage Agencies to Perform for the Public Good*

Since public non-profit agencies will underperform, what is one to do? A possible answer lies in providing agencies with economic incentives that align with society's wishes. This subsection explores how to do this by franchising the complete operation, perhaps even the ability to set fares, to for-profit agencies with carefully written contracts.[1]

The franchising process would be choreographed and monitored by a governmental master agency. The master agency would issue a request for proposals (RFP) with a well thought-out compensation formula based on performance benchmarks, but leaving open an additive constant K that would be determined by the bids. In exchange for this compensation the franchisee would be expected to design and operate the system; provide access to detailed operational data through an application programming interface (API); collect and keep all fares, and be responsible for all costs.

To see how this might work let us formulate the mathematical program of the profit maximizing agency. We shall show that it will be identical to the GDP (a good thing) if the compensation formula is set in a certain way.

To start, we first rewrite the GDP in shorthand, using $W = S - \$$ for society's net benefit and $\mathscr{X}$ for the feasible region of the decision variables ($\mathbf{x}$, $\mathbf{p}$) when the budget constraint is ignored. The GDP is then:

$$\text{(GDP)} \quad \max_{\mathbf{x},\mathrm{p}} \{W : \$ \leq (B + F); (\mathbf{x}, \mathbf{p}) \in \mathscr{X}\}.$$

The set $\mathscr{X}$ is allowed to include constraints that were not included in the original formulation of the GDP. Thus, the abbreviated formulation is even more general. For example, if we wish to deny the franchisee the ability to set fares, we can just specify their level $\mathbf{p_0}$ and include the constraint $\mathbf{p} = \mathbf{p_0}$.

Next, let us write in similar form the franchise's profit maximizing problem (PMP). We assume that the issued RFP requires the successful bidder to meet all the constraints of the GDP, other than the budget

[1]The role of unions in this type of arrangement deserves more serious consideration than what this short epilog allows.

constraint, or else face high penalties, i.e., the RFP enforces $(\mathbf{x},\ \mathbf{p}) \in \mathscr{X}$. Now, denote the agency's profit by P, where P is the difference between the agency's total revenue $R \equiv (I+F)$ and its cost \$ (i.e., $P = R - \$$), and write the PMP as follows:

$$\text{(PMP)} \quad \max_{\mathbf{x},\mathbf{p}}\{P : (\mathbf{x}, \mathbf{p}) \in \mathscr{X}\}.$$

We now explore how to set the compensation formula (i.e., the formula for I) so the PMP will have the same optima as the GDP. This is desirable because then if the agency maximizes profit it will also maximize society's benefit. The compensation formula is first derived assuming that $B = \infty$ (unlimited subsidies) and then $B < \infty$.

11.2.3.1. *Unlimited Subsidies*

If $B = \infty$, the feasible regions of the GDP and PMP problems match. Thus, these problems will have the same optima if their objective functions match up to an additive constant, i.e., if $P = W + K$, where K is an arbitrary constant.

This match can be achieved by adjusting the compensation, I. To derive the formula recall that $P = R - \$ = I + F - \$$ and $W = S - \$$. Hence, a match of the form $P = W + K$ is achieved if $I + F - \$ = S - \$ + K$, i.e., if we choose $I = (S - F) + K$. This is the desired formula. Note it does not include \$.

The variable K in the compensation formula is a fixed but arbitrary component that would be up for bid. The quantity in parenthesis is a performance-based adjustment that equals society's benefit minus the fare box revenue. The master agency would need to define this adjustment as accurately as possibly in terms of transparent and observable quantities such as fares, travel times and demand. For example, if the travel times $\mathbf{T}$ and fares $\mathbf{p}$ are observed but the demand is not, the formula could be $I = \underline{S}(\mathbf{p} + \mathbf{T}) - \mathbf{p} \cdot \underline{\mathbf{D}}(\mathbf{p} + \mathbf{T}) + K$, where the underlined symbols are the master agency's best approximation formulas of the gross benefit and demand functions. If reliable estimates of passenger counts $\boldsymbol{\alpha}$ are available then the more robust formula $I = \underline{S}(\mathbf{p} + \mathbf{T}) - \mathbf{p} \cdot \underline{\boldsymbol{\alpha}} + K$ could be used.

There is a special case where the result is even simpler. If the master agency wishes to model S as the sum of the consumer surplus $C(\mathbf{p} + \mathbf{T})$ and the fare box revenue F, as is often advocated in economics books, then the compensation formula reduces to $I = (S - F) + K = (C + F - F) + K = C + K$. As we now show, the variable component of the compensation, C, turns out to be a linear combination of the fares and travel times. We will also see that this combination represents the negative of the user's generalized cost of travel, which makes intuitive sense: the agency would bid for a nominal fixed compensation with the understanding that said compensation would be reduced by the (generalized) cost users experience due to insufficiently good service. This should work well since it gives the franchisee an incentive to provide good service at inexpensive rates.

To see why the variable part of the compensation is equal to the negative of the generalized cost, recall from economics that the gradient of C is $-\mathbf{D}$. This allows us to express C approximately as a linear function of the fares and travel times as we now show. [A first order expansion of $C(\mathbf{p}+\mathbf{T})$ in the neighborhood of the optimum, using $-\boldsymbol{\alpha}^{\boldsymbol{*}} \equiv -\mathbf{D}(\mathbf{p}^* + \mathbf{T}^*)$ for the gradient of C at that point, is: $C \approx C^* - (\mathbf{p} - \mathbf{p}^*) \cdot \boldsymbol{\alpha}^* - (\mathbf{T} - \mathbf{T}^*) \cdot \boldsymbol{\alpha}^*$. And this may be rewritten as the linear expression: $C \approx K' - (\mathbf{p} + \mathbf{T}) \cdot \boldsymbol{\alpha}^*$, where K' is the constant $(C^* + \mathbf{p}^* \cdot \boldsymbol{\alpha}^* + \mathbf{T}^* \cdot \alpha^*)$.] If this approximation for C is used the compensation formula would become: $I = C + K = [K + K'] - (\mathbf{p} + \mathbf{T}) \cdot \underline{\boldsymbol{\alpha}}^*$, where $\underline{\boldsymbol{\alpha}}^*$ is the master agency's estimate of the optimum demand. The term $[K + K']$ is the nominal part of the compensation, which is open for bid, and therefore can be replaced by a single constant, K. Using this notation, we can write the compensation formula as $I = K - (\mathbf{p} + \mathbf{T}) \cdot \underline{\boldsymbol{\alpha}}^*$. The second term (in boldface) is the performance-based adjustment. Note that it is a weighted sum across user classes of the observed user disutilities with more weight given to classes with high demand. Also note that the negative of this term represents the "total generalized user cost" as we pointed out.

11.2.3.2. *Finite Subsidies*

Let us now see what happens with a budget constraint. To satisfy this constraint, the RFP should be phrased exactly as before so the identity $P \equiv W$ continues to hold, but the compensation shall be capped at B (the maximum possible subsidy). We shall show that if the market for bids is efficient then the PMP arising from the capped RFP is equivalent to the GDP with a budget constraint.

The capped PMP is

$$(\text{CPMP}) \quad \max_{\mathbf{x},\mathbf{p}}\{P \equiv W : I \leq B; (\mathbf{x}, \mathbf{p}) \in \mathscr{X}\}.$$

Note the budget constraint is slightly different from that of the GDP, which is $\$ \leq B + F$. However, if the market for bids includes enough operating agencies to make it efficient (i.e., so that $P = 0$) then $P \equiv I + F - \$ = 0$, i.e., $I = \$ - F$. This allows us to express the budget constraint of the CPMP as $\$ - F \leq B$; or equivalently as $\$ \leq B + F$, which is the budget constraint of the GDP. Thus, under efficient markets CPMP $\equiv$ GDP, i.e., the proposed RFP structure induces the franchisee to operate for the good of society. In reality, the number of available high-quality operating agencies may be small so the ideal may not be reached, but the approach is appealing nonetheless since it nudges franchisees in the right direction.

11.2.3.3. *A Generalization to Include Externalities*

So far we have ignored externalities, but the approach can be extended to incorporate both the good and bad externalities that transit generates. The most salient good externality is the value to society of those trips attracted to transit from less desirable modes. This value, like the gross benefit S, should be a function of the demand and the generalized disutilities that transit produces, i.e., be of the form $G(\boldsymbol{\alpha}, \boldsymbol{\pi})$. The most important bad externalities are the environmental impacts associated with a transit operation. Their monetary value $E \geq 0$ should be a function of $\mathbf{x}$, which can usually be approximated as a weighted sum of key summary statistics, $\mathbf{y}(\mathbf{x})$, such as the vehicle-km and vehicle-hrs traveled by vehicle type and the lane-km of special lanes, i.e., $E \approx \mathbf{e} \cdot \mathbf{y}(\mathbf{x})$, where $\mathbf{e}$ is a vector of positive weights.

When considering externalities, the objective function of the GDP must be adjusted by their value, $(G - E)$, so it becomes $(S - \$) + (G - E)$; and the constraints do not change. Clearly, if we also add the adjustment term $(G - E)$ to our recommended compensation formula for an RFP, the equivalence between the GDP and the CPMP will be retained — and the successful franchisee will continue to work for the general good.

As an example, if the master agency wishes to use consumer surplus as a measure for user benefit and values the good externalities in proportion to the total demand attracted as $G = \mathbf{g} \cdot \boldsymbol{\alpha}$, then the compensation formula in the RFP would read: $I = K - (\mathbf{p} + \mathbf{T}) \cdot \underline{\boldsymbol{\alpha}}^* + \mathbf{g} \cdot (\boldsymbol{\alpha} - \underline{\boldsymbol{\alpha}}^*) - \mathbf{e} \cdot \mathbf{y}(\mathbf{x})$, where $\boldsymbol{\alpha}$ is the actual demand achieved by the franchisee. Note by putting yourself in the feet of a franchisee, how the two externality adjustments would induce you to increase ridership and reduce the environmental impacts of your design.

It is our belief that we do not see much incentive-based franchising in the real world despite its promise because the administrative organizations in many cities are compartmentalized and do not lend themselves to this form of transit management. Unions may also resist if potential franchisees do not use unionized labor — but negotiated compromises can be included as stipulations in the RFPs. Another barrier is that the necessary data to calculate the performance indicators may not be easy to collect and maintain. Fortunately, technology nowadays makes this “big data” activity a cinch. Thus, in technology-savvy cities where transportation functions are centralized under a single master agency, such as London in the U.K., there is no unsurmountable obstacle to performance-based, competitive franchising. In fact, "Transport for London" has been franchising bus operations for years, and more recently has begun to introduce performance-based incentives. This experiment is worth watching closely.

Homework

11.1. This exercise examines the economics of an infinitely long bus route. Its passengers are disaggregated into n homogeneous classes, $i = 1, 2, \ldots, n$, characterized by their door-to-door travel distances $\{l_1, l_2, \ldots, l_n\}$ [km] and their trip generation rates $\{\alpha_1, \alpha_2, \ldots, \alpha_n\}$ [pax/km-h]. For each class, trip origins are uniformly distributed in time and space. The operating agency is assumed to choose the station spacing s [km], the vehicle headway H [h], and the fare p_i [\$/pax] that is charged to passengers in each class $i = 1, 2 \ldots n$. We assume: (i) that $s \ll l_1 < \cdots < l_n$, such that each l_i is approximately a different multiple of s; (ii) that the vehicles are sufficiently large; (iii) that the vehicle cruising speed v [km/h], the dwell time per stop t_s [h], and the passenger walking speed v_w [km/h], are all constant; (iv) that headways are short so no schedule is advertised; and (v) that passengers do not have appointments at the destination. There are no significant infrastructure costs. The daily operating cost is a linear combination of the distance traveled and the number of stops made by all the buses during the day. The coefficients of this linear function are c_d[\$/bus-km] and c_s[\$/stop]. The disutility π_i for passengers in class i is the sum of the fare p_i and the average door-to-door travel time for the class, which we denote T_i. The demand for each class decreases exponentially with the disutility of the class, i.e., $\alpha_i = D_i \exp(-\pi_i)$, for some known value $D_i > 0$. Given all this information, please:

(a) Write expressions for the average agency cost per capita and the average passenger travel time T_i.

(b) Assuming that consumer surplus expresses the benefit to passengers and that externalities can be ignored, write a formula for the social benefit. (This must include a component for the agency.)

(c) Formulate the service design problems DP, GDP, and CDP' of Section 11.1, assuming a finite budget B[\$/km-h]. Solve each of these problems numerically with the following parameter values: $n=2$, $l_1=3, l_2=10, D_1=D_2=1, c_d=1, c_s=0.01, v=30, t_s=0.01, v_w=3$, $B=3$, $\mathbf{T_0}=(0.4, 0.8)$. Do you obtain the same design for DP and GDP? What design do you obtain if you insert $\boldsymbol{\pi}^*$ into CDP'? [*Hint*: you may use any readily available nonlinear optimization solver.]

(d) Now repeat (c) for the following two scenarios: (i) a flat fare such that $p_1 = p_2$; and (ii) a distance-based fare such that $p_1/l_1 = p_2/l_2$. Discuss the impact of the pricing strategy on the optimal system design, the ridership, and the social benefit.

(e) (Bonus question) How would you formulate and solve the design problem if $n \to \infty$ and the passenger travel distances are uniformly distributed in a finite interval?

11.2. Consider a bus shuttle that is to serve a homogeneous population. During the hours of service, passengers are expected to arrive at a constant rate, α [pax/h], which will depend on service quality. The operator can determine both the service frequency, x [buses/h], and the fare, p [\$/pax]. Each vehicle dispatch costs c_f [\$]. Passengers have appointments but service is regular and so frequent that passengers do not use the schedule. Thus, the average waiting time at both the origin and the destination is $1/x$. The demand rate α depends on the disutility, $\pi = p + \beta/x$, as follows: $\alpha = (D - \pi)/k$ for $\pi \in [0, D]$ and $\alpha = 0$ for $\pi > D$ [\$], where β [\$/h] is passengers' value of time, and k [\$-h/pax] is a positive constant. The agency covers its operating cost with a combination of the fare box revenue and a subsidy I [\$/h] that cannot exceed a limit B [\$/h].

(a) Assuming no externalities, write an expression for the social benefit function that includes the passengers and the operator. Then prove that the expression does not increase with the passenger disutility.

(b) Show that if there is no subsidy limitation and the agency is ideal, so it chooses the GDP solution, then it will provide free transit.

(c) Now assume that B is so low that $I = B$ and $B < c_f x$. Then, prove that the optimal solution to the GDP requires $p^* > 0$.

11.3. Consider again the bus shuttle problem 11.2. Find the combinations of frequency x and fare p, as well as the associated subsidy I (if applicable), that will maximize the three metrics listed below. Finding the closed-form solution is possible but tedious, so you may choose to solve the problem numerically with the following parameter values:

$\beta = 20$ \$/h, $c_f = 200$ \$, $k = 2$ \$-h/pax, $D = 100$ \$, $B = 2000$ \$/h. In addition, please give the resulting societal benefit, fare box revenue, and passenger waiting time.

(a) The total society benefit minus the system-wide cost.
(b) The fare box revenue.
(c) The number of riders.

Compare the results and discuss your findings.

Index

Lightning Source UK Ltd.
Milton Keynes UK
UKHW051213060320
359889UK00005B/49

9 789813 224087